W9-BEM-237

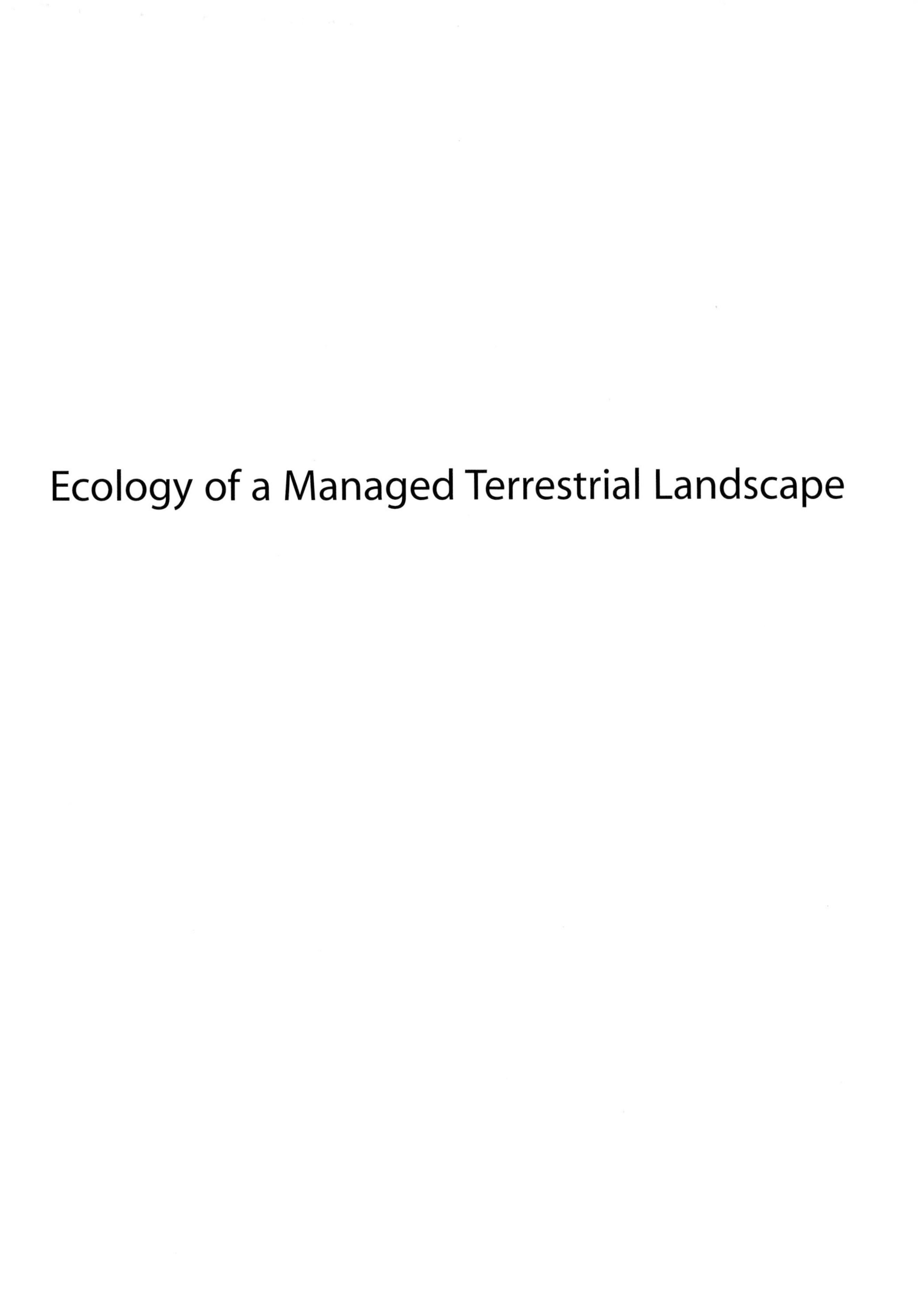

Ecology of a Managed Terrestrial Landscape

Edited by Ajith H. Perera, David L. Euler, and Ian D. Thompson

Ecology of a Managed Terrestrial Landscape:
Patterns and Processes of Forest Landscapes in Ontario

Published by UBC Press in cooperation with the
Ontario Ministry of Natural Resources

Printed in Canada on acid-free paper ∞

ISBN 0-7748-0749-0

Canadian Cataloguing in Publication Data

Main entry under title:
Ecology of a managed terrestrial landscape

"Published ... in cooperation with the Ontario Ministry of Natural Resources" – t.p.
Includes bibliographical references and index.
ISBN 0-7748-0749-0

1. Forest landscape management – Ontario. 2. Forest ecology – Ontario. 3. Forest policy – Ontario. I. Perera, A. (Ajith) II. Euler, David. III. Thompson, I.D. (Ian Douglas), 1949- IV. Ontario, Ministry of Natural Resources.

SD387.L35E36 2000 333.75'09713 C99-911210-4

This book was published with the help of a grant from the Ontario Ministry of Natural Resources.

UBC Press acknowledges the financial support of the Government of Canada through the Book Publishing Industry Development Program (BPIDP) for our publishing activities.

Canadä

We also gratefully acknowledge the support of the Canada Council for the Arts for our publishing program, as well as the support of the British Columbia Arts Council.

UBC Press
University of British Columbia
2029 West Mall, Vancouver, BC V6T 1Z2
(604) 822-5959
Fax: (604) 822-6083
E-mail: info@ubcpress.ubc.ca
www.ubcpress.ubc.ca

Contents

Preface

The dramatic change in the philosophy of forest management that has occurred over the last decade is a global phenomenon. There has been a consistent effort to shift the goal of forestry from sustaining multiple uses of forest products to sustaining the wide range of economic, cultural, social, and ecological values which forests provide. This broader focus demands a broader perspective on forest lands than was previously considered. Planning units have increased in size well beyond the individual forest stand, and planning issues require an increasingly long-term context. The "bigger-picture ecology" that is now required in policy development, land use planning, and forest management finds its conceptual foundation in the field of landscape ecology. The large spatial scales and long temporal scales inherent in landscape ecology supplement and complement the smaller scales addressed by autecology and community ecology.

Since the inception of landscape ecology as a discipline, the focus of study has been on areas where forest had been cleared for farming and urban development; that is, areas where forest cover occurred as small islands amid a vast "sea" of at least semi-permanently deforested land. Extrapolating the principles, knowledge, and experience gained in such non-forest landscapes to continuous forest landscapes in which the changes in forest cover are not likely to be permanent is of limited use. Forest landscape ecology needs to grow, as a discipline in its own right, in the direction of understanding the large-scale composition, structure, and function of predominantly forested regions, with the goal of providing an ecological basis for sustainable forest management. Forest landscape ecology is still in its early stages, and the sparse published literature reflects neither the breadth nor the depth of its full potential. This book is among the first comprehensive documents to develop and summarize a body of landscape ecological knowledge for a region where forests are the predominant land cover.

The province of Ontario, Canada, provides an excellent example of the new perspective and scale of forest management. The forested area of Ontario is one of the largest in the world and contains nearly 1.5 percent of the world's forests. The managed forest region in Ontario represents 70 percent of all the forested lands in the province and extends over more than 40 million ha, as almost continuous forest cover. The total contribution made by this managed forest region to the economy of the province exceeds $12 billion per year and includes the direct or indirect employment of more than 150,000 people in more than 40 communities dependent on the forest. The magnitude of the disturbances that occur in this forest zone has also been large: during some years, over 200,000 ha have been harvested, and much larger areas have been burned by wildfire. In recent years, Ontario has instituted a new broader ecological, economic, and social approach to forest management, through legislation and through numerous changes in the policy and planning processes governing land use and forest management.

Since the early 1990s, Ontario forest scientists have increasingly turned their attention to large-scale ecological studies. Concurrently, Ontario's primary land management agency, the Ontario Ministry of Natural Resources, began to adopt a larger-scale approach to forest policy, planning, and management. In this book, our intent is to provide "the bigger picture" of many aspects of Ontario's managed forest, including flora, fauna, disturbances, human use of resources, and forest management policy and planning. The information provided here will cause the reader to consider forest management from a different, very large-scale perspective.

This book is organized as an applied ecology text: first we consider the structure of the ecological systems, then their functions and processes, and finally applications of that knowledge. In accordance with this sequence, the chapters are grouped into three sections. Each chapter has been written independently, but with the intention of covering a specific part of the scope of this book. In keeping with the emphasis in forest landscape ecology on spatial and quantitative information, we have ensured, wherever possible, that the illustrations are spatially accurate.

The first of the three sections is intended to describe the managed forest landscape of Ontario and provide a context for the rest of the book. The reader is introduced to the topic of landscape ecology in a forestry context, the potential of this field of study to provide knowledge in support of forest management, and future challenges in the field. An overview is provided of the biogeographical setting of the province, beginning with the geology, climate, and soils. The forest vegetation of the managed forest regions of the province is described, and factors affecting the forest vegetation at the landscape scale are outlined. A comprehensive overview is provided of Ontario's forest vertebrates and their distributions, and changes in the range of selected species are reported. Finally, a detailed, quantitative description is provided of the spatial patterns of forest cover across the province and within the managed forest zone, with special reference to disturbances from forest fire and harvest practices.

The second section of the book examines processes and functions in the forest landscape of Ontario in detail, from various viewpoints. The role of past and present climate in shaping Ontario's forests is described, and forest changes that are predicted to occur in the future, in response to global climate change, are also discussed. A comprehensive discussion is provided of the patterns of large-scale disturbances caused by forest insects and diseases, with special reference to the dynamics of infestation by the spruce budworm (*Choristoneura fumiferana*). As wildfire is one of the most significant ecological processes in Ontario's forests, especially in the boreal region, an overview is provided of the forest fire regimes of the province, and methods of simulating wildfire and its consequences are examined. Large-scale spatial patterns of primary forest productivity across the province are described, and methods of estimating productivity are detailed. The section ends by synthesizing the response of forest vegetation and wildlife species to disturbances at large scales.

The third section of the book provides a provincial-scale perspective on current principles and challenges of forest policy and resource management in Ontario. The rich history of Ontario's forest policy development is detailed, and the reader is guided through historical changes in attitudes, perspectives, and values in forest management. This historical perspective is followed by an examination of recent developments in forest policy and legislation, with particular emphasis on Ontario's *Crown Forest Sustainability Act* (S.O. 1994, c. 25). This Act represented an important change in Ontario forest policy and, in part, reflected the new way of thinking about larger scales in forest management. Two current topics relevant to resource management at large scales are discussed: the process of strategic planning of land use, with particular reference to Ontario's forests, and the principles and potential value of adaptive management for forest management in Ontario. The final section ends with reflections on the present situation of landscape ecology and of forest management in Ontario, and suggestions of directions for future inquiry and research.

At the end of the text, we provide two valuable tools for the reader: a glossary of technical terms pertaining specifically to landscape ecology that may be unfamiliar to a first-time reader in this field, and a fold-out map of Ontario on which most of the place names used in the chapters are located with cartographic accuracy.

We are publishing this book mainly for two groups of readers. First, we address individuals who conduct forest research, develop forest policy and land use plans, and plan and implement forest management operations. Second, we expect the book to be instructive to graduate and senior undergraduate students in forestry and associated disciplines. We anticipate that this compilation of knowledge and critical discussion will fill a gap that currently exists in the literature of landscape ecology, with respect to the application of landscape ecological principles specifically to forested landscapes.

Ajith H. Perera
David L. Euler
Ian D. Thompson

September 1999

Acknowledgements

We acknowledge the financial and logistic support provided by the Ontario Ministry of Natural Resources, and in particular, Terry Taylor's managerial facilitation of this project. The Canadian Forest Service, the University of Waterloo, the University of Toronto, and Lakehead University provided in-kind support, in the form of their staff members' time spent as chapter authors and as technical editors. We are especially grateful to David Baldwin and Kent Todd for preparing the illustrations, Maureen Kuntz for editorial assistance, and Randy Schmidt, who provided project liaison from UBC Press. Special thanks go to Lisa Buse for her help in compiling the glossary, for editorial support, and for assistance in coordinating the review processes. The following individuals peer-reviewed chapters: Bill Baker, Ken Baldwin, Wayne Bell, Tom Clark, Steve Colombo, Bill Crins, Tom Crow, Peter Duinker, Richard Freitag, Andy Gordon, Guy Goudreau, Celia Graham, Art Groot, Eric Gustafson, Bob Haig, Steve Hounsell, Peter Kevan, Doug Larson, R.A. Lautenschlager, Mike Lavigne, Doug McCalla, Gray Merriam, John Middleton, Mike Moss, Brian Naylor, Simsek Pala, Bruce Pond, Rob Rempel, Clay Rubec, Taylor Scarr, Richard Sims, Brian Stocks, Roger Suffling, Charlie Van Wagner, and Dave Watton. Tom Crow and Andrew Jano reviewed the full book manuscript and provided valuable insight. This book could not have been completed without the dedication and perseverance of Susan Smith, who assisted the technical editors in numerous ways and edited the entire manuscript.

SECTION I:

Introduction to Landscape Ecology and Ontario's Managed Forest Region

This section describes the managed forest landscape of Ontario and provides contextual information for the subsequent sections. • Chapter 1 introduces the topic of landscape ecology in a forestry context and explores its applications in forest management. The following four chapters describe various aspects of the composition and structure of Ontario's managed forest landscapes. • Chapter 2 describes the geology, climate, and soils of the province. • Chapter 3 describes the forest vegetation of Ontario's managed forest regions, and • Chapter 4 the vertebrate species inhabiting these forests. Finally, • Chapter 5 provides a quantitative description of the present-day spatial patterns of forest cover that have resulted from almost a century of forest management practices and from a legacy of natural disturbances such as forest fire.

1 Landscape Ecology in Forest Management: An Introduction

AJITH H. PERERA* and DAVID L. EULER**

AN OVERVIEW OF LANDSCAPE ECOLOGY FUNDAMENTALS

Our purpose in this chapter is to provide an overview of landscape ecology, from development of the basic concepts to the adaptation of these concepts to the study of forested landscapes, with special reference to forest management. A comprehensive discussion of these themes is beyond the scope of the chapter, but key sources are cited for those readers who wish to explore landscape ecology in greater depth.

Ecology from a Landscape Perspective

From a very general perspective, a landscape is a large expanse of land. In land use planning, "landscape" is defined as a geographic unit which constitutes a mosaic of patches of various land cover types (Zonneveld 1995). Neither of these meanings incorporates ecological concepts, even in a generic sense, so neither provides a robust ecological definition of a landscape.

In the field of ecology, the world is organized into an ascending hierarchy of elements nested within each other: cells, organisms, populations, communities, ecosystems, landscapes, biomes, and finally the biosphere. In this framework, each level of elements bears a constant relation to the next in order of organization, but to define any one of the levels independently requires the specification of a particular point of view. Thus, a landscape can only be defined relative to a given organism (trees, mosses, or earthworms, for example) and is composed of whatever spatial units constitute ecosystems for that organism. The ecosystems, in turn, are composed of populations of the organism. This organism-centred definition is simple, generic, and fundamental to the study of landscapes. Allen and Hoekstra (1992) provide an excellent description of the ecological concepts behind the term "landscape."

Once a landscape is defined as an ecological system relative to the organism in question, the term "landscape ecology" can be defined as the study of the composition, structure, and function of that system (e.g., Forman and Godron 1986). "Composition" refers both to the types of elements composing the landscape (for example, forest ecosystems compose a forest landscape) and to the abundance of these elements, in terms of number, area, or biomass. "Structure" refers to the spatial relationship among these elements, which can be described by such criteria as proximity and clustering. "Function" refers to the primary processes (forest productivity, for example) occurring within components of the landscape, and to secondary processes that depend on the spatial interaction among those components (for example, the spread of a forest pest). Changes in a landscape include both short-term shifts in composition, structure and function (for example, forest succession after fire or harvest) and long-term adaptation (for example, an adaptation to climate change). Some have sought a simpler and a more specific definition of landscape ecology, such as "the study of the reciprocal effects of spatial pattern on ecological processes" (Pickett and Cadenasso 1995).

* Forest Landscape Ecology Program, Ontario Forest Research Institute, Ontario Ministry of Natural Resources, 1235 Queen Street East, Sault Ste. Marie, Ontario P6A 2E5

** Faculty of Forestry and the Forest Environment, Lakehead University, 955 Oliver Road, Thunder Bay, Ontario P7B 5E1

The Role of Scale in Landscape Ecology

As stated above, the biology of the organism defining a landscape determines the spatial scale of the landscape; similarly, the function rate of the organism defines the temporal scale on which a landscape functions. As a result, landscapes can have a wide range of spatial and temporal scales. For example, an earthworm landscape consists of only a few square metres and is characterized by process and change rates measured in days or hours. In contrast, a forest landscape may measure several square kilometres and is characterized by process and change rates measured in years.

A general notion exists that taking a landscape ecological perspective inherently involves large areas and a coarse level of detail, but, in fact, there is no fixed spatial or temporal scale associated with the terms "landscape," "landscape-scale," or "landscape-level." To avoid ambiguity in the use of landscape ecological concepts, one must specify both the organism defining the landscape and the explicit spatial and temporal scales identifying the scope of the study. Allen (1998) and King (1997) provide an extensive discussion of the concept of scale and its role in landscape ecology.

Concepts of Landscape Ecology

The original focus of landscape ecology was the spatial characteristics of land cover patches, and interactions among those patches, across landscapes dominated by human settlement. The principles developed in that context remain the same when applied to landscapes centred on organisms other than human beings, whether they are forest landscapes, for example, or beetle landscapes.

The island biogeography theory of MacArthur and Wilson (1967), which describes habitat isolation and consequent species responses, brought a substantial degree of rigour to landscape ecology with respect to inter-patch organism dynamics. This theory spawned the most popular topic of landscape ecology, habitat fragmentation (e.g., Harris 1984; Hansson et al. 1995); that is, the degree to which landscapes are composed of small patches isolated from others of their type. Concepts associated with fragmentation include "ecotones" (e.g., Hansen and di Castri 1992), or patch boundaries; "metapopulations" (e.g., Gilpin and Hanski 1991), or populations of populations; and "corridors" (e.g., Merriam and Saunders 1993), or paths by which flow can occur among patches. These topics remain the most studied aspects of landscape ecology.

Many topics other than landscape fragmentation have gained popularity since the mid-1980s. These include landscape disturbances (e.g., Turner and Dale 1991), post-disturbance changes (e.g., Baker 1989), and ecological flows (Gardner et al. 1992), the latter referring to the physical movement of biological and non-biological materials within landscapes. In addition, the application to landscapes of theoretical concepts such as hierarchy theory (e.g., Allen and Starr 1988), fractal geometry (e.g., Milne 1991), and neutral models (Gardner et al. 1987) fueled further conceptual studies and also provided insight in applied investigations.

All concepts of landscape ecology are based on an explicit understanding of the spatial patterns formed by constituent elements in a landscape. In fact, the quantitative analysis of landscape spatial structure has become one of the most studied facets of landscape ecology. The components forming the landscape are analyzed for such spatial criteria as heterogeneity, patchiness (the degree to which forest patches are small or isolated from others of their type), clustering (the degree to which the area of a given patch type is not dispersed), and proximity (the degree to which patches of the same type occur close to one another). For a recent review of the broad and rapidly expanding array of analytical techniques, the reader is referred to Gustafson (1998). Advances in landscape spatial analysis have been supported in recent years by the simultaneous progress achieved in the technologies of remote sensing, geographic information systems, and geostatistics.

The following sources provide in-depth information on the fundamental theories, concepts, and methods of landscape ecology, and their development: Forman and Godron (1986), Urban et al. (1987), Turner (1989), Turner and Gardner (1991), Hansen and di Castri (1992), and Bissonette (1997).

THE ECOLOGY OF FORESTED LANDSCAPES

A Brief History

Since Carl Troll coined the term "landscape ecology" in the 1930s, the major focus of the discipline has been landscapes composed of heavily settled or farmed areas with intermittent woodland patches.

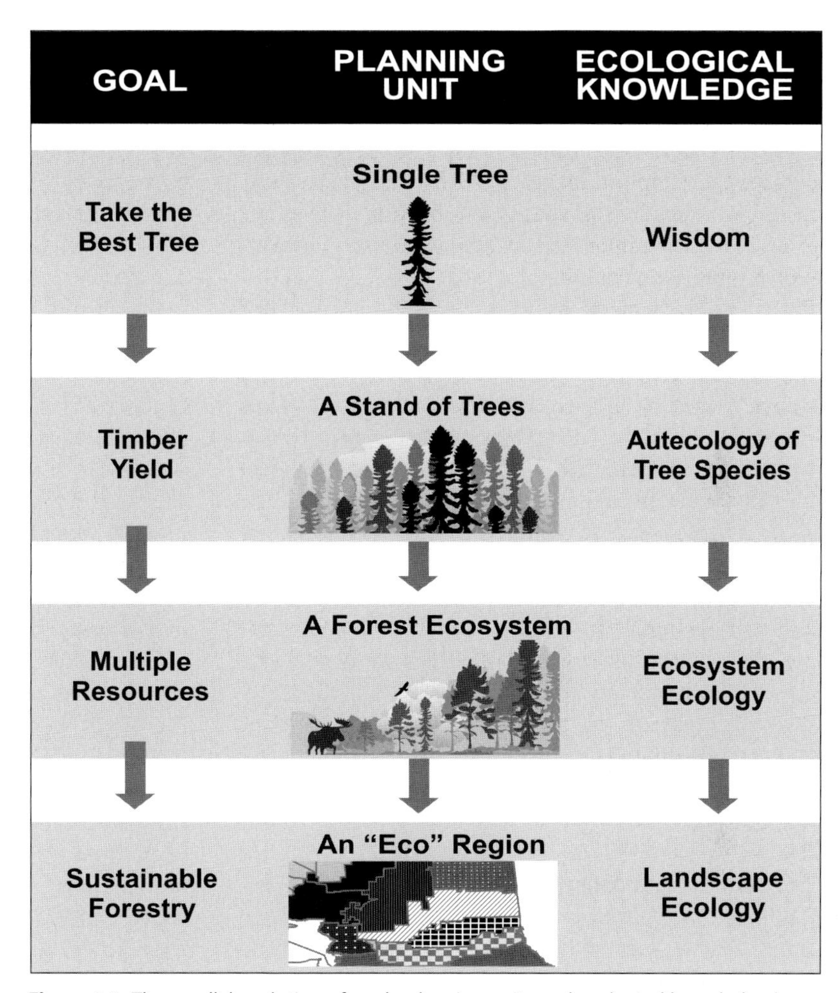

Figure 1.1 The parallel evolution of goals, planning units, and ecological knowledge in forest management.

Forest landscapes in which the dominant land cover is forest with intermittent patches of non-forested land did not become a focus of study until the mid-1980s.

The goal of forest management has also evolved from the harvest of individual trees to the ecological sustainability of forests. There has been a parallel evolution in the definition of the managed land unit, from individual trees to a broad unit of land that is ecologically meaningful, such as an ecoregion. Before ecological regions were adopted as planning units, knowledge of the ecology of individual tree species, termed "autecology," and knowledge of the ecology of communities and their environment, or "ecosystem ecology," provided sufficient scientific basis for management. The ecological patterns and processes of larger geographic units, such as ecoregions, however, extend beyond the spatial and temporal scope of traditional autecology or ecosystem ecology, and enter the realm of forest landscape ecology. It is logical that forest landscape ecology has emerged and developed as a field of study in parallel with the evolution of forest sustainability as the goal of forest management (Figure 1.1).

It is particularly appropriate that landscape ecology should develop and gain acceptance in Canada, and specifically in Ontario, where the areas of managed forest may be among the largest of any single nation or political region in the world. Not surprisingly, a recent literature survey showed an exponential increase in the number of mainstream publications on Canadian forest landscape ecology from the 1980s onward (Perera, unpub.).

Concepts and Applications of Forest Landscape Ecology

In theory, a forest landscape is an agglomeration of forest ecosystems, in which the ecosystems, in turn, are characterized by their different species composition and by structural qualities such as age and biomass. In practice, land cover types that can readily be identified (for example, types of forest stand or age classes) are often deemed to be surrogates for forest ecosystems. The following, therefore, is a working definition of a forest landscape: a large geographical unit dominated by a mosaic of forest cover types, perhaps interspersed with non-forest cover types such as wetlands and scrub, or even farms and settlements. Many other working definitions and descriptions of forested landscapes are proposed, such as an "agglomeration of watersheds" (Urban et al. 1987).

In the strictest sense, forest landscape ecology involves examining the relationships in spatial geometry among various land cover types at large spatial, temporal, and ecological scales. By definition, these scales lie above those of either individual communities of tree or wildlife species or the ecosystems containing those communities. Some studies of species populations, communities, and even physiological processes, however, have been identified with forest landscape ecology simply because they were conducted over large spatial extents.

Forest Landscape Composition and Spatial Patterns

All applications of forest landscape ecology are based on determining the compositional classification of a forest landscape. The classification may be as generic as broad land cover types, or as objective-specific as old-growth forest or moose habitat. Traditionally, aerial photographs provided the information needed to classify large forested landscapes. Now a whole array of information sources is available, including earth-observation and weather satellites. A range of techniques for classifying and interpreting the data has developed as well. For a good introduction to remote sensing in landscape ecology and natural resource management, we refer the reader to Quattrochi and Pelletier (1991) and Wilkie and Finn (1996).

A critical decision that must be made in classifying forest land cover composition is the selection of a data resolution and a classification scheme that are consistent with the research or management objective. This decision is a heuristic process, but it ensures that the landscape patterns and processes that emerge from the analyses are neither too coarse nor too detailed to test the intended hypothesis or make the required decision.

Once the compositional classes have been determined, the spatial structure of a forest landscape can be described in many ways. The main descriptors are heterogeneity (the diversity of land cover types within a given area), fragmentation (the degree to which forest patches are small or isolated from others of their type), and contagion (the degree to which forest patches are clumped). Related descriptors of landscape structure include edge (the interface between two land cover types), interior (the area of a forest patch that remains after discounting a buffer zone along the patch perimeter), and connectivity (the degree to which passage from one forest patch to another is unimpeded). What actually constitutes edge, interior, and connectivity is determined by the species in question; for example, the quality perceived as edge by moose might be quite different from the edge perceived by deer or marten in the same forest landscape. In the recent past, additional descriptors have emerged, including interspersion (the degree to which the land cover classes are juxtaposed), and lacunarity (the degree to which a landscape is either interrupted by gaps or continuous in texture).

A multitude of quantitative indices for measuring these characteristics have evolved since the mid-1980s (e.g., Krummel et al. 1987; O'Neill et al. 1988; Li and Reynolds 1995). These indices, termed "landscape metrics," are incorporated into both generic spatial-assessment software (e.g., Baker and Cai 1992; McGarigal and Marks 1995) and into specialized software packages for forest landscape analysis (e.g., Perera et al. 1997). The utility of the indices, like the descriptors they measure, varies with the objective of the study and depend on the land cover classification scheme and scale of analysis chosen. In other words, they depend on the extent of the landscape and the resolution of the source data. Perera et al. (1997) and Eng (1998) provide useful discussions on the application of these descriptors and indices in forest landscapes.

In addition to analysis of the spatial geometry within landscapes, forest landscape ecology includes the stratification of forest landscapes themselves according to geoclimatic variations among them. Identifying relatively homogeneous strata at a spatial scale higher than that of individual forest landscapes creates a hierarchical framework for analyzing any given area of forested land. Hills (1959) was the first to stratify forested landscapes into ecoregions, using forest productivity and composition as the indicators. The goal of that work was to identify and spatially delineate the geoclimatic variables that act as top-down constraints to the patterns and processes defining landscapes. An updated version of this concept, embedded in hierarchy theory, has been used to stratify forested landscapes according to forest productivity estimates derived from remote sensing data and geostatistical analytical techniques (Perera et al. 1996; Band et al., 1999).

Forest Landscape Function and Processes

Fundamental functions of forest landscapes, such as primary productivity and hydrology, are directly dependent on the composition and structure of the constituent elements of the landscapes. Methods for estimating forest primary productivity through physiological predictions have recently been extended to a landscape perspective (e.g., Running and Hunt 1993). It is now possible to predict forest primary productivity and hydrological processes by analyzing a combination of AVHRR data and physiological models (e.g., Band 1993) at spatial and temporal scales suitable for forest landscape management. The reader is referred to Running and Hunt (1993), Waring and Running (1998), and Band (2000, this volume) for further information on the description and assessment of the primary processes of forest landscapes.

Perhaps the most apparent process in a forest is disturbance from such causes as wildfire and insect pest infestation. The mechanisms by which these disturbances spread are of both theoretical and practical interest. Theoretical investigations into percolation models (e.g., Gardner et al. 1992; O'Neill et al. 1986) provide insight into the spread of disturbances based purely on landscape spatial structure. Forest disturbance by wildfire has received attention for several decades (e.g., Heinselman 1973). Li (2000, this volume) reviews the role of fire in landscapes and describes various modelling approaches that simulate fire processes. Large-scale insect pest disturbances in forest landscapes have been widely studied (e.g., Liebhold and Elkinton 1989; Holling 1992). Fleming et al. (2000, this volume) discuss this topic in detail, particularly with respect to infestation in Ontario forests.

Understanding disturbances caused by such agents as fire and insects within the forests themselves obviously provides a scientific basis for responding to the disturbance, perhaps by taking measures to suppress it. Of broader importance is the critical insight provided into temporal fluctuations in forest patterns and their associated processes. Understanding how forests change after disturbance provides the means of applying and assessing the premise of forest sustainability, which states that managing forests so that the results fall within the bounds of natural disturbance results will serve to sustain forest ecosystems.

Forest landscapes also change in response to changes in their external context; for example, in response to global changes in climate. Investigating and predicting this type of change, particularly at an international scale, is a topic of growing interest (e.g., Walker and Steffen 1996; Flannigan and Weber 2000, this volume). Disturbances that originate within the landscapes, such as wildfire and infestation, are also subject to change in response to such contextual changes (e.g., Gardner et al. 1996).

A multitude of ecological processes take place in forest landscapes, as in any ecological system. In generic terms, these processes include the flow of biological material (for example, propagules and populations) and of non-biological elements (for example, pollutants) among the land cover types. Biologists study the fluctuations in local populations of species caused by migration and localized extinction in response to changes in forest landscape structure (Voigt et al. 2000, this volume). The descriptors and indices of forest landscape spatial patterns are increasingly used in formulating and testing hypotheses of local population changes (e.g., Hansson 1994; Rempel et al. 1997). The long-term flow of propagules and germplasm across forest landscapes and the local adaptations that result are of interest to genecologists and conservation biologists (e.g., Goodman 1987). The flow of non-biological elements across forested landscapes is an important area of investigation (e.g., Merriam 1994) that has not yet been widely studied.

Forest landscape change also occurs in response to human activity. The conversion of forested land to settlement and agriculture is an important historical theme described extensively in both scientific and popular literature. Changes in landscape spatial patterns caused by forest harvest, and the consequences of these changes, are of increasing scientific interest. Franklin and Forman (1987) undertook pioneering studies in forested landscapes of the Pacific Northwest. Recent studies on this subject have been conducted by Li et al. (1993), Spies et al. (1994), Gluck and Rempel (1996), Gustafson and Crow (1994), Tang et al. (1997), and Eng (1998).

THE ROLE OF LANDSCAPE ECOLOGY IN FOREST MANAGEMENT

A Landscape Perspective for Foresters

The large spatial and temporal scales of landscape ecology provide a perspective that is not available from traditional forest ecosystem ecology. Numerous issues associated with the management goal of sustainable forests require explication at large spatial and temporal scales. One example is gap analysis; that is, the analysis of the disproportionate scarcity of certain ecological features in a management unit, relative to their representation in a larger region surrounding the management unit. Gap analysis at large scales is thus a prerequisite for developing and assessing biodiversity conservation plans in forest landscapes. Similarly, the resolution of socio-ecological issues, such as the quality and availability of old-growth forest and wilderness areas, requires long-term study at large spatial scales.

Ensuring the availability of habitat for large-bodied faunal species such as moose, for species with large habitat ranges such as migratory birds, or for species at the higher trophic levels such as lynx, requires planning at scales beyond a few forest stands. Likewise, concerns regarding species adaptations and the management of forest genetic resources cannot be addressed without knowledge of the structure of the forest landscape.

Only at large scales can natural forest disturbances be examined in a way that yields a comprehensive understanding of the effects of disturbance and their many implications for forest resource management. The most important implication relates to the spatial and temporal limits of perturbation thresholds; that is, the limits within which forest landscapes are capable of returning to their former state after a disturbance. In addition, the proper use of some of the information employed in forest management requires knowledge of the context of the information, which must be acquired from larger spatial and temporal scales. For example, the practice of silviculture requires information about species selection, site selection, and harvest rotations. These concepts are multi-scale issues, nested within the higher-level concepts of species adaptations, forest landscape stratification, and post-disturbance landscape change.

Forest landscape ecology provides a higher-level context for the silvicultural, physiological, and ecological information that constitutes the direct basis for forest management practices. The synoptic information and science provided by landscape ecology is, therefore, an essential foundation for sound forest management planning, from the development of forest policy to the design of strategies for forest land use.

Challenges for the Future

The forest management agencies of certain regions, such as the Pacific Northwest and Ontario, have taken a pioneering role in embracing landscape ecology as an important component in research, training, and decision-making. These initiatives have contributed to significant paradigm changes in forest management, of which the *Crown Forest Sustainability Act* of Ontario (1994) is a prime example.

Most forest managers and policy-makers, however, are only now discovering the value of landscape perspective. This discovery phase inevitably holds many challenges for forest managers, forest management agencies, and forest landscape ecologists, alike. First, forest managers have to endure an initial period of unfamiliarity with large spatial and temporal scales and discomfort with the coarse resolution inherent in landscape ecological information. The traditional approach of planning, managing, and monitoring of forest resources from the fixed perspective of the forest stand (that is, a single spatial and temporal scale) must yield to a multi-scale process. In this new, multi-scale approach, management issues and goals will be scaled in a hierarchically nested manner, and some decisions will be made at large spatial and temporal scales with information appropriate to those scales. Most forest managers have difficulty in accepting the concept that the bottom-up amalgamation of fine-resolution information and knowledge from community ecology and silviculture is an inefficient means of obtaining a large-scale perspective and does not necessarily elucidate patterns and processes emergent in landscapes. This hurdle can be overcome with education in the concept of ecological scale, as it applies to patterns, processes, and information sources, and in the simple logic of hierarchy theory.

To integrate a landscape perspective into their planning, forest management agencies will need to compile large-scale databases and to develop and adopt techniques supporting decision-making processes that operate at a large scale and are spatially explicit. This part of the task is made easier by progress in information technology, specifically in remote sensing, geographic information systems, database management software, and overall computing power. The more difficult part of the task, however, lies in generating the scientific knowledge and logic in forest landscape ecology that will enable forest managers to make intelligent use of the databases. One approach to satisfying this need is the practice of adaptive policy design and adaptive resource management, by which both policy and plans are applied as hypotheses and modified according to the results obtained. Adaptive management provides one possible framework for the effective application of currently available science.

The task faced by landscape ecologists is perhaps particularly daunting. Forest landscape ecology as a discipline is barely a decade old. At this stage in its development, it still relies on the direct importation of the ecological concepts, philosophies, and principles of landscapes from classical landscape ecology. That basic theory postulates landscapes in which forest occurs only as a network of small patches in a matrix of permanently non-forested land. A shift is required in this approach toward a true forest landscape ecology. The new approach will be a landscape in which forest is the dominant land cover and the mosaic of land cover patterns is more dynamic.

Another significant challenge is posed by the difficulty of conducting research into the patterns and processes of forest landscapes by traditional experimentation. At the large spatial and temporal scales involved, testing hypotheses by applying manipulative treatments, establishing controls, and making observations, will be a long-term, resource-exhausting process. This hurdle can be partly overcome by the judicious simulation of forest landscape patterns and processes in a spatially explicit manner, coupled with an adaptive approach to forest management. Integrating forest landscape analysis and monitoring into the process of forest management decision-making, as has been done in Ontario (OMNR 1996), will provide a steady stream of information in support of this learning process. In the short term, this approach will be hampered by the paucity of information on the mechanisms by which landscape ecological processes operate, and by the absence of both null models for defining the results of simulations and of readily available empirical data for validating the results.

Finally, we emphasize that landscape ecology will never solve all the problems of forest management planning in isolation; the fine-scale knowledge provided by community ecology, silviculture, and physiology will always be essential. Although it is unlikely that the true application value of the landscape perspective will be realized for some time, we enthusiastically believe that the top-down approach of forest landscape ecology provides unique and indispensable knowledge that readily complements traditional forest management science.

REFERENCES

Allen, T.F.H. 1998. The landscape "level" is dead: persuading the family to take it off the respirator. In: D. L. Peterson and V. T. Parker (editors). Ecological Scale. New York, New York: Columbia University Press. 35-54.

Allen, T.F.H. and T.B. Starr. 1988. Hierarchy: Perspectives for Ecological Complexity. Chicago, Illinois: University of Chicago Press. 310 p.

Allen, T.F.H. and T.W. Hoekstra. 1992. Toward a Unified Ecology. Complexity in Ecological Systems Series. New York, New York: Columbia University Press. 384 p.

Baker, W.L. 1989. A review of models of landscape change. Landscape Ecology 2(2): 111-133.

Baker, W.L. and Y. Cai. 1992. The r.le programs for multiscale analysis of landscape structure using the GRASS geographical information system. Landscape Ecology 7(4): 291-302.

Band, L.E. 1993. A Pilot Landscape Ecological Model for Forests in Central Ontario: Development of a Preliminary Landscape Ecological Model for the Management of Mixed Wood Forests in Central Ontario. Forest Fragmentation and Biodiversity Project Report No. 7. Sault Ste. Marie, Ontario: Ontario Ministry of Natural Resources, Ontario Forest Research Institute. 40 p.

Band, L.E. 2000. Forest ecosystem productivity in Ontario. In: A.H. Perera, D.L. Euler, and I.D. Thompson (editors). Ecology of a Managed Terrestrial Landscape: Patterns and Processes in Forest Landscapes of Ontario. Vancouver, British Columbia: University of British Columbia Press. 163-177.

Band, L.E., F. Csillag, A.H. Perera, and J.A. Baker. 1999. Deriving an Eco-regional Framework for Ontario through Large-Scale Estimates of Net Primary Productivity. Forest Research Report No. 149. Sault Ste. Marie, Ontario: Ontario Ministry of Natural Resources, Ontario Forest Research Institute. 30 p.

Bissonette, J.A. (editor). 1997. Wildlife and Landscape Ecology: Effects of Pattern and Scale. New York, New York: Springer-Verlag Inc. 410 p.

Pickett, S.T.A. and M.L. Cadenasso. 1995. Landscape ecology: spatial heterogeneity in ecological systems. Science 269: 331-334.

Eng, M. 1998. Spatial patterns in forested landscapes: implications for biology and forestry. In: Conservation Biology Principles for Forested Landscapes. J. Voller and S. Harrison (editors). Vancouver, British Columbia: University of British Columbia Press. 42-75.

Flannigan, M.D. and M.G. Weber. 2000. Influences of climate on Ontario forests. In: A.H. Perera, D.L. Euler, and I.D. Thompson (editors). Ecology of a Managed Terrestrial Landscape: Patterns and Processes in Forest Landscapes of Ontario. Vancouver, British Columbia: University of British Columbia Press. 103-114.

Fleming, R.A., A.A. Hopkin, and J.-N. Candau. 2000. Insect and disease disturbance regimes in Ontario's forests. In: A.H. Perera, D.L. Euler, and I.D. Thompson (editors). Ecology of a Managed Terrestrial Landscape: Patterns and Processes in Forest Landscapes of Ontario. Vancouver, British Columbia: University of British Columbia Press. 141-162.

Forman, R.T.T. and M. Godron. 1986. Landscape Ecology. New York, New York: John Wiley. 586 p.

Franklin, J.F. and R.T.T. Forman. 1987. Creating landscape patterns by forest cutting: ecological consequences and principles. Landscape Ecology 1(1): 5-18.

Gardner, R.H., B.T. Milne, M.G. Turner, and R.V. O'Neill. 1987. Neutral models for the analysis of broad-scale landscape pattern. Landscape Ecology 1(1): 19-28.

Gardner, R.H., M.G. Turner, V.H. Dale, and R.V. O'Neill. 1992. A percolation model of ecological flows. In: A.J. Hansen and F. di Castri (editors). Landscape Boundaries: Consequences for Biotic Diversity and Ecological Flows. Ecological Studies No. 92. New York, New York: Springer-Verlag. 259-269.

Gardner, R.H., W.W. Hargrove, M.G. Turner, and W.H. Romme. 1996. Climate change, disturbances and landscape dynamics. In: B. Walker and W. Steffen (editors). Global Change and Terrestrial Ecosystems. New York, New York: Cambridge University Press. 149-172.

Gilpin, M.E. and I. Hanski (editors). 1991. Metapopulation Dynamics. London, UK: Academic Press. 336 p.

Gluck, M.J. and R.S. Rempel. 1996. Structural characteristics of post-wildfire and clearcut landscapes. Environmental Monitoring and Assessment 39(1-3): 435-450.

Goodman, D. 1987. The demography of chance extinction. In: M. Soule (editor). Viable Populations for Conservation. Cambridge, UK: Cambridge University Press. 11-34.

Gustafson, E.J. 1998. Quantifying landscape spatial pattern: what is the state of the art? Ecosystems 1: 143-156.

Gustafson, E.J. and T.R. Crow. 1994. Modeling the effects of forest harvesting on landscape structure and the spatial distribution of cowbird brood parasitism. Landscape Ecology 9(4): 237-248.

Hansen A.J. and F. di Castri, editors. 1992. Landscape Boundaries: Consequences for Biotic Diversity and Ecological Flows. New York, New York: Springer-Verlag. 452 p.

Hansson, L. 1994. Vertebrate distributions relative to clear-cut edges in a boreal forest landscape. Landscape Ecology 9(2): 105-115.

Hansson, L., L. Fahrig, and G. Merriam (editors). 1995. Mosaic Landscapes and Ecological Processes. London, UK: Chapman and Hall. 452 p.

Harris, L.D. 1984. The Fragmented Forest: Island Biogeography Theory and the Preservation of Biotic Diversity. Chicago, Illinois: University of Chicago Press. 182 p.

Heinselman, M.L. 1973. Fire in the virgin forests of the Boundary Waters Canoe Area, Minnesota. Quarternary Research 3: 329-382.

Hills, G.A. 1959. A Ready Reference to the Description of the Land of Ontario and its Productivity. Preliminary Report. Maple, Ontario: Ontario Department of Lands and Forests, Division of Research. 142 p.

Holling, C.S. 1992. The role of forest insects in structuring the boreal landscape. In: H.H. Shugart, R. Leemans, and G.B. Bonan (editors). Analysis of the Global Boreal Forest. Cambridge, UK: Cambridge University Press. 170-191.

King, A.W. 1997. Hierarchy theory: a guide to system structure for wildlife biologists. In: J.A. Bissonette (editor). Wildlife and Landscape Ecology: Effects of Pattern and Scale. New York, New York: Springer-Verlag. 185-214.

Krummel, J.R., R.H. Gardner, G. Sugihara, R.V. O'Neill, and P.R. Coleman. 1987. Landscape patterns in a disturbed environment. *Oikos* 48(3): 321-324.

Li, C. 2000. Fire regimes and their simulation with reference to Ontario. In: A.H. Perera, D.L. Euler, and I.D. Thompson (editors). Ecology of a Managed Terrestrial Landscape: Patterns and Processes in Forest Landscapes of Ontario. Vancouver, British Columbia: University of British Columbia Press. 115-140.

Li, H., J.F. Franklin, F.J. Swanson, and T.A. Spies. 1993. Developing alternative forest cutting patterns: a simulation approach. Landscape Ecology 8(1): 63-75.

Li, H. and J.F. Reynolds. 1995. On definition and quantification of heterogeneity. *Oikos* 73(2): 280-284.

Liebhold, A.M. and J.S. Elkinton. 1989. Characterizing spatial patterns of gypsy moth regional defoliation. Forest Science 35: 557-568.

MacArthur, R.H. and E.O. Wilson. 1967. The Theory of Island Biogeography. Volume 1. Princeton, New Jersey: Princeton University Press. 203 p.

McGarigal, K. and B.J. Marks. 1995. FRAGSTATS: Spatial Pattern Analysis Program for Quantifying Landscape Structure. General Technical Report PNW-GTR-351. Portland, Oregon: USDA Forest Service. 122 p.

Merriam, G. 1994. Managing the Land: A Medium-Term Strategy for Integrating Landscape Ecology into Environmental Research and Management. Forest Fragmentation and Biodiversity Project Report No. 13. Sault Ste. Marie, Ontario: Ontario Ministry of Natural Resources, Ontario Forest Research Institute. 24 p.

Merriam, G. and D.A. Saunders. 1993. Corridors in restoration of fragmented landscapes. In: D.A. Saunders, R.J. Hobbs and P.R. Ehrlich (editors). Reconstruction of Fragmented Ecosystems: Global and Regional Perspectives. Natural Conservation No. 3. Chipping Norton, Australia: Surrey Beatty. 71-87.

Milne, B.T. 1991. The utility of fractal geometry in landscape design. Landscape and Urban Planning 21: 81-90.

O'Neill, R.V., D.L. DeAngelis, J.B. Waide, and T.F.H. Allen. 1986. A Hierarchical Concept of Ecosystems. Monographs in Population Biology No. 23. Princeton, New Jersey: Princeton University Press. 251 p.

O'Neill, R.V., J.R. Krummel, R.H. Gardner, G.Sugihara, B. Jackson, D.L. DeAngelis, B.T. Milne, M.G. Turner, B. Zygmunt, S.W. Christensen, V.H. Dale, and R.L. Graham. 1988. Indices of landscape pattern. Landscape Ecology 1 (3): 153-162.

Ontario Ministry of Natural Resources. 1996. Forest Management Planning for Ontario's Crown Forests. Toronto, Ontario: Queen's Printer for Ontario. 452 p.

Perera, A.H., D.J.B. Baldwin, and F. Schnekenburger. 1997. LEAP II: A Landscape Ecological Analysis Package for Land Use Planners and Managers. Forest Research Report No. 146. Sault Ste. Marie, Ontario: Ontario Ministry of Natural Resources, Ontario Forest Research Institute. 88 p.

Perera, A.H., J.A. Baker, L.E. Band, and D.J.B. Baldwin. 1996. A strategic framework to eco-regionalize Ontario. Environmental Monitoring and Assessment 39: 85-96.

Quattrochi, D.A. and R.E. Pelletier. 1991. Remote sensing for analysis of landscapes: An introduction. In: M.G. Turner and R.H. Gardner (editors). Quantitative Methods in Landscape Ecology: The Analysis and Interpretation of Landscape Heterogeneity. Ecological Studies No. 82. New York, New York: Springer-Verlag. 51-76.

Rempel, R.S., P.C. Elkie, and M.J. Gluck. 1997. Timber-management and natural-disturbance effects on moose habitat: landscape evaluation. Journal of Wildlife Management 61(2): 517-524.

Running, S.W. and E.R. Hunt, Jr. 1993. Generalization of a forest ecosystem process model for other biomes, BIOME-BGC, and an application for global-scale models. In: J. R. Ehleringer and C. B. Field (editors). Scaling Physiological Processes: Leaf to Globe. San Diego, California: Academic Press. 141-148.

Spies, T.A., W.J. Ripple, and G.A. Bradshaw. 1994. Dynamics and pattern of a managed coniferous forest landscape in Oregon. Ecological Applications 4: 555-568.

Statutes of Ontario. 1994. *Crown Forest Sustainability Act*. S.O. 1994, c. 25.

Tang, S.M., J.F. Franklin, and D.R. Montgomery. 1997. Forest harvest patterns and landscape disturbance processes. Landscape Ecology 12: 349-363.

Turner, M.G. 1989. Landscape ecology: the effects of pattern on process. Annual Review of Ecology and Systematics 20: 171-197.

Turner, M.G. and R.H. Gardner (editors). 1991. Quantitative Methods in Landscape Ecology. New York, New York: Springer-Verlag. 536.

Turner, M.G. and V.H. Dale. 1991. Modeling Landscape Disturbance. In: M.G. Turner and R.H. Gardner (editors). Quantitative Methods in Landscape Ecology. Ecological Studies No. 82. New York, New York: Springer-Verlag. 323-351.

Urban, D.L., R.V. O'Neill, and H.H. Shugart, Jr. 1987. Landscape ecology: a hierarchical perspective can help scientists understand spatial patterns. BioScience 37(2): 119-127.

Voigt, D.R., J.A. Baker, R.S. Rempel, and I.D. Thompson. 2000. Forest vertebrate responses to landscape-level changes in Ontario. In: A.H. Perera, D.L. Euler, and I.D. Thompson (editors). Ecology of a Managed Terrestrial Landscape: Patterns and Processes in Forest Landscapes of Ontario. Vancouver, British Columbia: University of British Columbia Press. 198-233.

Walker, B.H. and W.L. Steffen. 1996. GCTE science: objectives, structure and implementation. In: B. Walker and W. Steffen (editors). Global Change and Terrestrial Ecosystems. New York, New York: Cambridge University Press. 3-9.

Waring, R.H. and S.W. Running (editors). 1998. Forest Ecosystems: Analysis at Multiple Scales. New York, New York: Academic Press. 370 p.

Wilkie, D.S. and J.T. Finn. 1996. Remote Sensing Imagery for Natural Resources Monitoring – A Guide for First Time Users. Methods and Cases in Conservation Science. New York, New York: Columbia University Press. 295 p.

Zonneveld, I.S. 1995. Land Ecology: An Introduction to Landscape Ecology as a Base for Land Evaluation, Land Management, and Conservation. Amsterdam, The Netherlands: SPB Academic Publishing. 199 p.

2 Physical Geography of Ontario

DAVID J. B. BALDWIN,* JOSEPH R. DESLOGES,** and LAWRENCE E. BAND***

INTRODUCTION

Ontario, the second largest province of Canada, covers approximately 1 million km^2 and extends approximately from 42°N to 57°N latitude and from 75°W to 95°W longitude. The major patterns of geoclimatic features across the province are strongly interrelated and form the foundation for the biotic and natural disturbance processes of Ontario's many different landscapes. In this chapter, we describe the spatial distribution of the abiotic factors, including bedrock, surficial geology, climate, soils, and hydrology, that influence and interact with the biotic systems. The relationships among these elements, and the various biotic and natural disturbance processes at work across the province, are the subject of subsequent chapters in Sections I and II of this book. In keeping with the focus of this book on Ontario's northern managed forest landscape, we do not describe the southern, settled portion of the province, but refer the reader to well-known summaries of that region, such as Chapman and Putnam (1984).

GEOLOGY AND TERRAIN

The geology and terrain of Ontario are best understood by first examining the foundation of bedrock geology underlying the surficial deposits and landforms. Table 2.1 provides a summary of the geologic time scale relevant to Ontario's bedrock and surficial geology.

Table 2.1

Geologic periods relevant to Ontario's bedrock and surficial geology.

Eons	*Start - 10^6 Years Before Present*	*Era*	*Period*
Phanerozoic	1.65	Cenozoic	Quaternary
	0.011		– *Recent Epoch*
	5		– *Pleistocene Epoch*
	65		Tertiary
	135	Mesozoic	Cretaceous
	205		Jurassic
	250		Triassic
	290	Paleozoic	Permian
	320		Pennsylvanian
	360		Mississippian
	410		Devonian
	438		Silurian
	510		Ordovician
	570		Cambrian
Precambrian	2500	Proterozoic	
	4000	Archean	

(Adapted from Trenhaile 1998 and Webber and Hoffman 1970)

The bedrock geology of Ontario is variable in lithology, structure, and age, although approximately 61 percent of the province is underlain by Precambrian rock of the Canadian Shield (Thurston 1991). In the

* Forest Landscape Ecology Program, Ontario Forest Research Institute, Ontario Ministry of Natural Resources, 1235 Queen Street East, Sault Ste. Marie, Ontario P6A 2E5
** Department of Geography, University of Toronto, 100 St. George Street, Toronto, Ontario M5S 3G3
*** Department of Geography, University of North Carolina, CB# 3220 Chapel Hill, North Carolina, USA 27599

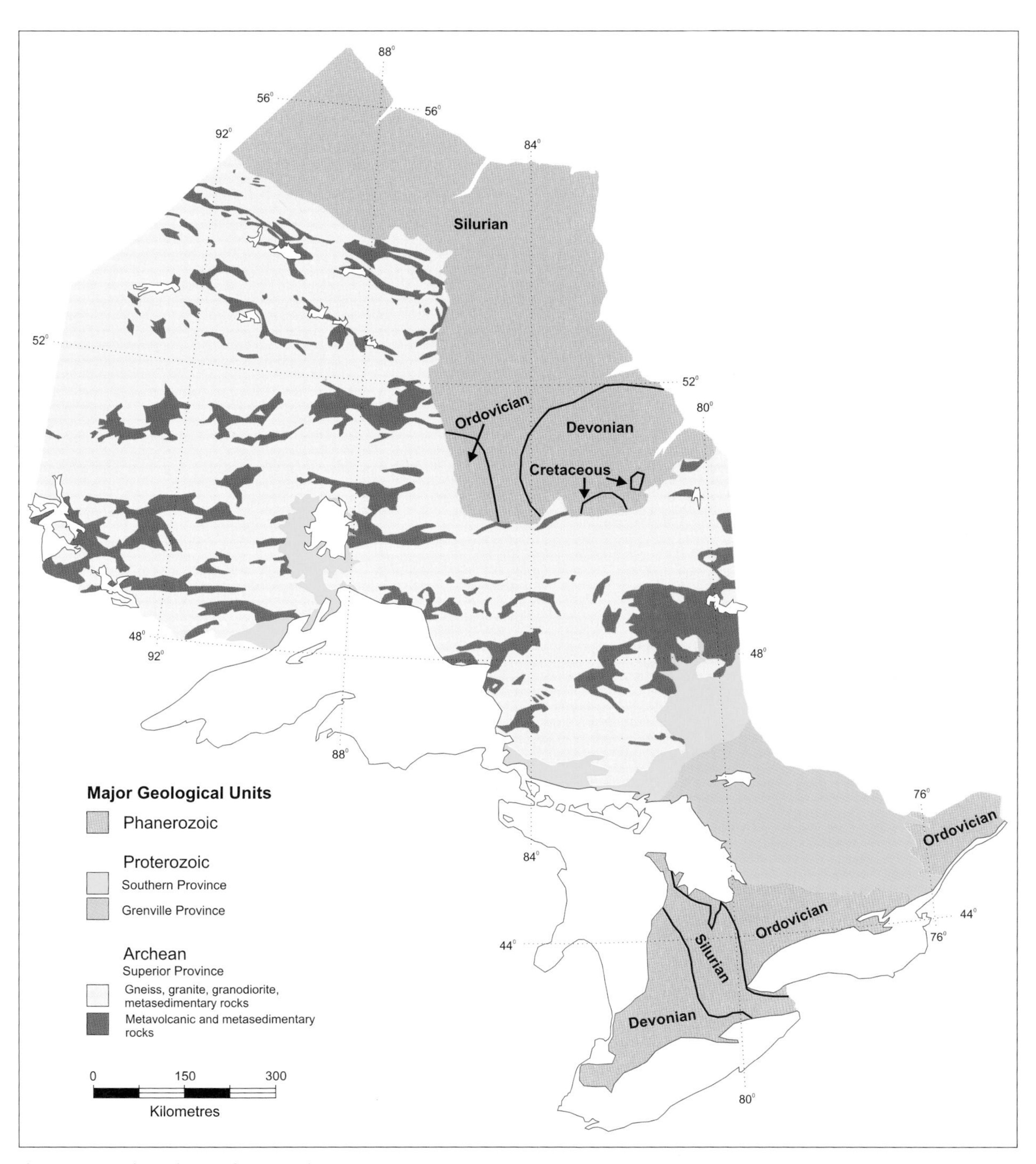

Figure 2.1 Geological map of Ontario showing major units.
(Adapted from Ontario Geological Survey 1991a-d and Webber and Hoffman 1970)

Phanerozoic age, sedimentary rocks developed in marine basins along the northern border of the Shield, forming the Hudson Bay lowlands, and in the Great Lakes Basins in the south (Figure 2.1). The Shield can be divided into three major geological and physiographic regions, from the oldest in the northwest to the youngest in the southeast. The northwestern region, known as the Superior Province, is more than 2.5 billion years old. This region, which can be described as lying north and west of the present city of Sudbury, is composed mainly of felsic intrusive rocks forming the rocky Severn and Abitibi uplands (Bostock 1970). The central region, known as the Grenville Province, is 1.0 to 1.6 billion years old. This region lies to the south of Sudbury, and is dominated by metasedimentary rocks that form the Laurentian highlands. The Penokean hills, a fold belt, and the Cobalt plain, an embayment, constitute the Southern Province, which is a narrow region approximately 1.8 to 2.4 billion years old extending from Sault Ste. Marie in the west to approximately Kirkland Lake in the east.

To the north of the Shield, in the area generally referred to as the Hudson Bay lowlands, the bedrock is composed of carbonate sedimentary formations. These formations date primarily from the Silurian period, but there are significant areas from the Ordovician and Devonian periods as well. Other sedimentary rocks occur near the city of Ottawa, in an area referred to as the Ottawa embayment, as well as throughout areas north of Lakes Erie and Ontario (Dyke et al. 1989). The clastic and marine carbonate bedrock of southern Ontario is interrupted by the Frontenac Axis, a southern extension of the Shield, which intersects the St. Lawrence Seaway east of Kingston. The Frontenac Axis has different forest cover and land use patterns than areas to either the west or east, owing to its uneven terrain and shallow, acidic soils, both characteristic of the Canadian Shield.

The topography of Ontario varies from flat plains to low, rolling uplands having 60 m to 90 m of relative relief, to dissected uplands with ridges, escarpments, and cuestas as high as 200 m above the adjacent terrain. This variation in topography originates from the bedrock structure; from extensive pre-Quaternary erosion, which is estimated at more than 6000 m since the last mountain-building episode (Card et al. 1972); and from Quaternary glaciations that both eroded and filled the pre-Quaternary surface. The significant relief that is present elsewhere in eastern North America is absent from Ontario. The highest point in the province, Maple Mountain near Temagami, is 693 m. The most rugged and fragmented surfaces occur in a band extending from the north shore of Lake Superior, across the Algoma highlands, through the Sudbury region and north of Manitoulin Island, and across the Madawaska highlands. This height of land forms a continental divide between the Great Lakes and Arctic drainage of Hudson Bay; north of this line, the elevation falls off monotonically (Figure 2.2). The large river basins that flow to the north across most of Ontario have low total relief and deranged drainage patterns from glaciation, both of which result in poor drainage.

All of Ontario underwent a set of major glacial advances and retreats during the last major glacial stade, up to 12,000 years BP. Northern parts of the province were still covered with ice 8,000 years BP (Hardy 1977, Ritchie 1989). The Hudson Ice, a distinct domain within the Laurentide Ice Sheet, had a number of separate lobes, one extending northeastward up the St. Lawrence River, a second westward into Lake Huron, and a third east-to-northeastward out of Lake Superior (Dyke et al. 1989). The retreat of the ice was neither uniform nor continuous. Isolated re-advances, such as the Cochrane surge, occurred during the period of final deglaciation. Figure 2.3 provides an overview of the major features and surficial materials across the province. During the retreat, streams formed along the margins of the ice sheets and created oblique, linear patterns of deposition. Glaciofluvial complexes that occur in various parts of the province (for example, the Oak Ridges complex in southern Ontario, and the Burntwood-Knife and Harricana-Lake McConnell complexes in central Ontario, each extending hundreds of kilometres) provide evidence of the massive scale of the convergent ice lobes and catastrophic meltwater discharges (Brennand and Shaw 1994) beneath the ice sheets and at their forward margins.

The retreating ice sheet fed a series of large meltwater lakes. Glacial Lake Algonquin covered much of the area from Sudbury to Huntsville. The Champlain Sea flooded well above the current levels of the Ottawa River Valley and of Lake Ontario. Lake Barlow-Ojibway covered a large area south of James Bay over what is now referred to as the Claybelt. The Tyrrell Sea lay over the Hudson Bay lowlands. In the northwest, Lake Agassiz covered virtually one-third of the province, joining with the Tyrrell Sea to the east for a

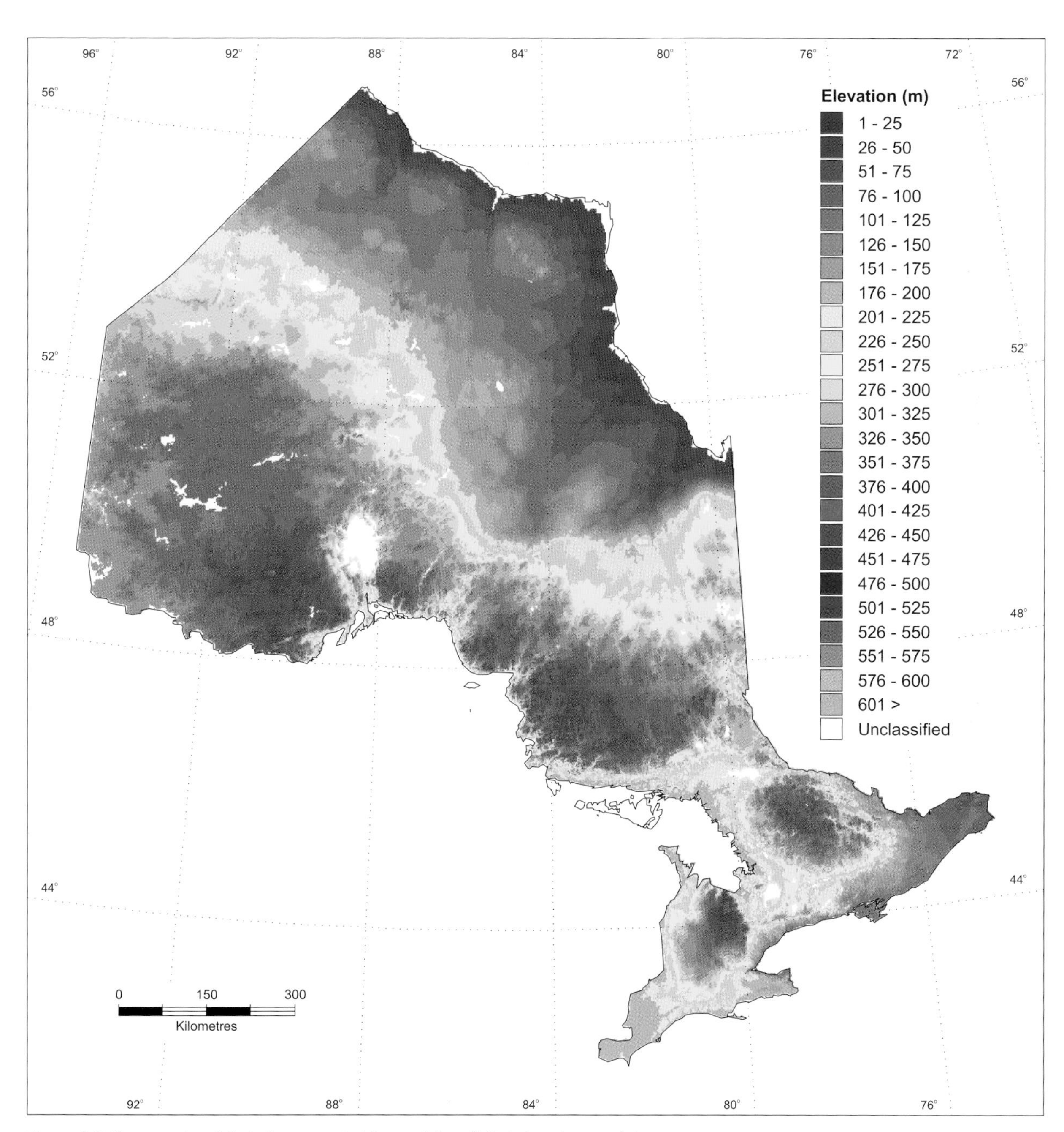

Figure 2.2 Topography of Ontario, generated from a 1-km digital elevation model.
(Data from Mackey et al. 1994)

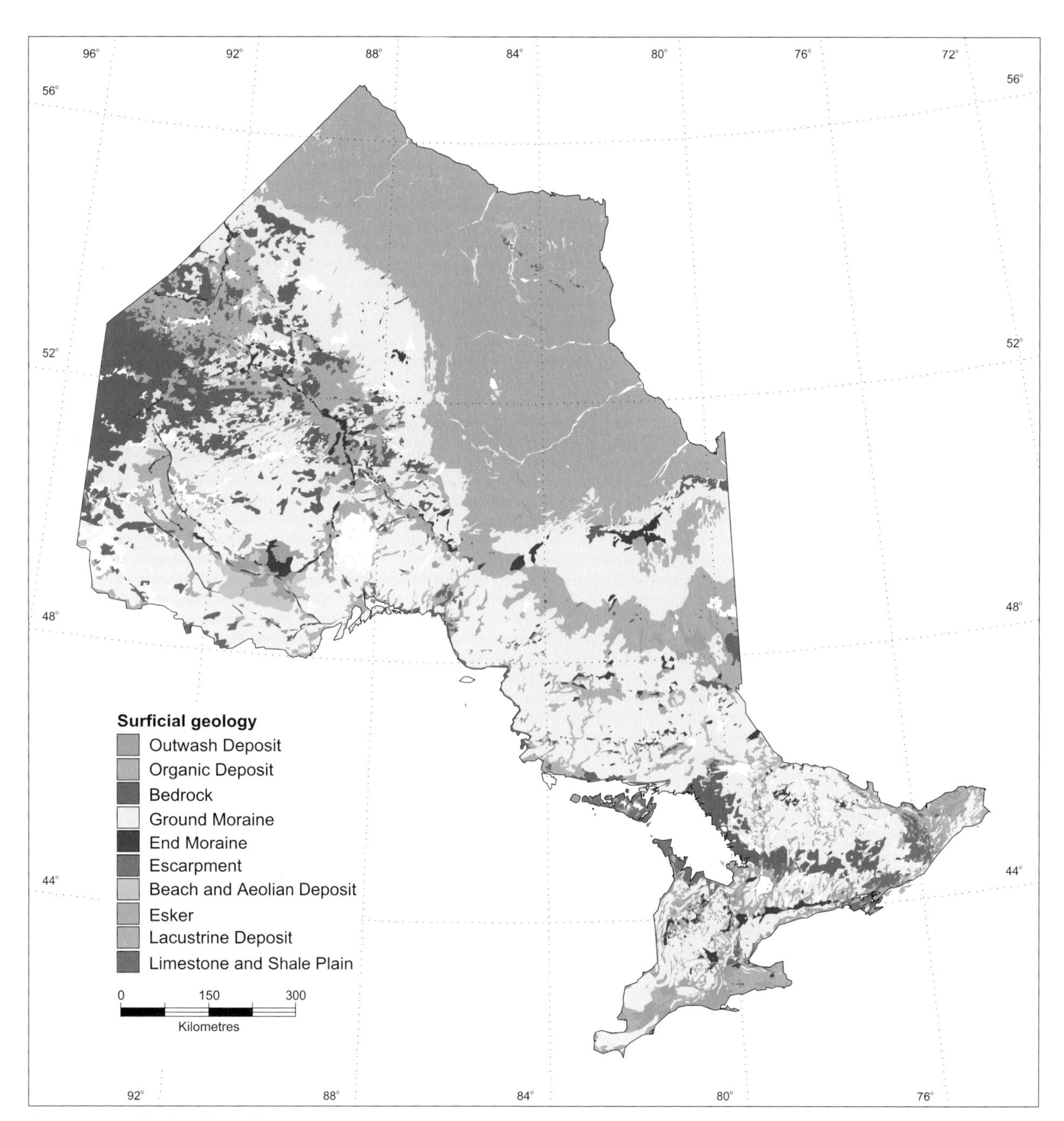

Figure 2.3 Surficial geology of Ontario.
(Adapted from Forest Landscape Ecology Program 1996)

short period (Teller 1985). Neither the postglacial lakes nor the Tyrrell Sea had fully receded until about 6000 BP. The general northward retreat of the Laurentide Ice Sheet left a predominately sandy to silty till cover over the southern part of the Shield and over southern Ontario, in contrast to the silty and clayey till and clay-rich lake and marine deposits that dominate the northern half of the province (Dredge and Cowan 1989). Low-lying and poorly drained glacial deposits have become covered with peat and other organic deposits (Figure 2.3).

Together, glaciation and postglacial deposition largely account for the present landscape of the province. The ice sheets sculpted or rounded the land and generally reduced its elevation, removed the original topsoils, and exposed the bedrock. Deposits laid down by glacial streams and lakes have strongly influenced soil development and, consequently, the composition of present-day forests (Thompson 2000, this volume).

Deposited till materials range from compact lodgement tills to surface ablation tills, and form both plains and high relief features such as moraines and drumlins. Kame deposits, interlobate moraines, and eskers, all composed of sand and gravel, are concentrated mainly in the southern half of the province, but also occur over large sections of the northern half (Dredge and Cowan 1989). The better-drained glacial and glaciofluvial landforms often stand above the more poorly drained substrate and thus provide locally distinct environments for drier ecosystems. Well-sorted sands occurring in long eskers and kame terraces, and other water-transported deposits, are often characterized by the development of distinct soils and forest cover complexes. Over most of the boreal and Great Lakes biomes, till deposits give rise to coarse soils, often with a considerable cobble component. Bedrock outcrops are common in the Shield, where parent materials were washed away or removed by glaciers.

CLIMATE

Most of Ontario's climate can be described as humid continental, with the notable exception of areas under the influence of Hudson Bay, which have a more maritime character. In general, Ontario's climate is affected by three major air sources: cold, dry, polar air from the north, the dominant factor during the winter months; Pacific polar air passing over the western prairies; and warm, moist, sub-tropical air from the Atlantic Ocean and the Gulf of Mexico (Webber and Hoffman 1970). The effect of these major air flows on temperature and precipitation depends on latitude, on proximity to major waterbodies, and, to a limited extent, on terrain relief.

Temperature

The most evident climate trend in Ontario is a well-defined gradient of temperature increase from north to south (Figure 2.4a). This pattern is modified by the influence of Hudson Bay and the Great Lakes and, to a lesser degree, by major topographic features. The cold, maritime climate of Hudson Bay and James Bay influences the northeast portion of the province, affecting areas as far south as Kirkland Lake. The result is a substantial reduction in growing-degree-days, relative to other locations in Canada at similar latitudes. For example, Winnipeg, Manitoba, lies at approximately the same latitude as Cochrane, Ontario, but has about 1000 more growing-degree-days. Lake Superior and Lake Huron moderate the winter temperature, "bending" the isolines around their shores (Figure 2.4b). Lake Superior also has a cooling effect on the northerly flow of warm summer air (Figure 2.4c). The impact of this cooling influence is felt in a sharp drop in growing-degree-days in areas north of Lake Superior, from the town of Marathon north to the settlement of Nakina (Figure 2.4d).

Local relief modifies the general temperature gradient at significant areas of increased elevation in the highlands near Caledon in southern Ontario, in the area of Algonquin Park, in the Algoma highlands northeast of Sault Ste. Marie, and in the highland area northwest of Thunder Bay (Figure 2.2). Pockets of lower temperature are evident at each of these heights of land (Figure 2.4a), and each has fewer growing-degree-days than the areas surrounding it (Figure 2.4d). The western portion of the province is influenced by the vast continental area of the Canadian prairies and the great plains of the US mid-west. In summer, warm air flowing from the southwest pushes up around the western shore of Lake Superior (Figure 2.4c), generating warm, dry conditions and increasing the number of growing-degree-days in this area, relative to eastern portions of the province at the same latitude.

Precipitation

The dominant precipitation trend in the province is an increase from northwest to southeast. This trend is modified significantly by strong lake and topographic effects in the central and southern portions of the province, particularly in areas to the lee of Lake Superior, Georgian Bay, and Lake Huron (Figure 2.5a).

Winter precipitation and snow accumulation are highly variable across the province (Figure 2.5b). The northern and western areas are influenced by the dry air of continental high-pressure zones. The southern and eastern areas are influenced by moister air from low-pressure areas. Winds accompanying the low-pressure conditions gather moisture while sweeping west to east across the Great Lakes and drop precipitation on the colder land mass. Several areas at the eastern ends of Lake Superior, Georgian Bay, and Lake Huron are called "snowbelts," as a result. The rise in elevation over the Algonquin highlands causes considerably greater snowfall in the area of Huntsville and Dorset than along the upper Ottawa River valley, 100 km to the east.

Summer precipitation patterns are more related to continentality than to lake effects. The climate of western areas of the province is dominated by continental high pressure, which reduces precipitation in early to mid-summer (Figure 2.5c). Precipitation is greatest away from the lakes, where air masses and storm cells build over land. The highest summer precipitation occurs in the central portions of the province and in the upper Ottawa Valley, northeast of North Bay. The annual water deficit (Figure 2.5d), derived from the Thornthwaite water balance equation, measures water storage as a function of the following: mean monthly temperature, total precipitation, latitude, and soil texture as a measure of water-holding capacity (Watson and MacIver 1995). Deficit values are particularly high in areas of southern Ontario just outside the lake-effect precipitation zones. These areas have relatively low precipitation, high summer temperatures, and well-drained soils. The water deficit is also high in the far west of the province, near Kenora, where a more continental regime of low precipitation prevails and where temperatures are relatively warm. Areas with low water deficits include the area immediately to the north of Sault Ste. Marie, which receives a combination of heavy winter precipitation and cool summer temperatures from the northward flow off Lake Superior. Low deficit values are also found in a cool area extending from Lake Superior toward Lake Nipigon. Here, the cool temperatures outweigh the limited precipitation values to keep moisture loss low.

Temporal Patterns in Precipitation and Temperature

Figure 2.6 contains climagraphs for key stations across the province. In the west (for example, at the town of Dryden) and in the north (for example, at Moosonee), the climate is more continental, winters are comparatively dry, and the annual range in temperature is wider. In the eastern and southern parts of the province (for example, around the cities of Toronto and Ottawa), precipitation is more evenly distributed throughout the year, and the annual temperature range is narrower.

Temporal trends in temperature and precipitation are not always clear, mainly for lack of consistent, long-term meteorological databases representing local climatic differences in various parts of the province. Several long-term records available from Environment Canada show a general warming trend from the late 1800s to about the mid-1900s. This trend reflects recovery from Little Ice Age conditions. Although there is no evidence of a similar trend in summer temperatures (Figure 2.7), winter temperatures at some stations have increased over the last couple of decades (Figure 2.8). No such trend is apparent in either annual or seasonal precipitation (Figure 2.9). It is not clear to what extent the apparent winter increases are caused by either short-term or long-term climate change, or to increasing urbanization in regions around the stations. The increases are apparently reflected in observations of satellite data made over the past two decades, which indicate a progressively earlier thaw and green flush and a longer active growing season in the subarctic regions of North America (Myneni et al. 1997). These observations may indicate a small increase in the length of the forest growing season in Ontario in recent years. It is unknown how this change may already be affecting the productivity of forests or may influence forest community dynamics in years to come. Climate change and its effects are discussed by Flannigan and Weber (2000, this volume).

SOIL DEVELOPMENT PATTERNS

The nature of soil development in Ontario depends on local combinations of climate, parent material, terrain, vegetation, and other organisms over time. Soils are formed by the physical and chemical weathering of bedrock and glacial parent material, and are

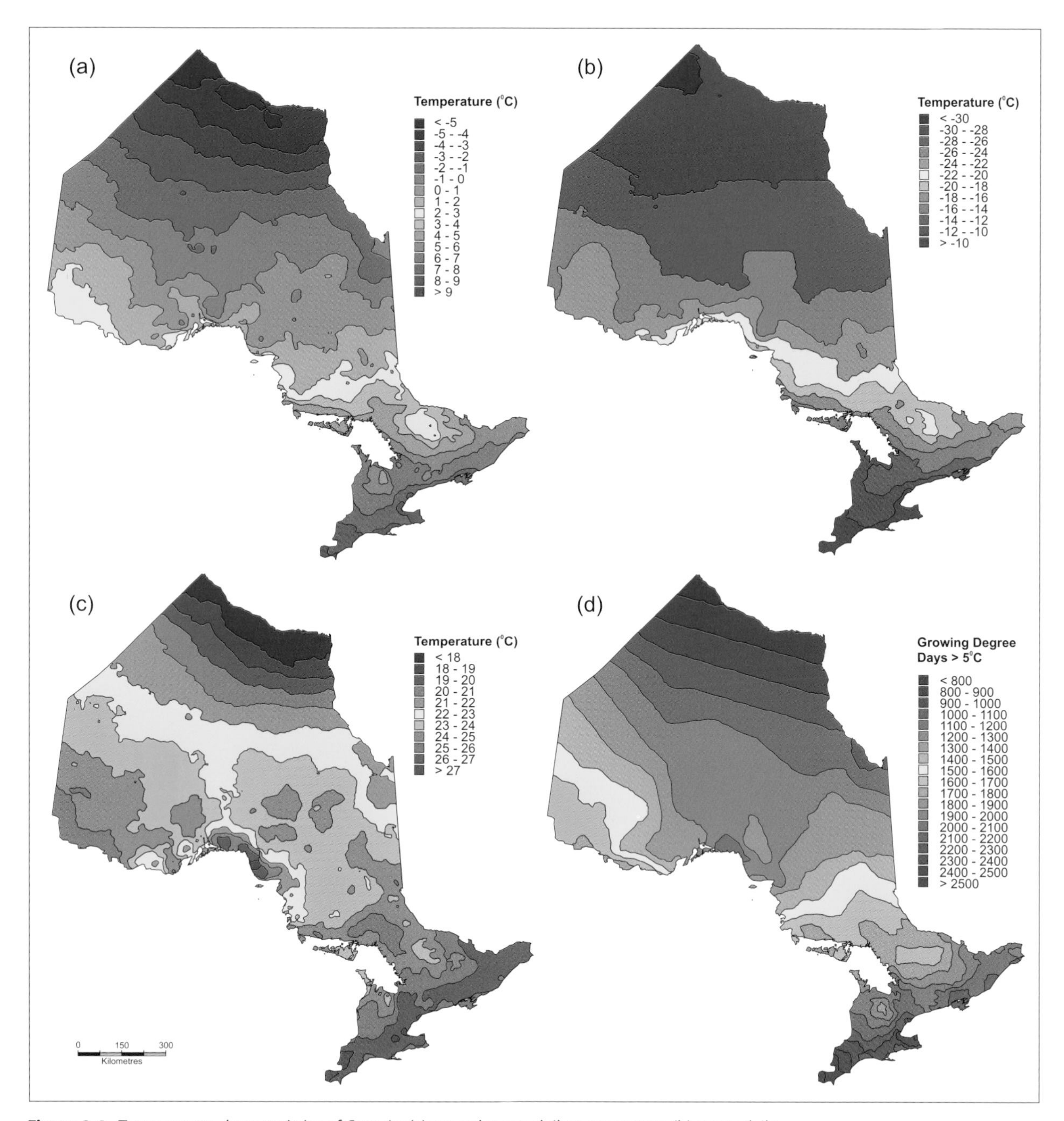

Figure 2.4 Temperature characteristics of Ontario: (a) annual mean daily temperature; (b) mean daily minimum temperature for January; (c) mean daily maximum temperature for July; and (d) growing-degree-days over 5°C. Temperatures interpolated from 30 years of data from more than 400 AES stations. (Adapted from Watson and MacIver 1995)

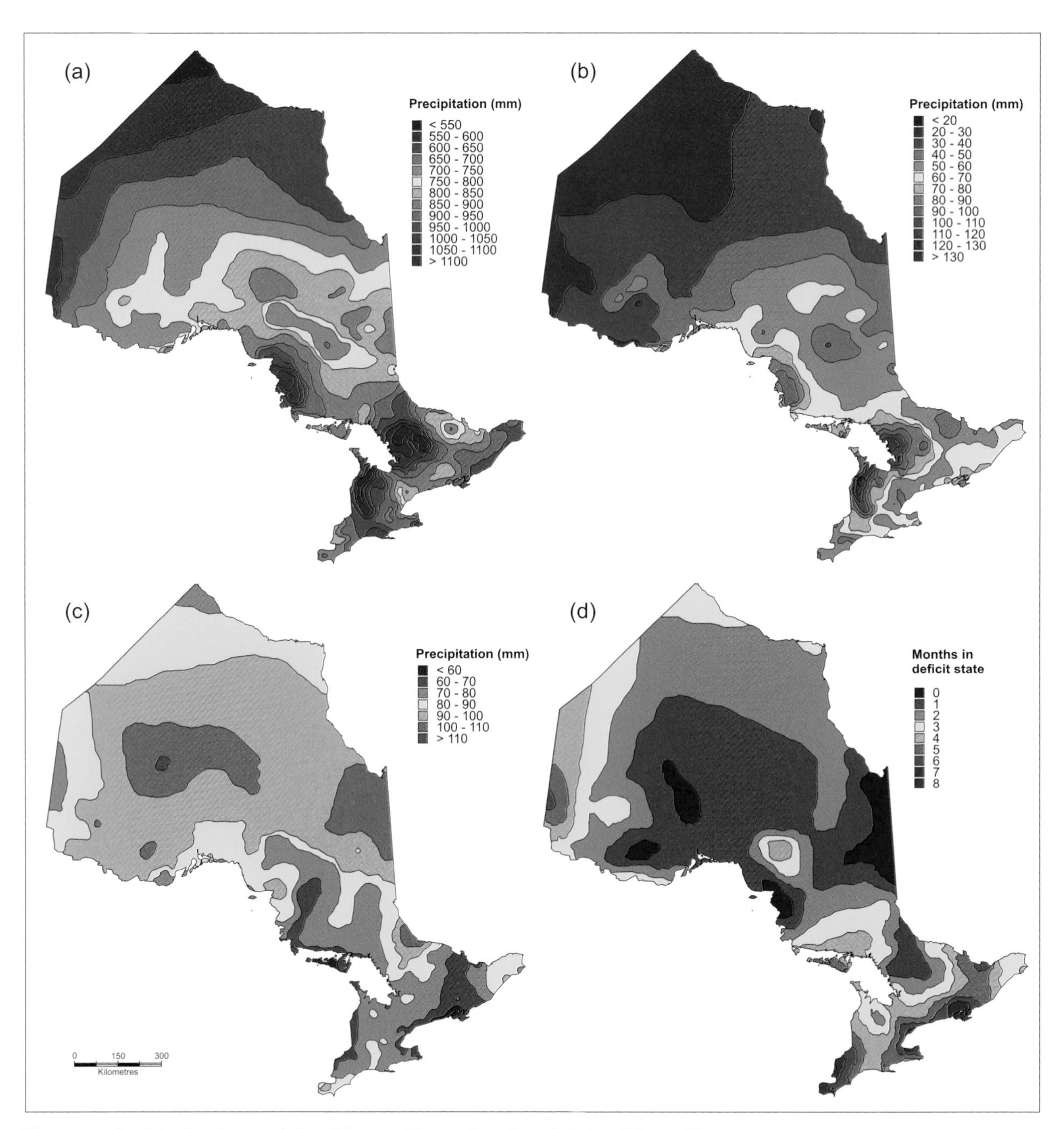

Figure 2.5 Precipitation characteristics of Ontario: (a) annual total precipitation; (b) monthly total precipitation for January; (c) monthly total precipitation for July; and (d) annual water deficit. Precipitation interpolated from 30 years of data from more than 400 AES stations.
(Adapted from Watson and MacIver 1995)

continually modified and shifted by water, wind, and gravity. Where glacial action has scoured away overlying deposits, the soils of Ontario closely reflect the underlying bedrock. Other soils reflect the tills and other morainic and lacustrine materials deposited by advancing and retreating ice sheets and their meltwater.

The Canadian System of Soil Classification (Agriculture Canada 1987) is a standard series of orders and component great groups by which soils can be identified and described. Six of the soil orders in this classification are predominant in Ontario. These are the organic and related organic cryosolic soils in northern parts of the province, brunisols in the northwest part of the Shield and south of the Shield, podzols over much of the central and southern Shield, luvisols in the Claybelt and over much of southern Ontario, and gleysols in poorly drained areas and in the Claybelt lacustrine deposits. Regosolic soils are dominant only in a thin band along the southwest shore of Hudson Bay. Figure 2.10 illustrates the soil orders and great groups that occur most extensively in Ontario, based on composition within the Soil Landscapes of Canada mapping units (Agriculture and Agri-Food Canada 1996). In the following section, we will examine these patterns of soil development in Ontario and the underlying factors of climate, geology, and terrain. The spatial relationship between soils and forest vegetation is examined by Thompson (2000, this volume).

Organic Soils

Organic soils dominate the Hudson Bay lowlands, the area of lacustrine deposits lying between Pickle Lake and Sandy Lake, and a band stretching from Espanola to Temagami, through Sudbury. They also occur as subdominant soils in poorly drained areas of otherwise sub-humid and arid landscapes (Clayton et al. 1977). As their name suggests, organic soils develop from organic deposits, and are defined as having more than a 30-percent component of organic material in specific portions or tiers of their profile (Agriculture Canada 1987).

Organic great groups are distinguished by the degree of decomposition of the organic material in the various tiers. The great group identified as fibrisols have a large percentage of well-preserved fibres in the upper and middle tiers of the soil profile. These soils are dominant in the vast bog and muskeg complexes of the Hudson Bay lowlands. Fibrisols have a

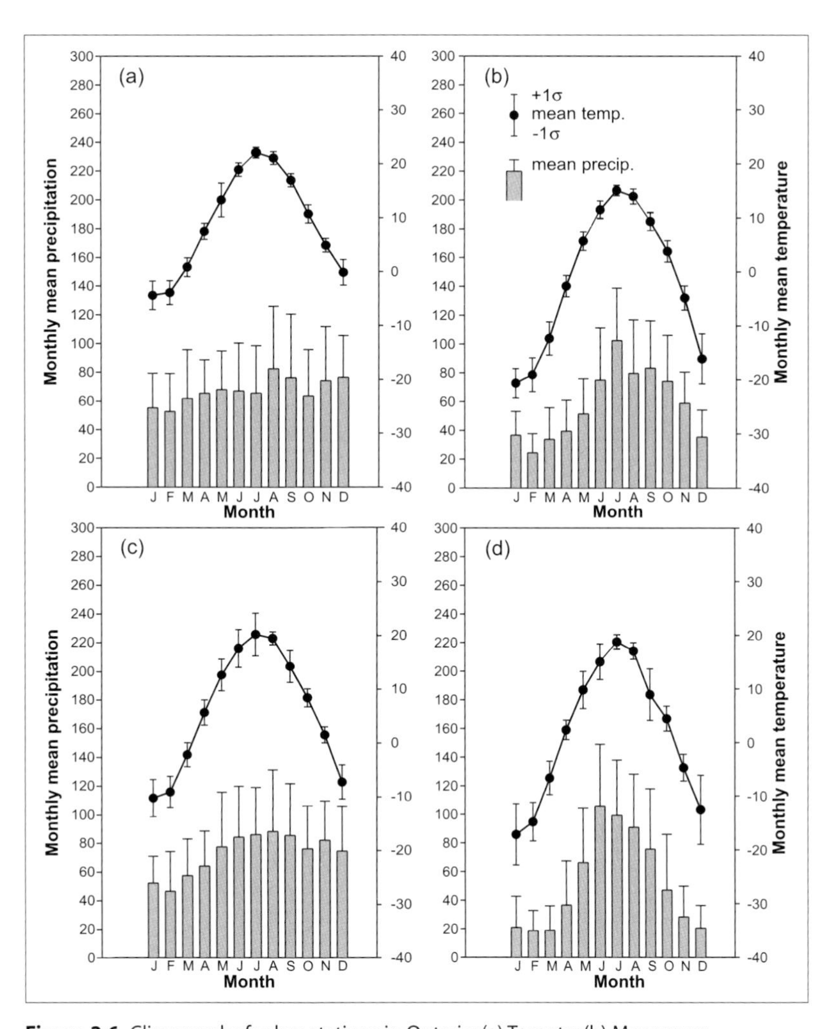

Figure 2.6 Climagraphs for key stations in Ontario: (a) Toronto; (b) Moosonee; (c) Ottawa; and (d) Dryden, derived from climate normals, 1961 to 1990.
(Data from Atmospheric Environment Service 1990)

particularly high water-holding capacity, and remain saturated for much of the year. Mesisols, the second great group, contain more material at a later stage of decomposition in the upper and middle tiers. These soils occur in a band around the wetter fibrisols along the south and west margins of the Hudson and James Bay lowlands, where the elevation declines into the lowland (Figure 2.2). They are also the dominant soils in the area of shallow tills and lacustrine deposits between Espanola and Temagami. They are intermixed with brunisols on the lacustrine deposits northeast of Pickle Lake, and with luvisols and gleysols in the Claybelt. Humisols contain highly decomposed organic material and are commonly referred to as muck soils. They dominate only in local patches, and are mapped as dominant only in a small area east of Lake St. Clair (Figure 2.10).

The water-holding capacity of organic soils, particularly fibrisols, moderates their soil temperature regimes, in comparison to surrounding or intermixed mineral soils. The latent heat involved in freezing and thawing the trapped water allows these soils to freeze and thaw later than other soils and thus to provide moderated growing conditions (Clayton et al. 1977).

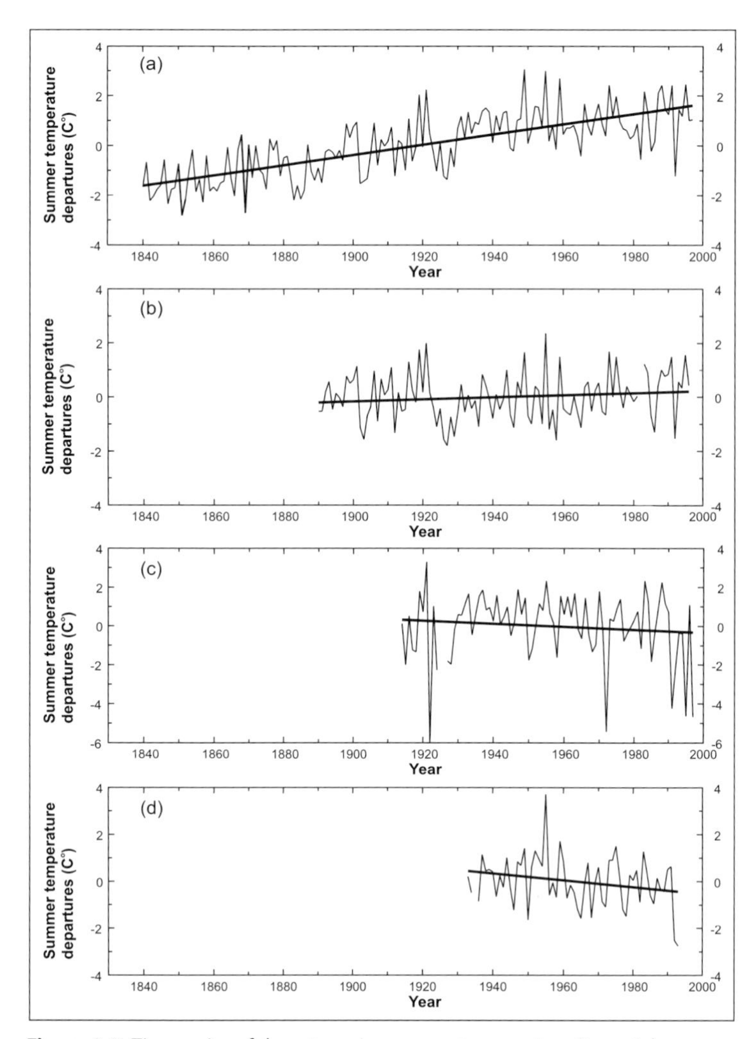

Figure 2.7 Time series of departures in summer temperature (June, July, and August) from period-of-record means for selected Ontario climate stations: (a) Toronto (downtown); (b) Ottawa; (c) Dryden; and (d) Moosonee. (Data from Atmospheric Environment Service 1998)

Cryosols

One great group of cryosolic soils, the organic cryosols, dominates a substantial area at the extreme northwest of Ontario (Figure 2.10). Cryosolic soils are characterized by the presence of permafrost within 1 m of the surface and by a significant degree of cryoturbation, or permafrost-related disturbance of the soil column, which causes various distinctive surface patterns and formations (Agriculture Canada 1987). The organic great group, in particular, develops on organic deposits in permafrost environments, such as the far northwest reaches of the province (Figure 2.3).

Brunisols

Brunisols dominate much of the Shield west of Lake Nipigon, from the border between Canada and the United States to latitude 53°N, and are interspersed with luvisols and podzols. They also dominate a large patch in the area of Chapleau and Gogama, on Manitoulin Island, in the Bruce Peninsula, and in a band surrounding the southern and eastern margins of the Shield. These soils are typically found in forested areas over coarse to medium glacial till or outwash, and on aeolian deposits (Figure 2.3). They occur on both imperfectly drained and well-drained sites, and are typically associated with rolling terrain. The layer development in these soils is distinct enough

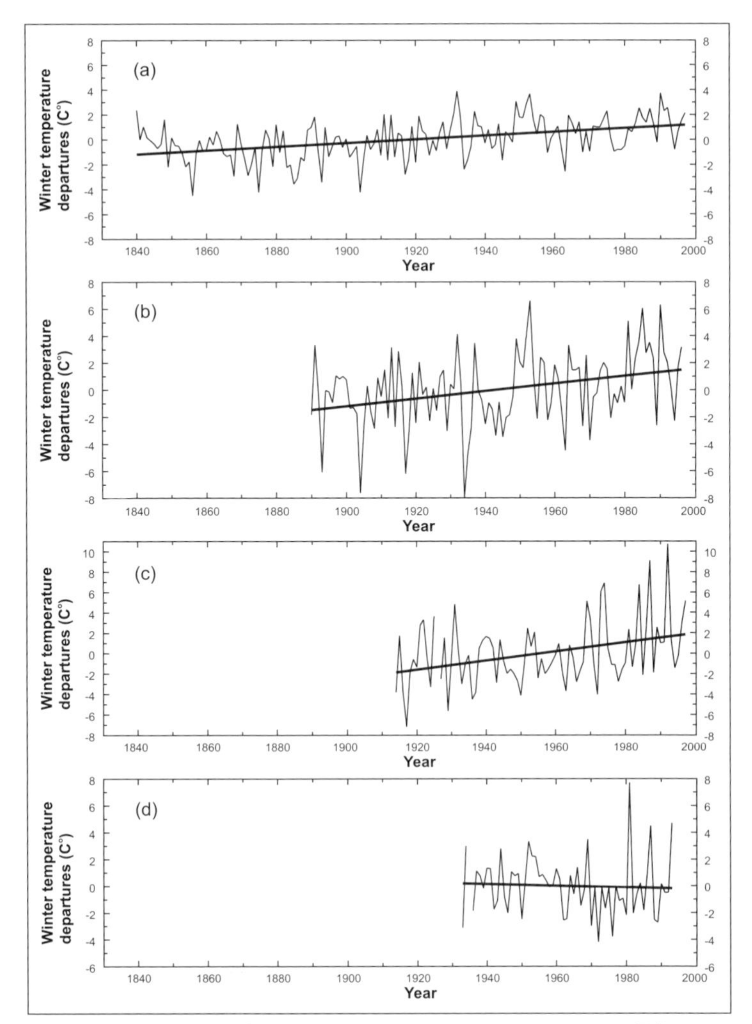

Figure 2.8 Time series of departures in winter temperature (January, February, and March) from the period-of-record means for selected Ontario climate stations: (a) Toronto (downtown); (b) Ottawa; (c) Dryden; and (d) Moosonee. (Data from Atmospheric Environment Service 1998)

to differentiate them from regosols, but leaching and weathering do not occur to the extent of forming a podzolic B layer (Clayton et al. 1977).

The brunisol great groups represented in Ontario include the melanic, eutric, and dystric types. Melanic brunisols develop on calcareous, rolling till and lacustrine deposits in eastern Ontario. Their distribution is closely associated with the limestone and shale plains surrounding the Shield, particularly on Manitoulin Island and in the Bruce Peninsula (Figure 2.3). Eutric brunisols also develop on calcareous till; in Ontario, their major distribution is associated with organic soils on a band of till between the organic deposits of the Hudson Bay lowlands and the lacustrine deposits of glacial Lake Agassiz. Dystric brunisols are more closely associated with acidic bedrock, with loamy to sandy acidic glacial till, and with outwash and lacustrine material. The Ontario brunisols are typically intermixed with luvisols and podzols in the western Shield area of the province.

Luvisols

Luvisolic soils are dominant in the lacustrine deposits of the Claybelt, intermixed with gleysolic and organic soils. They are also abundant in the lacustrine deposits near Kenora and Dryden and north of Opasquia Park. Southern Ontario soils are also dominantly luvisolic, intergrading with the band of melanic brunisols below the Shield in the north of the province and the gleysols of the Windsor and Niagara areas in the south.

Like brunisols, luvisols typically develop on relatively calcareous, forested glacial till and on glaciofluvial and glaciolacustrine deposits (Figure 2.3). The downward movement of forest litter and clay plays an important role in luvisol horizon development, but not to the extent of forming a podzolic B layer. Luvisols are typically intermixed with other forest soils, particularly the brunisols and podzols, but where transitions to grassland occur, gleysols are common co-dominants (Clayton et al. 1977).

Two great groups of luvisols are prominent in Ontario, the gray luvisols that develop in cool boreal and subarctic conditions, and the gray-brown luvisols found extensively in the south. The subarctic luvisols are typically unproductive; however, in boreal conditions, they are well suited to forest production (Thompson 2000, this volume). Large areas of luvisols support agriculture in southern Ontario and in the Claybelt.

Gleysols

Gleysolic soils are a dominant feature of the Claybelt,

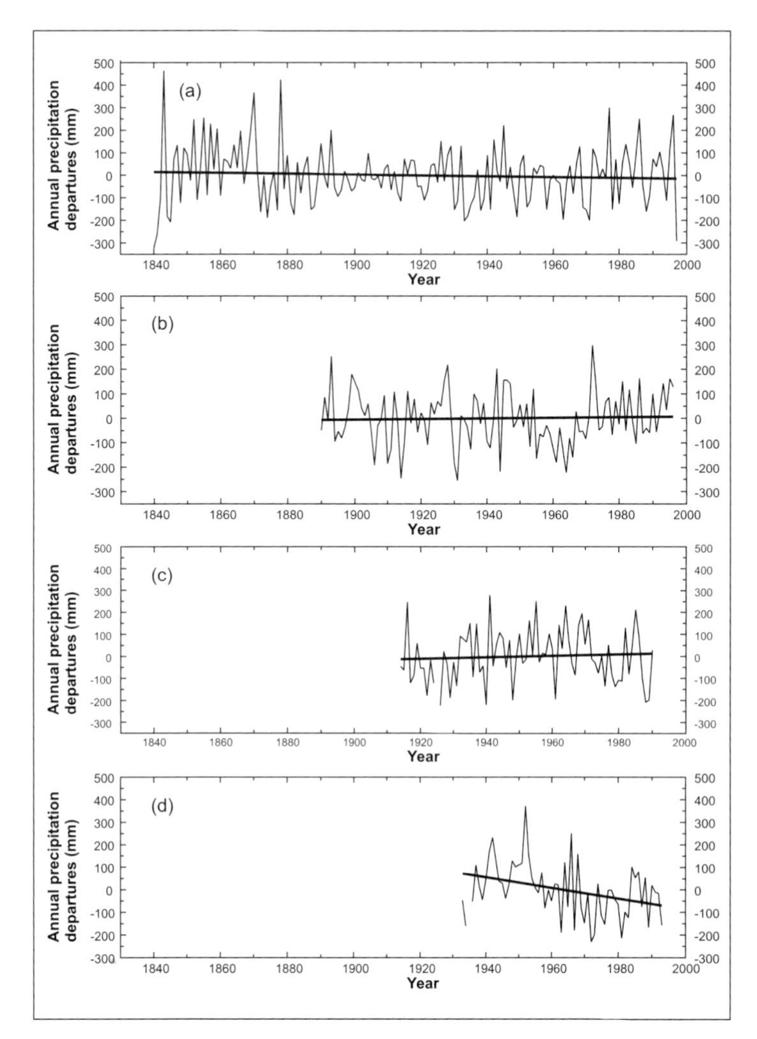

Figure 2.9 Time series of departures in annual precipitation from the period-of-record means for selected Ontario climate stations: (a) Toronto (downtown); (b) Ottawa; (c) Dryden; and (d) Moosonee.
(Data from Atmospheric Environment Service 1998)

and also occur in poorly drained areas north of Lake Erie and in the vicinity of Rainy River (Figure 2.11). These soils also form a thin band along the shore of James Bay, and are found in depressions, interspersed with other soils, in many other landscape complexes.

Gleysols typically develop on nearly level, calcareous tills, on lacustrine deposits, and on slowly permeable clay plains (Figures 2.2 and 2.3). These areas are typically poorly drained (Figure 2.11) or saturated with static water. It is this water which causes the gleyed and mottled layers, associated with reduction processes, that define these soils. Gleysols are often interspersed with gleyed versions of luvisols and organics at local sites of poor drainage occurring in otherwise semiarid surroundings (Clayton et al. 1977). Gleysols successfully support agriculture in southern Ontario and the Claybelt. In conditions where these soils usually develop, however, the lack of aeration and the proximity of groundwater to the surface often make them unproductive forest soils.

Podzols

Podzols are the dominant soils of the central and southern Shield in Ontario. They extend from north and east of Lake Superior, to the Ontario-Quebec border, and from the Claybelt to the southern limit of the Shield.

Podzols develop under forest stands on coarse-textured, stony, glacial tills and outwash, and on glaciofluvial sand lying on acidic parent material. The close association between these soils and acidic Shield conditions is evident in their correlation with the Frontenac Axis, which bisects plains of limestone and shale near Kingston (Figures 2.3 and 2.10). Forest litter is critical to the development of these soils. Strong weathering and leaching of organic matter and the presence of soluble aluminum and iron form the distinctive grayish podzolic B layer. Podzols develop in shallow layers in the Shield environment, and are more closely associated with the underlying bedrock parent material than many of the forest soils that develop on thicker glacial deposits (Clayton et al. 1977).

The dominant great group in Ontario, the humo-ferric podzols, typically develops on well-drained boreal sites, over coarse, iron-rich, acidic Shield areas (Figure 2.11). These soils are most commonly associated with exposed bedrock. Luvisols and brunisols, as well as organics and gleysols, also develop in depressions scattered across the Shield. Humid conditions, responsible for the substantial leaching of these soils,

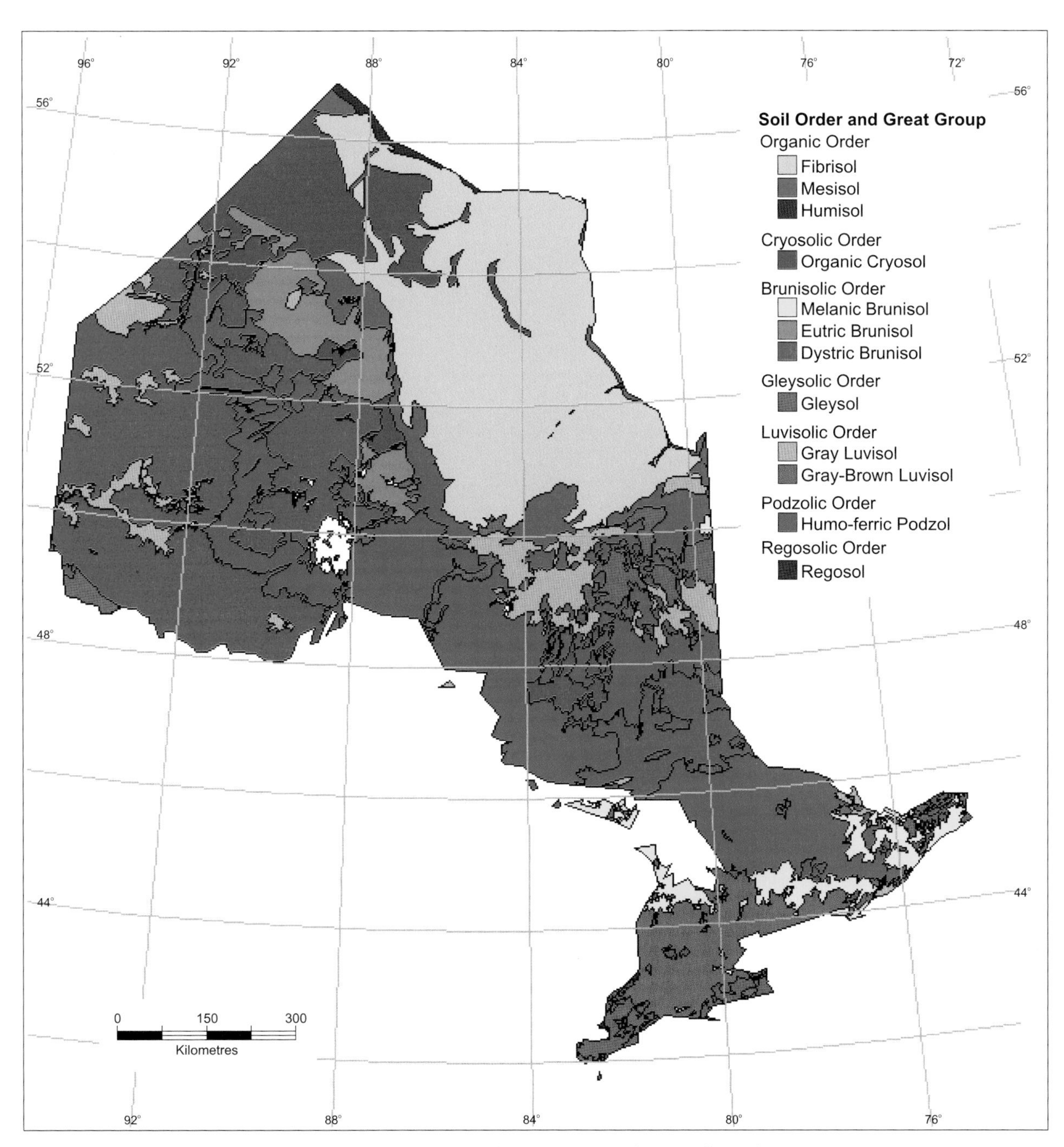

Figure 2.10 Dominant soil orders and great groups in Ontario, based on Soil Landscapes of Canada units.
(Data from Agriculture and Agri-Food Canada 1996)

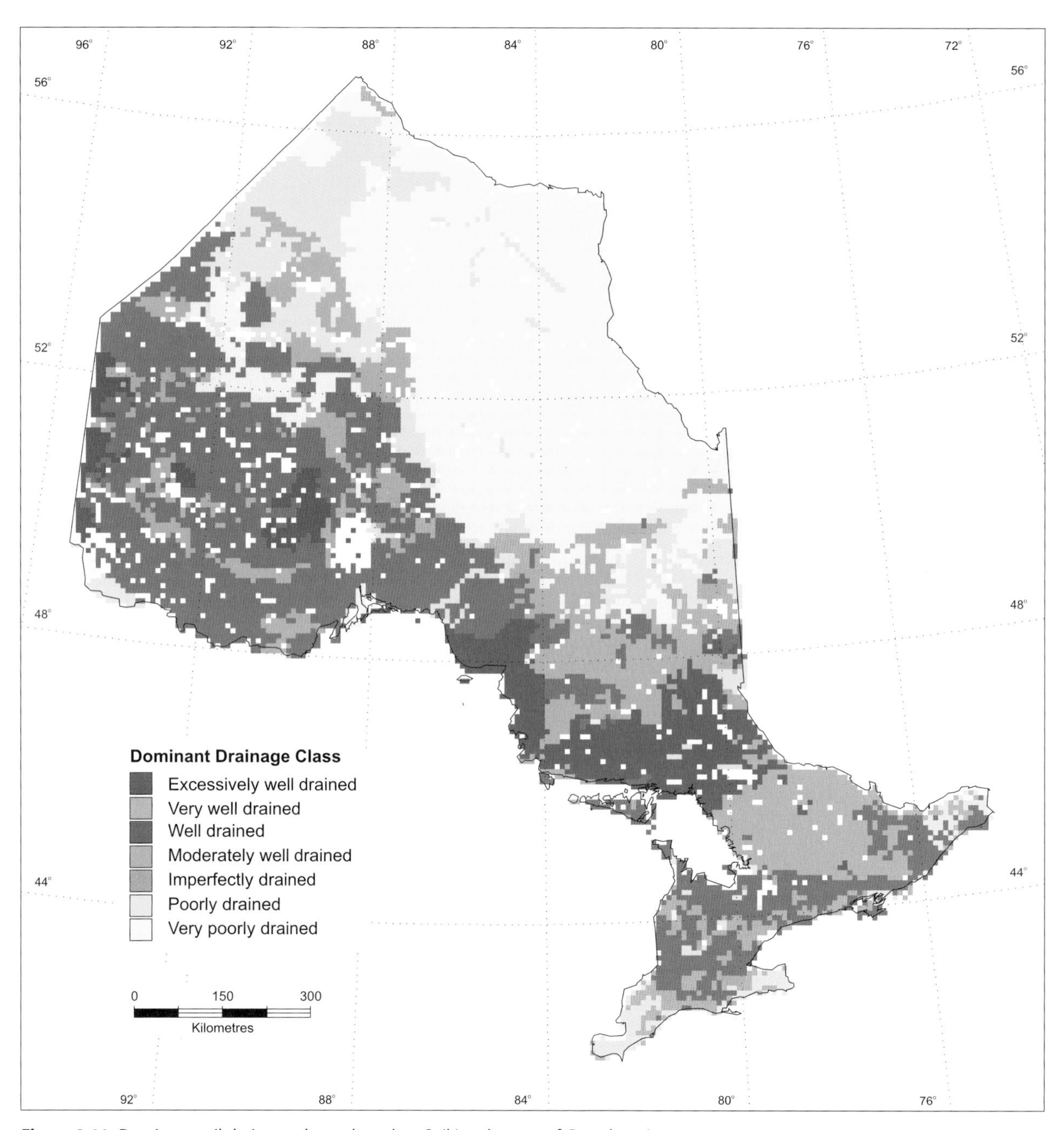

Figure 2.11 Dominant soil drainage classes based on Soil Landscapes of Canada units.
(Data from Agriculture and Agri-Food Canada 1996)

are also a determinant of their distribution (Figure 2.5a).

Regosols

Regosolic soils are the least prevalent of the major orders in Ontario; they dominate only a thin band along the Hudson Bay shoreline in the far northwest corner of the province. These are young soils, characterized by a lack of horizon development sufficient to meet the criteria of other soil types (Clayton et al. 1977). These soils directly reflect the parent material and may show cryoturbation, but not to the extent of the cryosolic order. Regosols are typically associated with tundra vegetation growing on materials of various grain sizes; however, the band of regosols in Ontario is developed on fine marine sediments.

The soils of Ontario are literally a product of their environment, in that they reflect either underlying bedrock units or patterns of surficial geology. The distribution of soil types and topography combine to determine drainage characteristics across the province.

HYDROLOGY

The drainage divide between the Great Lakes system and Hudson Bay is located close to the Great Lakes, so that most of the province drains towards the north (Figure 2.11). The area of individual drainage basins north of the divide can exceed 100,000 km^2 (Figure 2.12). Precipitation, evaporation, and runoff determine the annual water balance of a watershed. Strong variation in annual water balance occurs across the province, following gradients in precipitation, temperature, surface cover, and soils. The general trend from south to north of lower temperatures, shorter growing season, sparser canopy cover, and more predominantly clay-rich soils, is reflected in a corresponding increase in the ratio between annual runoff and annual precipitation. The extent of permafrost areas increases across the Hudson Bay lowlands and follows trends in soil drainage and temperature. Similarly, the combination of more poorly drained soils and lower temperatures is associated with a transition from forest to tundra and an increase in runoff ratio, despite lower overall amounts of precipitation.

Our comparison of the 30-year mean flow from 25 watersheds in Ontario with 30-year mean annual precipitation shows runoff ratios ranging from 30 to 40 percent in the watersheds of southern Ontario draining into Lake Ontario and Lake Erie. The watersheds draining to the north show runoff ratios well above 60 percent. Local variations in runoff ratio follow local variations in climate and soils as well as in the extent of forest and other land cover types.

Recent measurements of water balance recorded at sites in the boreal forest of northern Manitoba and central Saskatchewan show that mean daily evaporation during the growing season is roughly 1 mm per day, a rate which reflects energy limitations, stomatal limitations imposed by low root and soil temperatures, high vapour-pressure deficits, and sparse canopy cover (Sellers et al. 1995). These conditions are probably characteristic of the northern and western boreal biomes in Ontario as well, where they raise the potential for groundwater and soil-water runoff.

SUMMARY

The province of Ontario contains a range of natural landscapes that reflect variations in bedrock geology and climate, and particularly variations in surficial geology derived mainly from late-Quaternary glaciation. The landscapes are relatively young, in terms of toposequences and substrate, and are characterized primarily by the deranged drainage pattern characteristic of recent glacial activity. In addition to lakes and wetlands, the terrain is blanketed by interwoven patterns of well-drained sands, poorly drained clays, and extensive peat deposits, each with its characteristic forest associations. Across most of the province, surface water runoff is abundant, although there is a trend toward water deficit in the extreme western and southern regions.

Patterns of climate, topography, and hydrology are all strongly related at the provincial level. These broad patterns dictate the finer-scale patterns of soil development that result from both mesoscale and microscale climatic variation and from glacial, glaciolacustrine, and glaciofluvial deposition. The broad patterns outlined in this chapter control many of the processes occurring at the broadest landscape levels and provide the context for finer-scale landscape processes at work within landscape units. The nature of the relationships between landscape patterns and processes is examined in several of the chapters that follow.

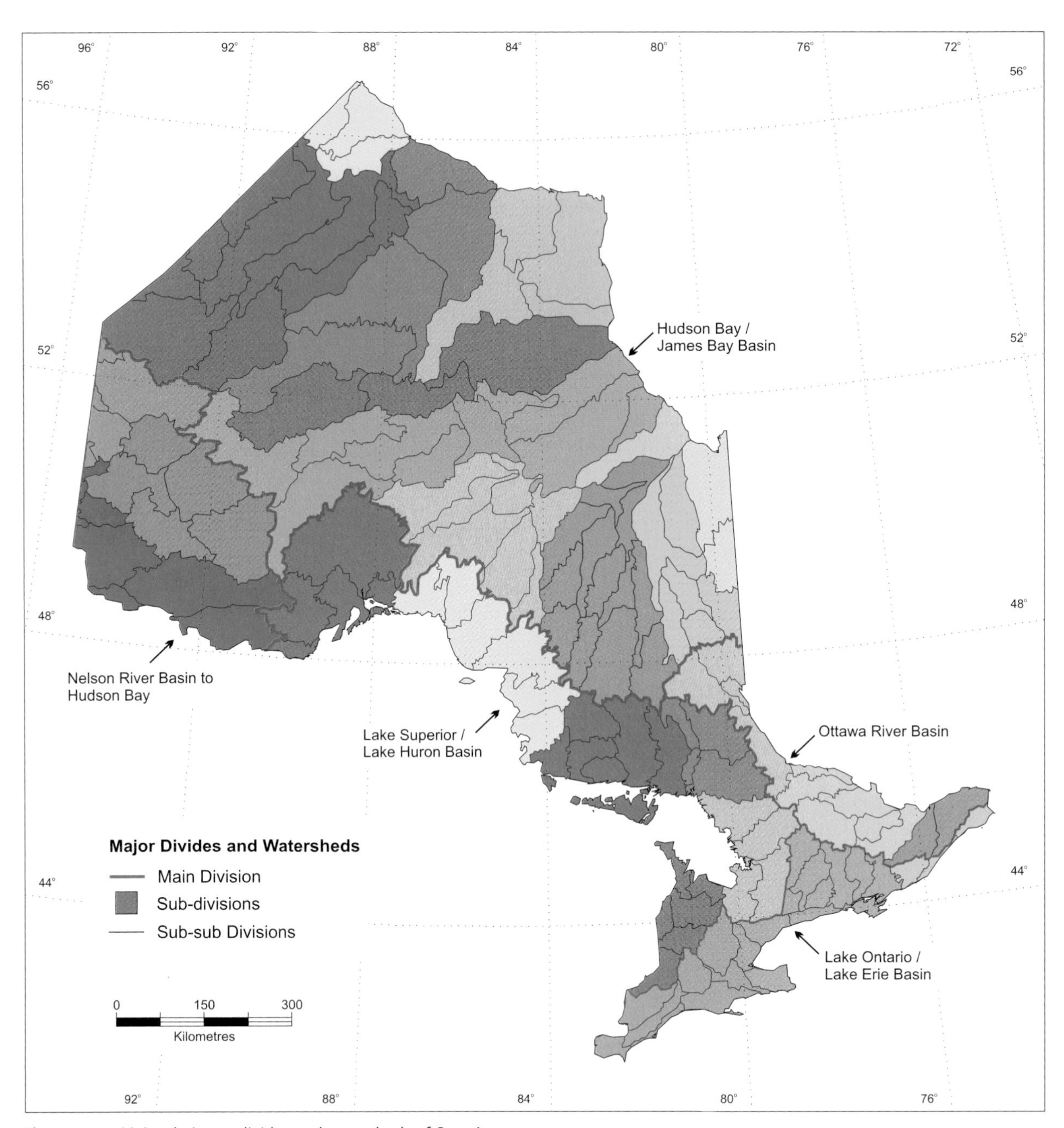

Figure 2.12 Major drainage divides and watersheds of Ontario.
(Data from Environment Canada 1990)

REFERENCES

Agriculture and Agri-Food Canada. 1996. Soil Landscapes of Canada. Version 2.2, National Soil Data Base. Ottawa, Ontario: Agriculture and Agri-Food Canada.

Agriculture Canada. 1987. The Canadian System of Soil Classification. Second Edition. Publication No. 1646. Ottawa, Ontario: Agriculture Canada, Research Branch. 164 p.

Bostock, H.S. 1970. Physiographic subdivisions of Canada. In: Geology and Economic Minerals of Canada. R.J.W. Douglas (editor). Economic Geology Report No. 1. Ottawa, Ontario: Geological Survey of Canada. 10-30.

Brennand, T.A. and J. Shaw. 1994. Tunnel channels and associated landforms, south-central Ontario: their implications for ice-sheet hydrology. Canadian Journal of Earth Sciences 31: 505-522.

Card, K.D., W.R. Church, J.M. Franklin, M.J. Frarey, J.A. Robertson, G.F. West, and G.M. Young. 1972. The southern province. In: Variations in Tectonic Style in Canada. Special Paper No. 11. Ottawa, Ontario: Geological Association of Canada. 335-380.

Chapman, L.J. and D.F. Putnam. 1984. Physiography of Southern Ontario. 1973. Special Volume No. 2. Toronto, Ontario: Ontario Geological Survey. 270 p.

Clayton, J.S., D.B. Ehrlich, J.H. Cann and I.B. Marshall. 1977. Soils of Canada. Volume 1, Soil Report. Ottawa, Ontario, Ontario: Canada Department of Agriculture, Research Branch. 243 p.

Dredge, L.A. and W.R. Cowan. 1989. Quaternary geology of the southwestern Canadian Shield. In: Quaternary Geology of Canada and Greenland. R.J. Fulton (editor). Geology of Canada, Publication No. 1. Ottawa, Ontario: Geological Survey of Canada. 214-235.

Dyke, A.S., J-S. Vincent, J.T. Andrews, L.A. Dredge, and W.R. Cowan 1989. The Laurentide ice sheet and an introduction to the Quaternary geology of the Canadian Shield. In: Quaternary Geology of Canada and Greenland. R.J. Fulton (editor). Geology of Canada, Publication No. 1. Ottawa, Ontario: Geological Survey of Canada. 178-189.

Flannigan, M.D. and M.G. Weber. 2000. Influences of climate on Ontario forests. In: A.H. Perera, D.L. Euler, and I.D. Thompson (editors). Ecology of a Managed Terrestrial Landscape: Patterns and Processes of Forest Landscapes in Ontario. Vancouver, British Columbia: University of British Columbia Press. 103-114.

Forest Landscape Ecology Program, Ontario Ministry of Natural Resources. 1996. Surficial Geology of Ontario: Digital Database and Viewing Software [CD-ROM]. Sault Ste. Marie, Ontario: Ontario Forest Research Institute, Ontario Ministry of Natural Resources.

Hardy, L. 1977. La déglaciation et les épisodes lacustres sur le versant québécois des basses terres de la baie de James. Géographie physique et Quarternaire 31: 261-273.

Mackey, B.G., D.W. McKenney, C.A. Widdifield, R.A. Sims, K. Lawrence, and N. Szcyrek. 1994. A New Digital Elevation Model of Ontario. NODA/NFP Technical Report TR-6. Sault Ste. Marie, Ontario: Canadian Forest Service. 31 p.

Myneni, R., C.D. Keeling, G. Asrar, C.J. Tucker, and R. Nemani. 1997. Increased plant growth in the northern high latitudes from 1981-1991. Nature 392: 323-325.

Ontario Geological Survey. 1991a. Bedrock Geology of Ontario, Northern Sheet. Map No. 2541. Scale: 1:1,000,000. Toronto, Ontario: Ontario Geological Survey.

Ontario Geological Survey. 1991b. Bedrock Geology of Ontario, West-Central Sheet. Map No. 2542. Scale: 1:1,000,000. Toronto, Ontario: Ontario Geological Survey.

Ontario Geological Survey. 1991c. Bedrock Geology of Ontario, East-Central Sheet. Map No. 2543. Scale: 1:1,000,000. Toronto, Ontario: Ontario Geological Survey.

Ontario Geological Survey 1991d. Bedrock Geology of Ontario, Southern Sheet. Map No. 2544. Scale: 1:1,000,000. Toronto, Ontario: Ontario Geological Survey.

Ritchie, J.C. 1989. History of the boreal forest in Canada. In: R.J. Fulton (editor). Quaternary Geology of Canada and Greenland. Geology of Canada. Publication No. 1. Ottawa, Ontario: Geological Survey of Canada. 508-512.

Sellers, P., F. Hall, H. Margolis, B. Kelly, D. Baldocchi, G. den Hartog, J. Cihlar, M.G. Ryan, B. Goodison, P. Crill, K.J. Ranson, D. Lettenmaier, and D.E. Wickland. 1995. The Boreal Ecosystem-Atmosphere Study (BOREAS): an overview and early results from the 1994 field year. Bulletin of the American Meteorological Society 76: 1549-1577.

Teller, J.T. 1985. Glacial Lake Agassiz and its influence on the Great Lakes. In: P.F. Karrow and P.E. Calkin (editors). Quaternary Evolution of the Great Lakes. Special Paper No. 30. Ottawa, Ontario: Geological Association of Canada. 1-16.

Thompson, I.D. 2000. Forest vegetation of Ontario: factors influencing landscape change. In: A.H. Perera, D.L. Euler, and I.D. Thompson (editors). Ecology of a Managed Terrestrial Landscape: Patterns and Processes in Forest Landscapes of Ontario. Vancouver, British Columbia: University of British Columbia Press. 30-53.

Thurston, P.C. 1991. Geology of Ontario: introduction. In: P.C. Thurston, H.R. Williams, R.H. Sutcliffe, and G.M. Scott (editors). Geology of Ontario. Special Volume No. 4. Toronto, Ontario: Ontario Geological Survey. 3-26.

Trenhaile, A. S. 1998. Geomorphology: A Canadian Perspective. Toronto, Ontario: Oxford University Press. 340 p.

Watson, B. G. and D.C. MacIver. 1995. Bioclimate Mapping of Ontario. Final report to the Ontario Ministry of Natural Resources. Unpublished. 55 p.

Webber, L.R. and D.W. Hoffman. 1970. Origin, Classification and Use of Ontario Soils. Publication No. 51. Toronto, Ontario: Ontario Department of Agriculture and Food. 58 p.

3 Forest Vegetation of Ontario: Factors Influencing Landscape Change

IAN D. THOMPSON*

INTRODUCTION

Ontario is the second largest province of Canada, covering 15 degrees of latitude and 20 degrees of longitude, and extending over more than 1 million km^2, of which 74 percent is forested (Ontario Ministry of Natural Resources [OMNR] 1996). The province has the most striking latitudinal zonation of forest regions of any political jurisdiction in North America east of the western cordillera. Four major forest regions occur along a latitudinal gradient of little more than 1000 km. These four regions are the Carolinian (or deciduous) forest; Great Lakes-St. Lawrence forest; boreal predominantly forest; and boreal forest and barrens (Rowe 1972) (Figure 3.1). A fifth region, consisting of true tundra and treed tundra, occurs at locations along the coast of Hudson Bay, such as at Cape Henrietta Maria. The focus of this chapter is the treed landscapes of the boreal forest and Great Lakes-St. Lawrence forest (hereafter referred to as the Great Lakes forest or region) which cover the entire central area of the province from Manitoba to Quebec, an area of approximately 64 million ha.

Figure 3.1 The forest regions of Ontario.
(Adapted from Rowe 1972)

Viewed from a height, the mixtures of forest types and age classes scattered across Ontario's forest landscape appear as a heterogeneous mosaic. It is the combination of different tree species in the forest patches, caused by many factors, which results in landscape pattern or structure. In the sections that follow, factors that control or influence the spatial pattern of forest types are examined. "Ecosystem" is a scalar concept; nevertheless, the "forest types" referred to in this

* Canadian Forest Service, Great Lakes Forest Research Centre, 1219 Queen Street East, Sault Ste. Marie, Ontario P6A 5M7

chapter are forest ecosystems, in the sense developed by Rowe (1961, 1962) and advanced by forest ecosystem classification systems used in Ontario (e.g., Sims et al. 1989; Chambers et al. 1997). The ultimate factors that control landscape structure act in a "top-down" manner, affecting ecological processes over large areas and long time-scales, while more proximate factors influence individual patch sizes and their rates of change over shorter time periods. Humans have played a large role in determining forest distribution and contemporary landscape structure across Ontario; however, the effects of logging on forest characteristics are not dealt with in detail here, except to introduce how factors that control landscape structure may have been altered in recent decades.

This chapter is intended to provide a broad overview of the forest vegetation of the province and to introduce the factors which determine the distribution of forest types. The task is large, so the level of detail that an individual reader seeks on a particular topic may not be found here. In some cases, the reader is guided to a more detailed discussion of the topic in subsequent chapters.

OVERVIEW OF THE BOREAL AND GREAT LAKES FOREST REGIONS

Most of the timber management in Ontario takes place in the boreal and Great Lakes forests (OMNR 1996). The landform features of both regions are dominated by glacial drift and postglacial marine and lacustrine sediments, as well as by exposed bedrock resulting from many Quaternary glacial events. As a further consequence of recent glaciation, the current forest associations are only about 7000 years old. Wetlands, areas of interrupted drainage, and numerous lakes and ponds are important components of the Ontario forest landscape and provide an irregular variety to its pattern. In the western part of the province, more than 25 percent of the land base is covered with water.

The Boreal Forest Region

The boreal forest (Figure 3.1) extends across the entire north-central area of the province, where it is underlain mostly by the Precambrian granite and gneiss bedrock of the Canadian Shield. The boreal forest has relatively few tree species, and these are adapted to a cold climate and frequent fires. This region is dominated by coniferous and mixedwood forests, occurring on coarse-textured and often shallow humo-ferric podzols, or brunisols, although luvisols are locally common in some areas (Baldwin 1991).

The local distribution of forest ecosystems is influenced by site conditions, as modified by disturbance events. The dominant natural disturbance is fire, coupled with and influenced by periodic local or widespread insect infestations and blowdown events. Because the climate differs from west to east, fire regimes across the province are not uniform, and forest characteristics vary regionally as a result. The approximate northern and southern limits of the boreal forest coincide with the July mean 13°C and 18°C isotherms (Ritchie and Hare 1971) and the average position of the summer and winter arctic air masses (Baldwin 1991). Annual productivity in the boreal forest is about 1.6 m^3 per hectare (Bickerstaff et al. 1981).

The most common dominant tree species associations are as follows: jack pine (*Pinus banksiana*) and black spruce (*Picea mariana*); jack pine and black spruce mixed with white birch (*Betula papyrifera*) and trembling aspen (*Populus tremuloides*); trembling aspen, white spruce (*Picea glauca*) and black spruce; trembling aspen and balsam fir (*Abies balsamifera*); and black spruce and balsam fir. Extensive stands of homogeneous black spruce dominate the landscape of the Claybelt in the east, but even here slight rises in topography provide moderately dry sites which support mixed forests with balsam poplar or trembling aspen and white birch. Wetter sites support either pure stands of black spruce or mixtures of black spruce with tamarack (*Larix laricina*) or, less commonly, eastern white cedar (*Thuja occidentalis*). The most abundant forest resource inventory working groups in the boreal forest, according to forest mapping conducted by the Ontario Forest Resources Inventory (FRI), are black spruce and jack pine, which account for 70 percent of the forest area (OMNR 1996).

The Great Lakes Forest Region

The Great Lakes region occurs in two distinct geographical areas of Ontario, one west, and the other east and southeast, of Lake Superior. These two areas are separated by an area of boreal forest that extends along the shore of Lake Superior from east of Thunder Bay to west of Lake Superior Provincial Park (Figure 3.1).

The forests of the Great Lakes region contain a greater variety of tree species and other plants than the boreal forests. These forests are generally mixedwoods and deciduous stands growing on a range of soil types, including podzols, luvisols, and brunisols (Rowe 1972). The Great Lakes region is characterized by a longer growing season and more precipitation than the boreal region. The major natural disturbances in the Great Lakes forests are fire and blowdown, but fewer fires occur than in the boreal region, and both disturbances tend to occur in smaller patches. Annual productivity in the Great Lakes forest is about 1.8 m^3 per hectare (Bickerstaff et al. 1981), approximately 13 percent greater than the annual productivity of the boreal forest.

Common tree associations of the Great Lakes forests are the following: mixed hardwoods (sugar maple [*Acer saccharum*], yellow birch [*Betula lutea*], and beech [*Fagus grandifolia*]); mixed hardwoods with white pine (*Pinus strobus*); trembling aspen with balsam fir; hemlock (*Tsuga canadensis*) with sugar maple and yellow birch; and white pine with red oak (*Quercus rubra*). The most abundant forest resource inventory working groups in the Great Lakes forest, based on FRI mapping, are sugar maple and poplars, which cover 45 percent of the forested areas (OMNR 1996). White pine was once a dominant cover species, but was heavily logged throughout the Great Lakes forest from the early 1700s. As a result, most of the stands once dominated by white pine are now gone (Aird 1985). A few examples of old-growth pine forests have been protected, including stands near Temagami, and at Bark Lake, near the town of Elliott Lake.

Transition Zone between the Boreal and Great Lakes Regions

A transition zone occurs between the two principal forest regions, where plant species from both may codominate in the landscape, depending on site type, history, and microclimate. The transition zone between the boreal and Great Lakes forests roughly corresponds to the 5°C mean annual isotherm east of Lake Superior and the 4°C mean annual isotherm to the west, or the area of 2000 to 2300 degree-days for both. Pastor and Mladenoff (1992) also suggest that there is a strong correspondence between the location of the boreal-Great Lakes forest ecotone and the line depicting 120 frost-free days, while Arris and Eagleson (1989) propose that the transition occurs at the -40°C annual minimum isotherm.

Only the most hardy and fire resistant of the true Great Lakes forest tree species, in particular red pine (*Pinus resinosa*) and white pine, occur at the extremes of the transition zone. While few of the Great Lakes species exist at the extreme northern edge of this ecotonal forest, the full complement of boreal species is present well south into the Great Lakes zone. Carleton and Maycock (1978) argued that the lack of Great Lakes species in the north may not be governed solely by climatic factors, but also by the greater probability of fire occurrence in boreal habitats than in areas to the south.

A BRIEF AUTECOLOGY OF MAJOR TREE SPECIES IN ONTARIO

There are about 24 tree species in the boreal region, of which only seven are abundant, while in the Great Lakes region there are as many as 45 tree species, of which fewer than 20 are abundant and others occur only in the eastern portion of the region (Table 3.1) (distributional data from Farrar 1995). The number of tree species declines latitudinally from a maximum of 79 species in the Carolinian or deciduous forest region in the extreme south of Ontario to a minimum of four or five species along Hudson Bay (Figure 3.2). The most rapid decline in species in the landscape occurs across the zone of transition forest, or ecotone, between the boreal and Great Lakes forests, west and east of Lake Superior, where there is a loss of about 30 species over a north-south distance of less than 200 km.

Many factors influence tree community assemblage, or landscape patches, at small scales. Responses by tree species to environmental influences and disturbances, and their ability to survive under local soil, moisture, and temperature regimes, is dictated by their autecology (Table 3.1). Individual tree species do best under certain site conditions and more poorly as these conditions decline from optimal in accordance with biophysical factors (Auclair and Goff 1971; Rowe 1972; Carleton and Maycock 1978). Plants may be unable to tolerate certain conditions at a given site, or some species may be better competitors on a given site type or under certain climatic conditions. Within both forest regions there are gradients in moisture and nutrient availability that influence which species occupy any site.

In Great Lakes forests, black spruce, white cedar, and black ash (*Fraxinus nigra*) are common on the wettest sites, while aspen, jack pine, red pine, and white pine prevail on the driest sites (Table 3.1). Sites with intermediate moisture levels maintain mixed forest stands usually dominated by sugar maple and beech. Eastern hemlock is a dominant species along

Table 3.1

A summary of autecological information for the important tree species of the boreal and Great Lakes forests in Ontario.

Tree species	*Forest biome*[1]	*Shade tolerance*[2]	*Fire tolerance*[3]	*Reproduction following fire*[4]	*Moisture tolerance*[5]	*Soil fertility*[6]	*Siterange Position*[7]	*Probable limiting factor*[8]	*Reproduction*
Black Spruce (*Picea mariana*)	B*, GL	L-M	L	H	H	L	L, (A)	450 GDD	Annual semi-serotinous cones
White Spruce (*Picea glauca*)	B, GL	M-H	L	L	L	M-H	U-M	450 GDD	Cone crops 2-3 yrs.
Jack Pine (*Pinus banksiana*)	B*, GL	L	L	H	L	L	U-M	1600 GDD	Annual semi-serotinous cones
Tamarack (*Larix laricina*)	B*, GL	L	L-M	L	H	H	L-M	1000 GDD	Cone crops 3-6 yrs.
Balsam Fir (*Abies balsamifera*)	B*, GL	H	L	L	M	M-H	U-M	500 GDD	Cone crops 2-4 yrs.
White Pine (*Pinus strobus*)	GL	M	H	H	M	L-M	M, (A)	2000 GDD	Cone crops 1-3 yrs.
Red Pine (*Pinus resinosa*)	GL	L	H	H	L	L	M-U	2000 GDD	Annual semi-serotinous cones
Eastern Hemlock (*Tsuga canadensis*)	GLe	H	L	M	M+	H	L-M	2300 GDD	Cone crops 2-3 yrs.
Eastern White Cedar (*Thuja occidentalis*)	GL, B	H	L	L	H	M-H	L	600 GDD	Cone crops 3-5 yrs.
Trembling Aspen (*Populus tremuloides*)	B, GL	L	L	M	M-L	M-H	M-U	600 GDD	Annual seeds, suckers
Balsam Poplar (*Populus balsamifera*)	B*, GL	L	H	H	M-H	H	L-M	500 GDD	Annual seeds, suckers
Large-toothed Aspen (*Populus grandidentata*)	GL	L	L	H	L	L-M	M-U	2200 GDD	Annual seeds, suckers
White Birch (*Betula papyrifera*)	B*, GL	L	L	H	L-M	M-H	M-U	1000 GDD	Seed crops 1-2 yrs.
Sugar Maple (*Acer saccharum*)	GLe	H	M	L	L-M	M	M-U	2000 GDD	Annual seeds
Red Maple (*Acer rubrum*)	GLe	M	L	M	M-H	M	A	2000 GDD	Annual seeds, suckers
Yellow Birch (*Betula alleghaniensis*)	GLe	M	L	H	M	M	M	2000 GDD	Annual seeds
Beech (*Fagus grandifolia*)	GLe	H	L	L	M	M-H	M	2300 GDD	Seed crops 2-3 yrs.
White Ash (*Fraxinus americana*)	GLe	M	L	L	L	H	U	2500 GDD	Seed crops 1-2 yrs.
Black Ash (*Fraxinus nigra*)	GL*, B	M	L	L	H	H	L	1700 GDD	Annual seeds
American Basswood (*Tilia americana*)	GLe	H	L	L	L	M-H	M-U	2500 GDD	Seed crops 1-2 yrs.

▶

◄ ***Table 3.1***

Tree species	Forest biome[1]	Shade tolerance[2]	Fire tolerance[3]	Reproduction following fire[4]	Moisture tolerance[5]	Soil fertility[6]	Siterange Position[7]	Probable limiting factor[8]	Reproduction
Red Oak (*Quercus rubra*)	GLe	M	L	L	M	L-M	M-U	2200 GDD	Seed crops 2-5 yrs.
American Elm (*Ulmus americana*)	GL	L	L	L	M-H	H	M-L	1900 GDD	Seeds crops 1-2 yrs.

Data are summarized from Fowells (1965), Hosie (1969), Ritchie (1987), and Sims et al. (1990).
[1] B = boreal forest; GL = Great Lakes forest, e = species occurs in the east, and either rarely or not at all west of Lake Superior
[2] L = low (only grows in full sun), M = medium (requires or can grow in partial shade), H = high (grows well under canopy)
[3] L = readily killed by fire, M = killed in hot fires, H = may survive all but major catastrophic fires
[4] L = does not reproduce well after fire, M = intermediate response to fire, H = well-adapted to reproduce following fire
[5] L = does not grow on wet sites, M = does best on fresh sites, H = can grow on wet sites
[6] L = low fertility tolerant, M = intermediate, grows on most soil types, H = requires high fertility soils
[7] L = lower slope, M = mid-slope, U = upper slope, A = all
[8] GDD = growing-degree-days >5.5°C (Ritchie 1987)
* most common in this biome

lakeshores on damper, acidic soils, but declines on higher ground as moisture decreases (Barnes 1991). Common species with a wide moisture tolerance include red maple and black spruce. In boreal forests, dry sites are dominated by jack pine and black spruce, wet sites by black spruce and sometimes tamarack, and intermediate sites by a mixture of most species, including aspen, balsam fir, and white spruce. Plant growth on sites that are extremely well drained may be nitrogen-limited, and it is predominantly the conifers, especially black spruce, hemlock, white pine, red pine, and jack pine, that are adapted to low nitrogen availability (Chapin et al. 1986). Nitrogen limitation is a common feature of cold, boreal systems, where the needle-dominated litter decomposes far more slowly than the leaf litter of the warmer Great Lakes systems (Pastor and Mladenoff 1992).

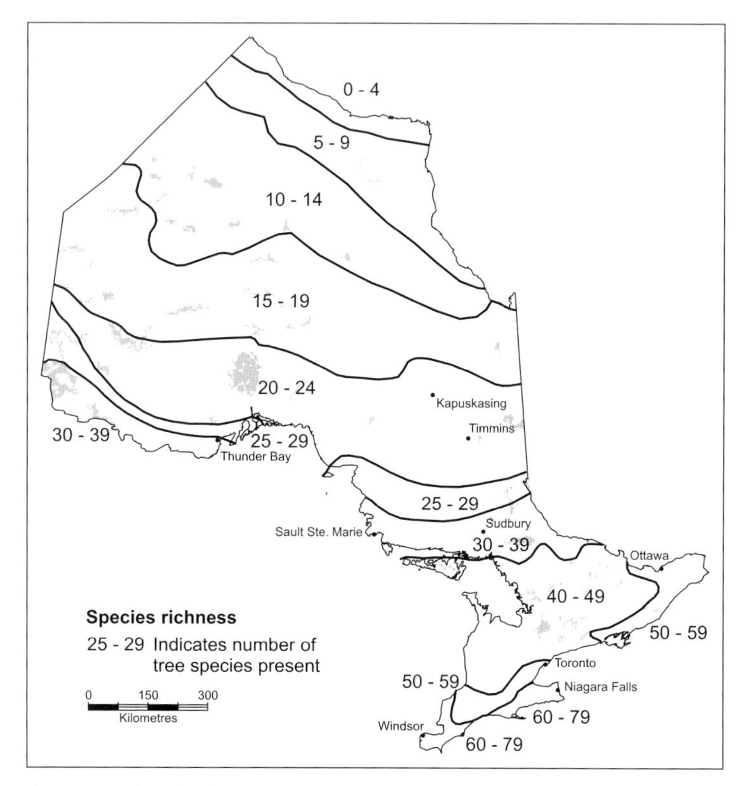

Figure 3.2 The distribution of tree species richness in Ontario.
(Data from Farrar 1995)

Certain tree species, including the pines, black spruce, the poplars, and white birch, are well adapted to fire as an agent of propagation. Other species tend to seed in under an established forest canopy and develop in shaded conditions. These shade-tolerant species include most of the southern hardwoods, balsam fir, and white spruce (Table 3.1). Fire does not favour these species by making it easier for them to occupy a site; in fact, balsam fir is an extremely shade-tolerant species which must develop under a canopy and can be eliminated from a stand by fire. Other species that have an intermediate tolerance to direct sunlight, such as red maple and yellow birch, grow in partially open conditions, often in gaps caused by fallen trees. Species with large seeds, such as beech, hickories (*Carya* spp.), hornbeams (*Ostrya* spp.), and oaks may not disperse as effectively within an area as do species with lighter, windborne seeds. Many of these heavy-seeded species occur only sporadically and always in mixed stands within Great Lakes forests. Many species cannot tolerate extreme cold and are thus confined to more southern areas. Several species, including red spruce (*Picea rubens*), white ash (*Fraxinus americana*), hickories, and basswood (*Tilia americana*), occur only in the most southern areas of the region (Fowells 1965; Ritchie 1987). Almost all Great Lakes species are thermally limited at the northern edge of their distributions (Table 3.1).

DEVELOPMENT OF FOREST LANDSCAPES IN ONTARIO

The degree and rate of change in forest patches across the landscape are dictated by the rate at which patches are formed as a result of forest disturbances and the intensity of the forest disturbances themselves (Baker 1992; Syrjänen et al. 1994). Many authors have suggested that landscapes exist in some form of spatiotemporal equilibrium, that forest ecosystems are predictable, and that ecosystem re-assembly is possible following disturbance. In fact, these ideas form the basis for the paradigm of sustainable development.

Three main theories of landscape equilibrium are used to explain how the structure of landscapes may change with time. First, the "steady-state shifting mosaic" concept advanced by Bormann and Likens (1979) suggests that, over a large enough area, all ecosystems constantly exist in their various successional forms, even though various places on the landscape change with disturbance. This concept is unlikely to apply to landscapes, such as Ontario's highly heterogeneous boreal forests, that are formed principally by large-scale disturbances. Secondly, the theory of "stationary landscape equilibrium with stochastic perturbation" (Loucks 1970) proposes that stability exists in landscapes as a result of predictable disturbance frequency. Data from most forest types suggest, however, that history and serendipity have as much to do with ecosystem development as predictable disturbance regimes, or even more. Thirdly, Suffling (1991, 1995) has advanced the argument, using data from Ontario, that landscapes are in constant disequilibrium. This is a view shared by Baker (1995), on the basis of considerable modelling of northern Minnesota forests. Under this theory, landscapes are constantly "catching up" to disturbance regimes that change at two temporal scales, the longer one on the order of several hundreds of years. Current climate data does not provide the basis for predicting that present forest ecosystems, which formed after disturbances at least 200 years ago, will recur at any particular point in the future, because forest response to disturbance is a complex process involving numerous interacting factors and time lags (Drake 1990; Bonan 1992).

KEY INFLUENCES ON FOREST VEGETATION AT LARGE SCALES

The development of forest vegetation across a landscape is influenced by three ultimate factors acting at large spatial and temporal scales: physical geography (including soils, bedrock, and topography), climate, and history. Physical geography and climate set fundamental limits on forest development, while history represents the net (or "combined," or "cumulative") effect of specific events of forest growth, change, and destruction over time. Coupled with these ultimate factors are proximate factors, or factors which act at more limited spatial and temporal scales to influence stand development. These include natural disturbances such as fire and wind, human intervention, periodic changes in weather, competition among species, and herbivory.

Climate, for example, clearly exerts a major influence on landscape structure by constituting the ultimate cause of moisture and temperature regimes (Suffling 1995) and by acting through processes that determine the presence or absence of tree species (Ritchie 1987; Bonan and Shugart 1989; Mackey et al. 1996). Historical perspectives have demonstrated and predicted the migration of tree species and the formation of species communities at certain periods, in response to long-term climate change. Evidence has been presented from Ontario (Liu 1990; Suffling 1995; Thompson et al. 1998) and from other areas in eastern Canada (e.g., Ritchie 1987; Payette 1993).

Effects of Ultimate Factors on Ontario Forest Vegetation at Large Scales

The role of climate in the development of forested landscapes is discussed by Flannigan and Weber (2000, this volume). For the most part, Ontario has a humid continental climate, except for areas near Hudson Bay which have a cold and maritime climate. A marked increasing temperature gradient exists from north to south and an increasing precipitation gradient from northwest to southeast across the province.

Long-term Climatic Effects on Forest Vegetation: Paleoecology of Ontario Forests

Paleoecological research implicates climate as a key factor in forest change during the postglacial period (Webb 1986; Ritchie 1987) and suggests that plant migration can generally track climate change, regardless of rapidity (Pitelka 1997). Several excellent paleoecological and paleobotanical studies have been conducted on forests in Ontario, and in nearby Quebec and northern Minnesota, which show that forest types and landscape structure have changed dramatically with warming since the final Holocene glaciation (Richard 1980; Ritchie 1987; Lewis and Anderson 1989; Liu 1990; Graumlich and Davis 1993; Payette 1993).

Virtually all of Ontario, and of Canada, underwent glaciation 12,000 years BP. Much of northern Ontario was still covered by ice 8000 years BP (Hardy 1977; Ritchie 1987), and was partially covered by large meltwater lakes that did not fully recede until about 6000 years BP. The most relevant data for northern Ontario come from Liu (1990), Lewis and Anderson (1989), and Ritchie (1987), who show that the transition zone between the Great Lakes and boreal forest species shifted steadily northward, except for short cooling periods, from 11,000 years BP to about 3000 years BP, and that a mixedwood forest of the Great Lakes type was already present near Sudbury by 9500 years BP. After the glaciers retreated, the boreal species moved north first because of their tolerance to colder temperatures. The first species to invade was white spruce, which is well suited to cold climates, and a spruce forest was present in south-central Ontario between 12,500 and 10,000 years BP (Ritchie 1987). Jack pine followed rapidly (Ritchie 1987; Liu 1990; Payette 1993). By 10,500 years BP, the spruce forest had given way to a forest dominated by jack pine and white pine on well-drained sites, and by a mixed deciduous forest containing virtually all of the common Great Lakes species, although in varying proportions (Ritchie 1987).

The evidence of early invasion by white spruce is inconsistent with the dominance of black spruce in the northern Ontario boreal forests today. Liu (1990) suggested that the edaphic characteristics immediately following glaciation favoured white spruce, and that black spruce was only able to invade after a longer period of soil development. A second possibility is that fire, which favours black spruce, occurred over several hundreds of years and eliminated white spruce from much of the earlier range of that species, particularly during a warmer period from 6500 to 6000 years BP (Payette 1993). The present distributions of white spruce and balsam fir occur in suitable areas that are free of fire, such as mesic sites with nutrient-rich soils (Foster 1984). Indeed, Mutch (1970) suggested that fire alone was responsible for the conifer domination of forests throughout the Holocene. The data gathered by Liu (1990) suggests that the Holocene period when forests were dominated by white spruce could have been short (less than 200 years) in some areas, or much longer (more than 1000 years) in areas such as southern Ontario (Ritchie 1987).

White birch was the first deciduous species to invade the boreal forest. It apparently arrived several hundreds of years after white spruce, and was followed quickly by poplars. However, the pollen record is poor for both these species, making the record of their migration unreliable (Ritchie 1987; Payette 1992). A warming climate subsequently favoured the development of mixed and deciduous forests dominated by white pine, which replaced the spruces and jack pine 1200 years or more after boreal species first invaded the areas currently occupied by Great Lakes forest. These new mixed forests were originally dominated by white pine, red pine, sugar maple, eastern hemlock, and beech, although there was apparently considerable variation in species mixtures and times of arrival across Ontario (Davis 1984; Ritchie 1987; Liu 1990). Hemlock-deciduous forests replaced the initial stands of pine and pine-oak on moister sites, and eventually became the dominant type on these sites, where they have remained for at least 2000 years (Davis et al. 1996).

The climate from 7000 to 3000 years BP appears to have been warmer than at present, so that Great Lakes forest types, dominated by white pine, prevailed as far north as Timmins before retracting to the present line, south of Gogama, about 2500 years ago (Liu 1990). Jacobson and Dieffenbacher-Krall (1995) showed a marked contraction in the distribution of white pine, a major tree species of the Great Lakes forest region, with the 0.5°C cooling trend and the associated reduction in fires over the past 1000 to 1500 years. During this period, white cedar also moved north, particularly into the Claybelt and other areas where postglacial lacustrine soils existed (Richard 1980; Liu 1990). The retreat of the Great Lakes forest during the subsequent cooling period (from 3000 BP until 1970) left isolated remnants within the boreal zone. These disjunct stands may play an important role in the re-expansion of Great Lakes forests northward which is probable under the current trend of global warming (Thompson et al. 1998), especially if the "long-jump and outlier" model of tree migration proposed by Clark et al. (1998) is correct. Good examples of extralimital white pine or red pine stands occur in Kenogami and Robb Townships near Timmins and at Kabinagagami Lake, south and west of Hearst.

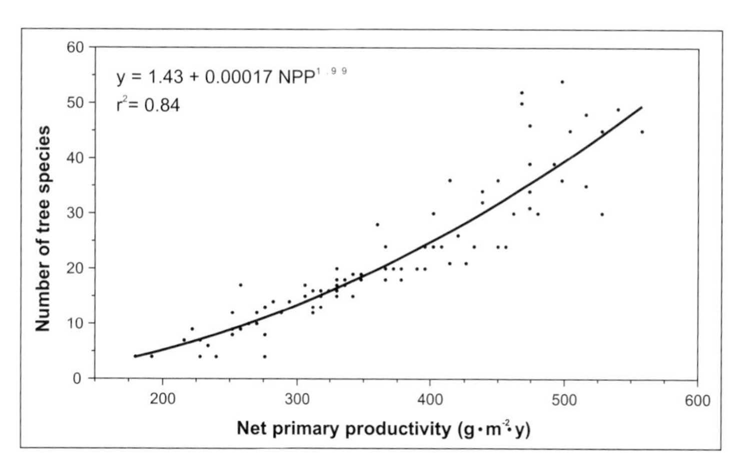

Figure 3.3 The relationship between tree species richness and net primary productivity.

Payette (1992) notes that, although climate and ecosystem processes can explain the migration and retraction of forest types, competition is an important factor in structuring plant communities. The current forests result not only from responses to climate and forest management, but also from the ability of individual species to compete, given specific historical sequences and chance events, and specific soil development and condition. We can learn from paleoecology how tree species responded to environmental changes in the past. Future forest responses to climate change may not be predictable on this basis, however, because the competitive regimes may have been altered by forest management, fire control, and stochastic events (Thompson et al. 1998; Pitelka 1997).

Climate, Primary Productivity, and Tree Species Distribution

Tree species richness correlates with many climatic variables, such as number of growing-degree-days, mean annual temperature, mean annual insolation, and total annual precipitation. Net primary productivity (NPP) is a variable that integrates many aspects of climate, such as length of growing season, soil temperature, and moisture regime (see Band 2000, this volume). Currie and Paquin (1987) found that annual evapotranspiration strongly described an observed variation with latitude and longitude in North American tree species richness. They suggested that this relationship may actually describe primary productivity, and that tree species richness is a predictable function of available energy.

Testing this hypothesis for Ontario, using the definition of a tree and the data on tree species distribution presented by Farrar (1995), reveals that a highly significant relationship exists between tree species richness and net primary productivity. For this comparison, net primary productivity (NPP) was derived from normalized difference vegetation index (NDVI) values, through use of the Regional HydroEcological Simulation System (RHESSys) model (Band et al. 1993). A multiple regression analysis related tree species richness to NPP and climate variables across 10,000-km² universal transverse mercator (UTM) grid blocks in the province ($r^2 = 0.84$, $F = 494.1$, df = 1, $P < 0.0001$) (Figure 3.3). In this model, NPP alone explained 87 percent of the variation in tree species richness across Ontario. Adding January minimum temperature to the model explained a further five percent of the variance. Another variable tested, annual water deficit, explained only minimal variance and was excluded from the final model. Other possibly important climate variables, such as annual rainfall and mean annual temperature, were correlated with primary productivity, so they were not included in the analysis individually.

At a broad level, primary productivity is an important variable which explains a high degree of the variance in the distribution of tree species in Ontario, and thus large-scale forest landscape patterns. Ultimately, climate controls where tree species and other plants can grow, acting through their responses to climate and their degree of adaptation to prolonged periods of cold weather. In particular, few deciduous species are found north of the line defining about 2000 growing-degree-days (Ritchie 1987) (Table 3.1), although Carleton (2000, this volume) reports some occurrences of southern species in riparian areas well north of this line.

A General Overview of Ontario's Physical Geography

Soils and geological setting are described in the previous chapter (Baldwin et al. 2000, this volume), so only certain key points are reiterated here with respect to forest development in the boreal and Great Lakes regions. Most of Ontario can be described as having either gently rolling or flat topography, primarily because of several glacial advances that have resulted in a peneplain. The bedrock geology is dominated by Precambrian rock of the Canadian Shield, consisting mostly of granites and gneisses, together with Phanerozoic sedimentary rocks that occur in basins across the Shield and over a large area of southern Ontario (Sado and Carswell 1987; Ontario Geological Survey 1991). Soils that have developed over these Shield rocks are acidic, except where they are modified by lacustrine or marine deposits. Periods of glaciation and postglacial processes were extremely important to the formation of the current forest landscape. Sculpting or rounding of the land generally reduced its elevation; the original topsoils were removed; extensive outwash and till plains were created; the bedrock was exposed in many areas, especially in the boreal region; glaciolacustrine sands and clays and glaciofluvial materials were deposited; and an interrupted drainage pattern was formed which persists throughout the province to the present day.

Physiographic and Edaphic Effects on Forest Landscapes

Topography and soil types exert strong influences on forest development, and the association between topography, soils, and forest types (or individual ecosystems) has been reasonably well studied (e.g., by Daubenmire 1968). Relationships exist among slope, soil type, microclimate, and soil moisture content, all of which can be used to infer local productivity (e.g., Brady 1984). Graumlich and Davis (1993) reported that substrate governs the distribution of pine and birch in the Great Lakes basin over a scale of hundreds of square kilometres. Knowledge of the associations between soil types and plants can be used to predict and model forest landscape structure. This relationship has been used in forest ecological classifications to describe forest ecosystems and their typical topographic locations (e.g., Sims et al. 1989; Johnson and Walsh 1997; Sims 1996).

Baldwin et al. (1990) and Sims (1996) reported common relationships found between toposequence and forest ecosystem types in the boreal forests of western Ontario. The patterns they described include the following: the occurrence of mixedwood forests on upper slopes, which are composed of deep soils of lacustrine origin, and the transition to increasingly conifer-dominated forests down the slopes; the distribution of black spruce with some white cedar in lowlands; the domination of jack pine and black spruce on shallow soils over bedrock or on sandy, well-drained soils; the occurrence of jack pine on upper slopes composed of deep, sandy or loamy ablation tills; and the admixture of an increasing component of black spruce and white spruce toward the lower slopes.

The soils of the Claybelt in the northeastern boreal forest, within the basin of the postglacial Lake Barlow-Ojibway, are dominated by clays and silty-clay gleisols rather than by the podzols or brunisols that are more typical of the boreal region. The combination of limited relief, clay soils, cold climate, and relatively poor drainage results in a forest cover dominated by black spruce, except where sandy soils from a glacial resurgence lie over the clay, supporting stands of mixedwoods and jack pine. Sims et al. (1998) proposed a toposequence model for the mixedwood forests of the central boreal region, consisting of mixed forests dominated by balsam fir and trembling aspen, with jack pine and black spruce on shallow soils over bedrock, and an increasing component of white birch and black spruce toward lower slopes and deeper soils. Throughout the boreal forest, the landscape pattern is affected by numerous wetlands that may be forested or not, depending on the proximity of the water table to the surface. Wetland forests, such as those that develop on bogs, are characterized by poor tree growth and are dominated by black spruce, tamarack, speckled alder (*Alnus incana rugosa*), and often white cedar.

In the Great Lakes region, soils are generally deeper than in the boreal zone, but till and postglacial lacustrine landforms remain the dominant features.

The marble belt, an area underlain by Precambrian calcareous rock in the central part of eastern Ontario, supports forest species requiring richer soils, such as ashes, basswood, and black cherry. Well-drained sand and silt deposits originating from the postglacial Champlain Sea in the upper Ottawa River valley, underlain in part by limestone, support forests that are dominated by white pine and aspen, and provide some of the true pineries that occur in the province. At the southeastern border of the Great Lakes region, north of Kingston, Precambrian bedrock exposure is common, resulting in forests more dominated by species favouring acidic soils. Similarly, across the northeastern edges of the transition forest between the boreal and Great Lakes regions, forests are very much controlled by bedrock, owing to the prevalence of outcrops and of shallow soils over rugged topography (Chambers et al. 1997).

West of Lake Superior, the western extension of the Great Lakes forest continues to about 100 km north of the Canada-United States border. Much of this area is underlain by glacial tills and postglacial sand and silt deposits, and there is only minor relief. Forests growing on well-drained soils are dominated by boreal species, including red pine and white pine, while transition forests occur on the more rugged topography near Atikokan. Forests dominated by deciduous species, more typical of the eastern Great Lakes forests, are common only in the area of Quetico Park, south of Atikokan, and occur on a range of soil conditions (Sims et al. 1989). Both the type and characteristic topography of soils influence the distribution of these forests (Barnes et al. 1982; Pregitzer and Barnes 1984; Chambers et al. 1997; Fassnacht and Gower 1998). A combination of red oak and white pine, sometimes including red pine, occurs on shallow soils along ridges, often containing bedrock outcrops, or on sandy-loam outwash plains. White pine and hardwoods, often with a component of trembling aspen or large-toothed aspen (*Populus grandidentata*), occur on shallow or sandy soils with good drainage or on deeper soils with a southerly exposure. Sugar maple and yellow birch occur on moderate slopes with deeper soils, while black spruce is found on imperfectly drained lowlands. Hemlock and hardwoods occupy richer bottomland or the moderately rich soils of end moraines and ground moraines. Finally, maple and red oak occur on deeper soils containing sand and gravel on level terrain, often including outwash plains. In combination, soil nutrients and moisture, soil depth, soil temperature, and topography are important factors underlying the formation of ecosystems and ecosystem processes in forest systems (Bonan 1992), and thus they influence forest landscape structure (Daubenmire 1968; Sims et al. 1989; Fassnacht and Gower 1998).

Historical Effects on Forest Vegetation at Large Scales

From the perspective of landscape ecology, history refers to the sequence of events that occurred at a particular location, as opposed to an individual recorded incident. Although history is not an ultimate constraint in the manner of climate or topography, to understand landscape dynamics, landscape ecologists must be historians to a certain extent. The discussion above on paleoecology illustrates how long-term historical changes have affected forest development and the presence, absence, and abundance of tree species across Ontario landscapes. Historical events act on species assemblages within forested systems at time scales ranging from many thousands of years to several hundreds. Events such as the Quaternary glaciations acted over thousands of years and have had a profound influence on Ontario's forests, one that endures to the present day. At shorter time scales, there have been local historical effects on forests, such as the change in natural fire frequency associated with the Little Ice Age in the mid-1700s (Johnson 1992) and the exceptional period of large numbers of fires from 1979 to 1981 (Van Wagner 1988). Understanding that these events occurred is important to gaining an appreciation of current forests.

Many anthropogenic historical events have affected forest development in Ontario. Aboriginal peoples burned forests in the Great Lakes region (Flader 1983; Clark and Royall 1996). The 350 years since European settlement began have seen numerous influences on forest landscape patterns, including logging, fire suppression, agricultural land clearing, the introduction of competitor and pathogenic species, and various silvicultural activities, among them the planting of forests not expected from natural succession. Other human influences include hydroelectric power installations, the construction and abandonment of roads, and the sum total of pollution effects. History introduces dynamism and serendipity into biological systems, or, as in the case of many anthropogenic influences, directed change to ecosystem trajectories, intended or otherwise. For example, widespread early logging of white pine in eastern Ontario, coupled with active fire suppression, especially in recent years, has resulted in a proliferation of tolerant hardwood forests in areas where mixedwood stands would otherwise have prevailed (e.g., Aird 1985). At the time of a catastrophic fire, the presence or absence of a seed crop in a particular conifer species can influence ecosystem development. If there were no cones in a given year, or if the conifer species was selectively removed from the system, deciduous species such as aspen or birch, which produce seeds annually or reproduce by suckering, may dominate the system. History plays an important role in how forest landscapes develop, and a knowledge of historical events contributes much to the understanding of ecosystem trajectories. A particular forest stand may have developed as the result of a particular sequence of many events over time; nevertheless, the cumulative importance of such historical effects on the assemblage and nature of current ecosystems is often overlooked.

Effects of Proximate Factors on Ontario Forest Vegetation

As was noted above, proximate factors are those which influence the development of forest vegetation at smaller scales. The principal influences of this type in Ontario are forest fires, windthrow, insect infestation, diseases, and human intervention in the form of logging and fire suppression.

Forest Fire

The role of fire in forest ecology is discussed further by Li (2000, this volume), who provides examples from spatiotemporal modelling. Fire is a major natural disturbance and an important factor in landscape structure in both the boreal and Great Lakes forests, but the temporal and spatial scales of the fires differ considerably between the two regions. At the landscape level, fire results in the development of a mosaic of patches differing in size, age, shape, and tree species composition (Pickett and White 1985; Heinselman 1981; Johnson 1992).

Fires in boreal forests are sometimes large, stand-destructive or "catastrophic" crown fires that may burn tens of thousands of hectares (fire class 3: Van Wagner 1983). Small fires that burn areas of less than 100 ha are the most frequent, but it is the largest fires that are primarily responsible for landscape structure (Heinselman 1973; Johnson 1992). For example, data compiled by OMNR for Hills Site Region 3E (Hills 1959, 1960) in the central part of eastern Ontario showed that 4 percent of all reported fires between 1920 and 1950 were responsible for 26 percent of the total area burned during those three decades. For site region 3S in the boreal northwest, 3 percent of the fires over the same period accounted for 40 percent of the area burned (OMNR, Central Region, unpub.). Suffling et al. (1988) showed that forest areas in western Ontario subjected to an intermediate level of fire disturbance (that is, an annual burn of 0.15 percent to 0.30 percent of the landscape) had greater landscape patch diversity than areas burned either more or less frequently.

Plotting the historical fires in Ontario from 1921 to 1996 (Figure 3.4) reveals some broad patterns that affect landscape structures in different areas of the province. The northwestern part of the province is subject to more fires than the eastern part, but several large fires have burned in the east. This pattern reflects the drier climate in the west, which results in frequent fires, but indicates that, when there are droughts in the east, the existence of large fuel loads that have accumulated through many moist years may result in very large fires. In boreal forests, the decade when the least area burned was 1951 to 1960, and the decade when the most area burned was 1921 to 1930.

Fires have been actively suppressed within Ontario's commercial forest lands since the late 1950s. Within the Great Lakes forests, this practice has resulted in the almost complete elimination of fire; in fact, the total of all the forest burned from 1960 to the present is less than the 50,000 ha burned in the single decade from 1940 to 1949, immediately before fire suppression began. Another clear pattern of fire distribution in the province is an association between patterns of access routes (roads and railways) and fires, both small and large, that are of anthropogenic origin. In fact, some of the most historically important fires, in terms of size and loss of life and property, were caused by trains, including the Matheson fire of 1916 and the Haileybury fire of 1922. North of the Claybelt, frequent cool, wet weather intruding from Hudson Bay well into north-central Ontario, coupled with wet soils, has resulted in a relatively low fire frequency.

The most common fires in Great Lakes mixedwood forests are surface fires (fire classes 1 and 2: Van Wagner 1983), which do not generate sufficient heat to destroy most large trees, and which rarely flare up into the canopy. These fires generate small openings in the forest and kill young, shade-tolerant trees and shrubs, and perhaps some individual larger trees, thereby altering succession beneath a mature forest canopy. Mature white pine and red pine are particularly adapted to resist this kind of fire (Wright and Bailey 1982). The rolling topography over much of the Great Lakes forest, with its numerous lakes and marshes, also affects fire size by restricting fire trajectories. Catastrophic fires are most common in boreal forests, where they result in high tree mortality and stand replacement. The intensity of both surface and crown fires is affected by the amount of fuel available and its distribution, by soil and fuel moisture, and by the rate of fire spread. These three criteria are controlled, in turn, by many other factors: the length of time since the last fire (that is, by the amount of fuel which has accumulated); weather, particularly wind and degree of accumulated moisture in the vegetation; forest management activity; damage or mortality caused by disease and insect infestation; the species composition and age of the forest itself; and, in the longer term, climate (Van Wagner 1983).

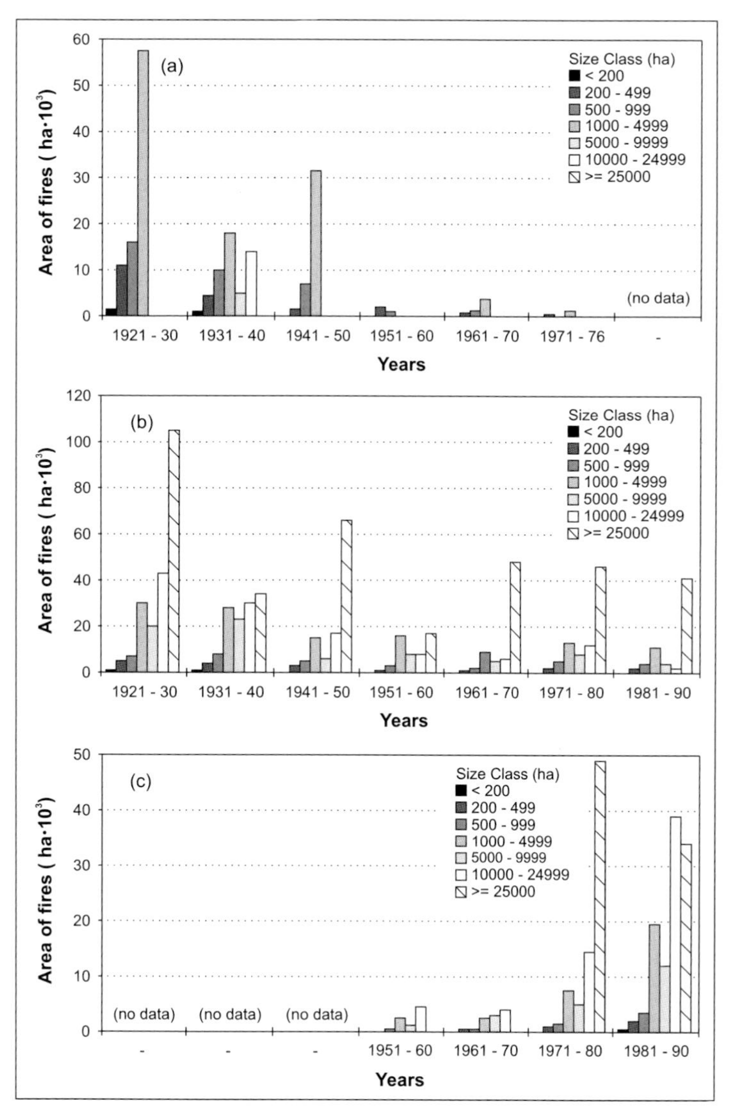

Figure 3.4 Size class distribution of forest fires from three forest biomes in Ontario: (a) Great Lakes-St. Lawrence forest region; (b) boreal forest region; and (c) Hudson Bay lowlands.

(Adapted from Forest Landscape Ecology Program, Ontario Ministry of Natural Resources 1998)

Natural fire rotation, the time it takes for all of a given landscape to burn completely at least once, is not uniform across the boreal forest region. Return times for fire in the northwestern and north-central parts of the province away from lakes Nipigon and Superior are shorter than in the northeastern areas, because of the moisture gradient previously mentioned and because of the difference between the common forest types in each region. In western Ontario, prior to the institution of fire suppression, stands dominated by jack pine and black spruce burned every 50 years, on average (Van Wagner 1978; Heinselman 1981); whereas, in eastern boreal forests, forests dominated by black spruce and jack pine burned about every 100 years (Cogbill 1985, Bergeron and Harvey 1997). Moister, black spruce-feather moss sites have fire return times of about 130 years (Cogbill 1985). In transition forests near Temagami, fire return times in black spruce communities average 64 years (Day and Carter 1991). Severe fires appear to result from several consecutive years that are drier than normal, and the fires can be particularly extensive if the dry periods coincide with general climatic warming. Cogbill (1985) suggests that this pattern of multiple, successive years of dry, warm conditions may be random. Van Wagner (1988) characterized fire frequency in Ontario as approximating a cycle of peak years every nine to ten years, with a decline in the area burned from 1920 to 1955, followed by an increase through the 1970s and 1980s. The trend of rising fire frequency has continued from 1990 to 1998 as well (Thompson et al. 1998).

In Great Lakes forests west of Lake Superior, Heinselman (1973) suggested a fire rotation period of 100 years, prior to fire suppression, but added that stands dominated by red and white pine may have lasted 150 to 350 years, surviving numerous ground fires every 20 to 40 years. In eastern white pine forests near Temagami, the estimated return time is 128 years (Day and Carter 1991). A study from a small area of pine-dominated mixedwood forest in Algonquin Park reported that low-intensity fires burn every 26 years and moderate-intensity fires every 47 years, but that stand-replacement fires occur at an interval of 200 years, on the average (Guyette and Dey 1995). Hemlock and hardwood forests have long fire cycles, at least 300 years and more likely 1200 to 1900 years (Henry and Swan 1974; Heinselman 1981; Frelich and Lorimer 1991).

The burning of a stand during a given fire rotation period does not preclude its burning again long before the average return time. Heinselman (1973), working in Minnesota, reported many areas of forests dominated by white pine that were burned over catastrophically two or three times within less than 100 years. The phenomenon of repeated burning must be taken into account in studies of how patch mosaics develop; for example, for the modelling of landscape use or the study of fire history. Although ground fires are clearly the most common type in Great Lakes forests, stand-replacement fires have affected large areas in the past, despite rolling topography and the high deciduous component, both of which normally reduce fire size. Richardson (1929) reported several enormous fires from early historical records; for example, a 5180-km^2 fire in the Lake Nipissing to Spanish River area in 1871, and another that burned most of the area from Temagami to the Quebec border, north along the Montreal River and south and west to Lake Superior and Lake Huron, around 1855, covering an area of at least 8000 km^2. Most information also suggests that there have been periods lasting from 5 to 50 years when large numbers of fires occurred, interspersed among intervals of little fire activity (Heinselman 1973; Stocks and Street 1983; Cogbill 1985). Heinselman (1973) took considerable care to point out that, although he calculated average fire return times in Minnesota, high variability exists in fire regimes within and among forest communities. What this means for both the Great Lakes and boreal forest regions in Ontario is that most areas have been burned many times since the re-establishment of forests in the early Holocene.

Van Wagner (1978) proposed that the age-class structure of forest stands across a landscape controlled by fire formed a negative binomial distribution, and illustrated this using data from Heinselman (1973) which is relevant to Great Lakes forests in Ontario. Exponential or Weibull models were supported in Suffling (1983) and Johnson and Larsen (1991), but not in Baker (1989, 1992). A problem in early models of landscape patch distribution and fire frequency was the incapacity to deal with the burning of patches several times during a fire rotation (Boychuk and Perera 1997). The latter authors found that spatiotemporal autocorrelation of disturbances resulted in a large variation in numbers of forest patches in various age classes in boreal forests. They proposed that landscape structure in fire-disturbed forests is governed by initial conditions and by long-term variability in burning rates. Modelling of fire frequencies is discussed in detail by Li (2000, this volume), and consequences for landscape structure under current fire regimes are detailed by Perera and Baldwin (2000, this volume).

Changes in landscape structure at the forest scale are relatively rapid during the first 20 years following a disturbance, as rapid succession by forest plant species takes place; however, succession slows in older forests, and a certain stability is attained which endures for many years. The successional pattern following fires is dependent on numerous factors, including the intensity of the fire, the season, soil moisture, and the availability of a seed supply. The pattern also differs among forest ecosystems, and both within and between the forest regions. Boreal forest landscapes tend to be dominated by even-aged stands as a result of repeated, often large, fires. In the Great Lakes forests, stands are generally uneven-aged because of long fire intervals, frequent ground fires, and the presence of many, common shade-tolerant species. Jack pine has thin bark and so is readily killed even by ground fires; however, sites dominated by jack pine usually return quickly to jack pine following a fire (Cayford et al. 1967; Cayford and McRae 1983), because the species seeds immediately from serotinous cones. Mixed stands can also occur on predominantly jack pine sites following a fire, depending on the original species composition, and on the season of the fire relative to the seeding of deciduous species (Cayford and McRae 1983; Gauthier et al. 1993). If fire is absent in jack pine or jack pine-aspen forests for more than 100 years, succession to mixed forests of black spruce, shade-tolerant hardwoods, and balsam fir is common (Day and Woods 1977).

Where the fire rotation period is less than about 150 years, balsam fir is seldom a dominant canopy species (Carleton and Maycock 1978; Payette 1992). Balsam fir is a secondary successional species on sites that remain unburned for long periods, becoming dominant only after 200 years or more (Carleton and Maycock 1978), following an even recruitment into the stand over about 150 years (Cogbill 1985). In the view of Clements (1916) on succession, balsam fir is a climax forest species. In central Ontario's boreal mixedwoods, large areas of the original forest of black spruce, white spruce, white birch, and aspen are being replaced by a simpler community of balsam fir and trembling aspen, in the absence of fire for 250 years, so that a highly uneven-aged and complex forest landscape results (I.D. Thompson and D.A. Welsh, unpub.). This is in contrast with succession in mixedwood forests, which initiate immediately following fire and produce stands which are extremely even-aged (Cogbill 1985). Balsam fir is readily killed by spruce budworm, so it often does not live to its potential maximum age.

Recruitment of dominant trees into black spruce stands that originate after fire continues for about 30 years. It appears that old stands are not colonized by new seedlings during their last 100 years (Cogbill 1985), although there may be considerable layering. In the western Quebec boreal zone, immediately adjacent to Ontario, fire in spruce forests was followed by a long period, perhaps up to 200 years, of domination of the forests by shade-intolerant hardwoods and mixedwoods, before conifers again became dominant (Bergeron and Harvey 1997). Similarly, in Ontario's Claybelt, mixed stands of black spruce and balsam fir can be converted through fire to pure black spruce, or to forests of aspen and balsam poplar (*Populus balsamifera*) that succeed to mixed forests of aspen or poplar with balsam fir (T.J. Carleton, unpub.; Kneeshaw and Bergeron 1998). Succession may vary with the type of disturbance, the availability of seed sources, and the serendipitous concurrence of seed production and disturbance, and is only somewhat dependent on time since fire. In general, however, shifts in stand composition at a site, through succession from one climax-type community to another, are exceptional, and boreal stand dynamics tend to be dominated by the differential aging of species within an unstable system.

In Great Lakes forests, where catastrophic fires are rare, pure stands of jack pine, red pine, and white pine, or a combination of these species, are usually self-replacing following either large or small fires, particularly if they occur on poor soils or on islands. In forests with a fairly even mixture of conifer and deciduous species, ground fires often occur in spring, when leaf litter from the previous year is dry. This type of fire travels slowly across the forest floor, doing little or no damage to larger trees and, if it is not severe, sparing younger trees altogether (Van Wagner 1983). Such fires favour the germination of many species, including red oak, white spruce, white birch, yellow birch, and white pine. This result is achieved by exposing mineral soils; reducing competition from shrubs such as hazel (*Corylus cornuta*), mountain maple (*Acer spicatum*), and balsam fir; and providing a flush of nutrients for seedlings (Heinselman 1981). In the absence of ground fires and stand-replacing fires, stands generally succeed to shade-tolerant tree species, such as balsam fir, eastern hemlock, and sugar maple, that can germinate in the absence of disturbance and exist for long periods beneath a canopy. Carleton et al. (1996) suggested that pine seedlings are out-competed by numerous shrubs, mosses, and forbs, so that little replacement of pine can be expected in mixed forests, without fire to reduce competitive effects.

It is likely that there are two conditions in which stand-replacing fires occur in Great Lakes forests. The first arises after a large blowdown has produced a large amount of fuel within a stand. A severe fire can occur in the first few years following that disturbance, prior to the decomposition of a high volume of fine fuels (Frelich and Reich 1996). The second possibility occurs if balsam fir is present in the understory and reaches a sufficient height to carry flames up into the canopy of taller pines. This situation would be exacerbated considerably should the balsam fir be dead or dying from attacks by spruce budworm, particularly in spring, prior to leaf flush in the understory (Stocks and Bradshaw 1981). In fact these latter conditions occurred during the 1948 Mississagi fire, near Thessalon, in which an area of more than 1.8 million ha was burned. Succession following such a large fire would ultimately enhance the presence of red pine and white pine in the new forests, if a seed source were present and a cone year occurred soon after the fire. The amount of red pine in the landscape is directly related to area burned (Flannigan 1993). Growth of white pine beneath a canopy often occurs following an initial stand dominated by white birch and aspens.

The structure of forested landscapes of Ontario is affected by fire to varying degrees, depending on location in the province relative to fire control activities during the past 50 years. Fires are affected, in turn, by weather, climate, insect infestations, fuel loads, and the species composition of the forest types. These factors change over time, interact in various ways, and are altered by humans; as a result, the locations and outcomes of fires are highly variable and difficult to predict. For example, climate change will result in altered fire regimes and altered landscape structure and species composition in Ontario forests (Suffling 1995). Change in landscape structure may result in myriad secondary effects, such as altered distributions of numerous animal species (Thompson et al. 1998). Fire has played a key role in the history of Ontario's forests in the past, and that role may become even more important in the future.

Wind Disturbance

Another common disturbance in Ontario forests is windthrow. This type of disturbance is important to stand and landscape structure in Great Lakes forests (Curtis 1959), where catastrophic fires are uncommon, but it is also an important agent of succession in boreal forests, in the absence of fire (Carleton and Maycock 1978; Kneeshaw and Bergeron 1998; Johnson 1992). In any boreal landscape, many stands will be of fire origin, but some will have escaped fire for a sufficiently long period for wind to begin to affect stand structure, species composition, and age distribution. Windthrow produces small gaps in the forest canopy by causing individual trees to fall to the ground, breaking the tops of adjacent trees as they fall. More violent storms may blow down large areas of forest. The extent of the damage that can be caused is indicated by the example of a 200-km^2 mixedwood forest dominated by old-growth white pine located north of Espanola, in which canopy gaps represented 7 percent of the total area (I.D. Thompson, unpub.). Runkle (1982) reported that 9 to 10 percent of an Appalachian deciduous and mixed forest occurred in gaps, the same proportion reported by Tyrrell and Crow (1994) in hemlock-hardwood forest in Wisconsin. Kneeshaw and Bergeron (1998) reported that an astonishing 40 percent of a boreal forest in western Quebec dominated by old-growth balsam fir was in gaps. The area of a given forest occurring as gaps varies with species composition and site, and with the age of the stand, as a reflection of the damage to individual trees by pathogens.

In Ontario boreal forests, black spruce are more susceptible to windthrow than are jack pine, trembling aspen, or white birch (Fleming and Crossfield 1983). Where fire has been absent for a considerable period, spruce budworm is largely responsible for killing balsam fir trees (Li 2000, this volume), thereby making the landscape susceptible to windthrow, and the amount of a boreal forest consisting of canopy gaps is directly related to the abundance of old balsam fir (Kneeshaw and Bergeron 1998).

Downbursts associated with severe thunderstorms, although uncommon, can cause large areas of blowdown which result in large forest gaps, sometimes over extensive areas. Several large blowdown events reportedly from this cause have been measured; for example, a single downburst near Gogama in 1973 demolished a forest area of about 1500 ha, and a severe storm near Ignace resulted in 21,000 ha of blowdown in black spruce-jack pine forest in 1988 (Flannigan et al. 1989). Dunn et al. (1983) described a huge blowdown of 334,000 ha in hemlock-hardwood Great Lakes forest of Wisconsin in 1977, which had resulted from 25 separate downbursts in one storm, and Canham and Loucks (1984) reported the blowdown of 3785 ha of mixedwood forest, also in Wisconsin. The latter authors calculated that the annual blowdown in the pre-settlement forest of Wisconsin exceeded 4000 ha, and that 17 to 25 percent of the forest was maintained in a successional state by wind.

Catastrophic stand replacement from windthrow has a cycle of at least 1400 years between events in red spruce (Cogbill 1996). Frelich and Lorimer (1991) reported that windthrow is the most important disturbance in northern hemlock-hardwood forests of the northeastern United States, where Dahir (1994) found that the average gap size was only 70 m^2. Surprisingly few data are available on gap sizes in boreal forests, but Kneeshaw and Bergeron (1998) reported that sizes followed a chi-square distribution in forests dominated by old balsam fir, with most gaps under 100 m^2 in size, but some ranging to 3000 m^2. In younger forests, these authors reported a negative exponential distribution of gap sizes, with most gaps under 50 m^2. Wind disturbance is a major factor in forest landscape structure and may act synergistically with insect infestation and fire.

Gap size influences which species can invade (Curtis 1959), and the frequency with which larger gaps occur is important to the ultimate species composition and structure of a stand, particularly in the absence of fire. Stands with small-gap dynamics are multi-aged, as young trees grow up in the gaps, surrounded by older trees (e.g., Frelich and Lorimer 1991). Larger forest gaps can be occupied by shade-intolerant species that could not otherwise become established in a closed-canopy forest (see Table 3.1 for the shade tolerance of various species). If gaps are small, however, then same-species replacement is often the normal successional sequence. This pattern is common in hemlock-hardwood forests, where most tree species are shade-tolerant (Dunn et al. 1983; Frelich and Lorimer 1991). In a deciduous forest in southern Quebec, Payette et al. (1990) reported that the gaps were generally small, existed less than 30 years, and were only occupied by shade-tolerant plants. A similar finding is reported by Brewer and Merritt (1977) in beech-maple forests in Michigan. In boreal forests, the most important tree species in canopy gaps is balsam fir, particularly in small gaps, because of its shade tolerance, high rate of seed production, and longevity of seed in the duff layer (Fowells 1965; Carleton and Maycock 1978; Kneeshaw and Bergeron 1998). Some large gaps remain open, however, and are dominated by dense shrubs for an extended period, because of the reduction in balsam fir success under competition (Ghent et al. 1957; Kneeshaw and Bergeron 1996). In boreal mixedwood stands, aspen regenerates well in canopy gaps and may replace itself in the canopy, in association with balsam fir (Paré and Bergeron 1995). White cedar is another species that can slowly invade gaps over time, in the absence of fire (Kneeshaw and Bergeron 1998).

Gap dynamics lead to uneven-aged stands and changes in species composition in many other forest types as well. In lowland black spruce stands of the Claybelt, Groot and Horton (1994) found even-aged forests that were of fire origin, while older, uneven-aged stands were either self-perpetuating or were becoming dominated by balsam fir and aspen, as these species took advantage of gaps created by blowdown. In nearby Minnesota, the absence of fire has made canopy gaps resulting from windthrow a mechanism by which the forest is changed from a relatively uniform-aged mixture of jack pine and aspen to a more uneven-aged mixed forest of balsam fir, white birch, white cedar, and black spruce (Frelich and Lorimer 1991). In the absence of fire, therefore, balsam fir is an increasingly important species in the boreal zone, where small-gap dynamics dominate forest renewal processes. Forests dominated by balsam fir are particularly susceptible to attack by spruce budworm, resulting in a highly patchy landscape.

Forest Insect Infestation

Certain insects affect the forest landscape pattern directly by causing tree mortality or by rendering the tree susceptible to disease, windthrow, or fire. Although there are many insects that can retard growth in Ontario's trees by burrowing into the bark, defoliating the tree, feeding on the seeds, or damaging branches and leaders, most insects do not cause immediate widespread mortality. Tree mortality can result from a suite of factors, including extreme cold, fungal attack, and drought, particularly if trees are already under stress from insect attack. The insects considered here are those which can alter the forest landscape pattern over large areas, as a result of an infestation that may last for several years. From 1977 to 1981 in Ontario, the average annual loss in tree biomass to insects and decay was more than 46 million m^3 (Gross 1985), and about 31 million m^3 were lost each year during the subsequent five-year period (Gross et al. 1992). A more detailed examination of the role of insects in changing Ontario landscapes is provided by Fleming et al. (2000, this volume).

The most important insect from a landscape perspective in Ontario is spruce budworm (*Choristoneura fumiferana*). This species is a cyclical conifer defoliator that occurs throughout both regions and has reached epidemic levels in various parts of the province over extended periods of time. Spruce budworm is the direct cause of extensive mortality in balsam fir and to a lesser extent in white spruce and sometimes of black spruce. This insect plays a particularly important role in shaping the patch dynamics of Ontario boreal forests for a period of several years after each infestation has receded, as trees become susceptible to secondary effects such as fire and windthrow, and as the patch composition changes through succession. The successional pattern that follows spruce budworm epidemics is dependent on two main factors: whether a fire occurs or not, and the nature of the tree species composition of both the understory and the overstory, as modified by local site conditions.

Epidemics of spruce budworm, or high periods in the budworm cycle, are associated with several years of dry weather, particularly in spring (Baskerville 1975). Fires often result, fuelled by dead and dying balsam fir. In budworm-killed balsam fir, fires remove fir from the stand, at least as a dominant species, and result in forests dominated by black spruce, white spruce, trembling aspen, and white birch (Baskerville 1975; Stocks 1985, 1987). The actual relative abundance of these species will depend on the availability of seed sources, in particular the occurrence of strong seed crops immediately before the fires or in surrounding areas after the fires.

Budworm can also remove young balsam fir from the understory of mixed conifer or mixed conifer-deciduous stands, changing the composition of the surviving stands and delaying the transition to balsam fir. After budworm has killed the balsam fir, there follows a relatively short period, about ten years, during which the stand remains susceptible to fire. If fire does not occur, the balsam fir will replace itself on the site (Ghent et al. 1957). In Great Lakes forests, balsam fir may occur in the understory, associated with mixedwoods dominated by white pine, or it may occur in stands dominated by balsam fir and white pine alone. The death of the balsam fir, in the absence of fire, can result in altered stand composition, as other shade-tolerant species receive a competitive advantage with increasing light after the fir dies. These other species may then replace balsam fir as the dominant understory component. One effect of fire suppression in Great Lakes forests has been the perpetuation of balsam fir in mixed forests, despite attacks by spruce budworm; however, the stands dominated by balsam fir still make up less than 10 percent of the region (OMNR 1996).

Jack pine budworm (*Choristoneura pinus pinus*) is an eruptive defoliator that feeds on jack pine throughout both of Ontario's forest regions. The species does not ordinarily kill its host, but severe infestations may do so (Nealis 1995); in fact, there has been widespread mortality for some years in the northern and western parts of the province (Rose and Lindquist 1984). The mortality rate in Ontario jack pine has been estimated by Gross et al. (1992) at one percent each year from the onset of defoliation, for a total mortality of more than 7 million m^3 from 1982 to 1987. The fact that large-scale defoliation and mortality caused by jack pine budworm render the affected stands particularly susceptible to catastrophic fire makes the insect an important factor in landscape dynamics. It is also important to the ecology of jack pine, because of the adaptation of that species to reproduction following fire (Nealis 1995).

Other important insect species in Ontario forests that may kill trees directly, but most often weaken them and invite secondary pathology, include two broadleaf defoliators, the gypsy moth (*Lymantria dispar*) in the Great Lakes forest and the forest tent caterpillar (*Malacosoma disstria*) in both the Great Lakes and boreal regions. Gypsy moth is not endemic to North America, but arrived in 1869 and had spread to Ontario by 1924 (Rose and Lindquist 1982). This insect prefers certain deciduous species as food (white oak, red oak, elm, and aspen) but also attacks many other trees, including most conifers, especially during the later larval stages (Gerardi and Grimm 1979). The most extensive defoliation by gypsy moth in Ontario occurred in 1985, when 241,000 ha were defoliated in south-central areas of the province (Roden and Surgeoner 1991). The affected area has declined each year since then, and by 1995 only about 20,000 ha were affected, mostly near Sudbury (Howse 1996).

Most of the trees killed by gypsy moth have been oaks, although other hardwoods and white pine have also been killed (Gross et al. 1992). No Ontario data exist on the forest succession that may occur following several years of defoliation by gypsy moth. Stand-level effects would only be expected if widespread mortality occurred. Because the gypsy moth attacks primarily tolerant hardwoods and aspen in Ontario, it is likely that stands would generally be self-replacing; however, oak forests in the USA killed by severe defoliation are often replaced by less susceptible, semi-shade-tolerant host species, such as red maple. The time since disturbance has been too short yet to determine if a similar pattern will be observed in Ontario.

Forest tent caterpillar appears to erupt on a 10- to 11-year cycle. Recent major infestations took place in Ontario in 1965, 1976, and 1987, but outbreaks are recorded as long ago as 1834 (Churcher and Howse 1989). Sugar maple, trembling aspen, and oaks are the preferred food trees of this insect. Trees are rarely killed, and new leaves are produced five to six weeks following defoliation. Loss of leaves for an extended period may confer a competitive advantage to semi-shade-tolerant species in the understory, such as spruces and red maple; however, no Ontario studies of succession following outbreaks of forest tent caterpillar have been conducted, so the ecological effects on stand and landscape structure are not known. Roland (1993) suggests that the amount of edge is important to the success of outbreaks, and that logging in boreal forests may be exacerbating rates of infestation. Forest tent caterpillar is not a major force in landscape-level changes in forest species composition and structure in Ontario.

Forest Disease: Stem Decay and Root Rot

The various fungal diseases pooled together under the names "stem decay" and "root rot" are important influences on the small-gap dynamics of Ontario forests. Stem decay and root rot are clearly small-scale decay processes that affect only a few trees within young to mature stands at any particular time. As a stand ages, however, the percentage of trees affected by various fungi can become considerable, and in the old-growth years, these diseases become major primary agents of stand decline (Whitney 1988; Basham 1991). Gross et al. (1992) reported an annual loss of forest biomass in Ontario of more than 8 million m^3 from root rot alone. Species composition, forest age, and site factors such as moisture and soil depth affect rates of decay and rot (Whitney 1988; Basham 1991). For example, root rot is more common in upland than in lowland black spruce (Whitney 1988). Aspens are particularly susceptible to fungal diseases, and the result is stem breakage during wind storms, which produces small forest gaps. Root rots are also important secondary effects to trees following mechanical damage, insect attack, or non-lethal fire.

ANTHROPOGENIC INFLUENCES ON FOREST VEGETATION AT LARGE SCALES

At intermediate scales, human intervention is important to the structure of forest landscapes. It is doubtful that the complex interrelationships of natural systems can be replicated by management (Hansen et al. 1991; Thompson and Welsh 1993; Hunter 1993). One of the goals of current forestry practices in many jurisdictions is to emulate natural processes; however, the concept that fire effects can be emulated through logging practices – a cornerstone of sustainable forest management – is unlikely to be valid, primarily because fire is a chemical process and logging is a physical process. Timber harvesting removes nutrients and structures from the stand and promotes alternative successional pathways to those occurring after fire (e.g., Carleton and MacLellan 1994; Baker et al. 1996; Carleton 2000, this volume). While there is evidence of attempts to manage forests sustainably in Ontario, the validity of these approaches to forest use and redevelopment will only come with growth of the coming forests, which will take 50 to 200 years, depending on the ecosystem. It is clear that human intervention has altered many forest ecosystems in North America (e.g., Davis 1996), including those in Ontario (Johnston and Elliot 1996), and that it has changed the forest landscape structure of Ontario (Gluck and Rempel 1996). A more detailed discussion of the impact of human intervention on Ontario forests is provided by Perera and Baldwin (2000, this volume) with respect to landscape metrics, by Carleton (2000, this volume) with respect to vegetation, and by Voigt et al. (2000, this volume) with respect to wildlife.

The most obvious alteration to forest landscapes by humans is by logging. Forest harvesting, coupled with fire suppression, has altered forest stand compositions across Ontario (Rogers 1978; Aird 1985; Hearnden et al. 1992; Carleton and MacLellan 1994) and in other nearby jurisdictions with similar forests (Baker et al. 1996; Davis 1996; Bergeron and Harvey 1997), reducing the presence of some species in favour of others. In general, species that thrive best following fire have declined, in favour of early successional species that are intolerant of both shade and fire. Species favoured by logging, and by lack of fire, include poplars and aspens, balsam fir, sugar maple, red maple, and beech. Species in decline because of long-term harvesting, past silvicultural practices, and the absence of fire, include the following: white pine, red pine, black spruce, white spruce, eastern hemlock, yellow birch, red spruce (Aird 1985; Hearnden et al. 1992; Davis 1996). White birch may also be included in this group, although it can sucker from stumps in areas where it is already present.

Reforestation strategies and policies have changed over time in Ontario, as have silvicultural practices. These changes have led to the redirection of forest successional trends and ultimately to changes in landscape structure (Perera and Baldwin 2000, this volume; Carleton 2000, this volume). Throughout the 1980s, for example, a major effort was made to plant coniferous species on clear-cut areas throughout the boreal forest. More recently, the practices of natural regeneration (termed a "free-to-grow" strategy) and of "careful logging" (which is intended to protect advanced growth of conifers in the understory [Groot 1984]), have become more common, concurrently with reductions in spending by governments on natural resources. Approximately 25 percent less area is currently planted than was planted in the early 1990s (Canadian Council of Forest Ministers 1997). The trend in the past has been toward a more homogeneous forest landscape and simpler forest species composition. A good example of these trends is provided by the hemlock-deciduous forest type in the Great Lakes region, in which logging of the hemlock component has resulted in large areas of predominantly even-aged sugar maple stands which often lack hemlock in the understory.

An important factor affecting the distribution of individual species, and ultimately the landscape pattern, is the reduction of seed sources by past logging over extensive areas. The five species of conifers (white pine, red pine, black spruce, white spruce, and eastern hemlock) which have dominated large areas of Ontario forest in the past have been particularly affected by logging, and thus have a diminished seed source, from a broad landscape perspective. Data from Aird (1985) and OMNR (1996) show that the present harvest of red pine and white pine has declined 75 percent from the maximum harvest of these species, which occurred around 1885. This decline corresponds to the reduction in forests dominated by these two species. Similarly, Baker et al. (1996) has modelled a predicted decline of about 80 percent in the amount of black spruce available in boreal Ontario over the next 50 years, as the forest is already substantially depleted of this species (Hearnden et al. 1992), especially on uplands. The lack of seed sources, coupled with altered disturbance regimes under predicted global warming, may further reduce the prominence of these species, thereby altering the complexity of forest patches and producing a more homogeneous landscape than might otherwise occur (Suffling 1995; Thompson et al. 1998).

Humans have also exerted a significant influence on the landscape through fire suppression. Fire-return times vary with the location of the ecosystem, but the average return time across Ontario is now 580 years, as opposed to less than 100 years under a natural system (Ward and Tithecott 1993). At Temagami, Day and Carter (1991) estimated that the fire-return time in white pine forests has now reached 1200 years, compared to the natural frequency of less than 100 years. In addition, the average size of fire in Ontario has been reduced substantially, and the mean annual area burned has declined by 88 percent, from about 700,000 ha per year prior to fire suppression to 81,000 ha per year (Ward and Tithecott 1993). Regardless of human efforts to reduce the effects of fire, however, some large fires escape control and reach a large size (Johnson et al. 1995). For example, a fire near Geraldton burned 44,000 ha in 1992, and another near Nipigon burned 113,000 ha in 1995.

Frelich and Reich (1995), working in Minnesota, suggested that reduced fire frequency has had the effect of altering successional pattern by creating multiaged forests where even-aged forests would be expected; for example, jack pine and aspen stands are converted to a mixture of aspen, black spruce, balsam fir, white birch, and white cedar. Fallen pines leave gaps in the canopy, through which secondary species enter the stand and eventually change the species composition and structure.

A further example of human intervention in natural processes can be found in the role which ungulates now play in altering successional pathways and climax forest types. In Ontario and the northeastern United States, populations of white-tailed deer (*Odocoileus virginianus*) are at historic highs, as a result of a period of mild winters and increased habitat availability, but also as a result of increased food supply from agriculture, the elimination of natural predators, strict deer-hunting regulations, and winter feeding programs (Voigt et al. 1992). In all areas where deer herbivory has been studied, the result has been the elimination of certain plant species, such as hemlock, and the alteration of forest succession (Anderson and Loucks 1979; Hough 1965; Frelich and Lorimer 1985; Alverson et al. 1988). These are examples of site-level effects which have, over time, come to influence landscape patterns. They illustrate the complex nature of forest ecosystems in settled areas.

SUMMARY

Ontario maintains a highly diverse flora. The two major forest regions, the boreal forest and the Great Lakes-St. Lawrence forest, are distinguished by different climatic regimes, which support largely conifer-dominated forests in the north, and deciduous and mixed forests in the south-central areas of the province. The boreal forest is part of the larger Holarctic boreal region, while the Great Lakes forest is unique to eastern North America. Paleoecological and climatic evidence supports Suffling's (1995) suggestion that the Ontario forests are not a landscape in equilibrium, but instead are continuously "catching up" to altered climate regimes and their associated disturbance patterns. Superimposed on this less than predictable pattern are both the anthropogenic influences of fire suppression and logging and the upper-level effects of broad-scale changes in climate. Other factors, too, have influenced Ontario's landscape, such as cumulative pollution effects, agricultural land use, hydroelectric power installations, and the construction and abandonment of roads. All of these contribute to creating different temporal and spatial distributions of forests and different forest successional patterns than would exist under a more natural disturbance regime.

Processes and factors that affect forest landscape structure in Ontario include those which act at the regional level (climate, terrain and soils, and history), those which act at the forest landscape level (fire, windthrow, and insect defoliation), and those which act at smaller scales within forest stands (various fungal pathogens and tree-falls). Although this is a hierarchy of scales, in which factors at the upper level impose constraints on the next, there are also interactions among the factors from each level which act to produce landscape structure. Humans appear to be altering processes at intermediate scales, by controlling fires, logging forests, and planting forests of types not expected from natural succession. The result is altered landscape structure at all scales that may have eventual effects on forest plant species, but this remains an important question for research.

An understanding of how past landscapes were formed and how processes have changed through history is needed to provide a perspective on the new landscape structure, one that is inevitably altered by humans. Contemporary forestry has the potential to alter Ontario forest landscape structure irrevocably through changes to patch size and distribution away from heterogeneity and towards homogeneity (e.g., Pickett and White 1985; Gluck and Rempel 1996), and through the creation of dynamics that differ from those under more natural regimes (Hunter 1990). How landscapes are being altered and what the subsequent effects on biological diversity may be is the subject of several of the chapters that follow.

REFERENCES

Aird, P.L. 1985. In Praise of Pine: The Eastern White Pine and Red Pine Harvest from Ontario's Crown Forests. Report PI-X-52. Chalk River, Ontario: Petawawa National Forestry Institute, Canadian Forest Service. 11 p.

Alverson, W.S., D.M. Waller, and S.L. Solheim. 1988. Forests too deer: Edge effects in northern Wisconsin. Conservation Biology 2: 348-358.

Anderson, R.C. and O.L. Loucks. 1979. White-tailed deer influence on structure and composition of *Tsuga canadensis* forests. Journal of Applied Ecology 16: 855-861.

Arris, L.L. and P.A. Eagleson. 1989. Evidence of a physiological basis for the boreal-deciduous forest ecotone in North America. *Vegetatio* 82: 55-58.

Auclair, A.N. and F.G. Goff. 1971. Diversity relations of upland forests in the western Great Lakes area. American Naturalist 105: 499-528.

Baker, J.A., T. Clark, and I.D. Thompson. 1996. Boreal mixedwoods as wildlife habitat: observations, questions, and concerns. In: C.R. Smith and G.W. Crook (editors). Advancing Boreal Mixedwood Management in Ontario: Proceedings of a Workshop, 17-19 October, 1996, Sault Ste. Marie, Ontario. Sault Ste. Marie, Ontario: Canadian Forest Service. 41-52.

Baker, W.L. 1989. Effect of scale and spatial heterogeneity on fire-interval distributions. Canadian Journal of Forest Research 19: 700-706.

Baker, W.L. 1992. Effects of settlement and fire suppression on landscape structure. Ecology 73: 1879-1887.

Baker, W.L. 1995. Long-term response of disturbance landscapes to human intervention and global climate change. Landscape Ecology 10: 143-159.

Baldwin, D.J.B., J.R. Desloges, and L.E. Band. 2000. Physical geography of Ontario. In: A.H. Perera, D.L. Euler, and I.D. Thompson (editors). Ecology of a Managed Terrestrial Landscape: Patterns and Processes in Forest Landscapes of Ontario. Vancouver, British Columbia: University of British Columbia Press. 12-29.

Baldwin, K.A. 1991. An introduction to the boreal forest in Ontario. Proceedings of the Entomological Society of Ontario 122: 73-86.

Baldwin, K.A., J.A. Johnson, R.A. Sims, and G.M. Wickware. 1990. Common Landform Toposequences of Northwestern Ontario. Report No. 3303. Sault Ste. Marie, Ontario: Canada-Ontario Forest Resource Development Agreement. 26 p.

Band, L.E. 2000. Forest ecosystem productivity in Ontario. In: A.H. Perera, D.L. Euler, and I.D. Thompson (editors). Ecology of a Managed Terrestrial Landscape: Patterns and Processes in Forest Landscapes of Ontario. Vancouver, British Columbia: University of British Columbia Press. 163-177.

Band, L.E., J.P. Patterson, R. Nemani, and S.W. Running. 1993. Forest ecosystem processes at the watershed scale: incorporating hillslope hydrology. Agriculture and Forestry Meteorology 63: 93-126.

Barnes, B.V. 1991. Deciduous forests of North America. In: E. Röhrig and B. Ulrich (editors). Ecosystems of the World, 7: Temperate Deciduous Forests. Amsterdam: Elsevier. 219-344.

Barnes, B.V., K.S. Pregitzer, and V.H. Spooner. 1982. Ecological forest site classification. Journal of Forestry 80: 493-498.

Basham, J.T. 1991. Stem Decay in Living Trees in Ontario's Forests. Information Report O-X-408. Sault Ste. Marie, Ontario: Forestry Canada. 64 p.

Baskerville, G.L. 1975. Spruce budworm: super silviculturalist. Forestry Chronicle 61: 138-140.

Bergeron, Y. and B. Harvey. 1997. Basing silviculture on natural ecosystem dynamics: an approach applied to the southern boreal mixedwood forest of Quebec. Forest Ecology and Management 92: 235-242.

Bickerstaff, A., W.L. Wallace, and F. Evert. 1981. Growth of Forests in Canada, Part 2: A Quantitative Description of he Land Base and the Mean Annual Increment. Information Report PI-X-1. Chalk River, Ontario: Petawawa National Forestry Institute, Canadian Forest Service. 136 p.

Bonan, G.B. and H.H. Shugart. 1989. Environmental factors and ecological processes in the boreal forest. Annual Review of Ecology and Systematics 20: 1-28.

Bonan, G.B. 1992. Soil temperature as an ecological factor in boreal forests. In: H.H. Shugart, R. Leemans, and G.B. Bonan (editors). A Systems Analysis of the Global Boreal Forest. Cambridge, UK.: Cambridge University Press. 126-143.

Bormann, F.H. and G.E. Likens. 1979. Pattern and Process in a Forested Ecosystem. New York: Springer-Verlag. 565 p.

Boychuk, D. and A. Perera. 1997. Modeling temporal variability of boreal landscape age-classes under different forest disturbance regimes and spatial scales. Canadian Journal of Forest Research 27: 1083-1094.

Brady, N.C. 1984. The Nature and Properties of Soils. New York: MacMillan Publishing. 750 p.

Brewer, R. and P.G. Merritt. 1977. Wind throw and tree replacement in a climax beech-maple forest. *Oikos* 30: 149-152.

Canadian Council of Forest Ministers. 1997. Compendium of Canadian Forestry Statistics 1996. Ottawa, Ontario: Canadian Forest Service. 234 p.

Canham, C.D. and O.L. Loucks. 1984. Catastrophic windthrow in the presettlement forests of Wisconsin. Ecology 65: 803-809.

Carleton, T.J. 2000. Vegetation responses to the managed forest landscape of central and northern Ontario. In: A.H. Perera, D.L. Euler, and I.D. Thompson (editors). Ecology of a Managed Terrestrial Landscape: Patterns and Processes in Forest Landscapes of Ontario. Vancouver, British Columbia: University of British Columbia Press. 179-197.

Carleton, T.J. and P. MacLellan. 1994. Woody vegetation responses to fire versus clearcut logging: a comparative study in the central Canadian boreal forests. Ecological Applications 6: 63-773.

Carleton, T.J. and P.F. Maycock. 1978. Dynamics of the boreal forest south of James Bay. Canadian Journal of Botany 56: 157-1173.

Carleton, T.J., P.F. Maycock, R. Arnup, and A.M. Gordon. 1996. In situ regeneration of *Pinus strobus* and *P. resinosa* in the Great Lakes forest communities of Canada. Journal of Vegetation Science 7: 431-444.

Cayford, J.H., S. Chrosciewicz, and H.P. Sims. 1967. A Review of Silvicultural Research in Jack Pine. Publication No. 1173. Ottawa, Ontario: Canadian Department of Forestry and Rural Development. 255 p.

Cayford, J.H. and D.J. McRae. 1983. The ecological role of fire in jack pine forests. In: R.W. Wein and D.A. MacLean (editors). The Role of Fire in Northern Circumpolar Ecosystems. Chichester, UK: John Wiley. 183-200.

Chambers, B.A., B.J. Naylor, J. Nieppola, B. Merchant, and P. Uhlig. 1997. Field Guide to Forest Ecosystems of Central Ontario. Field Guide FG-01. North Bay, Ontario: South Central Science Section, Ontario Ministry of Natural Resources. 200 p.

Chapin, F.S., P.M. Vitousek, and K. Van Cleve 1986. The nature of nutrient limitation in plant communities. American Naturalist 127: 48-58.

Churcher, J.J. and G.M. Howse. 1989. Another Outbreak of Forest Tent Caterpillars. Toronto, Ontario: Forestry Canada and Ontario Ministry of Natural Resources. 14 p.

Clark, J.S., C. Fastie, G. Hurtt and S.T. Jackson. 1998. Reid's paradox of rapid plant migration. Bioscience 48(1): 13-26.

Clark, J.S. and P.D. Royall. 1996. Local and regional sediment charcoal evidence for fire regimes in pre-settlement northeastern North America. Journal of Ecology 84: 365-382.

Clements, F.E. 1916. Plant Succession: An Analysis of the Development of Vegetation. Publication No. 242. Washington, DC: Carnegie Institute. 512 p.

Cogbill, C.V. 1985. Dynamics of the boreal forest of the Laurentian highlands, Canada. Canadian Journal of Forest Research 15: 252-261.

Cogbill, C.V. 1996. Black growth and fiddlebutts: The nature of old growth red spruce. In: M.B. Davis (editor). Eastern Old Growth Forests: Prospects for Rediscovery and Recovery. Washington, DC: Island Press. 113-125.

Cornell, H.V. and J.H Lawton. 1992. Species interactions, local and regional processes, and limits to the richness of ecological communities: a theoretical perspective. Journal of Animal Ecology 61: 1-12.

Currie, D.J. and V. Paquin. 1987. Large-scale biogeographical patterns of species richness of trees. Nature 329: 326-327.

Curtis, J.T. 1959. Vegetation of Wisconsin. Madison, Wisconsin: University of Wisconsin Press. 657 p.

Dahir, S.E. 1994. Tree mortality and gap formation in old-growth hemlock-hardwood forests of the Great Lakes regions [M.Sc. thesis]. Madison, Wisconsin: University of Wisconsin.

Daubenmire, R.F. 1968. Plant Communities: A Textbook of Plant Synecology. New York: Harper and Row. 300 p.

Davis, M.B. 1984. Holocene vegetational history of eastern United States In: H.E. Wright (editor). Late Quarternary Environments of the United States. London, UK: Longman Group. 166-181.

Davis, M.B. (editor) 1996. Eastern Old Growth Forests: Prospects for Rediscovery and Recovery. Washington, DC: Island Press. 383 p.

Davis, M.B., T.E. Parshall, and J.B. Ferrari. 1996. Landscape heterogeneity of hemlock-hardwood forest in northern Michigan. In: M.B. Davis (editor). Eastern Old Growth Forests: Prospects for Rediscovery and Recovery. Washington, DC: Island Press. 291-304.

Day, R.J. and G.T. Woods. 1977. The Role of Wildfire in the Ecology of Jack and Red Pine Forests of Quetico Provincial Park. Quetico Provincial Park Fire Ecology Study Report No. 5. Toronto, Ontario: Ontario Ministry Natural Resources. 79 p.

Day, R.J. and J.V. Carter. 1991. The Ecology of the Temagami Forest. Toronto, Ontario: Northeastern Region, Ontario Ministry of Natural Resources. 88 p.

Drake, J.A. 1990. Communities as assembled structures: do rules govern pattern? Trends in Ecology and Evolution 5: 159-164.

Dunn, C.P., G.R. Guntenspergen, and J.R. Dorney. 1983. Catastrophic wind disturbance in an old-growth hemlock-hardwood forest, Wisconsin. Canadian Journal of Botany 61: 211-217.

Farrar, J.L. 1995. Trees in Canada. Markham, Ontario: Fitzhenry and Whiteside; Canadian Forest Service. 502 p.

Fassnacht, K.S. and S.T. Gower. 1998. Comparison of soil and vegetation characteristics of six upland forest habitat types in north-central Wisconsin. Northern Journal of Applied Forestry 15: 69-76.

Flader, S.L. (editor). 1983. The Great Lakes Forest, An Environmental and Social History. Minneapolis, Minnesota: University of Minnesota Press. 336 p.

Flannigan, M.D. 1993. Fire regime and the abundance of red pine. International Journal of Wildland Fire 3: 241-247.

Flannigan, M.D., T.J. Lynham, and P.C. Ward. 1989. An extensive blowdown occurrence in northwestern Ontario. In: D.C. MacIver, H. Auld, and R. Whitewood (editors). Proceedings of the 10th Conference on Fire and Meteorology. Chalk River, Ontario: Petawawa National Forestry Institute, Canadian Forest Service. 65-71.

Flannigan, M.D. and M.G. Weber. 2000. Influences of climate on Ontario forests. In: A.H. Perera, D.L. Euler, and I.D. Thompson (editors). Ecology of a Managed Terrestrial Landscape: Patterns and Processes in Forest Landscapes of Ontario. Vancouver, British Columbia: University of British Columbia Press. 103-114.

Fleming, R.A., A.A. Hopkin and J.-N. Candau. 2000. Insect and disease disturbance regimes in Ontario's forests. In: A.H. Perera, D.L. Euler, and I.D. Thompson (editors). Ecology of a Managed Terrestrial Landscape: Patterns and Processes in Forest Landscapes of Ontario. Vancouver, British Columbia: University of British Columbia Press. 141-162.

Fleming, R.L. and R.M. Crossfield. 1983. Strip Cutting in Shallow-Soil Upland Black Spruce Near Nipigon, Ontario. Information Report O-X-354. Sault Ste. Marie, Ontario: Canadian Forest Service. 27 p.

Forest Landscape Ecology Program, Ontario Ministry of Natural Resources. 1998. Ontario's Forest Fire History: An Interactive Digital Atlas [CD-ROM]. Sault Ste. Marie, Ontario: Ontario Forest Research Institute, Ontario Ministry of Natural Resources.

Foster, D.R. 1984. Phytosociological description of the forest vegetation of southeastern Labrador. Canadian Journal of Botany 62: 899-906.

Fowells, H.A. 1965. Silvics of Forest Trees of the United States. Agriculture Handbook No. 271. Washington, DC: USDA Forest Service. 762 p.

Frelich, L.E. and C.G. Lorimer. 1985. Current and predicted effects of long-term deer browsing in hemlock forests in Michigan. Biological Conservation 43: 99-120.

Frelich, L.E. and C.G. Lorimer. 1991. Natural disturbance regimes in hemlock-hardwood forests of the upper Great Lakes region. Ecological Monographs 61:145-164.

Frelich, L.E. and P.B. Reich. 1995. Spatial patterns and succession in a Minnesota southern-boreal forest. Ecological Monographs 65:325-346.

Frelich, L.E. and P.B. Reich. 1996. Old growth in the Great Lakes region. In: M.B. Davis (editor). Eastern Old Growth Forests: Prospects for Rediscovery and Recovery. Washington, DC: Island Press. 144-160.

Gauthier, S., J. Gagnon, and Y. Bergeron. 1993. Population age structure of *Pinus banksiana* at the southern edge of the Canadian boreal forest. Journal of Vegetation Science 3: 783-790.

Gerardi, M.H. and J.K. Grimm. 1979. The History, Biology, Damage and Control of the Gypsy Moth. London, UK: Associated University Presses. 233 p.

Ghent, A.W., D.A. Fraser, and J.B. Thomas. 1957. Studies of regeneration in forest stands devastated by the spruce budworm. Forest Science 3: 184-208.

Gluck, M.J. and R.S. Rempel. 1996. Structural characteristics of post-wildfire and clearcut landscapes. Environmental Monitoring and Assessment 39: 435-450.

Graumlich, L.J. and M.B. Davis. 1993. Holocene variation in spatial scales of vegetation pattern in the upper Great Lakes. Ecology 74: 826-839.

Groot, A. 1984. Stand and Site Conditions Associated with Abundance of Black Spruce Advance Growth in the Northern Clay Section of Ontario. Information Report O-X-358. Sault Ste. Marie, Ontario: Canadian Forest Service. 15 p.

Groot, A., and B.J. Horton. 1994. Age and structure of natural and second-growth peatland *Picea mariana* stands. Canadian Journal of Forest Research 24: 225-233.

Gross, H.L. 1985. The Impact of Insects and Diseases on the Forest of Ontario. Information Report O-X-366. Sault Ste. Marie, Ontario: Canadian Forest Service. 96 p.

Gross, H.L., D.B. Roden, J.J. Churcher, G.M. Howse, and D. Gertridge. 1992. Pest-caused Depletions to the Forest Resource of Ontario, 1982-1987. Joint Report No. 17. Sault Ste. Marie, Ontario: Canadian Forest Service. 23 p.

Guyette, R.P. and D.C. Dey. 1995. A Presettlement Fire History in an Oak-Pine Forest Near Basin Lake, Algonquin Park, Ontario. Forest Research Report No. 132. Sault Ste. Marie, Ontario: Ontario Forest Research Institute, Ontario Ministry of Natural Resources. 7 p.

Haila, Y., K. Hanski, J. Niemelä, P. Puntilla, S. Raivo, and H. Tukia. 1994. Forestry and the boreal fauna: matching management with natural dynamics. *Annales Zoologica Fennici* 31: 187-202.

Hansen, A.J., T.A Spies, F.J. Swanson, and J.L. Ohmann. 1991. Conserving biodiversity in managed forests. Bioscience 41: 382-392.

Hardy, L. 1977. La déglaciation et les épisodes lacustres sur le versant québécois des basses terres de la baie de James. Géographie Physique et Quarternaire 31: 261-273.

Hearnden, K.W., S.V. Millison, and W.C. Wilson. 1992. A Report on the Status of Forest Regeneration. Toronto, Ontario: Ontario Independent Forest Auditor Commission. 116 p.

Heinselman, M.L. 1973. Fire in the virgin forests of the Boundary Waters Canoe Area, Minnesota. Quarternary Research 3: 329-383.

Heinselman, M.L. 1981. Fire intensity and frequency as factors in the distribution and structure of northern ecosystems. General Technical Report WO-26. In: H.A. Mooney, T.M. Bonnickson, N.L. Christensen, J.E. Lotan, and W.A. Reiners (editors). Fire Regimes and Ecosystem Properties. Washington, DC: USDA Forest Service. 7-57.

Henry, J.D. and J.M.A. Swan. 1974. Reconstructing forest history from live and dead plant material - an approach to the study of forest succession in southwest New Hampshire. Ecology 55: 772-783.

Hills, G.A. 1959. A Ready Reference to the Description of the Land of Ontario and its Productivity. Preliminary Report. Maple, Ontario: Ontario Department of Lands and Forests. 142 p.

Hills, G.A. 1960. Regional site research. Forestry Chronicle 36:401-423.
Hosie, R.C. 1969. Native trees of Canada. Canadian Forestry Service, Queen's Printer, Ottawa. 380 p.
Hough, A.F. 1965. A twenty-year record of understory vegetation change in a virgin Pennsylvania forest. Ecology 46: 370-373.
Howse, G.M. 1996. Gypsy moth in Canada, 1995. In: Proceedings, 1995 Annual Gypsy Moth Review: The Gypsy Moth: A Changing Horizon. Acme, Michigan: USDA Forest Service and Michigan Department of Agriculture. 36.
Hunter, M.L. 1990. Wildlife, Forests, and Forestry: Principles of Managing Forests for Biological Diversity. Englewood Cliffs, New Jersey: Prentice-Hall. 370 p.
Hunter, M.L. 1993. Natural fire regimes as spatial models for managing boreal forests. Biological Conservation 65: 115-120.
Jacobson, G.L. and A. Dieffenbacher-Krall. 1995. White pine and climate change: insights from the past. Journal of Forestry 93: 39-42.
Johnson, E.A. 1992. Fire and Vegetation Dynamics: Studies from the North American Boreal Forest. Cambridge, UK: Cambridge University Press. 129 p.
Johnson, E.A. and C.P.S. Larsen. 1991. Climatically induced changes in fire frequency in the southern Canadian Rockies. Ecology 72: 194-201.
Johnson, E.A., K. Miyanishi, and J.M.H. Weir. 1995. Old-growth, disturbance, and ecosystem management. Canadian Journal of Botany 73: 918-926.
Johnson, J.A. and S.A. Walsh 1997. Building air photo interpretation keys to the NOW FEC S-types and V-types in the Roslyn Lake Study Area: a case study. NODA Technical Report TR-40. Sault Ste. Marie, Canadian Forest Service. 39 p.
Johnston, M.H. and J.A. Elliot. 1996. Impacts of logging and wildfire on an upland black spruce community in northwestern Ontario. Environmental Monitoring and Assessment 39: 283-297.
Kneeshaw, D.D. and Y. Bergeron. 1996. Ecological factors affecting the abundance of advance regeneration in Quebec's southwestern boreal forest. Canadian Journal of Forest Research 26: 888-898.
Kneeshaw, D.D. and Y. Bergeron. 1998. Canopy gap characteristics and tree replacement in the southeastern boreal forest. Ecology 79: 783-794.
Lewis, C.F.M. and T.W. Anderson. 1989. Oscillations of levels and cool phases of the Laurentian Great Lakes caused by inflows from glacial Lake Agassiz and Barlow-Ojibway. Journal of Paleolimnology 2:99-146.
Li, C. 2000. Fire regimes and their simulation with reference to Ontario. In: A.H. Perera, D.L. Euler, and I.D. Thompson (editors). Ecology of a Managed Terrestrial Landscape: Patterns and Processes in Forest Landscapes of Ontario. Vancouver, British Columbia: University of British Columbia Press. 115-140.
Liu, K.-B. 1990. Holocene paleoecology of the boreal forest and Great Lakes-St. Lawrence forest in northern Ontario. Ecological Monographs 60:179-212.
Loucks, O.L. 1970. Evolution of diversity, efficiency, and community stability. American Zoologist 10: 17-25.
Mackey, B. G., D. W. McKenny, Y.-Q. Yang, J. P. McMahon, and M. F. Hutchinson. 1996. Site regions revisited: a climatic analysis of Hills' site regions for the province of Ontario using a parametric method. Canadian Journal of Forest Research 26: 333-354 + errata.
Mutch, R.W. 1970. Wildland fires and ecosystems - a hypothesis. Ecology 51: 1046-1051.
Nealis, V. 1995. Population biology of the jack pine budworm. In: W.J.A. Volney, V.G. Nealis, G.M. Howse, A.R. Westwood, D.R. McCullough, and B.L. Laishley (editors). Jack Pine Budworm Biology and Management. Information Report NOR-X-342. Edmonton, Alberta: Canadian Forest Service. 55-72.
Noss, R.F. 1990. Indicators for monitoring biodiversity: a hierarchical approach. Conservation Biology 4: 355-363.
Ontario Geological Survey. 1991. Bedrock Geology of Ontario, Explanatory Notes and Legend. Map Number 2545. Toronto, Ontario: Ministry of Northern Development and Mines.
Ontario Ministry of Natural Resources. 1996. Forest Resources of Ontario 1996. Toronto, Ontario: Queen's Printer for Ontario. 86 p.
Paré, D. and Y. Bergeron. 1995. Above-ground biomass accumulation along a 230-year chronosequence in the southern portion of the Canadian boreal forest. Journal of Ecology 83: 1001-1007.
Pastor, J. and D.J Mladenoff. 1992. The southern boreal-northern hardwood forest border. In: H.H. Shugart, R. Leemans, and G.B. Bonan (editors). A Systems Analysis of the Global Boreal Forest. Cambridge, UK: Cambridge University Press. 216-240.
Payette, S. 1992. Fire as a controlling process in North American boreal forest. In: H.H. Shugart, R. Leemans, and G.B. Bonan (editors). A Systems Analysis of the Global Boreal Forest. Cambridge, UK: Cambridge University Press. 144-169.
Payette, S. 1993. The range limit of boreal tree species in Quebec-Labrador: an ecological and palaeoecological interpretation. Review of Paleobotany and Palynology 79: 7-30.
Payette, S., L. Filion, and A. Delwaide. 1990. Disturbance regime of a cold temperate forest as deduced from tree-ring patterns: the Tantaré Ecological Reserve, Quebec. Canadian Journal of Forest Research 20: 1228-1241.
Perera, A.H. and D.J.B. Baldwin. 2000. Spatial patterns in the managed forest landscape of Ontario. In: A.H. Perera, D.L. Euler, and I.D. Thompson (editors). Ecology of a Managed Terrestrial Landscape: Patterns and Processes in Forest Landscapes of Ontario. Vancouver, British Columbia: University of British Columbia Press. 74-99.
Pickett, S.T.A. and P.S. White (editors). 1985. The Ecology of Natural Disturbance and Patch Dynamics. New York: Academic Press. 208 p.
Pitelka, L.F. 1997. Plant migration and climate change. American Scientist 85: 464-473.
Pregitzer, K.S. and B.V. Barnes. 1984. Classification and comparison of upland hardwood and conifer ecosystems of the Cyrus H. McCormick Experimental Forest, Michigan. Canadian Journal of Forestry 14: 362-375.
Richard, P. 1980. Histoire postglaciaire de la végétation au sud du Lac Abitibi, Ontario et Québec. Géographie Physique et Quarternaire 31:77-94.
Richardson, A.H. 1929. Forestry in Ontario [booklet]. Toronto, Ontario: Ontario Department of Forestry. 73 p.
Ritchie, J.C. 1987. Postglacial Vegetation of Canada. Cambridge, UK: Cambridge University Press. 178 p.
Ritchie, J.C. and F.K. Hare. 1971. Late Quarternary vegetation and climate near the arctic treeline of northwestern North America. Quaternary Research 1: 331-342.
Roden, D.B. and G.A. Surgeoner. 1991. Survival, development time, and pupal weights of larvae of gypsy moth reared on foliage of common trees of the Great Lakes region. Northern Journal of Applied Forestry 8: 126-128.

Rogers, R.S. 1978. Forests dominated by hemlock: Distribution related to site and post-settlement history. Canadian Journal of Botany 56: 843-854.

Roland, J. 1993. Large-scale forest fragmentation increases the duration of tent caterpillar outbreak. *Oecologia* 93: 25-30.

Rose, A.H. and O.H. Lindquist. 1982. Insects of Eastern Hardwood Trees. Technical Report No. 29. Ottawa, Ontario: Canadian Forest Service. 304 p.

Rose, A.H. and O.H. Lindquist. 1984. Insects of Eastern Pines. Publication No. 1313. Ottawa, Ontario: Canadian Forest Service. 127 p.

Rowe, J.S. 1961. The level-of-integration concept and ecology. Ecology 42: 420-427.

Rowe, J.S. 1962. Soil, site, and land classification. Forestry Chronicle 38: 420-432.

Rowe, J.S. 1972. Forest Regions of Canada. Publication No. 1300. Ottawa, Ontario: Canadian Forest Service. 172 p.

Runkle, J.R. 1982. Patterns of disturbance in some old growth mesic forest of eastern North America. Ecology 63: 1533-1546.

Sado, E.V. and B.F. Carswell. 1987. Surficial Geology of Northern Ontario. Ontario Geological Survey Map No. 2518. Toronto, Ontario: Ontario Ministry of Northern Development and Mines.

Sims, R.A. 1996. The derivation of spatially referenced ecological databases for ecosystem mapping and modeling in the Rinker Lake Research area of Northwestern Ontario. NODA File Report No. 33. Sault Ste. Marie, Canadian Forest Service. 48 p.

Sims, R.A., H.M. Kershaw, and G.M. Wickware. 1990. The Autecology of Major Tree Species in the North Central Region of Ontario. Publication No. 5310. Thunder Bay, Canada-Ontario Forest Resource Development Agreement/ Ontario Ministry of Natural Resources. 126 p.

Sims, R.A., W.D. Towill, K.A. Baldwin, and G.M. Wickware. 1989. Forest Ecosystem Classification for Northwestern Ontario. Ottawa, Ontario: Canada-Ontario Forest Resource Development Agreement. 191 p.

Stocks, B.J. 1985. Forest fire behavior in spruce budworm-killed balsam fir. In: Recent Advances in Spruce Budworm Research. Proceedings, CANUSA Spruce Budworm Research Symposium, September 16-20, Bangor, Maine. Ottawa, Ontario: Canadian Forest Service. 198-209.

Stocks, B.J. 1987. Fire potential in the spruce budworm-damaged forests of Ontario. Forestry Chronicle 63: 8-14.

Stocks, B.J. and D.B. Bradshaw. 1981. Damage caused by spruce budworm: fire hazard. In: J. Hudak and A.G. Raske (editors). Review of the Spruce Budworm Outbreak in Newfoundland - Its Control and Forest Management Implications. Information Report N-X-205. St. John's, Newfoundland: Canadian Forest Service. 57-58.

Stocks, B.J. and R.B. Street. 1983. Forest fire weather and wildfire occurrence in the boreal forest of northwestern Ontario. In: R.W. Wein, R.R. Riewe, and I.R. Methven (editors). Resources and Dynamics of the Boreal Zone. Ottawa, Ontario: Association of Canadian Universities for Northern Studies. 249-265.

Suffling, R. 1983. Stability and diversity in boreal and mixed temperate forests: a demographic approach. Journal of Environmental Management 17: 359-371.

Suffling, R. 1991. Wildland fire management and landscape diversity in the boreal forest of northwestern Ontario during an era of global warming. In: S.C. Nodvin and T.A. Waldrup (editors). Fire and the Environment: Ecological and Cultural Perspectives. Proceedings of an International Symposium, March 20-24,1990, Knoxville, Tennessee. General Technical Report SE-69. Asheville, North Carolina: USDA Forest Service, Southeastern Experiment Station. 97-106.

Suffling, R. 1995. Can disturbance determine vegetation distribution during climate warming? A boreal test. Journal of Biogeography 22: 501-508.

Suffling, R., C. Lihou, and Y. Morand. 1988. Control of landscape diversity by catastrophic disturbance: a theory and a case study of fire in a Canadian boreal forest. Environmental Management 12: 73-78.

Syrjänen, K., R. Kalliola, A. Puolasmaa, and J. Mattsson. 1994. Landscape structure and forest dynamics in subcontinental Russian European taiga. *Annales Zoologici Fennici* 31: 19-34.

Thompson, I.D. and D.A. Welsh. 1993. Integrated resource management in boreal forest ecosystems: impediments and solutions. Forestry Chronicle 69:32-39.

Thompson, I. D., M. D. Flannigan, B. M. Wotton, and R. Suffling. 1998. The effects of climate change on landscape diversity: an example in Ontario forests. Environmental Monitoring and Assessment 49: 213-233.

Tyrrell, L.E. and T.R Crow. 1994. Structural characteristics of old-growth hemlock-hardwood forests in relation to age. Ecology 75: 370-386.

Van Wagner, C.E. 1978. Age-class distribution and the forest fire cycle. Canadian Journal of Forest Research 8: 220-227.

Van Wagner, C.E. 1983. Fire behaviour in northern conifer forests and shrublands. In: R.W. Wein and D.A. MacLean (editors). The Role of Fire in Northern Circumpolar Ecosystems. Chichester, UK: John Wiley. 65-80.

Van Wagner, C.E. 1988. The historical pattern of annual burned area in Canada. Forestry Chronicle 64:1 82-185 and erratum 64: 319.

Voigt, D.R., J.A. Baker, R.S. Rempel, and I.D. Thompson. 2000. Forest vertebrate responses to landscape-level changes in Ontario. In: A.H. Perera, D.L. Euler, and I.D. Thompson (editors). Ecology of a Managed Terrestrial Landscape: Patterns and Processes in Forest Landscapes of Ontario. Vancouver, British Columbia: University of British Columbia Press. 198-233.

Voigt, D.R., M. Deyne, M. Malhiot, B. Ranta, B. Snider, R. Stefanski, and M. Strickland. 1992. White-tailed deer in Ontario: Background to a Policy. Draft Document. Toronto, Ontario: Ontario Ministry of Natural Resources, Wildlife Policy Branch. 83 p.

Ward, P.C. and A.G. Tithecott. 1993. The Impact of Fire Management on the Boreal Landscape of Ontario. Publication No. 305. Sault Ste. Marie, Ontario: Ontario Ministry of Natural Resources. 12 p.

Webb, T. 1986. Is the vegetation in equilibrium with climate? An interpretive problem for late Quarternary pollen data. *Vegetatio* 67:75-91.

Whitney, R. D. 1988. Root Rot Technology Transfer. Sault Ste. Marie, Ontario: Canadian Forest Service. 35 p.

Wright, H. A. and A. W. Bailey. 1982. Fire Ecology: US and Southern Canada. New York: John Wiley. 501 p.

4 Forest Vertebrates of Ontario: Patterns of Distribution

IAN D. THOMPSON*

INTRODUCTION

The Great Lakes and boreal forests (Rowe 1972) of Ontario provide habitat for approximately eight species of reptiles, 14 species or subspecies of amphibians, 60 species of mammals, and 150 species of birds (Banfield 1974; Cook 1984; Cadman et al. 1987). A further 140 species of birds, 15 species of reptiles and amphibians, and 2 species of mammals occupy habitats in waterbodies and wetlands which are associated with or dependent on forests. Neither these latter species nor fish are considered in this chapter. By definition, insects are also beyond the scope of the chapter, although they constitute the greatest animal diversity in Ontario's forests, occurring in several tens of thousands of species. Insects which have contributed to landscape-level change in Ontario forests are described in a later chapter (Fleming et al. 2000, this volume).

Animal communities in forested systems are affected by a number of factors related to habitat suitability and availability that may differ among species and among forest types. All animals respond to ultimate factors such as climate, which affect their distribution within the province. Animal communities are also structured through proximate and ecosystem-dependent processes which reflect the distribution of other animal species acting as competitors or predators within food webs. Most animal species have particular habitat preferences within forests, and the links between these preferences and forest processes is dependent to a large extent on body size (Harestad and Bunnell 1979; Holling 1992). Habitat supply is governed by a range of factors which are scale-dependent spatially and temporally, relative to the animal species, so that animals could be grouped according to their responsiveness to factors operating at various scales. For example, populations of woodland caribou (*Rangifer tarandus*) respond to broad-scale forest patterns, such as those created by fire, and to broad-scale forest processes such as photosynthesis. Red-backed voles (*Clethrionomys gapperi*), on the other hand, respond to change at a much smaller scale, probably at the level of the individual forest stand, in such factors as the amount of moss cover and the production of fungal fruiting bodies.

At one time, animal species and communities responded to landscape patches formed as a result of natural disturbances, within the broad constraint of climate. An important question of the present day is whether animal species are responding to changes in landscape structure caused by human intervention. In this chapter, broad species distribution patterns are described relative to hierarchies of processes. A later chapter (Voigt et al. 2000, this volume) examines the individual factors operating in species distribution and provides answers about the responses of particular species to changes in landscape structure.

BROAD SPECIES DISTRIBUTION PATTERNS IN ONTARIO

There are latitudinal and longitudinal gradients in the occurrence of vertebrate species in the province

* Canadian Forest Service, Great Lakes Forest Research Centre, 1219 Queen Street East, Sault Ste. Marie, Ontario P6A 5M7

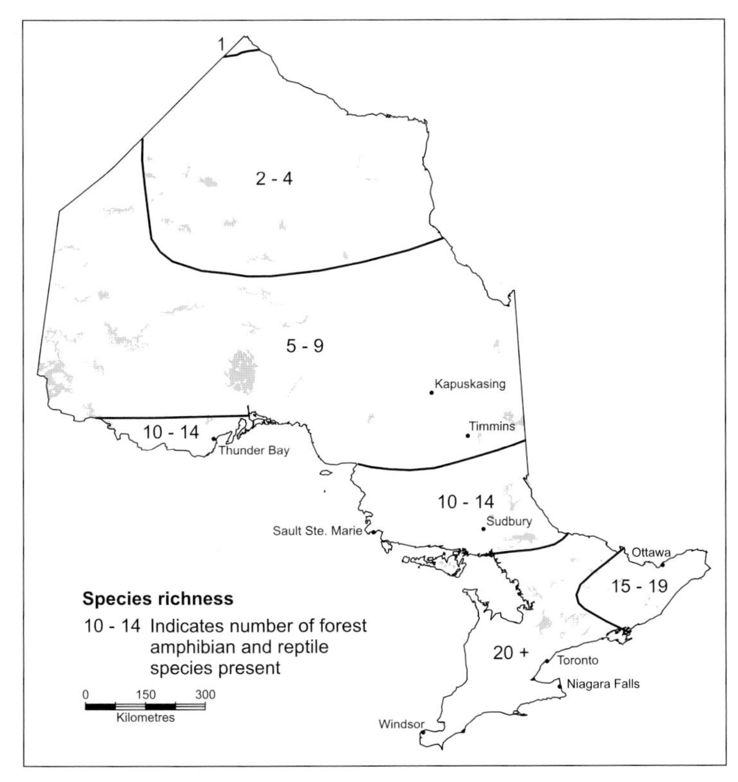

Figure 4.1 The distribution of forest amphibian and reptile species in Ontario. (Data from Cook 1984)

(Figures 4.1 to 4.3, based on species data from Cook 1984; Cadman et al. 1987; Dobbyn 1994). The highest forest animal richness in Ontario occurs in an area of about 25,000 km² in the east-central portion of the province, bounded approximately by Highway 11 on the west, Highway 17 on the north and northeast, Highway 41 to the southeast, and Highway 7 on the south. Here there are approximately 110 species of forest birds, 49 mammal species, and 22 forest-dwelling amphibian and reptile species. This area, which includes Algonquin Provincial Park, is located within the Great Lakes forest and lies at about 45°N latitude, or farther to the south than most of Canada. The area supports a broad array of forest types, including many boreal-type stands, yet it is close to the Carolinian forest region. It has reasonably abundant precipitation, a relatively warm climate (and thus high forest productivity), and a relatively low human presence, certainly in comparison to the immediately adjacent southern region. Fully one-third of the area lies within parks or other protected areas. These factors combine to produce a range of habitat conditions attracting a high number of species. Many of the species that breed in this eastern area of the province are not found west of Lake Superior. The converse, however, is not true; there are no species of terrestrial forest vertebrates which occur exclusively in the west of the province.

The uneven distribution of species across Ontario likely has its origin in the distribution of many species in southeastern deciduous and mixed forests during the early Holocene period, when the Wisconsin glaciation ended (e.g., Mayr 1946). Many Great Lakes bird species, for example, prefer hardwood forests, including orioles (*Icterus* spp.), the yellow-throated vireo (*Vireo flavifrons*), and several species of flycatcher (*Empidonax* spp.). The number of vertebrate species declines as one moves away from the area of highest species richness near Algonquin Park and proceeds northward and eastward to the area between the Ottawa River valley and Lake Ontario. In the most southerly area of the province, high human populations and a history of land clearing for intensive agriculture appear responsible for a decline in the number of species. Proceeding northward from Algonquin Park, one finds species numbers declining from the Great Lakes forests, to the ecotonal forests, and into the boreal forest. The number of species is further reduced in the Hudson Bay lowlands, and the lowest number of species is found along the coast of Hudson Bay in the extreme north of the province (Figures 4.1 to 4.3). The data on species occurrence for the northwestern portion of the province is less reliable than the data for more accessible areas, particularly in the south, because fewer studies and observations have been conducted in this region.

Disproportionate Representation of Vertebrates in Ontario

The degree of responsibility which a jurisdiction bears for the conservation of a species can arguably be linked to the extent of the range of that species within the jurisdiction, relative to the area of some larger jurisdiction. The ratio of the two areas is used to identify species which a given jurisdiction has a disproportionately large responsibility to maintain. This concept is particularly important for rare species. Within the two

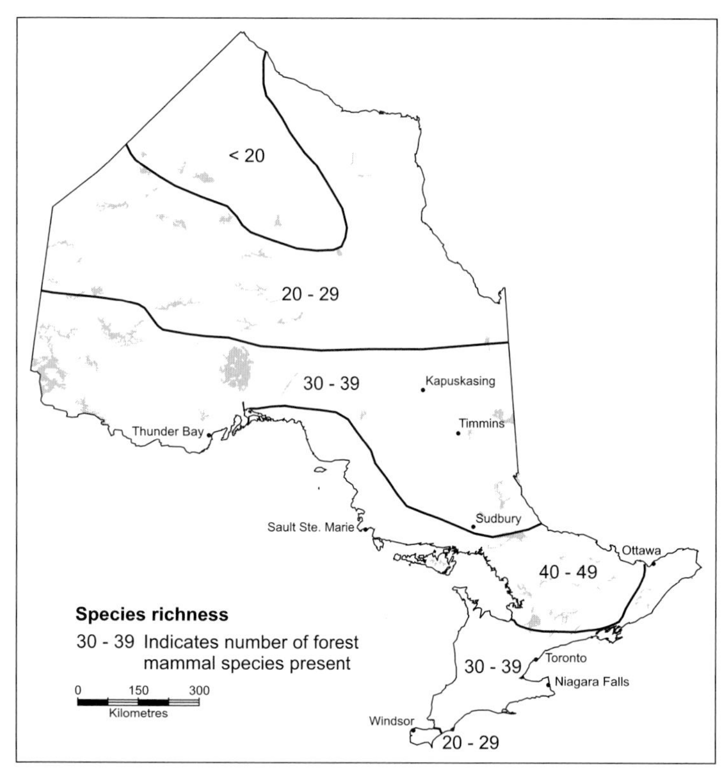

Figure 4.2 The distribution of forest mammal species richness in Ontario. (Data from Dobbyn 1994)

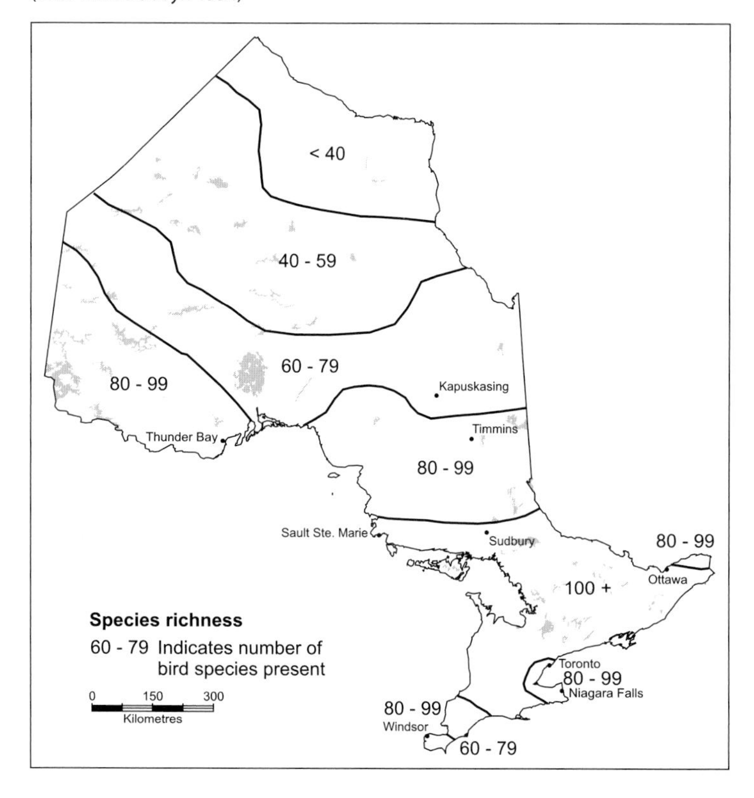

◄ **Figure 4.3** The distribution of forest bird species richness in Ontario. (Data from Cadman et al. 1987)

forest regions considered here, Ontario has a disproportionate responsibility in Canada for the conservation of several forest species, including the following: the southern bog lemming (*Synaptomys cooperi*), rock vole (*Microtus chrotorrhinus*), white-footed mouse (*Peromyscus leucopus*), southern flying squirrel (*Glaucomys volans*), red-shouldered hawk (*Buteo lineatus*), broad-winged hawk (*B. platypterus*), wood thrush (*Hylocichla mustelina*), chestnut-sided warbler (*Dendroica pensylvanica*), Connecticut warbler (*Oporornis agilis*), four-toed salamander (*Hemidactylium scutatum*), red-backed salamander (*Plethodon cinereus*), wood turtle (*Clemmys insculpta*), and ringneck snake (*Diadophus punctatus*). There are several other rare reptiles and amphibians for which Ontario maintains the only range in Canada and hence bears full Canadian conservation responsibility, but none is strictly a forest species. Two examples are the eastern Massasauga rattlesnake (*Sistrurus catenatus catenatus*) and Fowler's toad (*Bufo woodhousei fowleri*).

Among the forest-dependent species, the red-shouldered hawk and the wood turtle are listed provincially as "vulnerable," and southern flying squirrel is listed nationally as "vulnerable." The wood thrush and the Connecticut warbler are species that have declined during the past 20 years, and estimates for wood thrush populations suggest that less than a third of the original population remains (Sauer et al. 1997). Four of the twelve forest species for which Ontario has a high conservation responsibility, based on range in Canada, are either rare or in decline. The Connecticut warbler is the only one of these that occurs in the boreal forest; the others are Great Lakes forest species.

PATTERNS IN FOREST VERTEBRATE SPECIES RICHNESS

Species Richness in the Forest Landscape

Species richness is a scalar concept. The factors and processes affecting the number of species differ with scale, from a large geographic region of North America, to a smaller jurisdiction such as Ontario, a discrete area such as a specific Hills ecoregion, or finally a local area in which site-specific factors govern the presence or absence of a species within habitat patches. Across Ontario, there are clear patterns of species occurrence. At the scale of large regional landscapes, climate is an important factor affecting productivity within forests. Along with soil conditions and topography (Bonan 1992), it likely has a key influence on the occurrence of vertebrate species over broad areas of North America (Pianka 1966; Currie 1991), and likewise across Ontario.

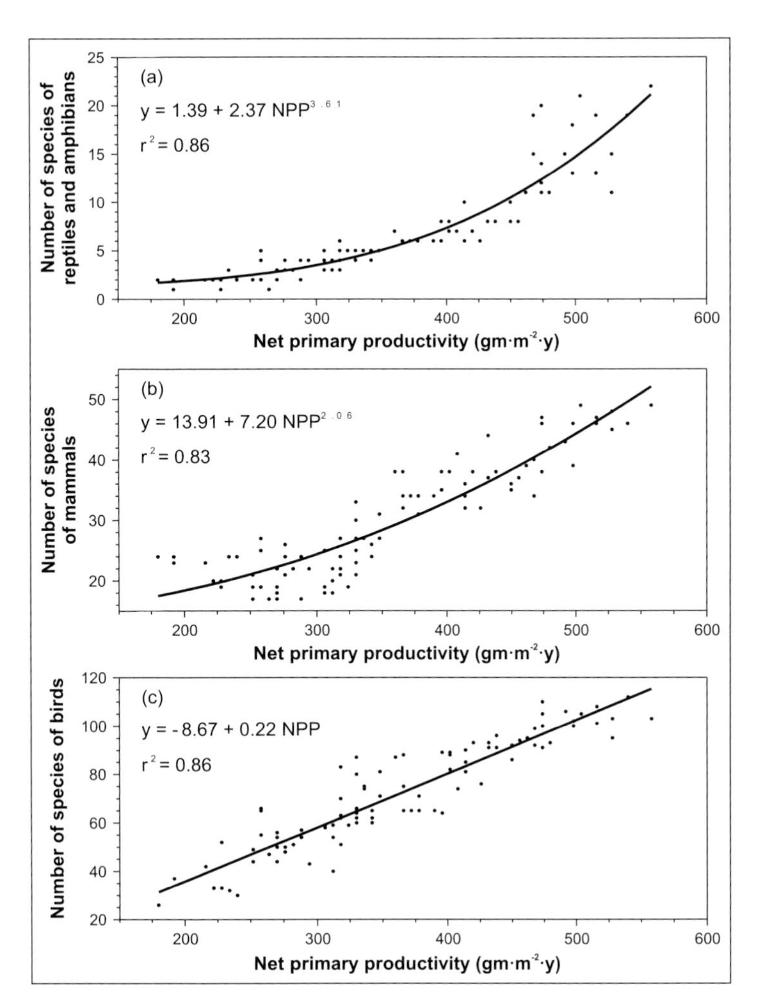

Figure 4.4 The relationship between net primary productivity (NPP) and forest vertebrate species richness in Ontario for (a) reptiles and amphibians; (b) mammals; and (c) birds.

Net primary productivity (NPP) in forests is directly related to climate (Bonan and Shugart 1989; Bonan 1992) and becomes lower as one moves northward in Ontario (Bickerstaff et al. 1981). The effects of climate on species distributions were examined for this chapter by assessing the relationships between species abundance and NPP. The results of analyses of species richness by 10,000-km^2 Universal Transverse Mercator (UTM) grid blocks, using NPP and January minimum temperature as independent variables, are shown in Figure 4.4a, b, and c. In deriving these curves, data from approximately Highway 7 south to Lake Ontario and Lake Erie were excluded, because of the lack of forest cover in these highly agricultural and urbanized landscapes. Species that do not use forests for some aspect of their life history were omitted (for example, most waterfowl, many sparrow species, European hare [*Lepus europaeus*], bullfrog [*Rana catesbeiana*]). Numbers of species per UTM block reported in Cadman et al. (1987) and Dobbyn (1994) varied with the effort of observers, and with the available data used to compile species distributions. In cases where the number of species in a given block was low compared to adjacent UTM blocks, an average number of species from the surrounding blocks was applied.

Species distributions were all strongly correlated to NPP (Figure 4.4). The best models for mammals (Figure 4.4b) and reptiles and amphibians (Figure 4.4a) were curvilinear, indicating a greater than expected increase in species richness, and hence in primary productivity, with decreasing latitude. In the case of amphibians and reptiles (Figure 4.4a), the regression model was improved somewhat by the addition of January minimum temperature, which explained an additional five percent of the variance. January temperature is likely a good predictor of frost depth in soils and freezing depth in lakes, and may thus influence winter survival of hibernating reptiles and amphibians. Cook (1984) also suggested that unpredictability in local climate would have an impact on the local species richness of amphibians and reptiles. In birds, the response of species richness to primary productivity was strongly linear, suggesting a somewhat different response to productivity than for the previous groups (Figure 4.4c). That difference is probably related to the migratory life history of most bird species, which enables them to arrive on the breeding grounds when annual productivity is highest.

These models clearly suggest the importance of latitudinal change in forest cover and productivity to vertebrate species distribution in Ontario. Currie (1991) found similar results for a much larger data set: he tested species richness of vertebrate classes for all of North America, using potential evapotranspiration (PET) as the independent variable. PET is basically a climatic variable that is influenced by local conditions of temperature and soil humidity. In Ontario, where there is a clear moisture gradient between east and west, PET would not be expected to correlate strongly with species distribution; furthermore, no relationship necessarily exists between PET and NPP. Primary productivity is a strong predictor of vertebrate species richness, because it integrates energy, actual evapotranspiration, nutrients, and water availability. The remarkably high amount of variation explained by the curves depicted in Figure 4.4 supports the contention that climate, as reflected by NPP, is a key influence on species richness at provincial and regional landscape scales.

Latitudinal and longitudinal patterns in species occurrence

The only two species of amphibian in the far north of Ontario are the boreal (or striped) chorus frog (*Pseudacris triseriata*) (not a true forest species), and the wood frog (*Rana sylvatica*) (Figure 4.1). Two other extreme-northern occurring amphibians, American toad (*Bufo americanus*) and northern leopard frog (*Rana pipiens*), are found along much of the Hudson Bay coast but do not extend to the far northwestern border with Manitoba (Cook 1984). Most Ontario species of reptiles and amphibians are absent in the boreal region, where there are only eight forest amphibian species (three salamanders, four frogs, and one toad), one snake, and one turtle (a non-forest species). The eastern garter snake (*Thamnophis sirtalis*) is the only reptile occurring throughout the boreal forest and into the Hudson Bay lowlands.

Within the Great Lakes forests and the transition zone to the boreal forest, there are five species of forest reptiles and amphibians that only occur east of Lake Superior (Table 4.1), and two others whose distributions end near Thunder Bay (yellow-spotted salamander [*Ambystoma maculatum*] and eastern redback salamander). The differences in distribution among these species contribute to the fact that species richness is higher in eastern Ontario than in western Ontario, and also to the conclusion reached by Thompson et al. (1995) that the western and eastern old-growth white pine forests maintain distinct faunas.

Mammals also follow a provincial pattern of decline in species abundance with latitude and longitude. As with amphibians and reptiles, certain species of mammals occur primarily or only east of Lake Superior. One difference in the species distribution of mammals, compared to birds as well as amphibians and reptiles, is the high species richness that extends westward along the eastern shore of Lake Superior (Figure 4.2). For example, in the area of Pukaskwa National Park near Marathon, there are only five fewer species of mammals than in the vicinity of Algonquin Park. Among the mammals not found north and west of Algonquin Park are the southern flying squirrel, red bat (*Lasiurus borealis*), eastern pipistrelle (*Pipistrellus flavius*), and small-footed bat (*Myotis leibii*). Four other forest mammals also occur only in eastern Ontario (Table 4.1).

Some mammals have a northerly distribution and are not found (or are very rare) in southern forests; these include the woodland caribou, wolverine (*Gulo gulo*), heather vole (*Phenacomys intermedius*), rock vole (although some records exist south of the boreal region), arctic shrew (*Sorex arcticus*), and northern bog lemming (*Synaptomys borealis*). The latter species is only found in the Hudson Bay lowlands, along with several arctic and marine species such as the arctic fox (*Alopex lagopus*) and polar bear (*Ursus maritimus*), which are not included in Figure 4.2. The woodland caribou has a disjunct distribution in Ontario: remnant herds live in isolated locations along the coast of Lake Superior and on islands in that area, completely separated from the main distribution of the species about 150 km to the north. Bergerud (1974) suggested that these herds occupy undisturbed, isolated habitats that are difficult to hunt, have not been logged, and maintain a low predator density.

Differences in species composition between the boreal and Great Lakes forest species is well exemplified by birds. Most of the Great Lakes forests maintain 80 to 100 avian species, but the boreal bird fauna consists of fewer than 80 species throughout the forest region and only about 40 species in the Hudson Bay lowlands. The number of bird species declines considerably toward the southwest, with its highly agricultural landscape, almost as significantly as to the far north (Figure 4.3). There are fourteen species that breed primarily in Great Lakes forests, and only in the east (Table 4.1). A further nine species occur almost exclusively in the Great Lakes forests but are sporadically found elsewhere in the province as well. These include Cooper's hawk (*Accipter cooperii*); the northern saw-whet owl (*Aegolius acadicus*); red-headed woodpecker (*Melanerpes erythrocephalus*); great-crested flycatcher (*Myiarchus crinitus*); white-breasted nuthatch (*Sitta carolinensis*); warbling vireo (*Vireo gilvus*); pine warbler (*Dendroica pinus*); the indigo bunting (*Passerina cyanea*), which prefers edges but breeds in deciduous and mixed forests; and the northern oriole (*Icterus galbula*). In boreal forests, there are nine species of forest-dependent birds which breed there exclusively or are uncommon in Great Lakes forests. These are the sandhill crane (*Grus canadensis*), which breeds in treed or open bogs; great gray owl (*Strix nebulosa*); boreal owl (*Aegolius nebulosa*); boreal chickadee (*Parus hudsonicus*); palm warbler (*Dendroica palmarum*); Connecticut warbler; Wilson's warbler (*Wilsonia pusilla*); pine grosbeak (*Pinicola enucleator*); and the three-toed woodpecker (*Picoides tridactylus*).

Table 4.1

Great Lakes forest species occurring only in eastern Ontario (east of Lake Superior).

Amphibians:	
Four-toed salamander	*(Hemidactylium scutatum)*
Two-lined salamander	*(Eurycea bislineata)*
Birds:	
Red-shouldered hawk	*(Buteo lineatus)*
Yellow-billed cuckoo	*(Coccyzus americanus)*
Eastern screech owl	*(Otus asio)*
Willow flycatcher	*(Empidonax traillii)*
Blue grey gnatcatcher*	*(Polioptila caerulea)*
Wood thrush	*(Hylocichla mustelina)*
Loggerhead shrike*	*(Lanius ludovicianus)*
Yellow-throated vireo	*(Vireo flavifrons)*
Blue-winged warbler	*(Vermivora pinus)*
Golden-winged warbler	*(V. chysoptera)*
Cerulean warbler	*(Dendroica cerulea)*
Rufous-sided towhee*	*(Pipilo erythrophthalmus)*
Orchard oriole*	*(Icterus spurius)*
Reptiles:	
Wood turtle	*(Clemmys insculpta)*
Redbelly snake	*(Storeria occipitomaculata)*
Ringneck snake	*(Diadophis punctatus)*
Mammals:	
Opossum*	*(Didelphis virginiana)*
Hairy-tailed mole	*(Parascalops breweri)*
Red bat	*(Lasiurus borealis)*
Eastern pipistrelle	*(Pipistrellus flavius)*
Small-footed bat	*(Myotis leibii)*
Woodland vole*	*(Pitymys pinetoreum)*
Southern flying squirrel	*(Glaucomys volans)*
Gray fox*	*(Urocyon cinereoargentateus)*

* rare in true Great Lakes forest zone

The extreme southern area of the province, now dominated by non-forested landscapes, has lost many of its species through human activities. Several species are known to have been extirpated from this area, including the passenger pigeon (*Ectopistes migratorius*); wild turkey (*Meleagris gallopavo*) (recently reintroduced); timber rattlesnake (*Crotalus horridus*); blue racer (*Coluber constrictor*), which occurs on Pelee Island; fisher (*Martes pennanti*); river otter (*Lontra canadensis*); and northern cricket frog (*Acris crepitans*).

Species Richness in Forest Types

While regional differences in species richness across Ontario are affected by variables that have an impact over broad landscapes, forest-level differences in species assemblages, or beta diversity, are controlled by factors relating to the functional ecology of the forest system. Species richness in a given forest is dependent on many factors, as reported by Tilman (1982), Askins et al. (1987), Drake (1990), Currie (1991), Cornell and Lawton (1992), and Ricklefs and Schluter (1993). Some factors are abiotic; these include local climate, as related to relative productivity and survivorship; habitat heterogeneity, or the variety of patch types; the amount of particular habitat types in a landscape, or the suitability of forest patches for certain species; the history which the forest has undergone; and the predictability and degree of disturbance. Biotic factors include competition among species, parasitism, and predation.

Major exogenous natural processes that affect landscapes at the spatial scale of individual forests include fires and large blowdowns. Disturbances are important factors in maintaining the wildlife diversity that has evolved under past disturbance dynamics in forests (Levin 1992). These disturbances act predictably through the creation of diverse habitats in time and space. Declines in populations of moose (*Alces alces*) and white-tailed deer (*Odocoileus virginianus*) as forests age provide an example of the necessity for stand renewal to maintain species that require early successional forests. Haila et al. (1994) suggested that three predictions can be made with respect to wildlife in landscapes subject to disturbance. The first is that the most abundant species must have generalized habitat requirements, as evidenced by a high degree of spatial heterogeneity in the landscapes they occupy. Secondly, the types of species living in the different successional stages of the forests must roughly parallel the r-K continuum (see Voigt et al. 2000, this volume). Thirdly, continuity of habitat in time and space is critical to the survival of some species, particularly those not well adapted to dispersal.

McLaren et al. (1998) proposed a number of indicator species for broad forest types, including mixedwood, upland conifer, and tolerant hardwood forests, as identified at the "forest level," which they defined as 1,000 ha to 10,000 ha. The species they chose were primarily those with a range of intermediate body sizes, such as the snowshoe hare (*Lepus americanus*) and flying squirrels (*Glaucomys* spp.), but they also included some forest songbirds, because initial habitat selection by these species is related to the suitability of forested landscapes.

Species Richness in Forest Patches

Alpha diversity refers to differences in numbers of species among patches of habitat. In forests, small-bodied species are primarily affected by local habitat conditions, or patch factors. Forest processes acting at local scales and affecting alpha diversity include small blowdowns, small fires, and canopy gaps from the toppling of individual dead trees. Species richness within forest types (or individual patches) is related to patch suitability as habitat (that is to patch type, size, and age), and to individual within-patch structures such as snags and fallen logs.

There has been some testing in Ontario of the value of forest ecosystems, as defined by various classification systems (e.g., Sims et al. 1989), as predictors of community associations of small-bodied vertebrate species. Welsh and Lougheed (1996) showed that there was a high degree of specialization among neotropical migrant songbirds by Forest Ecosystem Classification (FEC) types, or groups of types, in the Claybelt. Small mammal communities near Kapuskasing differed among four closely related ecosystems dominated by mature black spruce (I.D.Thompson, unpub.). Martell (1983a) demonstrated how forest age and successional stage influenced the structure of Ontario boreal rodent communities. Naylor and Bendell (1983) found that small rodent species richness was affected by age and vegetation diversity in various jack pine ecosystems near Gogama.

Individual structures within stands are important in affecting habitat suitability for some species. For example, although pileated woodpeckers (*Dryocopus pileatus*) occupy forests in Ontario ranging widely in type and age, their occurrence in a particular stand appears related to the availability of large-diameter aspen or poplar that can be excavated for nests (Spytz 1993; Kirk and Naylor 1996). Woodpeckers may act as keystone species in forest ecosystems because of the large number of species, including breeding bats, owls, and chickadees, that require the tree cavities excavated by woodpeckers for parts of their life history (Thompson and Angelstam 1999). Thus, species interactions at small scales are important to species richness as well. McLaren et al. (1998) proposed certain species as indicators of processes underway at the stand or local habitat scale which reflected such interrelationships among species and site-specific ecosystem functions.

Patch quality has an important influence on whether a given forest stand is occupied by a particular species and on the level of a productivity of the species at the site (Van Horne 1983). Some patches are occupied for long periods, while other, less suitable patches may be used only temporarily and then abandoned. This difference is the basis for the theory of metapopulations, which postulates that some habitats, or patches, are net producers of a species, while others must receive a constant immigration of individuals to remain occupied (Pulliam 1988). In boreal Alberta and Ontario, red-backed voles prefer certain forest types and ages, and the value of these patches differs in terms of how long individuals survive there and how productive they are (Martell 1983b; Bondrup-Nielsen 1987). Similar inferences were made for marten (*Martes americana*) by Thompson (1994) for central boreal Ontario. A more detailed discussion of factors that influence species richness and of metapopulation theory is provided by Voigt et al. (2000, this volume).

An Example of the Effects of Patch Age: Relationships of Animal Communities to Old, White Pine Ecosystems at Two Scales

The age of a forest is an important local factor affecting species occurrence. Many species are adapted to successional forests in Ontario (Mönkkönen and Welsh 1994), because much of the landscape is in young age classes at any point in time (Van Wagner 1978). Old forests, however, are structurally more diverse than younger stands and may support more specialist species (Hunter 1990).

One forest community that has declined considerably across Ontario is forest dominated by old white pine stands (Aird 1985). Studies in old-growth white pine from western, north-central and south-central Ontario showed that the communities of animals that live in white pine forests were not the same throughout the province, and that these communities were affected by the age and type of stand (Thompson et al. 1995). At the regional level, communities of small mammals, carabid beetles, songbirds, ants, and salamanders in the western part of the province were found to be different from those in the east. At the local patch scale, while all of the species among the various taxa studied were found in other forest types as well as pine, some species were consistently most abundant in old-growth pine stands, a finding suggesting that this forest may be their optimal habitat. For example, two species of carabid beetles and some salamanders were more abundant in old pine forests than in either mature pine forests or other forest types. The results of this study have clear implications for forest management with respect to maintaining white pine locally, and across its range in the province.

The conclusion that old pine forests maintain faunal communities distinct from those of other forest types was confirmed and extended through more intensive studies conducted north of Espanola that concentrated on species at the patch level for bats, salamanders, carabid beetles, ants, and winter resident birds. The bat communities were substantially different among the forest types. Two *Myotis* species, the little brown bat (*Myotis lucifugus*) and the long-eared bat (*Myotis septentrionalis*), together with the hoary bat (*Lasiurus cinereus*) and the silver-haired bat (*Lasionycterus noctivagans*), occurred significantly more often in old pine than in mature pine or other mixedwood forest types (Jung et al. 1999). There appear to be at least four carabid beetle species that are only found in old-growth pine forests (A. Applejohn et al., unpub.). Among the resident birds, red crossbills (*Loxia curvirostra*) and white-winged crossbills (*Loxia leucoptera*) were recorded more frequently in white pine forests than in mixedwoods where pine was absent, and pine siskins (*Carduelis pinus*) were recorded only in pine-dominated stands (I.D. Thompson, unpub.). The conclusion is that white-pine-dominated mixedwoods in general, and old-growth pine forests in particular, maintain discrete animal communities. The landscape-scale implications of this research, such as the question of how much old pine is needed in a given area to maintain the full complement of species, and in what patch sizes, are as yet unresolved.

CHANGES IN FOREST VERTEBRATE DISTRIBUTION

Biological systems involve the transfer of energy through various trophic levels. A key quality of biological systems is that they exist in a hierarchy of self-organizing systems that maintain an inherent stability through various cooperative relationships in time and space. These cooperative relationships are what structure ecological communities and are referred to as functional attributes of the system. Although animals live off energy produced within the system, they also play an important role at various levels in the return of energy and elements to be recycled within the system. Human disturbance alters the energy flows by altering pathways, through many means, but primarily through a reduction in energy assimilation and transfer and a reduction in recycling capability (Bonan 1992). Often, thresholds exist in ecological systems such that, if the ecosystems are sufficiently disturbed, they will move to another state (Holling 1992); for example, boreal black spruce forests that are logged with sufficient damage to the site can become, and remain, alder swamps or alder grasslands. Such changes can have cascading effects on components supported by the ecosystem, including animal communities (Haila et al. 1994). For animal communities, one of the major human impacts is the reduction of total available habitat in the landscape (Andren 1994). In Ontario's forested landscapes, habitat loss occurs through the change of ecosystems to different states, and not through forest fragmentation. The latter is an ephemeral condition in these forest landscapes and is not an important factor in changes to animal populations or species richness (Voigt et al. 2000, this volume).

Wherever they are practised, logging and fire suppression have produced a landscape pattern different from that expected from natural disturbances (Hansson 1992; Andren 1994), and Ontario is no exception (Gluck and Rempel 1996; Perera and Baldwin 2000, this volume). In addition to changing the species composition of patches, logging also influences patch size, patch shape, and patch distribution, particularly in boreal forests where clear-cutting is more common than in the Great Lakes forests. The changes in landscape pattern influence ecological processes that, in turn, affect animal community structure through habitat availability and suitability, dispersal capability and success, altered predator-prey dynamics, delayed non-linear species responses to habitat loss, and differential influence on habitat specialists and generalists (Hansson 1992; Andren 1994; Doak and Mills 1994; Haila et al. 1994; Mönkkönen and Welsh 1994; Kareiva and Wennergren 1995). There is evidence from many areas of North America for declines in animal species as a result of logging. The species affected include the spotted owl (*Strix occidentalis*) in the Pacific Northwest of the United States and in British Columbia (Murphy and Noon 1991), caribou throughout eastern Canada (Bergerud 1974), marten in Newfoundland (Thompson 1991), and certain forest passerine bird species in eastern North America (Askins et al. 1987; Dickerman 1987; Sauer et al. 1997), among many others.

In Ontario, concern has been raised for several species, some of which are associated with old-growth forest, because of reduced habitat availability as a result of timber harvesting. The species of concern include the woodland caribou, marten, wood turtle, wood thrush, red crossbill, and many species from the Carolinian forest region. The latter forest region has been severely reduced through conversion to urban areas and agriculture. Although no species in the province has become extinct, except the passenger pigeon, and none are endangered solely as a result of habitat loss through logging, long-term continuous declines in some common species are disturbing. For example, data from the North American Breeding Bird Survey show a 20-year monotonic decline in wood thrushes (Sauer et al. 1997) that may be related to such landscape-level processes as ecosystem change and habitat loss. The effects of human activities may only become apparent over the long term. A more detailed examination of the effects of landscape changes on wildlife are found in Voigt et al. (2000, this volume).

Recent Changes in the Range of Selected Ontario Forest Species

There is no recent comprehensive literature documenting change in species ranges in the province, but a number of past accounts of changing distributions of mammals and birds have been published (Peterson and Crichton 1949; de Vos and Peterson 1951; Peterson 1957; Snyder 1957; de Vos et al. 1959; de Vos 1962, 1964; Outram 1967). Some of the distributions reported in those papers are identified as "extreme" or "maximum," but more often range limits are reported without clarification of the supporting evidence. For example, the northern limit of white-tailed deer reported by Peterson (1957) was based on an article by Cross (1937), who cited no data for the limit in question. Although the 1940s were characterized by a warming trend (Thomas 1957), it is unlikely that deer were common as far north as was suggested by Cross. Voigt et al. (1992) showed a more realistic line for the breeding range of deer in 1940, based on weather patterns, habitats, the clearing of land associated with rail, road, and hydro-power developments, the proximity of agriculture in Manitoba, and early agriculture-based attempts at northern settlement during a warming period from around 1910 to the 1930s. (The harsh winters of the 1950s subsequently caused failures in most northern agriculture, and the abandoned farmlands gave rise to an increase in successional habitats.) Distributions mapped in the past must be used with care, because they are based on incidental sightings and may differ from actual breeding range.

Atlases of species distributions are now available for mammals and birds in Ontario (Cadman et al.

1987; Dobbyn 1994;). The atlas of birds presents a one-time record of species distribution. The atlas of mammals displays temporal distribution modestly; for example, certain species are shown as occurring in a particular block before 1969 only if they have not been recorded there in later years. More detailed temporal data are available for those species listed under the *Ontario Endangered Species Act*, because of increased efforts to locate them. For the ungulates, good count data are available, including long-term aerial inventories of moose and two province-wide surveys of caribou. As summarized by Voigt et al. (1992), several data sources are available for white-tailed deer, including hunting records and pellet survey data.

Fur-trapping records that are available from 1947 onward for some management districts of the Ontario Ministry of Natural Resources (OMNR) may help in determining species distributions over time and likely population trends. The areas of the OMNR management districts have changed several times during the past 40 years, however, and trapping data are fraught with variance (Thompson 1988). Despite these problems, long-term trends can be determined and valid spatial comparisons made, because most of the variance in trapping data occurs between time periods rather than between areas. A major contribution to the knowledge of species distributions in Ontario comes from the recollections of former and present employees of OMNR who worked as field personnel. Many have been keen naturalists who maintained personal records of the distributions or occurrence of various wildlife species in their areas. The knowledge of these individuals is often the only source of information available on changes in the distribution and populations of many species between 1960 and 1990. The information sources identified above have been compiled to produce the maps in Figures 4.5 to 4.7, which show the most likely breeding range of selected species at various times.

The reported distributions of species shown in these figures represent contiguous breeding range within which there is a high probability that the species is, or was, common or abundant. Boundary lines depicted in the figures should be viewed as bands several kilometres wide. Species may be rare or uncommon, or occur intermittently, beyond these boundaries. For several species, notably the large ungulates, there are small isolated populations that have existed over many years beyond the illustrated limits. For example, moose occur in the Alfred Bog and Larose Forest near Ottawa, woodland caribou occur along the Lake Superior shoreline from Pukaskwa Park to an area near White River, as well as on the Slate Islands, and white-tailed deer are found near Timmins and Cochrane. City-dwelling populations of raccoons (*Procyon lotor*) and black squirrels that are not part of the larger wild distribution, such as those in Thunder Bay, were not considered as part of the animals' range for the maps. The maps also show the estimated species distributions for 1970 and/or 1980 where sufficient data exist. The years since 1950 represent the longest time period for which there were data that can be used with confidence, but where information was available from previous decades, it is noted in the text. Voigt et al. (2000, this volume) has examined these changes in range in light of several possible hypotheses that may explain why the species have undergone such large changes in their distributions.

White-Tailed Deer

The breeding range of white-tailed deer has advanced to the north across the province at least twice, first after it reached a northward maximum between 1940 and 1950, and again in recent years following a southward range retraction that started in the 1960s and continued through the 1980s (Figure 4.5a). Peterson (1957) and de Vos (1964) showed deer throughout the boreal zone of central Ontario during the period from 1930 to 1961. Much of this range was apparently estimated by straight line without supporting records, particularly north of Lake Nipigon, and may have represented scattered northward range extensions both to the extreme west near the Manitoba border and north of the Claybelt. A review of a series of reports entitled *Reports of the Minister of Lands and Forests of the Province of Ontario* issued from 1920 to 1960 suggested that deer were common in the Red Lake to Sioux Lookout areas, but not east of Thunder Bay (named "Port Arthur" at that time), nor north of Lake Superior or south of Lake Nipigon; no deer were reported near Lake Opasatika (south of Kapuskasing) in 1925, and none were seen along the Albany River by a survey party in 1930. The Minister's report from 1956 states that there were two distinct areas important for deer in the province, one west of Lake Superior and the other southward from the North Bay

District Consistent with the above information, Voigt et al. (1992) showed deer distribution in 1940 as non-continuous north of Lake Superior.

There was subsequently a southward decline in deer distribution across the province that may have started in the late 1950s and continued until the late 1970s. King (1976) reported that high mortality occurred in deer during the winters of 1958-59, 1959-60, and 1970-71, and that by 1960 few deer remained north of Lake Nipissing and Sault Ste. Marie. After about 1980, deer distribution again moved steadily northward, although deer were not as plentiful along the northern edge of their distribution in 1998 (Figure 4.5a) as they were during the early 1950s. Throughout the period when deer occurred well south of their former range, they continued to survive in some isolated northern populations such as those in the eastern Claybelt and near Thunder Bay.

Deer in Algonquin Park are of particular interest because that population has shown large changes that have been relatively well recorded. Wilton (1984) reported that deer declined throughout the period from 1958 to 1974. Since 1974, and contrary to trends elsewhere in the province, deer numbers have not increased in Algonquin Park, and while summer density is unknown, less than 200 animals now overwinter in the Park (N. Quinn and M. Wilton, unpub.). Most animals move out of the Park into deer yards to the south, for example, to the large yard near Round Lake and Wilno.

Woodland Caribou

The southern limit of caribou range in Ontario has declined since the arrival of European settlers in the province (Bergerud 1974). That decline continues northward today (Darby and Duquette 1986), so that caribou are presently found north of regions of high moose density (Figure 4.5b). The probable historical range of caribou at the turn of the century was as far south as Lake Nipissing in the east, and south of the Canadian border into Wisconsin and Minnesota in the west (de Vos and Peterson 1951; de Vos 1964). In the early 1900s, caribou were hunted by Indians for the sum of one dollar per carcass to supply wood-cutting camps north of Lake Nipissing with fresh meat. Although caribou ranged throughout the area between Longlac, Lake Nipigon, and Pukaskwa Park, accounts suggest that by the 1950s and 1960s, they occurred in small herds using large areas, and were not regularly distributed throughout this region (Figure 4.5b). A small number of animals that had occupied an area near Caramat-Hillsport-Longlac, survived at least until 1980 (G. Eason, pers. comm.). An animal was shot in this area in 1997. Armstrong et al. (1998) reviewed historical information and concluded that the range of caribou in 1950 did not include the area east of Thunder Bay between Lakes Nipigon and Superior, as depicted by de Vos (1964) and Peterson (1957). The 1995 line (Figure 4.5b) identifies the zone of current continuous caribou range. There are probably 15,000 to 20,000 caribou in the province north of that line. At least three relict populations continue to survive to the south: a herd of 10 to 30 animals in the area of Pukaskwa National Park, a small herd on Pic Island, and a larger population on the Slate Islands that has ranged from 80 to over 600 during the past 20 years. Transplanted caribou have now survived their first decade on Michipicoten Island (G. Eason, pers. comm.).

Fisher

Fisher exist in at least two distinct populations in the province, as originally depicted by de Vos (1952). One population is an eastward extension of the western North American population that extends into Ontario along the border with Manitoba and Minnesota, as far east as Sioux Lookout and Thunder Bay. The other population is centred near Algonquin Park and is continuous with the population in Quebec, extending north to Cochrane, and at some points as far west as Geraldton and Wawa. It is possible that these two populations overlap south of Lake Nipigon, whenever fisher are abundant, as has apparently been the case from 1996 to 1998.

The changes in range of the fisher in eastern Ontario have been among the most dramatic and dynamic of all species reported here (Figure 4.5c). A province-wide decline in fisher occurred from 1964 to 1975 (Table 4.2). The decline began initially in the White River area in the mid-1960s, and had spread north through Manitouwadge and Geraldton by 1967. This was followed by a decline in northwestern Ontario that started in about 1974, and finally one in the Cochrane region by 1976. The decline in the west was not as severe as in the east. In the Claybelt, fisher had declined in the Longlac area at least five years before the decline at Cochrane. Fisher remained common at Kirkland Lake, even during the periods of lowest

Table 4.2

Registered number of fisher caught per 100 km² on traplines in Ontario Ministry of Natural Resources districts from 1955 to 1997.

Year	*Coch.*	*Chap.*	*Ft.Fr.*	*Ger.*	*Gog.*	*Kap.*	*Ken.*	*K.L.*	*N.Bay*	*Pem.*	*P.Snd.*	*S.L.*	*SSM*	*Sud.*	*Th.B.*	*W.R.*	*RedL.*
55-56	0.64	0.75	1.10	0.13	0.80	0.49	0.51	0.78	0.35	1.97	1.18	0.98	0.69	0.58	0.88	0.45	0.28
57-58	0.68	0.62	1.11	0.17	0.91	0.99	0.74	1.12	0.86	1.86	1.20	0.43	0.63	0.62	0.58	0.52	0.11
58-59	0.65	0.54	1.13	0.12	0.49	0.52	0.66	1.04	0.62	0.98	0.93	0.22	0.35	0.54	0.73	0.26	0.10
59-60	0.61	0.54	1.20	0.14	0.67	0.45	0.59	1.16	1.41	2.03	1.15	0.21	0.55	0.65	0.85	0.34	0.15
61-62	0.53	0.49	1.19	0.09	0.44	0.42	0.72	0.90	2.09	2.49	0.78	0.36	0.44	0.73	0.92	0.16	0.25
62-63	0.29	0.30	1.45	0.07	0.16	0.28	1.04	0.75	2.24	2.40	0.73	0.66	0.27	0.55	0.85	0.11	0.43
63-64	0.72	0.64	1.68	0.07	0.62	0.42	1.21	1.76	2.13	2.69	0.99	1.01	0.43	0.77	1.21	0.21	0.45
64-65	0.40	0.36	1.33	0.04	0.41	0.13	0.71	1.36	1.97	1.99	0.95	0.59	0.29	0.75	0.30	0.08	0.33
65-55	0.62	0.25	1.49	0.08	0.28	0.26	0.78	2.20	2.14	2.83	1.22	0.66	0.20	0.88	0.60	0.11	0.26
67-68	0.21	0.10	1.47	0.02	0.26	0.09	0.52	1.33	1.28	2.13	1.82	0.19	0.06	0.46	0.43	0.02	0.14
72-73	0.50	0.26	12.7	0.03	–	0.09	7.27	2.94	3.99	4.01	1.58	1.84	0.20	2.87	2.48	–	–
73-74	0.18	0.09	8.03	0.01	0.49	0.05	3.37	1.95	6.55	3.11	1.11	0.36	0.00	2.88	0.98	–	1.07
76-77	0.16	0.08	12.1	0.01	0.40	0.03	4.45	1.24	1.88	1.00	1.11	0.30	0.03	1.18	0.89	–	1.10
77-78	0.14	0.04	6.11	0.01	0.57	0.03	4.46	1.40	2.89	1.62	1.81	0.31	0.00	1.27	0.50	–	1.57
78-79	0.08	0.05	6.17	0.05	0.34	0.01	4.38	0.80	3.71	2.55	4.27	0.39	0.03	2.01	0.38	–	1.92
79-80	0.08	0.06	8.11	0.01	0.62	0.01	6.47	1.04	3.89	4.23	5.86	0.42	0.00	1.98	0.20	–	3.16
80-81	0.03	0.08	8.26	0.03	0.32	0.01	6.05	0.52	2.31	4.17	5.58	0.40	0.00	1.32	0.47	–	2.66
81-82	0.08	0.11	6.37	0.02	0.18	0.02	5.59	0.96	3.15	7.76	0.00	0.29	0.06	1.46	0.44	–	1.39
82-83	0.21	0.09	7.65	0.05	0.30	0.08	5.40	1.30	3.80	8.16	0.00	0.56	0.03	2.16	0.43	–	1.79
83-84	0.25	0.07	6.13	0.03	0.22	0.01	5.33	1.20	3.95	5.32	0.00	0.24	0.03	2.38	0.52	–	1.36
84-85	0.21	0.08	6.19	0.01	0.12	0.01	3.40	1.07	3.15	5.52	0.00	0.10	0.13	1.56	0.27	–	0.72
85-86	0.16	0.06	6.57	0.02	0.22	0.14	2.99	0.93	3.99	4.53	5.97	0.14	0.20	3.03	0.28	–	0.72
87-88	0.13	0.08	3.65	0.04	0.31	0.10	3.03	1.25	4.28	5.45	7.08	0.04	0.39	2.43	0.37	–	0.27
88-89	0.15	0.28	2.91	0.03	0.48	0.09	2.63	1.14	4.43	4.49	8.54	0.04	0.29	1.96	0.15	–	0.22
90-91	0.13	0.25	1.55	0.02	0.23	0.08	0.58	0.21	3.08	5.27	7.97	0.04	0.42	1.57	0.20	–	0.09
91-92	0.10	0.19	0.73	0.01	0.15	0.04	0.42	0.21	2.17	3.87	4.19	0.01	0.24	1.18	0.13	–	0.07
92-93	0.06	0.30	0.52	0.01	0.27	0.03	0.28	0.21	2.12	5.01	5.29	0.02	0.38	1.16	0.15	–	0.07
93-94	0.07	0.21	0.77	0.01	0.21	0.04	0.48	0.21	2.49	2.18	5.00	0.00	0.30	0.85	0.14	–	0.03
94-95	0.09	0.11	1.27	0.01	0.18	0.07	0.70	0.21	3.66	6.10	7.22	0.01	0.38	1.58	0.15	–	0.09
95-96	0.08	0.19	0.69	0.02	0.36	0.10	0.46	0.21	3.00	5.34	6.23	0.02	0.21	1.45	0.10	–	0.05
96-97	0.07	0.32	1.36	0.01	0.10	0.07	0.44	0.21	3.07	8.45	8.01	0.01	0.28	1.39	0.15	–	0.05

Some variance in numbers, particularly for Kenora and Fort Frances, may be explained by uncertainty over the district boundaries, as a result of various corporate reorganizations. White River was no longer a separate district after 1970. Dashes indicate that no data were available. No data were available for the period from 1969 to 1971. (Data from J.A. Baker, OMNR, unpub.)

Coch. = Cochrane; Chap. = Chapleau; Ft.Fr. = Fort Frances; Ger. = Geraldton; Gog. = Gogama, Kap. = Kapuskasing; Ken. = Kenora; K.L. = Kirkland Lake; N. Bay = North Bay; Pem. = Pembroke; P.Snd. = Parry Sound; S.L. =- Sioux Lookout; SSM. = Sault Ste. Marie; Sud. = Sudbury; Th.B. = Thunder Bay; W.R. = White River; RedL. = Red Lake population levels elsewhere in the eastern part of the province, but fisher catches were extremely low there in the early 1990s. Most information suggests that, during periods of low populations, fisher occurred in small numbers across much of the same range that they occupy in years when they are abundant. An exception to this pattern seems to be an area from Wawa to Kapuskasing on the east, across to Longlac to the north and west, and south to Thunder Bay on the west. Fisher appear to have been virtually absent from this extensive area during almost 30 years from 1968 to 1995, including the area around White River, where they had been abundant until about 1960 (B. Snider and G. Eason, pers. comm; I.D. Thompson, unpub.; Table 4.2). Mean incidental captures of fewer than 30 animals per year from 1975 to 1982 (a period of low quotas) indicate that scattered individuals apparently continued to survive in the Cochrane and Timmins areas, but that they were clearly uncommon.

Catches again rose in the Cochrane District in 1982 and in Timmins after 1987, but numbers did not support catch rates similar to those of the late 1960s (Table 4.2). Fisher have become abundant in many areas of southeastern Ontario, most noticeably over the past three years (1995 to 1997). In the southeast, they are common north of Kingston and Tweed, and along the more wooded areas near the Highway 7 corridor to near Peterborough (B. Snider, pers. comm.). In the northeast, their range has extended once again north of Cochrane, west of Hearst, and as far as Longlac in the east. However, the fact that they are still uncommon (although present and increasing in 1998) north and west of Wawa, White River, and Manitouwadge, also suggests that the current population expansion has not reached levels observed in the early 1960s. Because numbers of fisher are increasing across the province, locating the line depicting current range was difficult (Figure 4.5c).

Perhaps the most puzzling aspect of the various expansions and contractions of fisher range has been the lack of temporal consistency across the province. Based on trapping statistics, fisher have been considerably more abundant in western Ontario than in the east during the past 25 years (Table 4.2). In the northwest, the exact northern boundary of the area where fisher can be considered common has varied. Fisher occur, in low numbers, as far north as the Hudson Bay lowlands (de Vos 1952; Dobbyn 1994), as does the population in Manitoba (Douglas and Strickland 1987). Range fluctuations have been minor in the west compared to the east, and large areas of range have not been completely abandoned, even during periods of low population. There appear to have been three population lows in the west over the past 40 years, around 1956, 1967, and 1987, but the population has always remained abundant in areas around Fort Frances and Kenora. During years of reduced numbers and range, fisher were still common in the northwest near Pikangikum, but declined near Sioux Lookout to the east and Sandy Lake to the north. Following the 1987 to 1991 low, they again became common north of Sandy Lake in 1998, but are not yet abundant north of Sioux Lookout, although they are being caught as far north and east as Lake St. Joseph. Fisher to the north and east of Thunder Bay declined in the early 1970s and have not yet reached their former abundance. In 1998, they were uncommon, but increasing, near Nipigon, east to Terrace Bay and north to Geraldton.

There has been considerable concern at various times during the past 80 years that fisher might become extremely rare or even extinct in various areas of the province (Seton 1929; de Vos 1952; Douglas and Strickland 1987). Around 1900, fisher were common in the James Bay area and inland from Attawapiskat to Fort Hope (de Vos [1952] quoting Hudson's Bay Company fur buyers), but became rare there by 1920. Apparently a decline of similar magnitude to that seen in the mid-1970s in the east occurred in fisher populations at least once previously in the mid-1930s, when they became rare throughout the province (de Vos 1952; M. Novak, pers. comm). Following that decline, the population did not recover in the east until the period from 1945 to 1950 (Peterson and Crichton 1949; de Vos 1952), but unfortunately no reliable data exist for western Ontario. Fisher were re-introduced into the Parry Sound area during the period from 1957 to 1963 and on the Bruce Peninsula and Manitoulin Island during the period from 1979 to 1982 (Douglas and Strickland 1987). The eastern part of the Parry Sound District and the Pembroke area, however, still supported a substantial harvest during the late 1950s (Table 4.2), so a core area of preferred habitat may exist southeast of Lake Nipissing and in the northern areas of the Ottawa River valley.

Moose

Moose have expanded their range southward in eastern

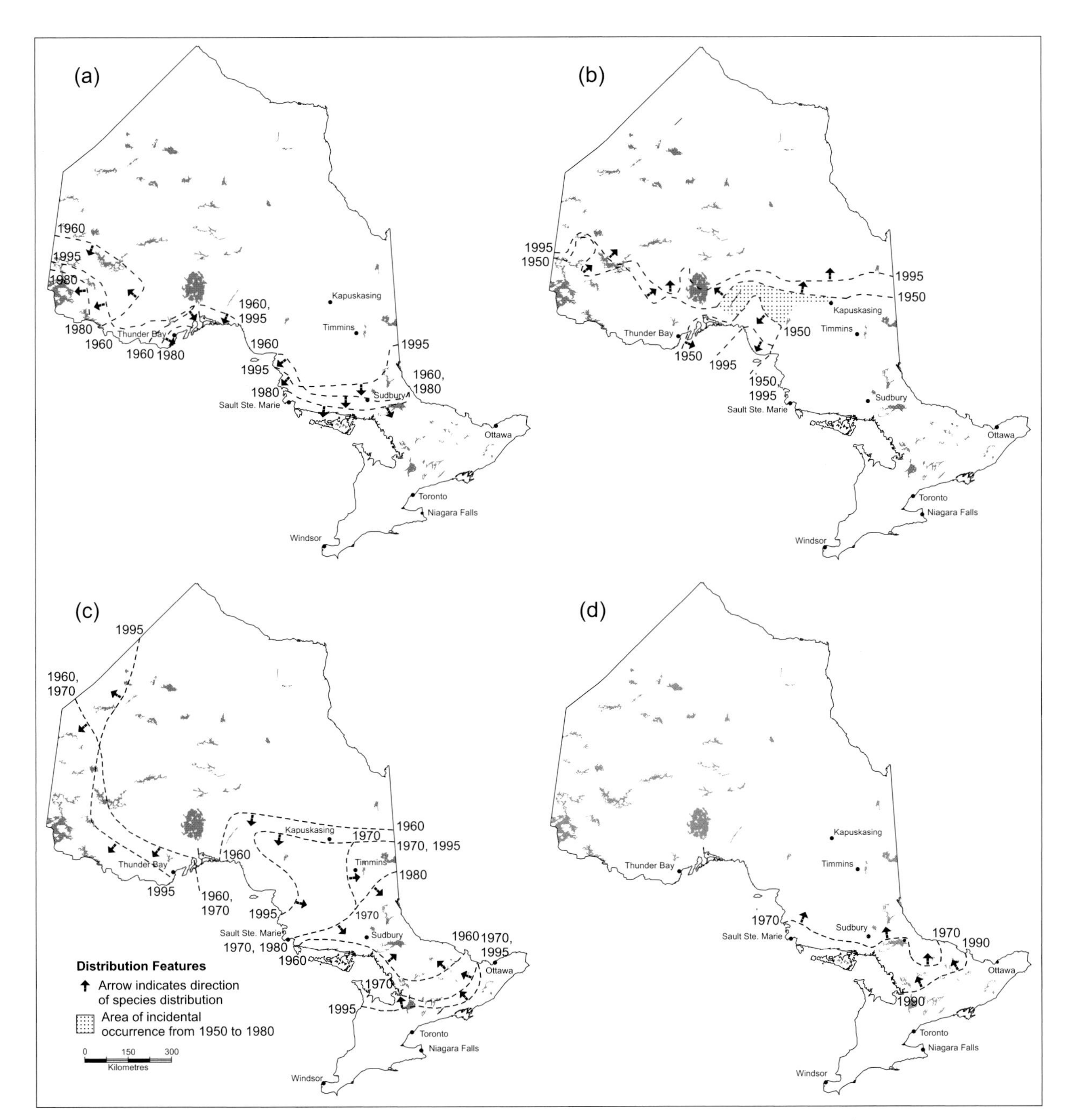

Figure 4.5 Changes in the range of (a) white-tailed deer, (b) woodland caribou, (c) fisher, and (d) moose in Ontario from either 1950 or 1960 to 1995.
(Data from Voigt et al. 1992, Armstrong et al. 1998, deVos 1952, and OMNR, unpub.)

Ontario during the past 30 to 40 years. That increase in moose density appears to be centred in Algonquin Park (M. Wilton, pers. comm.). Aerial survey data from the former (that is, pre-1970) Tweed and Lindsay Districts of OMNR, covering an area from Pembroke to Georgian Bay, indicated that few moose occurred south of North Bay in the late 1950s (the approximate density was 0.01/km^2), and virtually none occurred outside of Algonquin Park. Pimlott et al. (1969) reported a density of 0.2/km^2 in Algonquin Park during the same period. The 1960 distribution (Figure 4.5d) encompasses the Algonquin population, which, although low at that time, was self-sustaining. Moose numbers have increased steadily since the early 1970s in the Algonquin area, where the present density of 0.4/km^2 to 0.6/km^2 is considered to be one of the highest in the province (Wilton 1984, and pers. comm.). Concurrent with the increase in moose within the Park has been an extension of their range west to Georgian Bay, east to Deep River and Pembroke along the Ottawa River, and south along a line near Eganville, Bancroft, and Minden (Figure 4.5d).

Bobcat

Bobcats (*Lynx rufus*) have occurred in Ontario since the late 1940s, and their range has not changed appreciably in the past 40 years. They occupy two core areas: eastward from Sault Ste. Marie to Blind River and Sturgeon Falls, and in the area near Fort Frances and Rainy River. There are a few individuals scattered in eastern Ontario between Kingston and Pembroke, and near Sarnia (Dobbyn 1994). The species has not expanded its range south along Georgian Bay, or either north or east of Rainy River. In fact, in the west, bobcat range appears to have declined somewhat south from Kenora (Figure 4.6a).

Porcupine

Porcupine (*Erethizon dorsatum*), like fisher, have undergone protracted population fluctuations and large changes in range, particularly in eastern Ontario (Figure 4.6b). Porcupine have never occurred in high numbers as far north as fisher, their major predator. In western Ontario, there appear to be two distinct populations, one between Rainy River and Kenora in the west, and the other near Thunder Bay, probably as the result of expansion northward from Minnesota. This latter population has increased eastward to form a continuous range, in recent years, with the eastern population, which has concurrently expanded to the west and the north. Porcupines were common in the Thunder Bay to Nipigon area during the 1940s, but had become rare to absent there by 1960. The population centred on Rainy River has recently expanded eastward as far as Atikokan and northward to Dryden. In the east, the range of porcupines prior to 1970 was unclear. They have apparently always been uncommon in the Timmins-Chapleau areas, but were abundant in the eastern Claybelt throughout the 1950s. The number of porcupines declined during the 1940s and 1950s, and fell significantly in the 1970s, at which point they reached their lowest distribution in the province. A population continued to exist after 1970 in the southern portion of the Claybelt, in the Haileybury area, but porcupines were uncommon in continuous forest at Temagami. An expansion in distribution in the mid-1990s has been extremely widespread, even including an area north of Geraldton, and may be without precedent for much of the central portion of the province (Figure 4.6b).

Grey Squirrel

Grey squirrels (*Sciurus carolinensis*) have moved considerably northward since their range was last documented by de Vos (1964) (Figure 4.6c). This species represents a case of continuous historical range expansion. Grey squirrels first appeared in the Blind River area in the mid-1970s, but were in Sault Ste. Marie in the late 1960s. This pattern suggests that grey squirrels invaded the eastern end of the north shore of Lake Huron from Michigan and the western end from an area lying further to the southeast in Ontario. They are now common along the line from Sault Ste. Marie, through Blind River and Sudbury, to North Bay; however, they are rare in transition forests north of these areas, that is, through the ecotone between the boreal and Great-Lakes forests. No grey squirrels are found at Wawa, they are only rarely seen in Lake Superior Park, and they do not occur to any extent to the north of Espanola or at Temagami. In the west, they are only common in the extreme southwest area near Fort Frances and Rainy River, where they were rare in 1960 (de Vos 1964).

Raccoon

In 1960, raccoons had not yet reached the North Bay area, and were found only south of the French River in the east (Figure 4.6d). In 1970, they occurred from

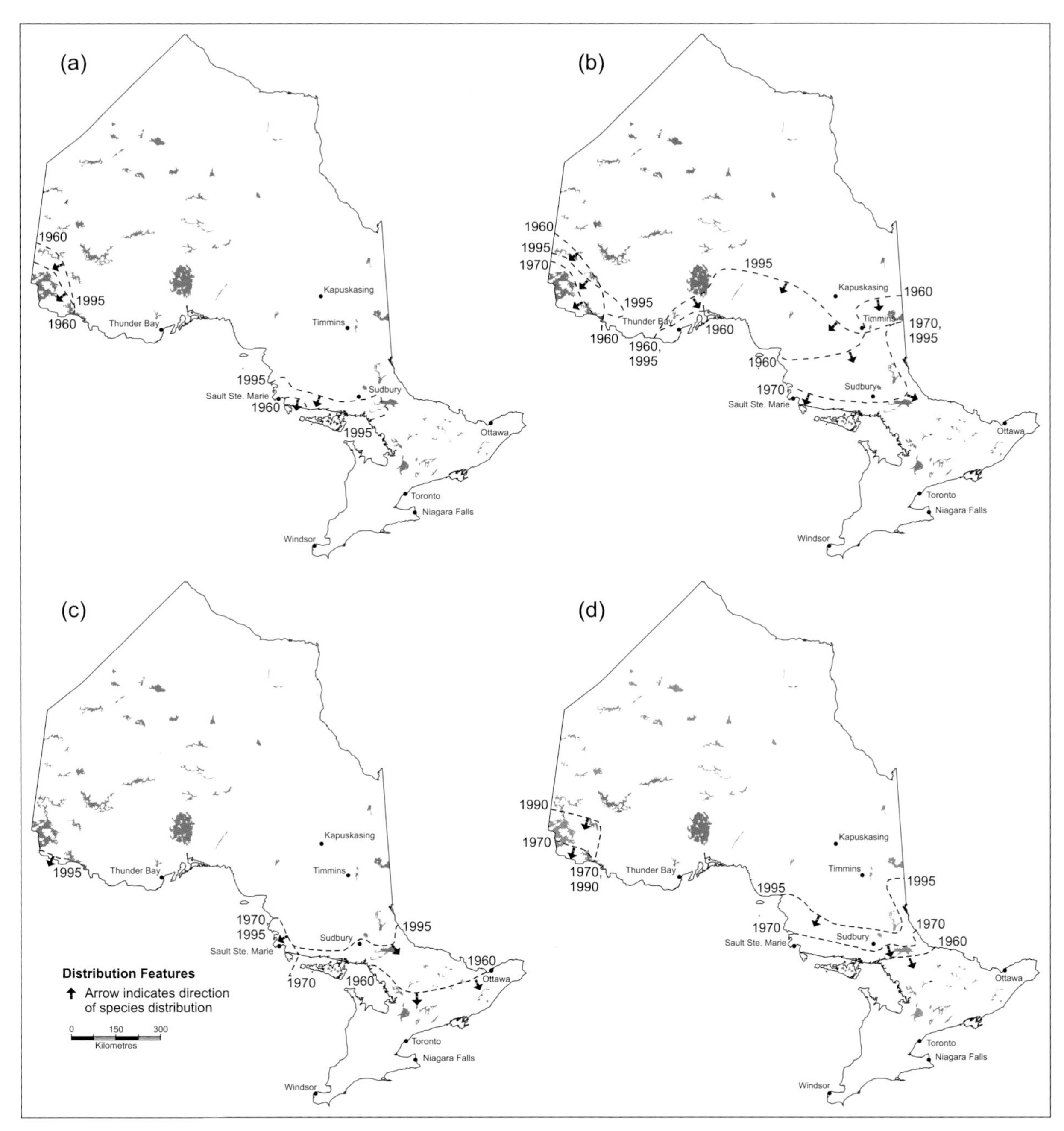

Figure 4.6 Changes in the range of (a) bobcat, (b) porcupine, (c) grey squirrel, and (d) raccoon in Ontario from 1960 to 1995. Figures (a), (b) and (c) represent data from OMNR (unpub.); Figure (d) represents data from de Vos (1964).

Rainy River to Fort Frances in the west. De Vos (1964) suggested that they were only found in southernmost Ontario in 1960 and not yet in western Ontario, but Simkin (1966) reported a population in the Rainy River area, and several long-time residents of the Algonquin area have reported that raccoons were reasonably common there in the early 1960s. Raccoons have continued their northward range expansion during the past four decades. They were common in Sault Ste. Marie and Blind River by the mid- to late 1970s and reached the eastern areas of the Claybelt north of Temagami around the same time. They tend to be uncommon in forested areas at the northern fringes of their range and are most abundant in agricultural regions along Lake Huron, west of North Bay, and in the Claybelt. Raccoons are currently rare at Wawa, Chapleau, and Timmins. In the west, they are common along the western areas of Highway 17 from Dryden to Kenora. There are occasional astounding records of northerly movement by raccoons, some listed by Simkin (1966), from locations as far north as Sandy Lake and Big Trout Lake, as well as other observations at various places along the Canadian National Railways (CNR) line in the northwest of the province.

Sandhill Crane

There are probably three distinct populations (and two subspecies) of sandhill cranes that occur in Ontario: a northern population that breeds in the Hudson Bay lowlands and northern boreal forests, a population from Lake of the Woods to Lake St. Joseph in the west (Lumsden 1971; Cadman et al. 1987), and a more recently established population that appears to have moved into the province from Michigan (Tebbel and Ankney 1982) (Figure 4.7a). The western population may have expanded to the north recently, but movements of this population have been poorly documented since 1971. Pedlar and Ross (1997) suggested that there may have been no range expansion or increase in population of the Algoma population after 1985, despite increased sightings of the birds by local naturalists during the period from 1985 to 1996, but they also suspected that their helicopter survey technique was insensitive to any increase. Regardless, sandhill cranes have clearly expanded their range in the Algoma area since 1970, when the first recorded breeding occurred (Cadman et al. 1987), to a line that is now north of Wawa, Chapleau, and, in the east, Sudbury. The northeastern population, which is a different subspecies than the southern and western birds

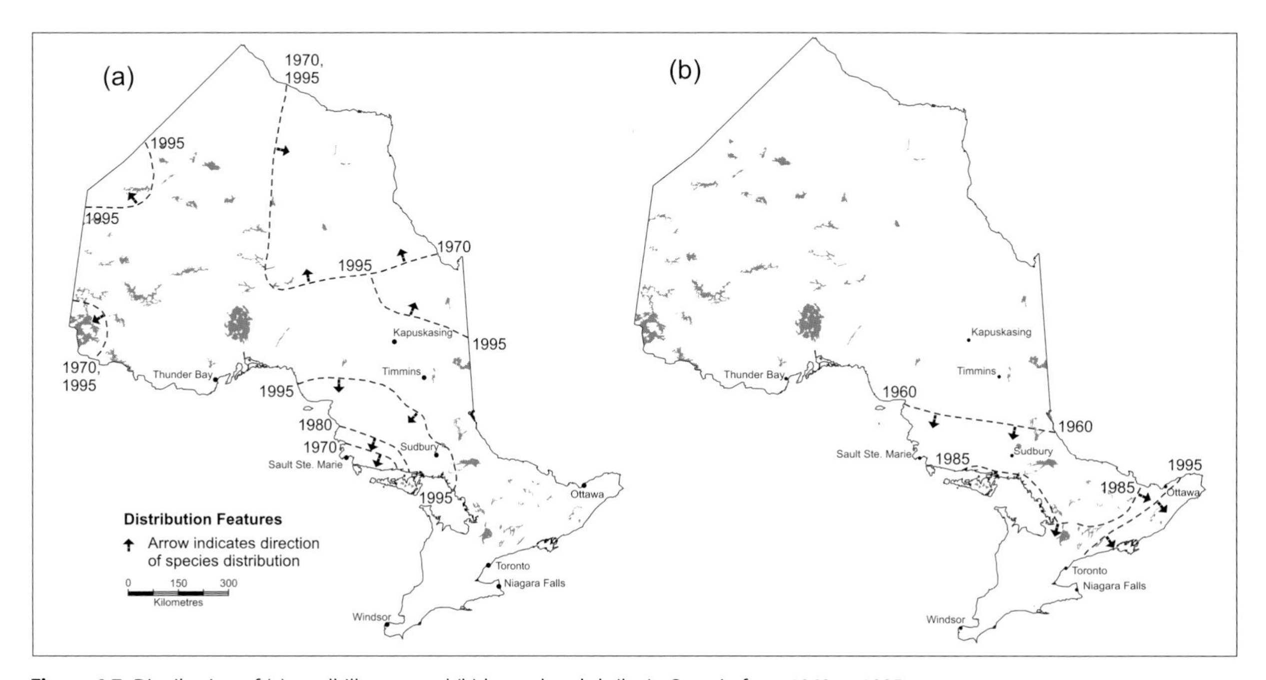

Figure 4.7 Distribution of (a) sandhill crane and (b) loggerhead shrike in Ontario from 1960 to 1995.
(Data from Tebbel 1982 and Cadman 1985, 1990, respectively)

(Lumsden 1971), has expanded south into the Cochrane and Timmins area.

Loggerhead Shrike

While loggerhead shrike (*Lanius ludovicianus*) is not a forest species, changes in the populations have been affected by changes in forest cover. In the 1960s, loggerhead shrikes were found in suitable habitat south of a line from north of Sault Ste. Marie to North Bay (Figure 4.7). By the time Cadman (1985) documented their status, from a concern over declining numbers, their range had shrunk to well south of Algonquin Park, but still included agricultural areas east of Blind River. By 1995, there were very few loggerhead shrikes remaining in areas around Napanee, Smiths Falls, and Carden Township (Johns et al. 1994). Only 18 breeding pairs were observed in 1997 (M.D. Cadman, unpub.).

SUMMARY

The distribution of Ontario's forest vertebrates is determined first by the location of their ancestral populations following the last glacial period. This fact is illustrated by the number of Great Lakes forest region species that are absent from similar forests west of Lake Superior. Relationships between species richness and net primary production for all groups of vertebrates indicate that faunal distributions, at the level of broad landscapes across Ontario, reflect available energy in the systems. Exactly where animal species individually settle in a landscape depends on proximate factors and processes, many of which are affected by human interventions in that landscape. Community structure is influenced by large-scale and long-term factors.

Animals have historically altered their distributions in Ontario in response to numerous factors, but how they have responded to anthropogenic changed habitats has yet to be resolved. Perspective on the influence of all of these factors is gained through an appreciation of history, which provides a measure of understanding about how contemporary landscape structure may have been altered by humans. While no species has become extinct, except the passenger pigeon, there have been fundamental shifts in species distributions and local abundances across the province, including numerous extirpations from southern Ontario. Continuous declines in some species, such as the wood thrush and loggerhead shrike, suggest that changes in forest habitats are, indeed, occurring. Further discussion of how animals respond to various sources of disturbance, particularly human activity, together with an assessment of the specific reasons for change in the ranges of certain Ontario forest vertebrate species, are provided by Voigt et al. (2000, this volume).

ACKNOWLEDGEMENTS

The following individuals, who are either former or current OMNR personnel, contributed their knowledge of changes in species distributions: Evan Armstrong, Jim Baker, Al Bisset, Neil Dawson, Gord Eason, Mike Eliuke, Linda Ferguson, Mick Gauthier, Greg Gillespie, Fred Johnston Sr., Scott Jones, Ken Koski, Dave Lyons, Vic Miller, Ben Miron, Brian Naylor, Milan Novak, Don Perry, Norm Quinn, Gerry Racey, Bruce Ranta, Will Samis, Barry Snider, Dennis Voigt, Milan Vukelich, Bernie Wall, Randy Wepruk, Mike Wilton, and Orville Wohlgemuth. Mike Cadman and Ken Ross of the Canadian Wildlife Service provided information on migratory bird species. Darcy Ortiz of the Canadian Forest Service, prepared many of the figures, that were later finalized by Kent Todd and Dave Baldwin of OMNR.

REFERENCES

Aird, P.L. 1985. In Praise of Pine: The Eastern White Pine and Red Pine Harvest from Ontario's Crown Forests. Report PI-X-52. Chalk River, Ontario: Petawawa National Forestry Institute, Canadian Forest Service. 11 p.

Andren, H. 1994. Effects of habitat fragmentation on birds and mammals in landscapes with different proportions of suitable habitat: a review. *Oikos* 71:355-366.

Armstrong, E., G. Racey, and N. Bookey. 1998. Landscape-level considerations in the management of forest-dwelling woodland caribou in northwestern Ontario. Unpublished abstract. Paper to the North American Caribou Conference, March, 1998, Whitehorse, Yukon.

Askins, R.A., M.J. Philbrick, and D.S. Sugeno. 1987. Relationship between the regional abundance of forest and the composition of bird communities. Biological Conservation 39: 129-152.

Banfield, A.W.F. 1974. The Mammals of Canada. Ottawa, Ontario: National Museum of Natural Sciences. 438 p.

Bergerud, A.T. 1974. Decline of caribou in North America following settlement. Journal of Wildlife Management 38: 757-770.

Bickerstaff, A., W.L. Wallace, and F. Evert. 1981. Growth of Forests in Canada, Part 2: A Quantitative Description of the Land Base and the Mean Annual Increment. Information Report PI-X-1. Chalk River, Ontario: Petawawa National Forestry Institute, Canadian Forest Service. 136 p.

Bonan, G.B. 1992. Soil temperature as an ecological factor in boreal forests. In H.H. Shugart, R. Leemans and G.B. Bonan (editors). A Systems Analysis of the Global Boreal Forest. Cambridge, UK: Cambridge University Press. 126-143.

Bonan, G.B. and H.H. Shugart. 1989. Environmental factors and ecological processes in the boreal forest. Annual Review of Ecology and Systematics 20: 1-28.

Bondrup-Nielsen, S. 1987. Demography of *Clethrionomys gapperi* in different habitats. Canadian Journal of Zoology 65: 277-283.

Cadman, M.D. 1985. Status Report on the Loggerhead Shrike in Canada. Unpublished report. Committee on the Status of Endangered Wildlife in Canada, Ottawa, Ontario.

Cadman, M.D. 1990. Update status report on the loggerhead shrike in eastern Canada. Ontario rare bird breeding program, Fed. Ont. Naturalists. Unpublished report to COSEWIC. 33 p.

Cadman, M.D., P.F.J. Eagles, and F.M. Helleiner. 1987. Atlas of the Breeding Birds of Ontario. Waterloo, Ontario: University of Waterloo Press. 617 p.

Cook, F.R. 1984. An Introduction to Canadian Amphibians and Reptiles. Ottawa, Ontario: National Museum of Natural Sciences. 200 p.

Cornell, H.V. and J.H Lawton. 1992. Species interactions, local and regional processes, and limits to the richness of ecological communities: a theoretical perspective. Journal of Animal Ecology 61: 1-12.

Cross, E.C. 1937. The white-tailed deer of Ontario. Rod and Gun in Canada 38: 14-16 + 32.

Currie, D.J. 1991. Energy and large-scale patterns of animal- and plant-species richness. American Naturalist 137: 27-49.

Darby, W. R. and L.S. Duquette. 1986. Woodland caribou and forestry in Northern Ontario, Canada. *Rangifer*, Special Issue 1: 87-93.

De Vos, A. 1952. The Ecology and Management of Fisher and Marten in Ontario. Wildlife Series, Technical Bulletin No. 1. Toronto, Ontario: Ontario Department of Land and Forests. 90 p.

De Vos, A. 1962. Changes in the distribution of mammals and birds in the Great Lakes area. Forestry Chronicle 38: 108-113.

De Vos, A. 1964. Range changes of mammals in the Great Lakes region. American Midland Naturalist 17:210-231.

De Vos, A., A.T. Cringan, J.K. Reynolds, and H.G. Lumsden. 1959. Biological investigations of traplines in northern Ontario. Wildlife Series, Technical Bulletin No. 8. Toronto, Ontario: Ontario Department of Lands and Forests. 48 p.

De Vos, A. and R.L. Peterson. 1951. A review of the status of woodland caribou in Ontario. Journal of Mammalogy 32: 329-337.

Dickerman, R.W. 1967. The old northeastern subspecies of red crossbill. American Birds 41: 189-194.

Doak, D.F. and L.S. Mills. 1994. A useful role for theory in conservation. Ecology 75: 615-626.

Dobbyn, J. 1994. Atlas of the Mammals of Ontario. Don Mills, Ontario: Federation of Ontario Naturalists. 120 p.

Douglas, C.W. and M.A. Strickland. 1987. Fisher. In: M. Novak, J.A. Baker, M.E. Obbard, and B. Malloch (editors). Wild Furbearer Management and Conservation in North America. Toronto, Ontario: Ontario Trappers Association and Ontario Ministry of Natural Resources. 511-529.

Drake, J.A. 1990. Communities as assembled structures: do rules govern pattern? Trends in Ecology and Evolution 5: 159-164.

Fleming, R.A., A.A. Hopkin, and J.-N. Candau. 2000. Insect and disease disturbance regimes in Ontario's forests. In: A.H. Perera, D.L. Euler, and I.D. Thompson (editors). Ecology of a Managed Terrestrial Landscape: Patterns and Processes in Forest Landscapes of Ontario. Vancouver, British Columbia: University of British Columbia Press. 141-162.

Gluck, M.J. and R.S. Rempel. 1996. Structural characteristics of post-wildfire and clearcut landscapes. Environmental Monitoring and Assessment 39: 435-450.

Haila, Y., K. Hanski, J. Niemelä, P. Puntilla, S. Raivo, and H. Tukia. 1994. Forestry and the boreal fauna: matching management with natural dynamics. *Annales Zoologici Fennici* 31: 187-202.

Hansson, L. 1992. Landscape ecology of boreal forests. Trends in Ecology and Evolution 7: 299-302.

Harestad, A.S. and F.L. Bunnell. 1979. Home range and body weight - a reevaluation. Ecology 60: 389-402.

Holling, C.S. 1992. Cross-scale morphology, geometry, and dynamics of ecosystems. Ecological Monographs 62: 447-502.

Hunter, M.L. 1990. Wildlife, Forests, and Forestry: Principles of Managing Forests for Biological Diversity. Englewood Cliffs, New Jersey: Prentice-Hall. 370 p.

Johns, B., E. Telfer, M. Cadman, D. Bird, R. Bjorge, K. De Smet, W. Harris, D. Hjertaas, P. Laporte, and R. Pittaway. 1994. National Recovery Plan for the Loggerhead Shrike. Report No. 7. Ottawa, Ontario: Recovery of Nationally Endangered Wildlife. 32 p.

Jung, T.S., I.D. Thompson, R.Titman, and A.P. Applejohn. 1999. Habitat selection by forest-dwelling bats in relation to stand type and structure. Journal of Wildlife Management. In press.

Kareiva, P. and U. Wennergren. 1995. Connecting landscape patterns to ecosystem and population processes. Nature 373: 299-302.

King, D.R. 1976. Estimates of white-tailed deer population and mortality in central Ontario, 1970-72. Canadian Field Naturalist 90: 29-36.

Kirk, D.A. and B.J. Naylor. 1996. Habitat Requirements of the Pileated Woodpecker with Special Reference to Ontario. Technical Report No. 46. North Bay: Ontario Ministry of Natural Resources, South Central Science and Technology Unit. 49 p.

Levin, S.A. 1992. The problem of pattern and scale in ecology. Ecology 73: 1943-1967.

Lumsden, H.G. 1971. The status of the sandhill crane in northern Ontario. Canadian Field Naturalist 85: 285-293.

Martell, A.M. 1983a. Changes in small mammal communities after logging in north-central Ontario. Canadian Journal of Zoology 61: 970-980.

Martell, A.M. 1983b. Demography of southern red-backed voles and deer mice after logging in north-central Ontario. Canadian Journal of Zoology 61: 958-969.

Mayr, E. 1946. History of the North American bird fauna. Wilson Bulletin 58: 1-68.

McLaren, M.A., I.D. Thompson, and J.A. Baker. 1998. Selection of vertebrate wildlife indicators for monitoring sustainable forest management in Ontario. Forestry Chronicle 74: 241-248.

Mönkkönen, M. and D.A. Welsh. 1994. A biogeographical hypothesis on the effects of human caused landscape changes on the forest bird communities of Europe and North America. *Annales Zoologici Fennici* 31:61-70.

Murphy, D.D. and B.D. Noon. 1991. Coping with uncertainty in wildlife biology. Journal of Wildlife Management 55: 773-782.

Naylor, B.J. and J.F. Bendell. 1983. Influence of habitat diversity on the abundance and diversity of small mammals in jack pine forests. In: R.W. Wein, R.R. Riewe, and I.R. Methven (editors). Resources and Dynamics of the Boreal Zone. Ottawa, Ontario: Association of Canadian Universities for Northern Studies. 295-307.

Outram, A.A. 1967. Changes in the mammalian fauna of Ontario since Confederation. Ontario Naturalist, September: 19-21.

Pedlar, J.H. and R.K. Ross. 1997. An update on the status of the sandhill crane in northern and central Ontario. Ontario Birds 15: 4-13.

Peterson, R.L. 1957. Changes in the mammalian fauna of Ontario. In F.A. Urquhart (editor). Changes in the Fauna of Ontario. Toronto, Ontario: University of Toronto Press. 43-58.

Peterson, R.L. and V. Crichton. 1949. The fur resources of the Chapleau District. Canadian Journal of Forest Research 27: 68-84.

Perera, A.H. and D.J.B. Baldwin. 2000. Spatial patterns in the managed forest landscapes of Ontario. In: A.H. Perera, D.L. Euler, and

I.D. Thompson (editors). Ecology of a Managed Terrestrial Landscape: Patterns and Processes in Forest Landscapes of Ontario. Vancouver, British Columbia: University of British Columbia Press. 74-99.

Pianka, E.R. 1966. Latitudinal gradients in species diversity: A review of concepts. American Naturalist 100: 33-46.

Pimlott, D.H., J.A. Shannon, and G.B. Kolenosky. 1969. The Ecology of the Timber Wolf in Algonquin Provincial Park. Research Report Number 87. Maple: Ontario Department of Lands and Forests. 92 p.

Pulliam, H.R. 1988. Sources, sinks, and population regulation. American Naturalist 132: 632-651.

Ricklefs, R.E. and D. Schluter (editors). 1993. Species Diversity in Ecological Communities: Historical and Geographical Perspectives. Chicago, Illinois: University of Chicago Press. 414 p.

Rowe, J.S. 1972. Forest Regions of Canada. Publication No. 1300. Ottawa, Ontario: Canadian Forest Service. 172 p.

Sauer, J. R., J. E. Hines, G. Gough, I. Thomas, and B. G. Peterjohn. 1997. The North American breeding bird survey results and analysis. Version 96.3. Laurel, Maryland: Patuxent Wildlife Research Center.

Seton, E.T. 1929. Lives of Game Animals. New York, New York: Doubleday, Doran. 378 p.

Simkin, D.W. 1966. Extralimital occurrences of raccoons in Ontario. Canadian Field Naturalist 80: 144-146.

Sims, R.A., W.D. Towill, K.A. Baldwin, and G.M. Wickware. 1989. Forest Ecosystem Classification for Northwestern Ontario. Ottawa, Ontario: Canada-Ontario Forest Resource Development Agreement. 191 p.

Snyder, L.L. 1957. Changes in the avifauna of Ontario. In: F.A. Urquhart (editor). Changes in the Fauna of Ontario. Toronto, Ontario: University of Toronto Press. 26-42.

Spytz, C.P. 1993. Cavity-nesting bird populations in cutover and mature boreal forest, northern Ontario [M.Sc. Thesis]. Waterloo, Ontario: University of Waterloo. 88 p.

Statutes of Ontario. *Endangered Species Act*. R.S.O. 1990, c. E15.

Tebbel, P.D. and C.D. Ankney. 1982. The status of sandhill cranes in central Ontario. Canadian Field Naturalist 96: 163-166.

Thomas, M.K. 1957. Changes in the climate of Ontario. In: F.A. Urquhart (editor). Changes in the Fauna of Ontario. Toronto, Ontario: University of Toronto Press. 59-75.

Thompson, I.D. 1988. Habitat needs of furbearers in relation to logging. Forestry Chronicle 64: 251-261.

Thompson, I.D. 1991. Could marten become the spotted owl of eastern Canada? Forestry Chronicle 67: 136-140.

Thompson, I.D. 1994. Marten populations in uncut and logged boreal forests in Ontario. Journal of Wildlife Management 58: 272-280.

Thompson, I.D. and P. Angelstam. 1999. Species in forest biodiversity management. In: M.L. Hunter (editor). Management of Forest Biodiversity. Oxford, UK: Cambridge University Press. 434-459.

Thompson, I.D., A. Applejohn, T.S. Jung, and L.A. Walton. 1995. A Short-Term Study of the Faunal Associations in Old White Pine Ecosystems. Forest Fragmentation and Biodiversity Program, Technical Report No. 21. Sault Ste. Marie, Ontario: Ontario Ministry of Natural Resources. 43 p.

Tilman, D. 1982. Resource Competition and Community Structure. Princeton, New Jersey: Princeton University Press.

Van Horne, B. 1983. Density as a misleading indicator of habitat quality. Journal of Wildlife Management 47: 893-901.

Van Wagner, C.E. 1978. Age-class distribution and the forest fire cycle. Canadian Journal of Forest Research 8: 220-227.

Voigt, D.R., M. Deyne, M. Malhiot, B. Ranta, B. Snider, R. Stefanski, and M. Strickland. 1992. White-Tailed Deer in Ontario: Background to a Policy. Draft Document. Toronto, Ontario: Ontario Ministry of Natural Resources, Wildlife Policy Branch. 83 p.

Voigt, D.R., J.A. Baker, R.S. Rempel, and I.D. Thompson. 2000. Forest vertebrate responses to landscape-level changes in Ontario. In: A.H. Perera, D.L. Euler, and I.D. Thompson (editors). Ecology of a Managed Terrestrial Landscape: Patterns and Processes in Forest Landscapes of Ontario. Vancouver, British Columbia: University of British Columbia Press. 198-233.

Welsh, D.A. and S.C. Lougheed. 1996. Relationships of bird community structure and species distributions to two environmental gradients in the northern boreal forest. Ecography 19: 194-208.

Wilton, M.L. 1984. How the moose came to Algonquin. *Alces* 23: 89-106.

5 Spatial Patterns in the Managed Forest Landscape of Ontario

AJITH H. PERERA* AND DAVID J.B. BALDWIN*

INTRODUCTION

The focus of study of landscape ecology, since its inception, has been land cover patterns and the ecological implications of those patterns (Forman and Godron 1986). Spatial patterns of forest land cover are important, either explicitly or implicitly, to many ecological processes, such as species population dynamics. These patterns have a direct importance to such economic concerns as the selection of sites for timber mills, and to such social concerns as the conservation of wilderness. Turner (1989) presented an excellent early discussion on the relevance of land cover spatial patterns to a range of ecological processes, including disturbance, species dynamics, and nutrient flows. Forman (1995a) has provided a more recent overview of the concepts and terminology of land cover patterns, of techniques for describing these patterns, and of the socio-ecological implications of the patterns themselves.

In Ontario, there are many important reasons for assessing spatial patterns in the extensive forest cover of the province. Forest cover patterns reflect the combined influence of relatively static factors, such as latitude, precipitation, landforms, soils, and terrain, and of spatial and temporal trends in species adaptation and distribution. Forest patterns reflect the effects of disturbance, whether natural or caused by human activity; conversely, land cover patterns also constitute context variables in the study of species patterns or disturbance regimes. Spatial indices of forest cover patterns have become valuable metrics for the regional-level monitoring of ecological processes (e.g., Turner 1989). Finally, the *Forest Management Planning Manual for Ontario's Crown Forests* in use in the province (OMNR 1996) prescribes that forest managers assess and monitor spatial indices of forest cover at a hierarchy of scales, from the level of forest management planning to sub-regional and regional levels.

Our goal in this chapter is to provide an overview of the spatial characteristics of Ontario's managed forest landscape and to discuss the ecological implications of these characteristics. Land cover studies in Ontario began with the development by Hills (1959) of an ecological land classification scheme that attempted to explain the spatial distribution of forests through geoclimatic variables. Subsequently, Rowe (1972) provided a detailed description of the distribution of Ontario's major forest cover types. Neither Hills nor Rowe provided a description which was spatial, in the true sense; that is, neither was based on the study of strictly spatial attributes.

That fact is not surprising, because the concepts, techniques, and source data for quantifying spatial patterns of land cover have only been developed and become widely known within the last decade (O'Neill et al. 1988; Turner and Gardner 1991; McGarigal and Marks 1995; Gustafson 1998). Several spatial land cover studies have been conducted over specific sites in Ontario in recent years. Suffling (1988) examined the landscape diversity in a plot-transect in northwestern Ontario and concluded that intermediate-level

* Forest Landscape Ecology Program, Ontario Forest Research Institute, Ontario Ministry of Natural Resources, 1235 Queen Street East, Sault Ste. Marie, Ontario P6A 2E5

disturbances produce the highest levels of landscape diversity. Perera and Baldwin (1993) studied the spatial characteristics of red and white pine forests in western and central Ontario. They found that these forests occurred mainly in small but not isolated patches, which had low interior areas only at high buffer widths. Gluck and Rempel (1996) analyzed the structural differences between post-wildfire and clear-cut landscapes in a western Ontario study area and concluded that clear-cuts produced larger and more irregular patches, with less interspersion, than patches produced by fire. In the same study area, Rempel et al. (1997) examined the trends in moose population in relation to the spatial characteristics of the disturbances.

No comprehensive studies of land cover patterns have been conducted, however, for all of Ontario or for broad regions of the province. As a result, the discussion in this chapter draws heavily on a study we have recently concluded on the overall spatial patterns of forest cover (Perera and Baldwin, in press). The land cover types we used were based on a series of supervised classifications of Landsat Thematic Mapper (TM) data (classifications, that is, based on preassigned class definitions rather than on the distinction of differences per se) which provided continuous land cover data over the entire 1.1 million km^2 area of Ontario (e.g., Spectranalysis Inc. 1994). We implicitly assumed that the broad categories of land cover (Table 5.1) obtained from this source are functionally distinct as well. By this, we mean that the cover types respond differently to climate, terrain, and disturbance factors, and that they each have a different effect on such ecological processes as productivity, tolerance of and recovery from disturbances, and species habitat supply. In other words, we viewed these cover types as the constituent elements of a landscape. Our choice of land cover classes and of spatial scale (for example, of data resolution) could arguably be too coarse or too fine, relative to different objectives of ecological studies or management planning. Throughout the following discussion, the reader must always keep in mind the spatial scale involved and should guard against interpreting our findings at any other scale.

Terms used to describe spatial patterns are not specifically defined in this chapter, although the context in which they are used may make their meaning evident. For clarification, the reader may refer to the glossary provided in this book and to the introductory chapter on landscape ecology (Perera and Euler 2000, this volume).

ECOREGIONAL PATTERNS IN ONTARIO

A prerequisite to studying spatial patterns in forest landscapes is to stratify the landscapes. This step is particularly important in examining patterns over very large areas such as Ontario. Termed "ecoregionalization systems," the various approaches taken to stratification provide the context for studying land cover patterns. The resulting strata, generically known as "ecoregions," are distinguished according to broad, yet spatially explicit, environmental features such as climate, geology, and terrain; therefore, they explain elements and processes that exert control over finer land cover patterns. In deriving ecoregions, one attempts to attain the greatest ecological homogeneity within the ecoregional units, and the greatest heterogeneity among the units. Ecoregional systems thus provide a better ecological basis for examining landscape patterns than such divisions as administrative units, which may reflect no ecological distinctions.

Hills (1959) pioneered ecoregionalization in North America, when he developed an ecological land classification system for Ontario (Figure 5.1). Hills's objective was to classify the province into zones of biomass productivity, in the context of forest timber production, focusing on the relationship of climate and landforms to the occurrence and succession of forest vegetation. The resulting system is implicitly hierarchical and is developed from the most local, or lowest, level upward. Site regions constitute the highest level of this hierarchy, site districts the intermediate level, and landform units the lowest level, and each level is encompassed in the level above and defined by the level below.

By combining field observations with a rudimentary mapping of broad relief features, broad bedrock classes, and depth and texture classes of unconsolidated materials, Hills established landform units based on associations between landforms and vegetation succession patterns, so that the effect of landforms was assessed as the local effect of slope and aspect. These landform units were then amalgamated into site districts. Hills proposed that climate exerts a higher-level control, in the form of continental effects, in determining forest vegetation, succession, and biomass productivity. Consequently, he used climatic gradients to combine the site districts to form site regions, but also took into account soil development gradients and land use categories.

The Hills ecoregionalization system has played a very important role in many facets of forest and resource management in Ontario, ranging from the design of the provincial system of protected areas, to the development of a site classification system for forest management. A major drawback to this system, however, has always been the absence of documentation on the methods and information used to derive it. Many attempts have been made to modify the Hills system. Both the Canadian system of terrestrial ecoregions (Wiken 1977, later mapped by Wickware and Rubec 1989) (Figure 5.2a) and the revised

Table 5.1

Description of land cover classes across Ontario and within the forest management zone.

Provincial Landcover Class	*FMZ Landcover Class*	*Landcover Class Description*
Dense Forest	Dense Deciduous Forest	Continuous canopy composed of approximately 80% deciduous species
	Dense Coniferous Forest	Continuous canopy composed of approximately 80% conifer species. This class also includes dense conifer swamp, not including tamarack cover (included in the Fen class)
	Mixed Forest, Mainly Deciduous	Continuous canopy (dense forest) with deciduous component between 50 and 80%
	Mixed Forest, Mainly Coniferous	Continuous canopy (dense forest) with coniferous component between 50 and 80%
Sparse Forest	Sparse Forest	Includes a range of sparse / poor forest conditions including sparse conifer (conifer stands on bedrock, black spruce in depressions) sparse deciduous (bedrock bearing vegetation, shrub cover along riverbanks, overgrown agricultural fields)
	Cut, Burn, Scrub	Includes areas disturbed by or regenerating following recent harvest or fire (within last 30 years). Canopy ranges from nil to denser than the sparse forest class (but not as dense as Dense and Mixed Forest classes)
Non-Forest Natural	Coastal Mudflats	Includes both vegetated and very sparsely vegetated mudflats primarily near Hudson Bay
	Coastal Marshes	Includes both intertidal marsh (subject to regular tidal influence) and supertidal marsh (subject to exceptionally high tides only)
	Inland Marshes	Includes coastal marshes beyond saltwater influence near Hudson Bay, shallow / ephemeral, cattail and deep water marshes in the south
	Hardwood Swamp	Associated mainly with rivers, lake shores, and old lakebeds
	Miscellaneous Wetlands	Includes most wetland types through the central portion of the Forest Management Zone
	Fen	Includes dense graminoid cover, shrub-rich and treed fens and fens with pools (pools comprising up to 20% of the surface)
	Open Bog	Ranges from lichen-rich peat plateaus to bogs with varying amounts of shrub cover but little or no tree cover
	Treed Bog	Represents treed bogs with forest cover less dense than the sparse forest classes. This class occurs on nutrient poor, poorly drained areas, often as a narrow ring around open bogs
Non-Forest Anthropogenic	Built-up Areas, Bedrock	Includes all urban settlement (excluding agriculture), mine tailings, and extraction sites and prominent bedrock outcrops with virtually no vegetation
	Agriculture	Agricultural activity including row crops and open fields with bare plowed soil
	Pasture and Abandoned	Includes areas that showed grass cover in spring through to summer including alvar on shallow soils over limestone. These areas show a lack of regular field or row patterns

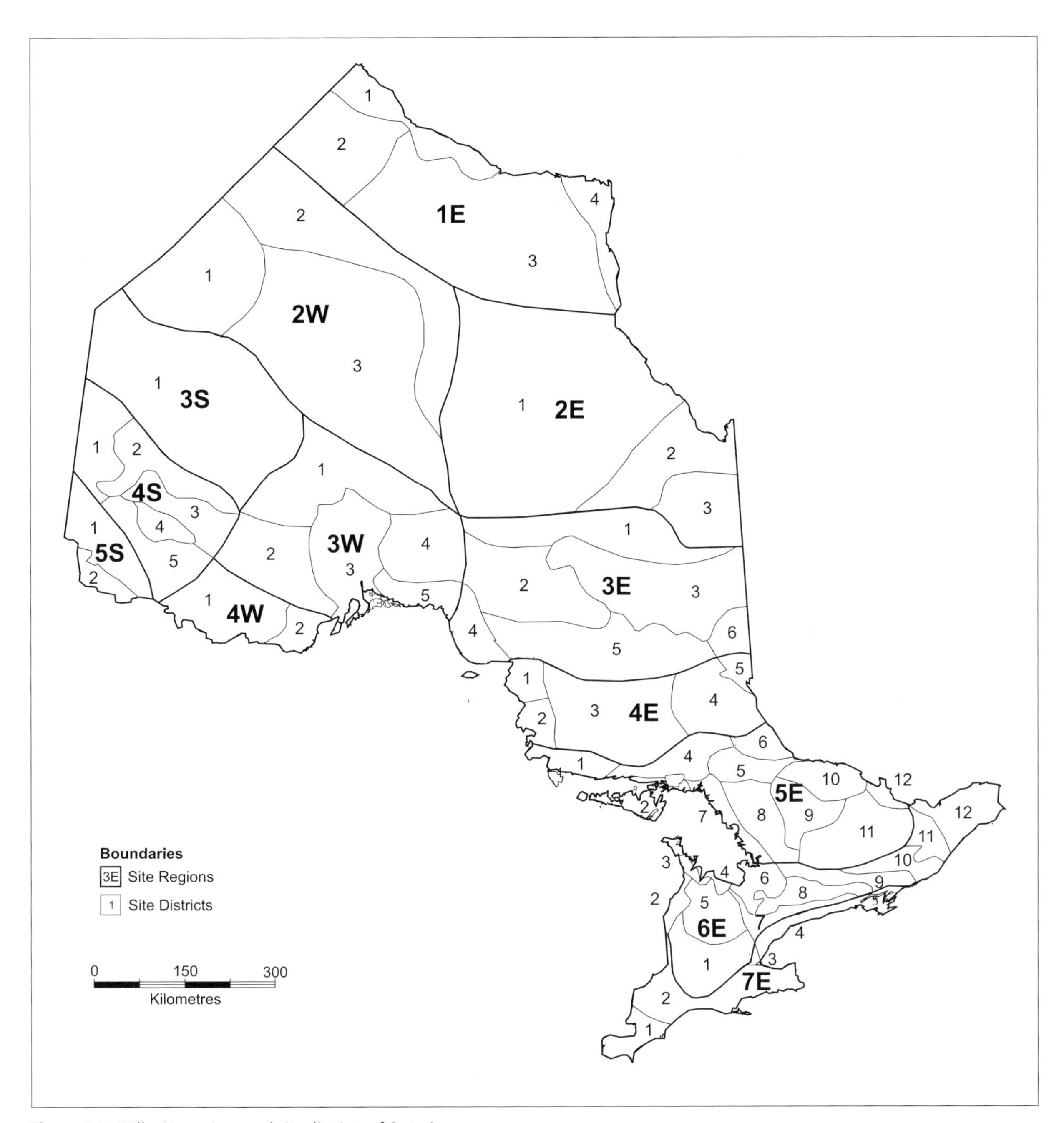

Figure 5.1 Hills site regions and site districts of Ontario.
(Adapted from Hills 1959)

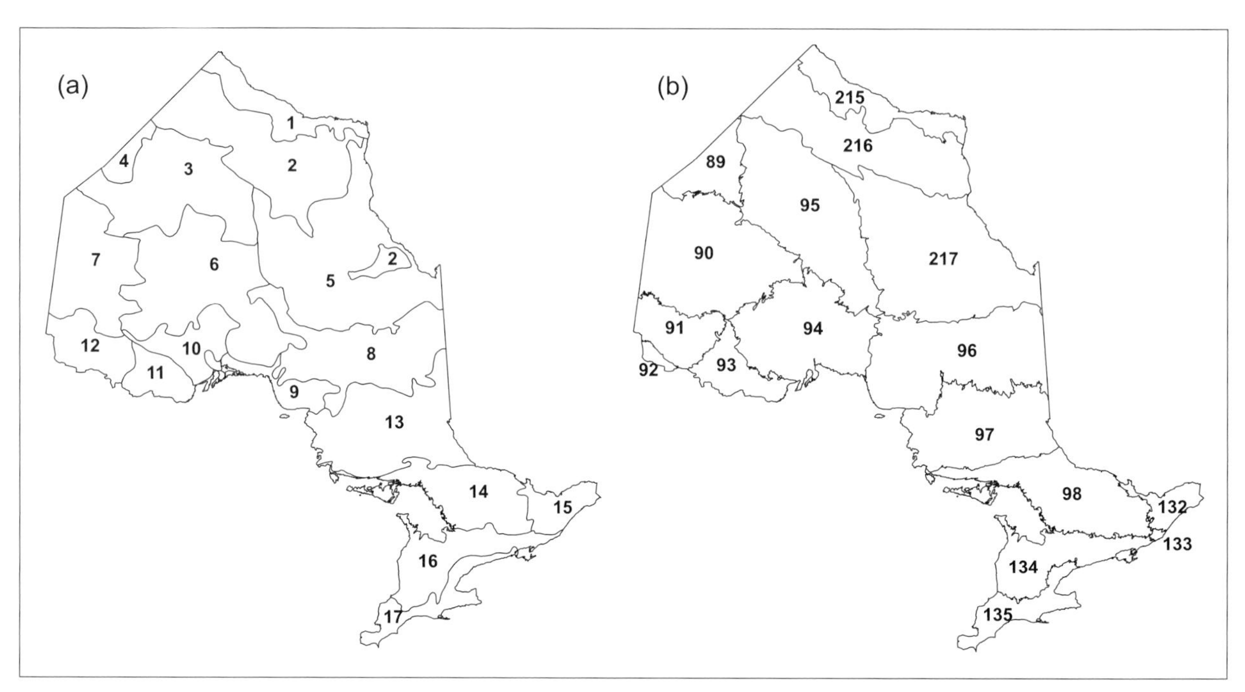

Figure 5.2 Federal ecoregionalization systems influenced by Hills (1959): (a) Terrestrial Ecoregions of Ontario; and (b) Revised Terrestrial Ecoregions of Ontario.
(Adapted from Wickware and Rubec 1989 and Ecological Stratification Working Group 1995, respectively)

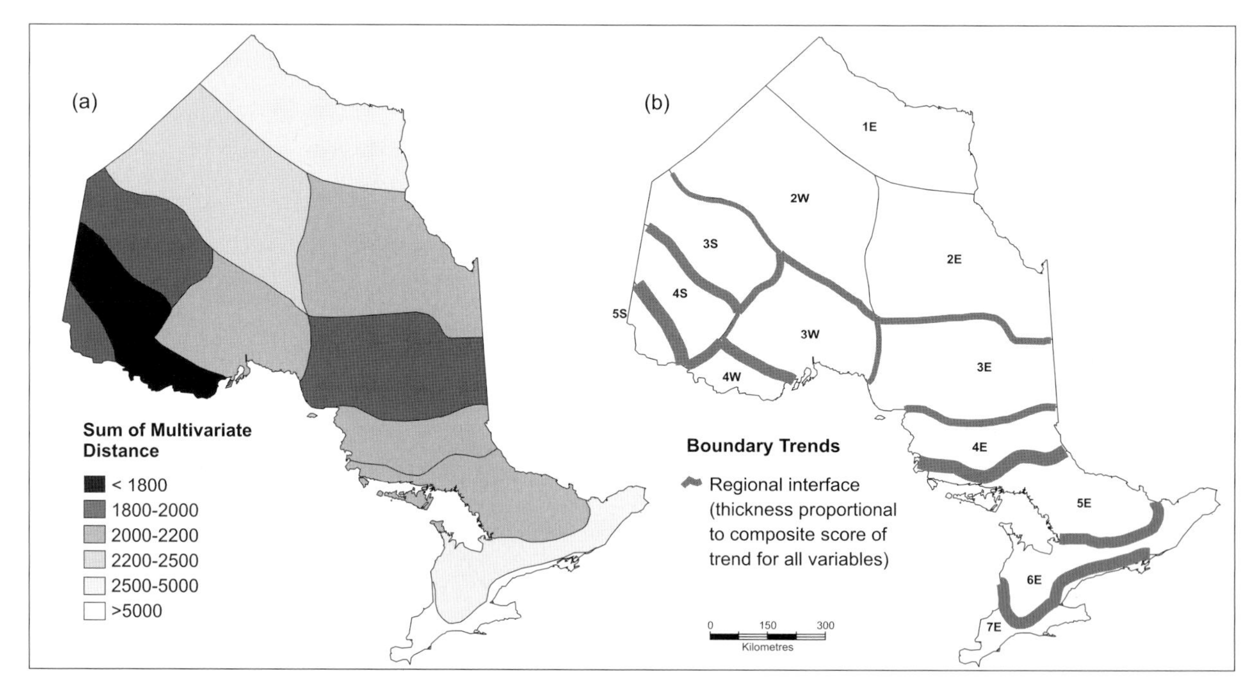

Figure 5.3 Examples of ecoregional quantification: (a) total pairwise multivariate distance (based on a suite of geoclimatic variables) from each region to all others, mapped by region; and (b) analysis of means of a suite of geoclimatic variables and multivariate distances across a gradient for each regional interface. Interface widths shown proportional to strength of boundary cumulative score.
(Adapted from Baldwin et al. 1998)

terrestrial ecoregions (Ecological Stratification Working Group 1995) (Figure 5.2b) attempted to improve the precision of the Hills units through the use of information derived from satellite data. But these too suffer from the same basic shortcoming: a lack of clear documentation of quantitative methods and data that prevents the systems from being more than qualitative and makes it impossible to reproduce or test them independently.

For any of these systems to be used to their full effect, their units and boundaries must be quantified; only then will the spatial patterns they exhibit make sense and their underlying geoclimatic properties be understood. A recent study provides this fundamental information for several Ontario ecoregional frameworks (Baldwin et al., in press). Advances made since these systems were first published in the techniques of spatial analysis and in the development of provincial spatial geoclimatic databases have now provided the means to quantify these qualitative systems (Figure 5.3). These methods include techniques to quantify both the regional units (Figure 5.3a) and the trends at work across their boundaries (Figure 5.3b).

Although their work was intended to serve the specific needs of timber management, both Hills and his successors recognized the need to delineate areas that would represent more than simply where different forest types grow. By defining units of similar geoclimatic attributes or existing vegetation, these systems attempt to define regions that encompass broad ecological processes. A more recent system, the Hierarchical Ecoregional Framework (HEF), has been developed on this same premise, but has explicitly linked the definition of regions to net primary productivity, as quantified at an explicit scale (Perera et al. 1996; Band et al. 1999). The units of HEF (Figure 5.4) were constituted by identifying the factors that explain the greatest degree of variation across an independent spatial representation of primary productivity at each level of the hierarchy. Band (2000, this volume) provides a more detailed discussion of this process. The HEF system provides a number of benefits, including a spatially explicit methodology, an explicit foundation in ecological principles, and a quantitative basis for reproducing, testing and updating the units.

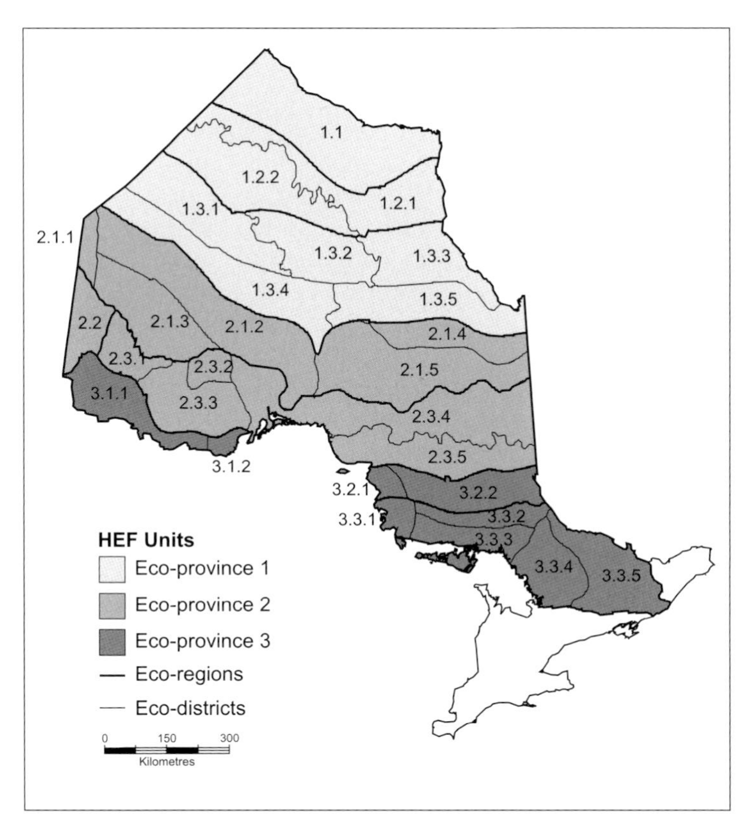

Figure 5.4. The Hierarchical Ecoregional Framework. (Adapted from Perera et al. 1996)

LAND COVER PATTERNS AT A PROVINCIAL SCALE

Our spatial assessment of Ontario land cover patterns (Perera and Baldwin, in press) was based on three distinct zones in the province: the northern zone (north of latitude 52°N), the zone of extensive forest management (between 45°N and 52°N), and the southern zone (south of 45°N). Ontario's forests occupy 40 percent of the northern zone, over 80 percent of the forest management zone, and about 20 percent of the southern zone (Figure 5.5a).

Ontario's extensively managed forest zone covers more than 500,000 km^2, or approximately half of the land base of the province. The forest in this zone is predominantly dense, in contrast to the predominantly sparse forest found in the area north of 52°N (Figure 5.5b). (Dense forest, in this classification, is defined as forest in which the tree canopy forms a ground cover greater than 50 percent.

Sparse forest is defined as having a canopy cover of less than 50 percent.) The spatial characteristics of the complex of land cover classes differ from one zone to another. The forest patches in the far north exist in a matrix of natural wetlands (marshes, bogs, and fens), while in the heavily settled area of southern Ontario, they are dispersed within a matrix of agriculture and developed land. The patchiness of the forest cover is low in the forest management zone (Figure 5.6a). (A landscape is considered "patchy" if the mean patch size is small and the distance to a like neighbour, or "isolation," is high. Patches are deemed small or large and inter-patch distance high or low, only in comparison to other scenarios, not by absolute values.) The variation in forest patch size across the province shows a distinct trend (Figure 5.6b). The many small patches found in the northern and southern zones exhibit a low degree of size variation, while there is greater size variation among the larger patches of the forest management zone. Forest patch size variability is highest in the transition areas between the forest management zone and either of the other two zones.

The dense forest classes also demonstrate a lower degree of patchiness in the forest management zone than in the northern or southern zones. This fact is apparent in a larger mean patch size (Figure 5.7a), lower inter-patch isolation (Figure 5.7b), and a higher percentage of patch interior that remains after a 240-m buffer is discounted (Figure 5.7c). The spatial complexity of dense forest patches is also higher in the forest management zone, especially in comparison to the southern zone (Figure 5.7d). (Patches are deemed complex when the shape index [Patton 1975] is high.)

LAND COVER PATTERNS IN THE FOREST MANAGEMENT ZONE

The forest management zone is subject to extensive disturbance; for example, in 1995, a total area of more than 2000 km^2 was harvested (Canadian Forest Service 1997), and a further area of 2000 km^2 was burned by forest fires (Perera et al., in press). Over the 45-year period from 1951 to 1995, nearly 2 million ha were burned (Figure 5.8) and a further 6.6 million ha were harvested by clear-cutting (Figure 5.9).

The extent of human activity in the forest management zone, in terms of agriculture, mining, and urbanization, is low, representing only 4 percent of the land area. This activity is concentrated in few centres of agriculture and settlement. About 50 percent of the land cover of the forest management zone is covered by a combination of dense coniferous forest and dense deciduous-coniferous mixed forest (Figure 5.10). The dense coniferous forest is dominated by jack pine and black spruce. The dense mixed forest is dominated by black spruce and aspen in the north, and by red pine, white pine, and maple in the south. Sparse forest and dense deciduous forest cover constitute 25 percent of the land area, of which 11 percent consists of recently-harvested forest, burns, and scrub vegetation. A wide range of wetland types, including coastal and inland marshes, coastal mudflats, swamps, fens, and bogs, occupy 10 percent of the zone.

To analyze the spatial pattern of land cover across the forest management zone, we used a grid of 10-km by 10-km cells (Perera and Baldwin, in press). Each 100-km^2 cell was analyzed for the spatial characteristics of its land cover types, including patchiness, amount of edge, interior, isolation, complexity, contiguity, and diversity. Seventeen land cover types were distinguished in the classification, derived from Landsat TM data, that was used for the forest management zone. These classes are nested within the provincial land cover classes (Table 5.1). The average values for a 100-km^2 cell in the forest management zone are illustrated in Figure 5.11.

All cover types in the average 100-km^2 unit occur in patches from 5 to 20 ha in size (Figure 5.11a), with the exception of dense conifer forest and recently disturbed forest, which occur in larger patches. The variation in patch size is higher in the forest cover types than in any other cover type (Figure 5.11b). The forest patches also have more linear edge than any other cover type, and the mixed conifer class has the most among the forest types (Figure 5.11c). The amount of patch interior remaining after application of a 120-m-wide edge buffer is highest in the recently disturbed forest and dense conifer classes, but also disproportionately high in the coastal marsh and open bog wetland types (Figure 5.11d). The wetland patches are generally more isolated from like neighbours than are the forest patches (Figure 5.11e). Patches of all cover types have the same general probability of being intermixed (55 to 70 percent), with the exception of the coastal wetland patches, in which the probability of intermixing with other land cover types is lower (Figure 5.11f).

For a closer examination of variation in spatial patterns, it was necessary to stratify the forest management zone. We used the ecological land classification

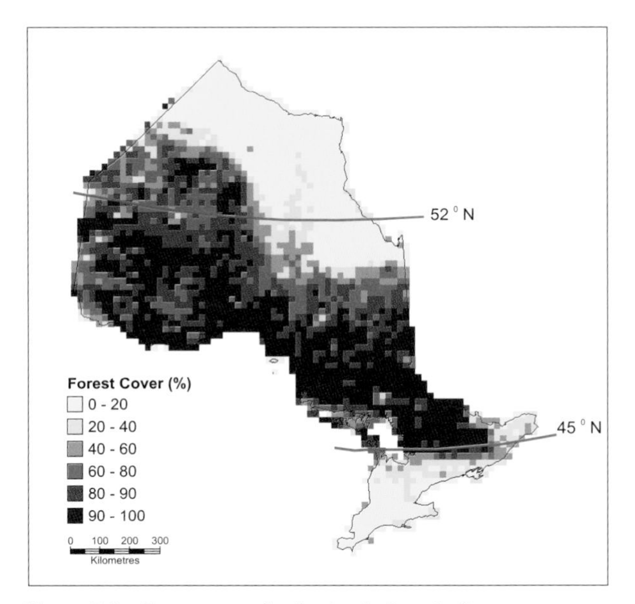

Figure 5.5a Forest cover distribution in Ontario. Percentage forest cover, based on a 20-km by 20-km grid.

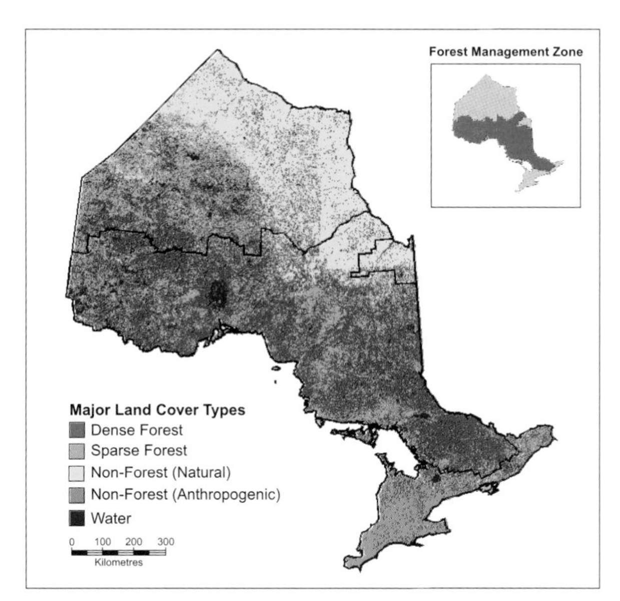

Figure 5.5b Major land cover types of Ontario.

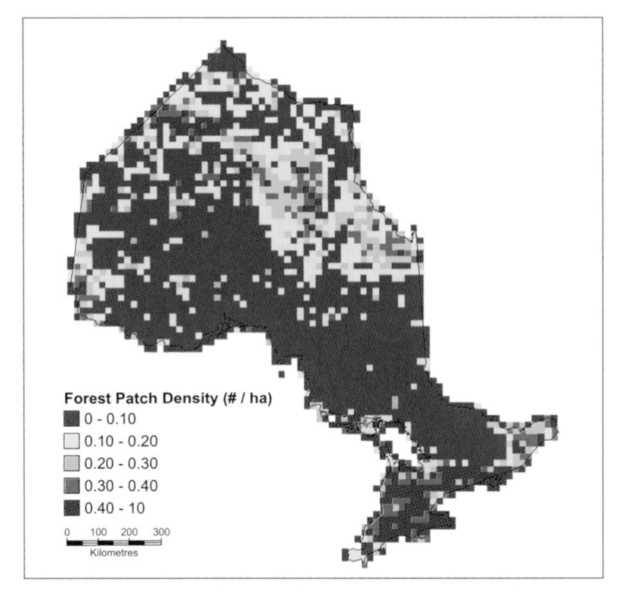

Figure 5.6a Forest (dense and sparse) patch density of Ontario.

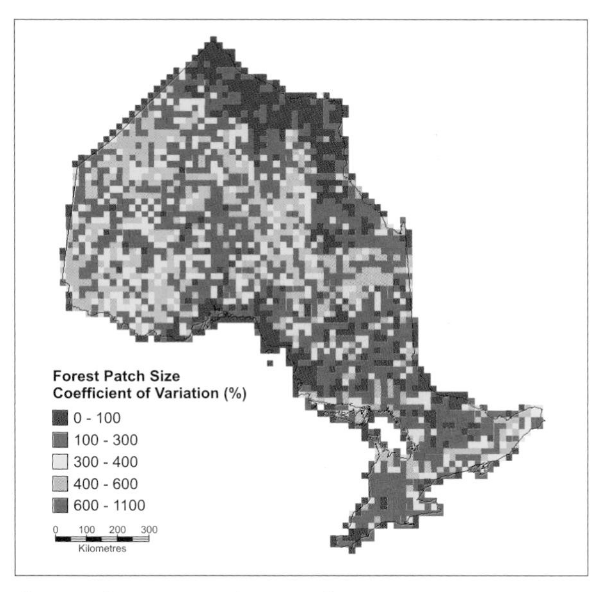

Figure 5.6b Variation in forest patch size in Ontario.

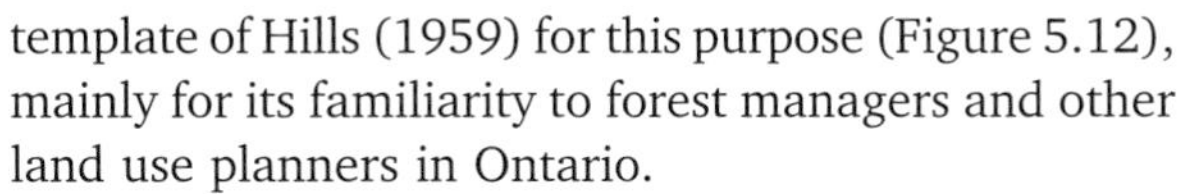

template of Hills (1959) for this purpose (Figure 5.12), mainly for its familiarity to forest managers and other land use planners in Ontario.

We found that the dominant land cover is different in each ecoregion, a result which indicates the presence of sub-zonal variations within the forest management zone (Table 5.2). Ecoregions 2W and 3S are dominated by dense conifer forest, and ecoregion 2E is co-dominated by dense conifer and fens. Ecoregions 3E, 4E, and 3W are co-dominated by the dense conifer cover, mixed conifer, and mixed deciduous types. Dense deciduous forests are a significant component of only three ecoregions, 4W, 5S, and 5E. Ecoregion 4S has a high component of recently disturbed forest.

Overall, the diversity of land cover composition in all ecoregions is similar, as measured by the richness

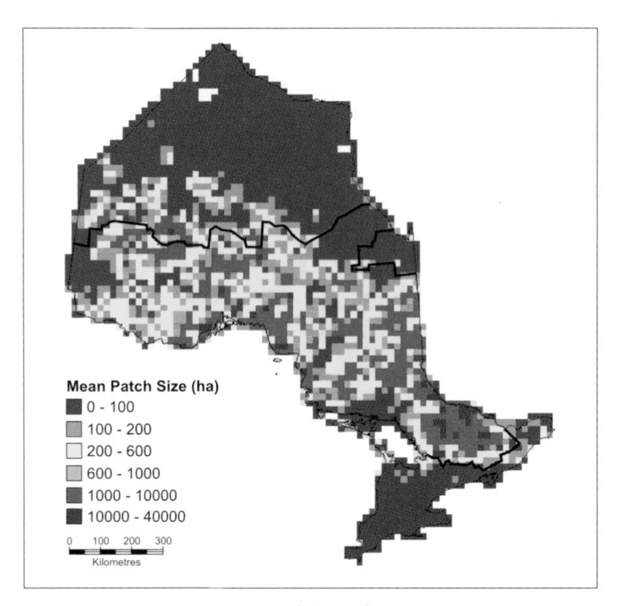

Figure 5.7a Mean patch size of dense forest in Ontario.

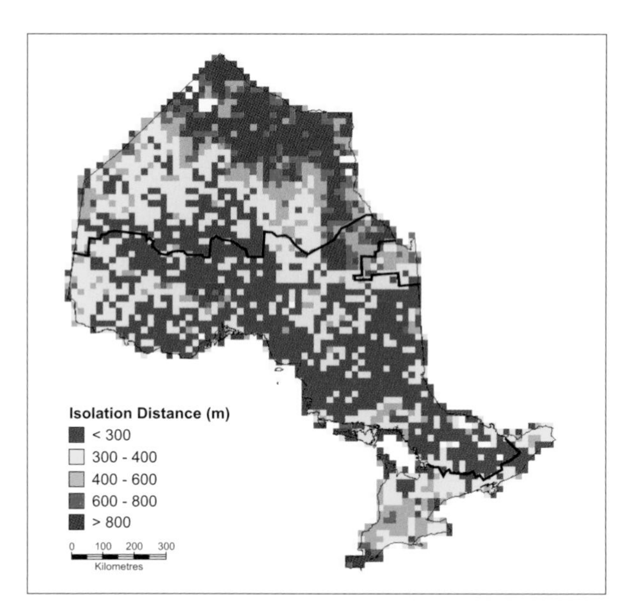

Figure 5.7b Isolation distance of dense forest patches in Ontario.

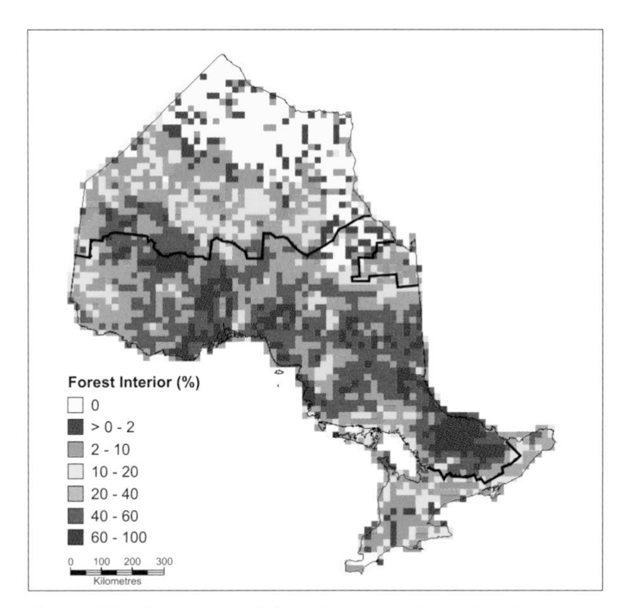

Figure 5.7c Percentage of dense forest patch interior, given a 240-m-wide edge.

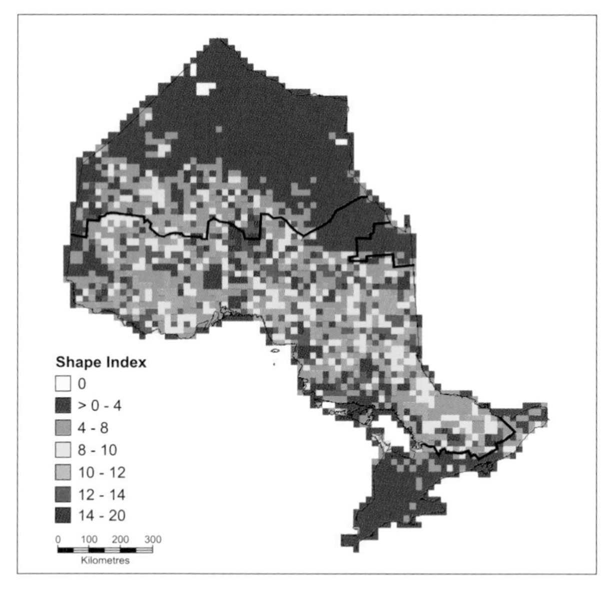

Figure 5.7d Complexity of dense forest patches in Ontario.

of cover types, by the evenness of their area apportionment, and by heterogeneity indices. The spatial clustering of the cover types, as measured by an index of contagion (Li and Reynolds 1993), was also similar among the ecoregions in the forest management zone. Only ecoregion 3S represented a minor exception: here, spatial contagion was higher and compositional evenness lower, because of a predominance of dense conifer forest cover.

Spatial Patterns of the Forest Cover

Having examined general land cover patterns across the province, and then the patterns within the forest management zone, we will now focus on the spatial

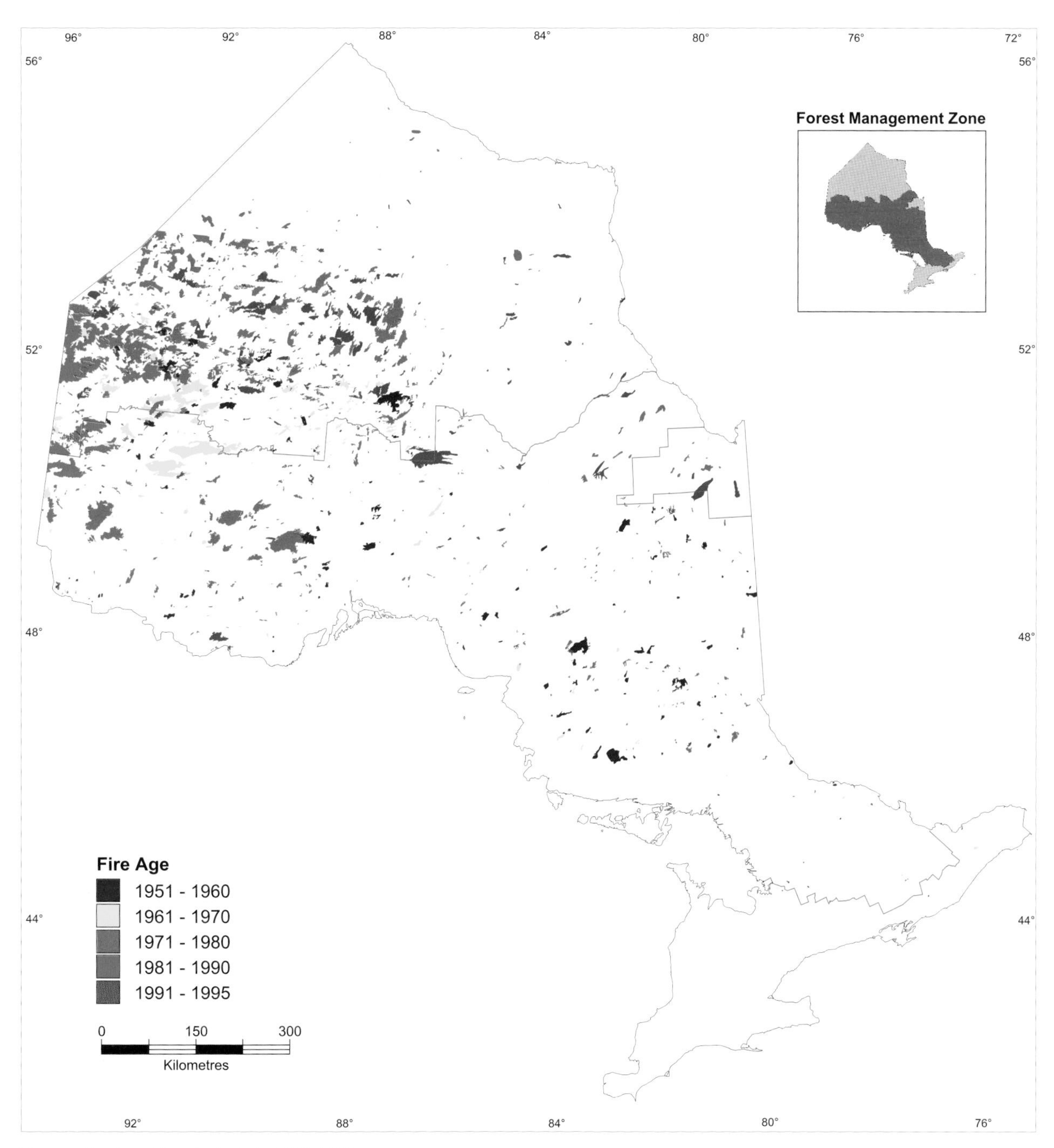

Figure 5.8 Forest fires in Ontario, 1951-1995.
(Data from Ontario Ministry of Natural Resources, Forest Landscape Ecology Program, 1998)

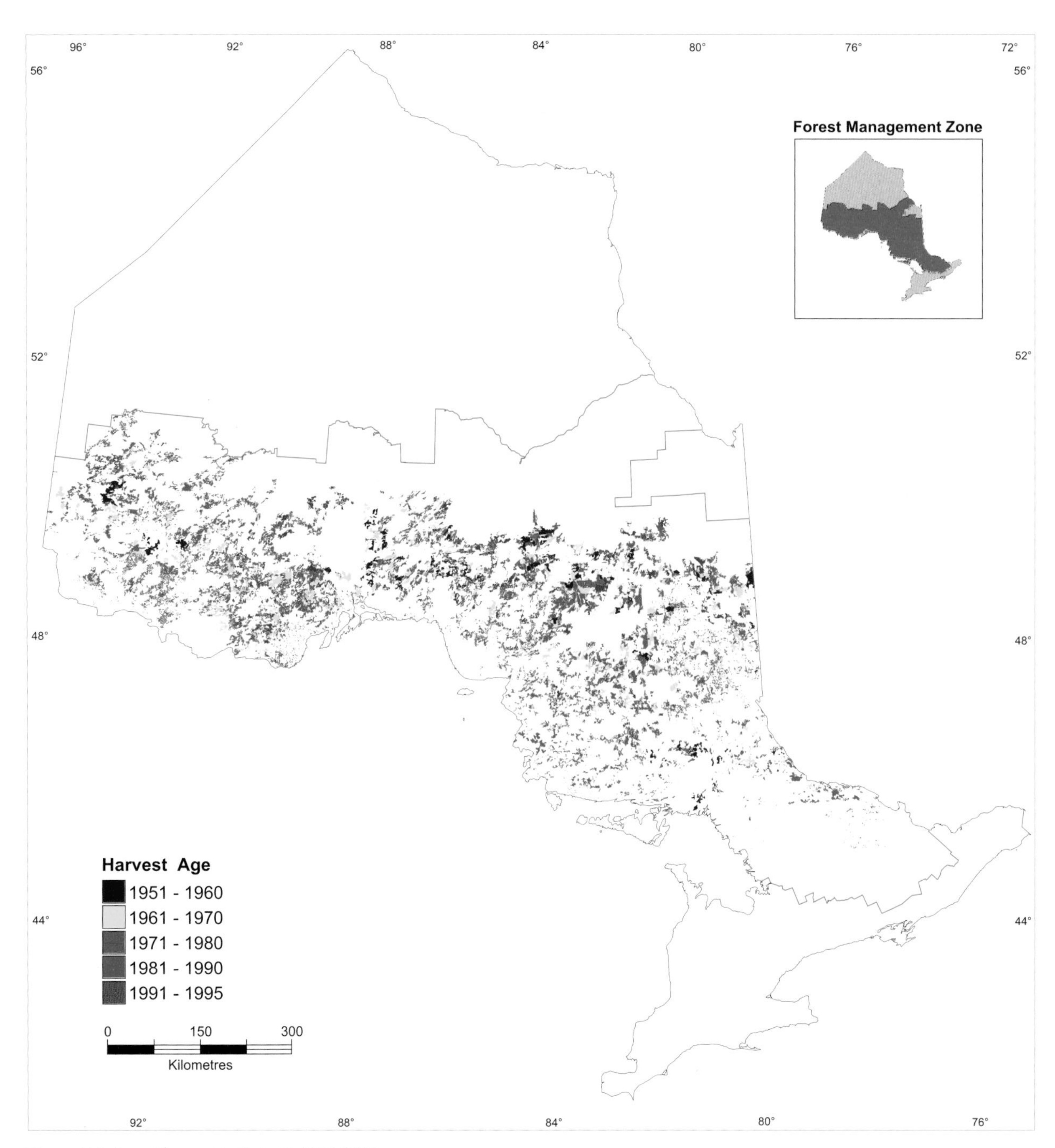

Figure 5.9 Forest harvest in Ontario, 1951-1995.

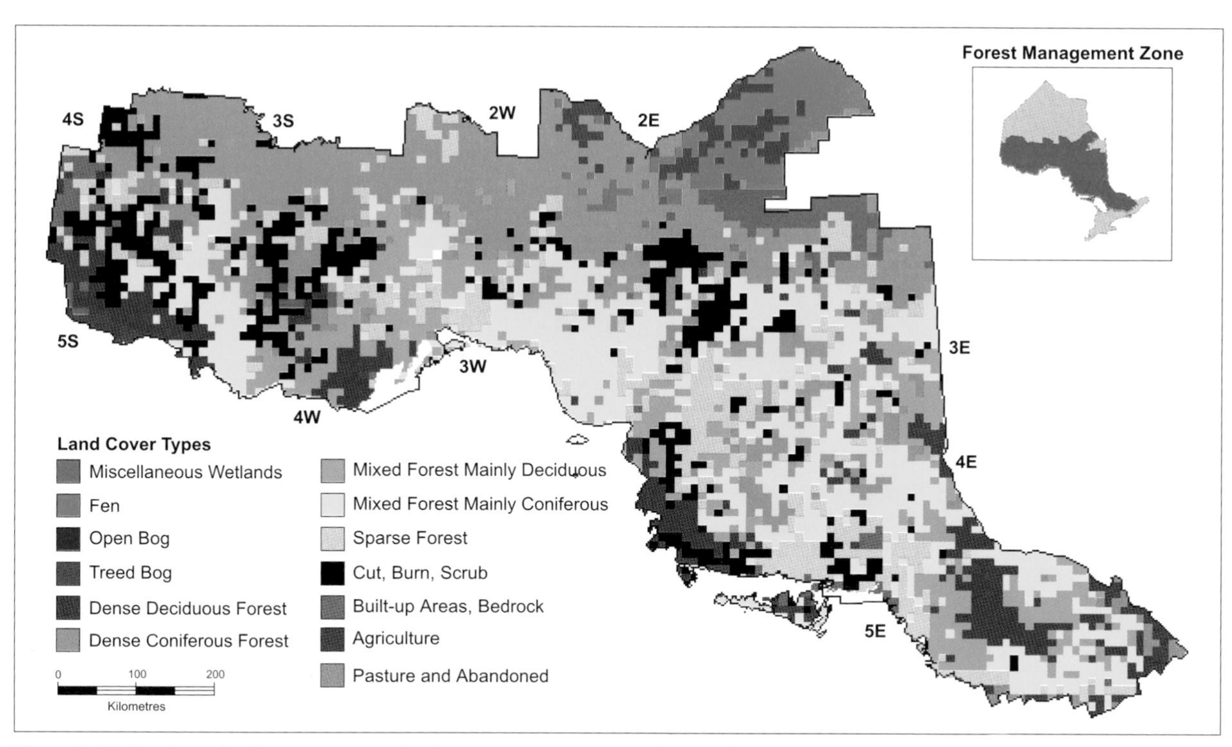

Figure 5.10 Dominant land cover types in the forest management zone, by a 10-km by 10-km grid.

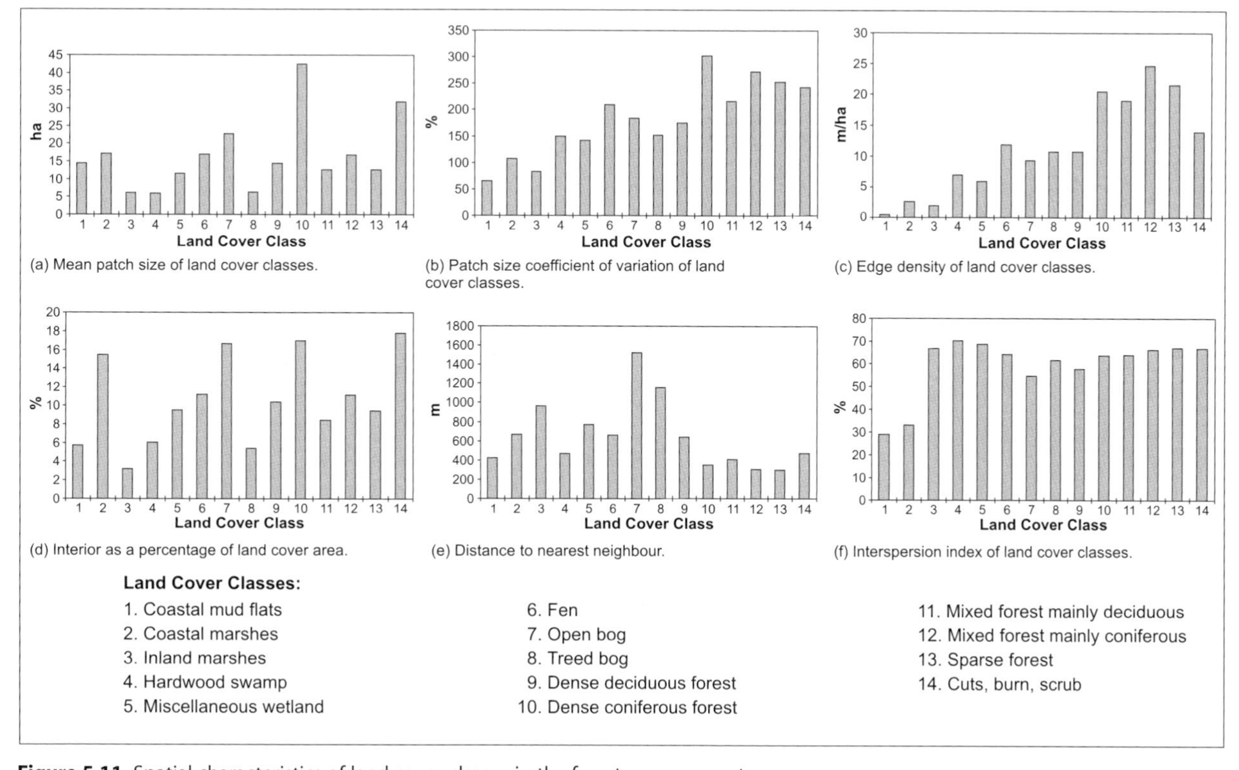

Figure 5.11 Spatial characteristics of land cover classes in the forest management zone.

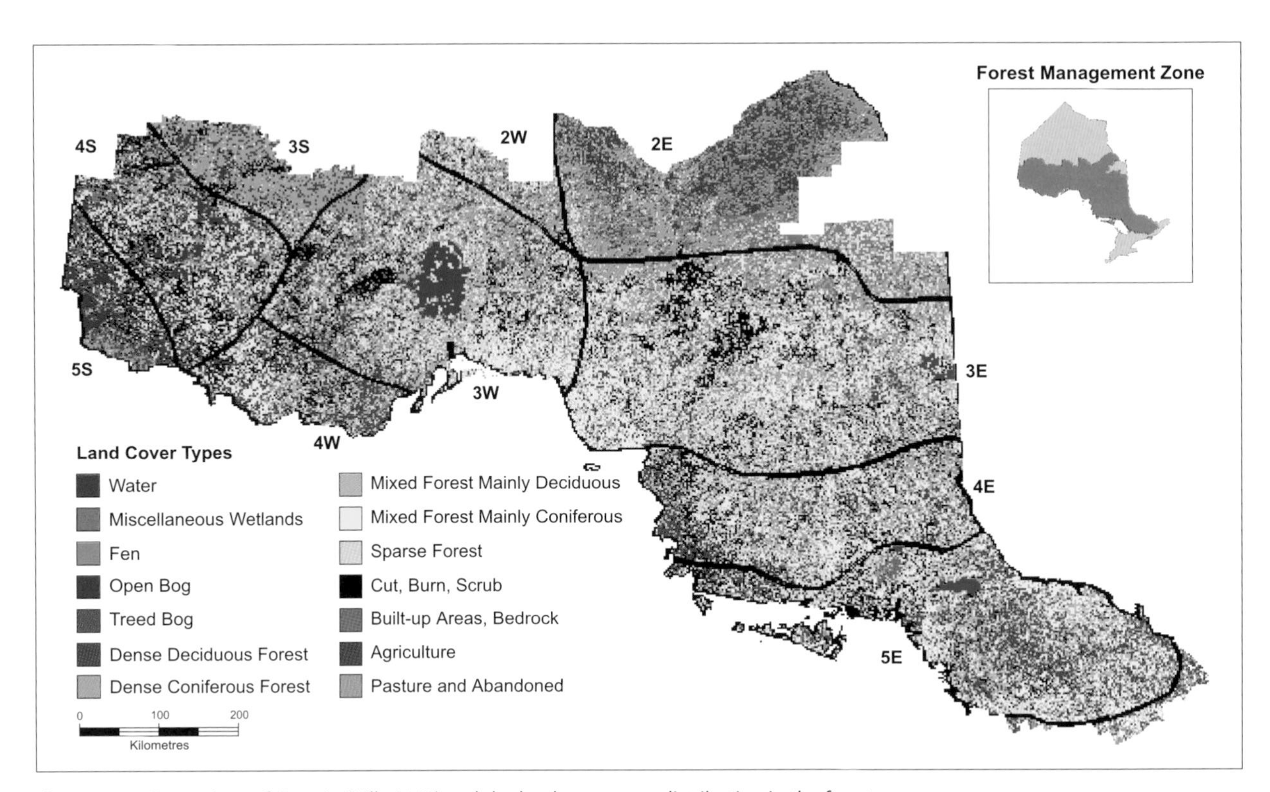

Figure 5.12 Ecoregions of Ontario (Hills 1959) and the land cover type distribution in the forest management zone.

Table 5.2

Dominant land cover types in the Hills ecoregions (1959) within the forest management zone in Ontario.

	% Area by Class										
Eco-region	*DDF*	*DCF*	*MDF*	*MCF*	*SF*	*CB*	*MWL*	*FN*	*OB*	*TB*	*BU*
2E	27.72						9.76	25.08	13.79	10.12	
2W		46.59	8.52	6.19	22.18			6.35			
3E		19.25	25.57	13.29	17.68	11.91					
3S		67.14		7.48	4.10	10.14				2.91	
3W		24.56	18.84	19.43	16.69	9.81					
4E		13.68	23.22	18.79	15.31	12.63					
4S	8.25	19.84	17.32		16.75	24.73					
4W	22.73		21.15	20.30	8.63	13.19					
5E	15.36		22.62	18.91	15.22	9.70					
5S	27.22		8.79		11.22	19.68					10.98

Note: DDF=dense deciduous forest; DCF=dense conifer forest; MDF=mixed deciduous forest; MCF=mixed coniferous forest; SF=sparse forest; CB=cuts,burns, scrub; MWL=miscellaneous wetlands; FN=fen; OB=open bog; TB=treed bog; BU=built-up areas, bedrock.

patterns within the forest cover itself in the forest management zone. The forest cover classes used in this description are dense deciduous forest, dense conifer forest, mixed conifer forest, mixed deciduous forest, and sparse forest.

Patchiness

As stated above, a patchy landscape is one composed of patches that are relatively small and relatively distant from others of their type. In this respect, the most notable aspects in the forest cover of the forest management zone are the larger patch size and low degree of isolation of the dense conifer type in the northern ecoregions, 2E, 2W, and especially 3S (Figures 5.13a and 5.13b). The larger and less isolated patches of dense deciduous forest occur not in the north but in the southern ecoregions, 4E, 4W, 5E, and 5S. With respect to the mixed conifer and mixed deciduous forest classes, there are no major differences in either patch size or isolation among the ecoregions, although the degree of isolation is somewhat higher in the mixed deciduous patches of ecoregions 2E, 4S, and 5S, and in the mixed conifer patches of ecoregion 3S. The patches of recent disturbance from harvest or fire are large in all ecoregions and isolated in all ecoregions except 5E, 4E, and 5S.

Spatial Complexity

The shape index and the degree of interspersion (McGarigal and Marks 1995) indicate the spatial complexity. When the shape index is high, patches are considered complex. When the percentage of interspersion, or juxtaposition, is high, the landscape itself is more complex.

Patch complexity in the dense conifer forest cover is extremely high in the northern ecoregions 2E, 2W, and 3S (Figure 5.13c). The opposite is observed for dense deciduous forest patches. The land cover patches of recent disturbances (that is, those less than 20 years old) from either fire or forest harvest are more complex in shape than the patches of any other land cover type. The probability that one forest cover type will be intermixed with any other forest type is similar for all forest cover classes in all ecoregions (Figure 5.13d).

Interior and Edge

The proportion of patch area left after a 120-m-wide buffer along the patch perimeter is discounted measures the patch interior, while the amount of edge is based on the proportion of the edge, or boundary, of the patch that is "hard" edge (Perera and Baldwin, in press). Hard edge indicates the proportion of a patch perimeter that borders anthropogenic, non-forest cover types.

Owing to its large mean patch size, dense conifer forest has more interior in the northern ecoregions than elsewhere, and more than the dense deciduous forest in the southern ecoregions. Overall, the percentage of interior in individual forest patches ranges from 5 to 15 percent (Figure 5.13e). In all ecoregions, dense deciduous forest patches have more hard edge (that is, more patch perimeter bordering on agriculture and settlement) than any other forest cover type (Figure 5.13f). Of course, all forest patches in those ecoregions containing a higher degree of settlement (that is, ecoregions 5S, 5E, and 4W) have more hard edge.

Spatial Patterns of the Forest Age Classes

Based on the history of forest fire and forest harvest from 1951 to 1995 (Figures 5.8 and 5.9), we categorized the forest cover of the forest management zone into three broad age classes: forest less than 15 years old (that is, forest disturbed at some time between 1981 and 1995), forest between 15 and 45 years old (that is, forest disturbed at some time between 1951 and 1980), and forest more than 45 years old (that is, forest which was undisturbed in 1951 and remained so in 1995). We do not imply that the third age class was never disturbed or that it resulted from only natural influences and not human influence. The age classes are general categories; a forest patch identified as belonging to a given age class may include a wide range of age conditions and cover-class spatial patterns.

The oldest class of forest (more than 45 years of age) overwhelmingly dominates the forest cover in all ecoregions except 4S, where the two younger age classes (15 to 45 years of age and under 15 years of age) represent more than 40 percent of the forest cover (Figure 5.14a). The relative amount of the youngest and intermediate classes is similar among all ecoregions except 2W, which contains a higher proportion of the youngest class. Generally, mean forest patch size (Figure 5.14b) is smallest in the youngest

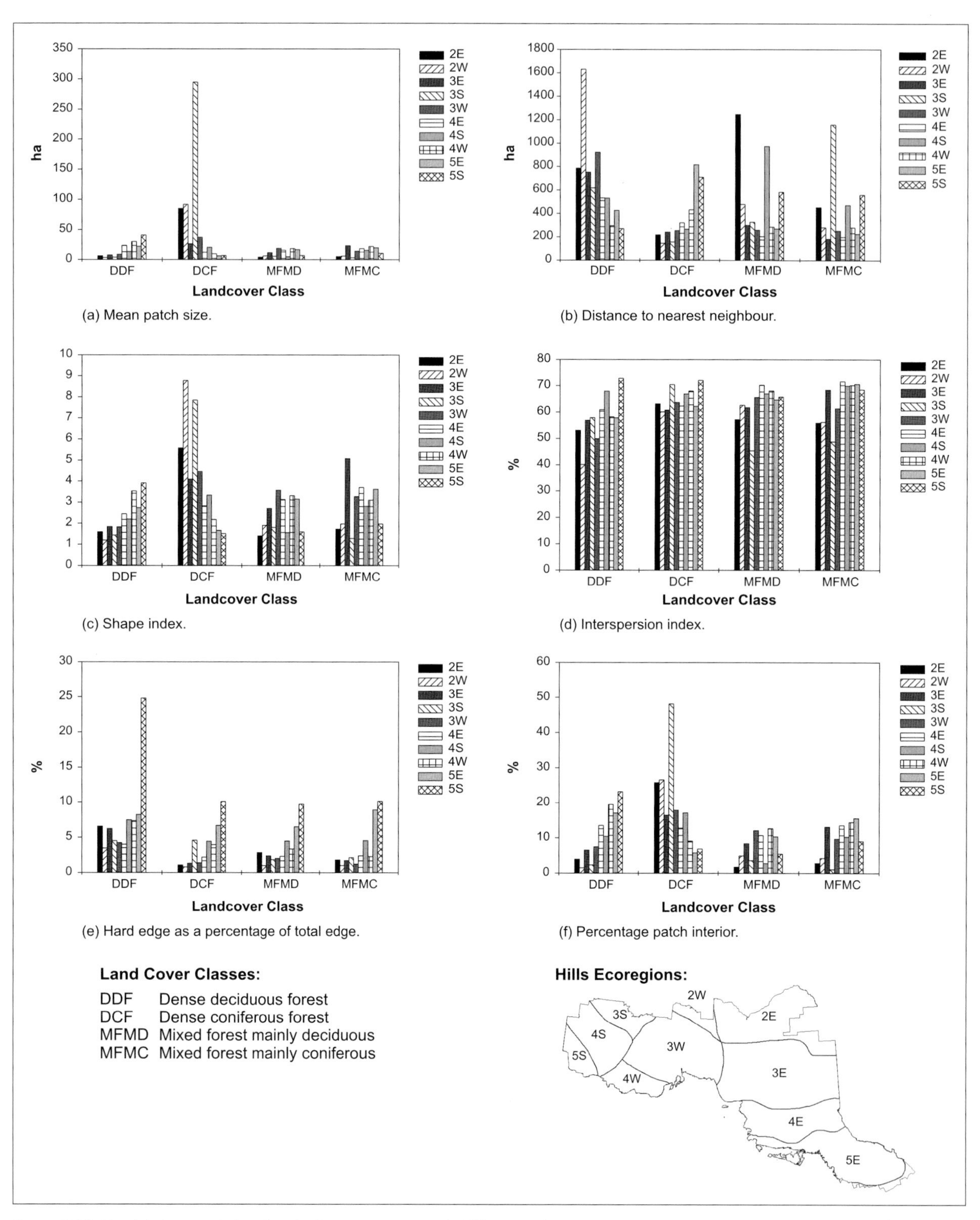

Figure 5.13 Spatial characteristics of major forest cover types in Hills ecoregions.

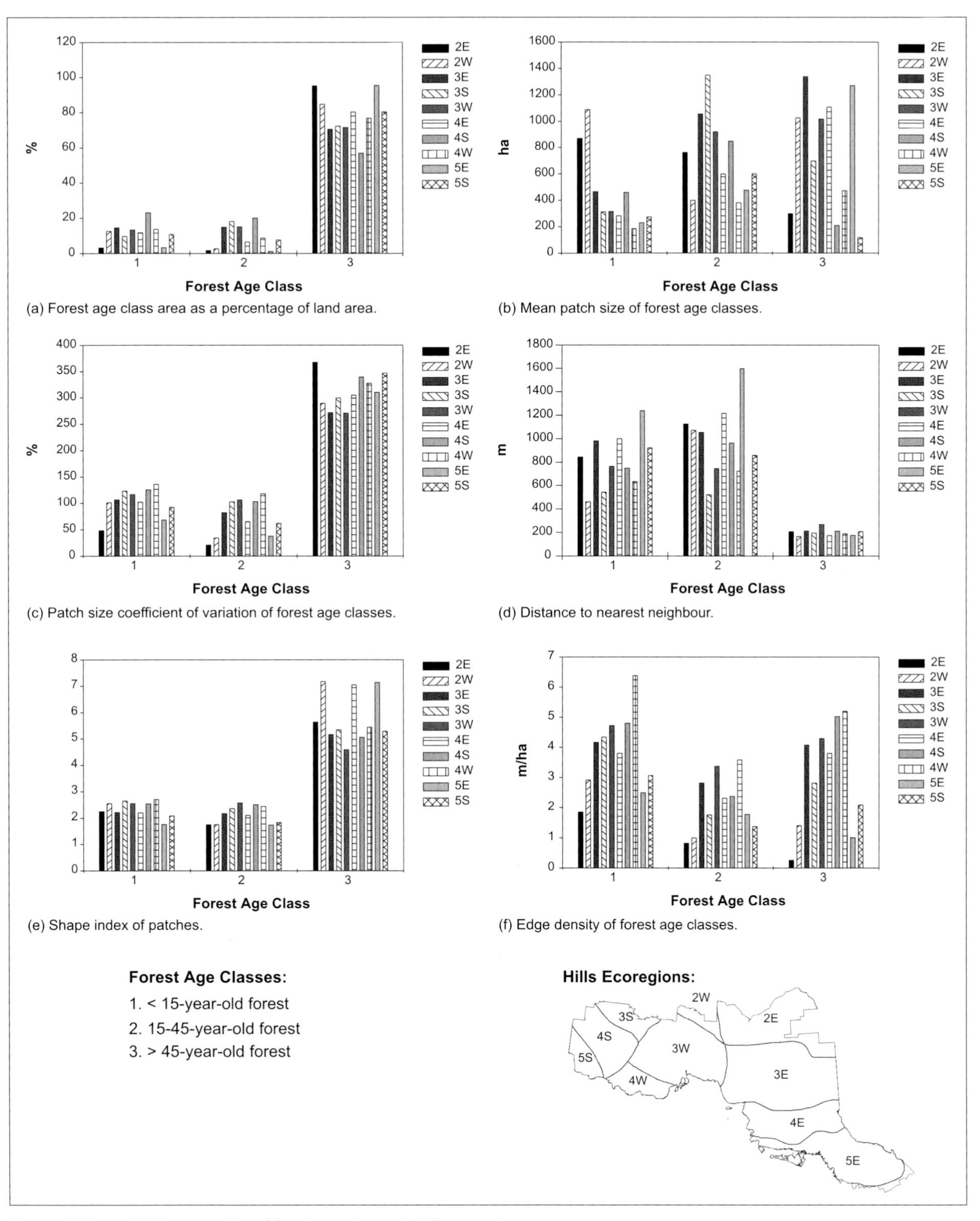

Figure 5.14 Spatial characteristics of forest age classes in Hills ecoregions.

forest age class, but in the two northern ecoregions, 2E and 2W, the youngest age class contains the largest mean forest patch size. The two classes younger than 45 years of age have a considerably lower patch-size variation (Figure 5.14c) than the older forest, but the patches are more isolated than in the older forest (Figure 5.14d). With respect to shape complexity, the patches of the youngest age class have simpler shapes (Figure 5.14e) than patches of the other two age classes. The amount of edge associated with the youngest age class is considerably higher than that of the two older classes in all ecoregions (Figure 5.14f); this is most notable in ecoregion 4W. Comparing Figures 5.15a and 5.15b reveals areas where new edge has been created within the most recent 15-year period. Only ecoregions 2E and 2W appear altogether free of new edge, but some pockets of ecoregions 4W, 3E, and 5E have relatively little.

The forest cover in the oldest forest category is dominated by the mixed conifer type in ecoregions 3E and 4E, and by the dense conifer type in ecoregions 2E, 2W, 3S, and 3W. The dominance of dense deciduous forest in this age class is limited to small areas on the southern parts of the forest management zone, in ecoregions 4W, 5E, and 5S.

Spatial Patterns of Forest Harvest and Forest Fire

The disturbance history depicted in Figures 5.8 and 5.9 is limited to disturbances having an area of 200 ha or more. At this coarse scale, many minor disturbances may have been excluded, and small, non-disturbed areas may have been included. The clear-cut areas, furthermore, include forest roads, landings, or other parts of the area of harvest operations where the forest cover has been removed. While these areas may not be covered by the silvicultural definition of forest harvest, they are included in the area where forest cover has been disturbed during the process of harvest.

The area disturbed by forest fire and forest harvest by clear-cutting represents 20 percent of the area of the forest management zone. The proportion of the land area disturbed by forest fire during the period from 1951 to 1995, as compiled by 100-km^2 cells, is illustrated in Figure 5.16a. The same information is presented for forest harvest in Figure 5.16b. It appears that the forest fires are spatially more clustered, while the forest harvest areas are dispersed across the forest management zone.

The southeastern and the northeastern parts of the forest management zone, ecoregions 5E, 2E, and 2W, were not generally subject to a high degree of disturbance from either fire or forest harvest. Ecoregion 5E had almost no disturbance by fire (less than 0.25 percent of the area). Similarly, harvesting represents only 0.4 percent of ecoregion 2E and only 1.9 percent of ecoregion 2W. The most disturbed ecoregion is 4S, where 15 percent of the land was burned and a further 25 percent harvested over the 45-year period. More of ecoregion 3S was burned (17 percent) than of any other ecoregion. The ecoregions most heavily harvested since 1951 include 3E (24 percent), 3W (20 percent), 4W (18 percent), and 4E (17 percent). Two of the four largest parks in Ontario, Quetico and Pukaskwa, were free of forest harvest over the 45-year period (Figure 5.16a). On the other hand, the other two largest parks, Lake Superior and Algonquin, were relatively free of wildfire during the same period (Figure 5.16b).

The distribution of disturbances from fire and forest harvest is not the same for all landform types. Within broad classes of surficial geology, which are outlined by Baldwin et al. (2000, this volume), we examined the cumulative area disturbed by forest fires and clear-cuts over the period from 1951 to 1995 (Figure 5.17). We calculated the relative occurrence of fire and harvest in each surficial geology class, after weighting the values shown in Figure 5.17 with the difference in the total area of disturbance from each source (2 million ha burned and 6.6 million ha harvested) and the relative area coverage of each surficial geology class. The most abundant surficial geology class, ground moraine, showed an equal tendency to be burned or harvested over the 45-year period. End moraines, outwash plains, lacustrine deposits, and beach and aeolian deposits were harvested twice as much as they were burned. Conversely, bedrock outcrops were burned seven times more than they were harvested, and organic deposits were burned 50 times more than they were harvested.

Burns and clear-cuts create edges in forest patches, along which non-disturbed areas are juxtaposed against disturbed areas. In Ontario, the issue of edge and its positive effects on certain wildlife species became the focus of attention in the 1970s, with the result that forest management policies were changed to create more edge for species such as moose (Euler 1981). The issue of edge is an ongoing scientific and policy debate about the long-term effects of edge on

Figure 5.15a Amount of edge (m) in the <15-year age class in the forest management zone, calculated on a 10-km by 10-km grid.

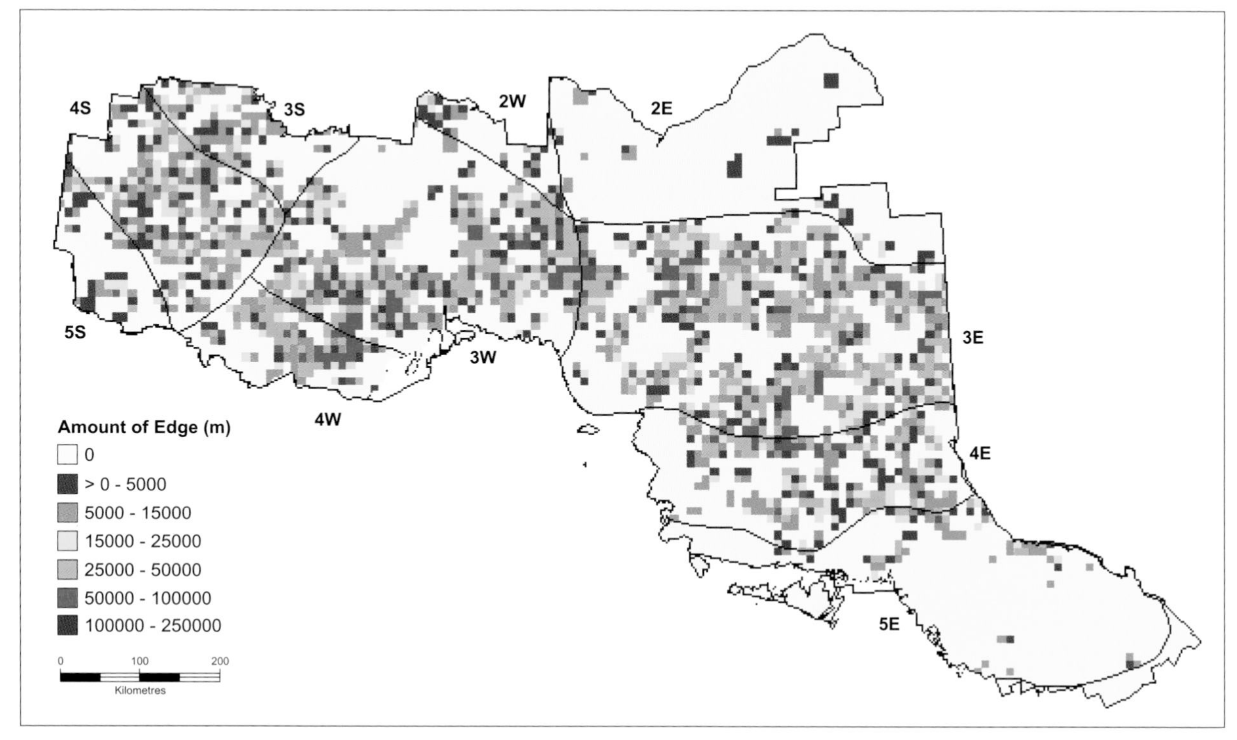

Figure 5.15b Amount of edge (m) in the 15- to 45-year age class in the forest management zone, calculated on a 10-km by 10-km grid.

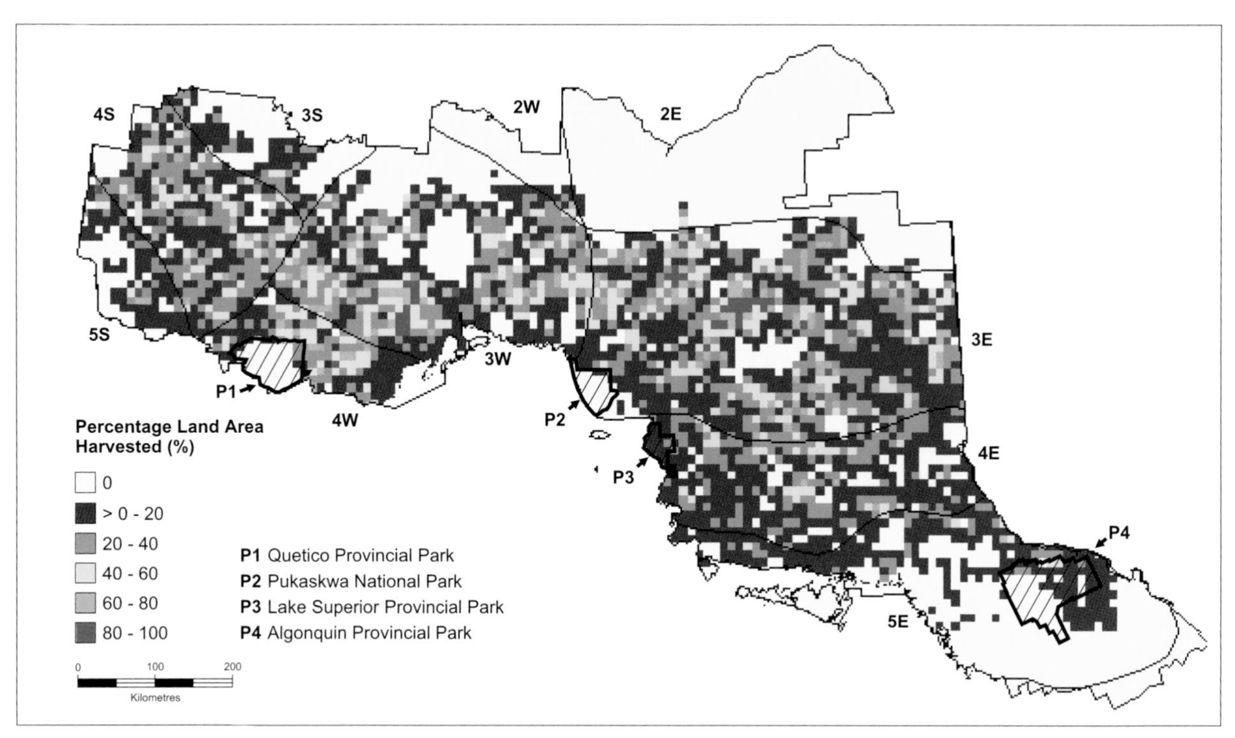

Figure 5.16a Percentage of land area harvested in the forest management zone from 1951 to 1995, calculated on a 10-km by 10-km grid.

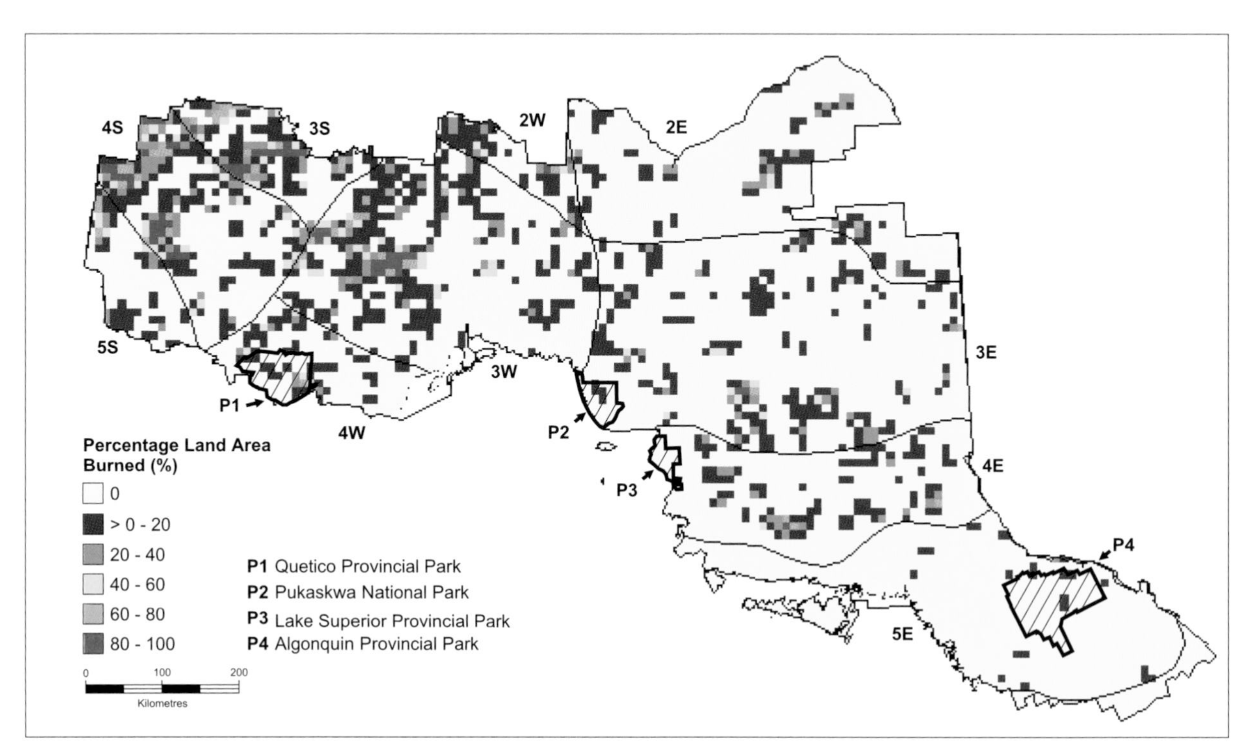

Figure 5.16b Percentage of land area burned in the forest management zone from 1951 to 1995, calculated on a 10-km by 10-km grid.

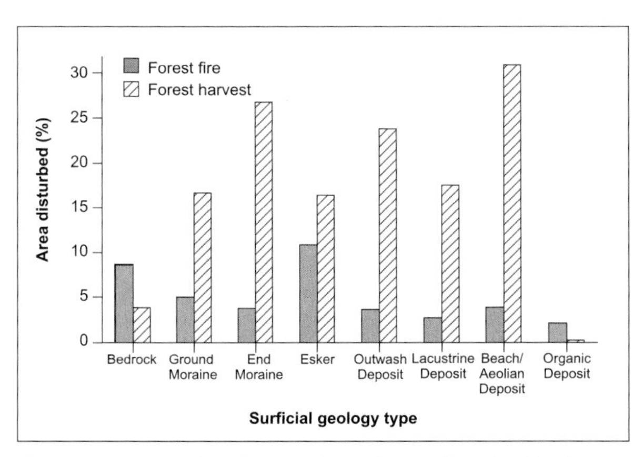

Figure 5.17 Area of surficial geology classes disturbed by forest fire and forest harvest in the forest management zone from 1951 to 1995.

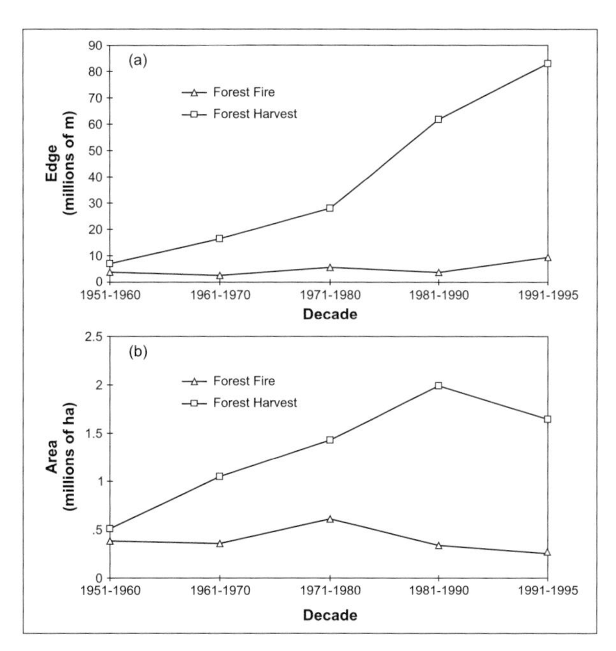

Figure 5.18 Comparison of forest fire and forest harvest in the forest management zone from 1951 to 1995 for (a) amount of edge created and (b) total area of disturbance.

biodiversity (Rempel et al. 1997). As illustrated in Figure 5.18a, the total amount of edge created by forest harvest (termed "cut edge") has increased several-fold over the last 25 years, in comparison to the amount of edge created by forest fire (termed "fire edge"). The same figure shows that total cut edge has sharply increased over fire edge since the 1980s. There are two reasons for this increase. The more obvious lies in a progressive increase in harvested area, compared to the area of forest fires (Figure 5.18b). From 1951 to 1995, the total area harvested by clear-cutting in each decade increased from 0.5 million ha (1951 to 1960) to almost 2.0 million ha (1981 to 1990). During only five years, from 1991 to 1995, the area of total harvest exceeded 1.5 million ha. In comparison, the total area of forest fires over 200 ha in size fluctuated somewhat, but did not greatly exceed 0.5 million ha per decade.

The second reason for the disproportionate increase in cut edge is the trend toward smaller clear-cuts. From the 1980s to 1995, there was a definite trend in forest harvest towards smaller clear-cut blocks. This trend is evident from increases in both the proportion of area harvested in cuts smaller than 500 ha (Figure 5.19a) and the proportion of the total number of clear-cuts represented by the smaller cut blocks (Figure 5.19b). For example, during the period from 1991 to 1995, clear-cuts smaller than 500 ha accounted for more than 50 percent of the total number of cut blocks, compared to less than 20 percent during the period from 1951 to 1960. The proportion of the forest area harvested in clear-cuts larger than 5000 ha decreased from over 60 percent in the decade from 1951 to 1960, to less than 20 percent from 1991 to 1995. From 1951 to 1995, however, there were no significant changes in the size-class distribution of forest fires, in terms of either the total fire extent (Figure 5.19c) or the number of forest fires (Figure 5.19d), beyond a very slight increase in smaller fires. Not only do the smaller clear-cuts have very low ratios of patch area to perimeter, and yield more edge as a result, but they also give rise to other small clearings for landings and logging roads, which create a substantial amount of edge themselves.

Figure 5.20 illustrates the spatial distribution of edge created by forest fires and clear-cuts over the 45-year period from 1951 to 1995. There is wide variation among the ecoregions in the total amount of edge created and in its source (Figure 5.20a). In some ecoregions, for example, 2W and 2E, forest fires accounted for over 65 percent of the edge. In other ecoregions, the proportion of edge created by forest harvest is much higher: for example, 68 percent in ecoregion 3S, 84 percent in 4S, 88 percent in 3W, 91 percent in 4W, 95 percent in 3E and 4W, and 97 percent in 5E.

Quetico Park in ecoregion 4W and Pukaskwa Park in ecoregion 3E are almost free of cut edge, in contrast with Lake Superior Park in ecoregion 3E and Algonquin Park in ecoregion 5E (Figure 5.20b).

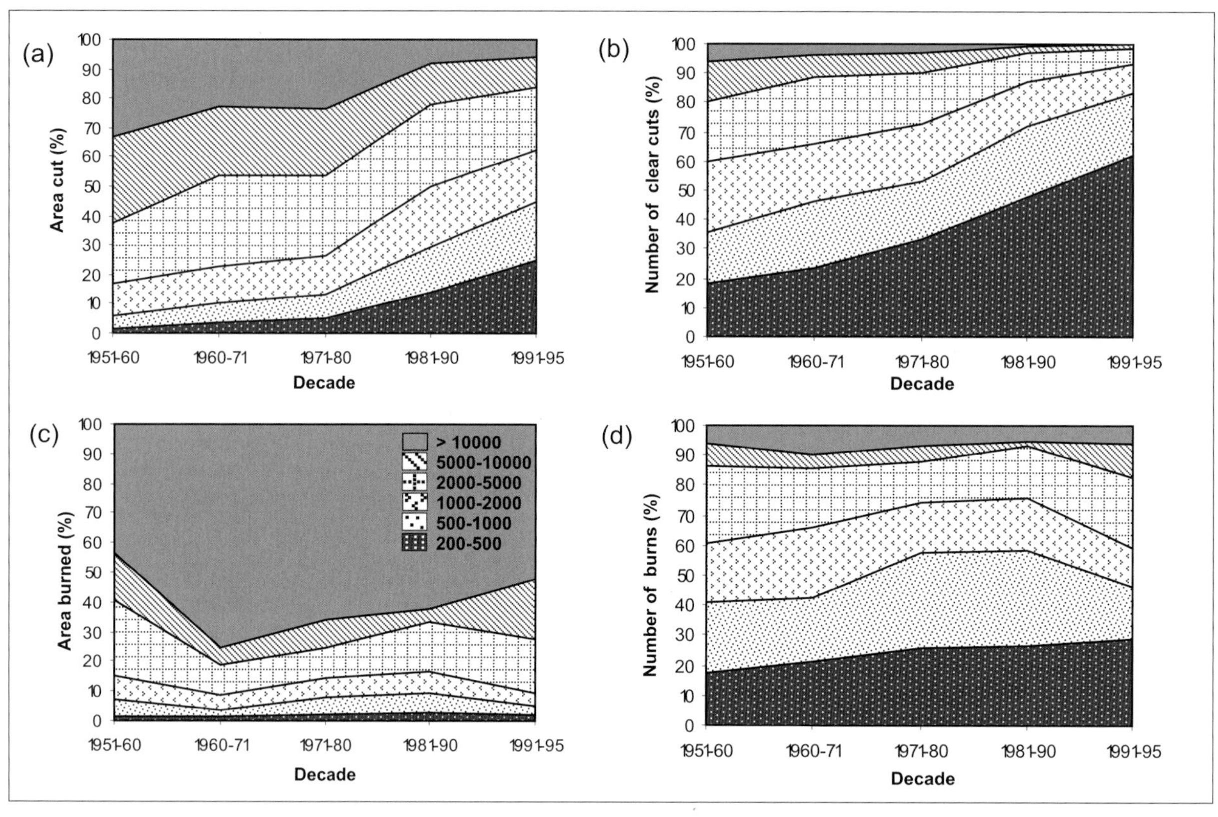

Figure 5.19 Area and number of clear-cuts and fires from 1951 to 1995 by decade and size category: (a) cut area; (b) number of cuts; (c) fire area; and (d) number of fires. All values are the percentage of the total number or area for each decade and size category.

ECOLOGICAL IMPLICATIONS AND FUTURE DIRECTIONS

In the preceding discussion, we have assessed the spatial characteristics of Ontario's forested lands, both by general composition type and broad age class, within main physiographic zones, ecoregions, and units of underlying surficial materials. We have concluded by reporting clear trends in the spatial patterns formed by forest fires and clear-cut harvesting. In the following section, we cautiously interpret our findings, place them in the perspective of the knowledge of Ontario's forest landscape ecology that is required, and point the way to important areas of future investigation.

Interpreting Ontario's Land Cover Patterns

The forests in Ontario's forest management zone are not fragmented, but comprise almost continuous cover; therefore, the classical concepts of landscape ecology that are popularly expounded (e.g., Forman and Godron 1987) are not directly relevant. Concepts such as fragmentation, isolation, corridors, and interior, for example, were formulated specifically for heavily settled or farmed landscapes. As a result, it is not possible to extrapolate the vast body of knowledge in landscape ecology directly to the managed forests of Ontario. The same is true of studies in forested landscapes of the Pacific Northwest (e.g., Franklin and Forman 1987; Ripple et al. 1991; Spies et al. 1994; Tang et al. 1997; Eng 1998), the northern lake states (e.g., Mladenoff et al. 1993), or Fennoscandia (e.g., Haila et al. 1994). These results cannot be directly extrapolated to Ontario's managed forests because of differences in socio-economic and ecological context. The most apparent difference is that of scale; Ontario's forested landscape is very large and continuous, and undergoes very large-scale disturbance, compared to either the Pacific northwest or Fennoscandia.

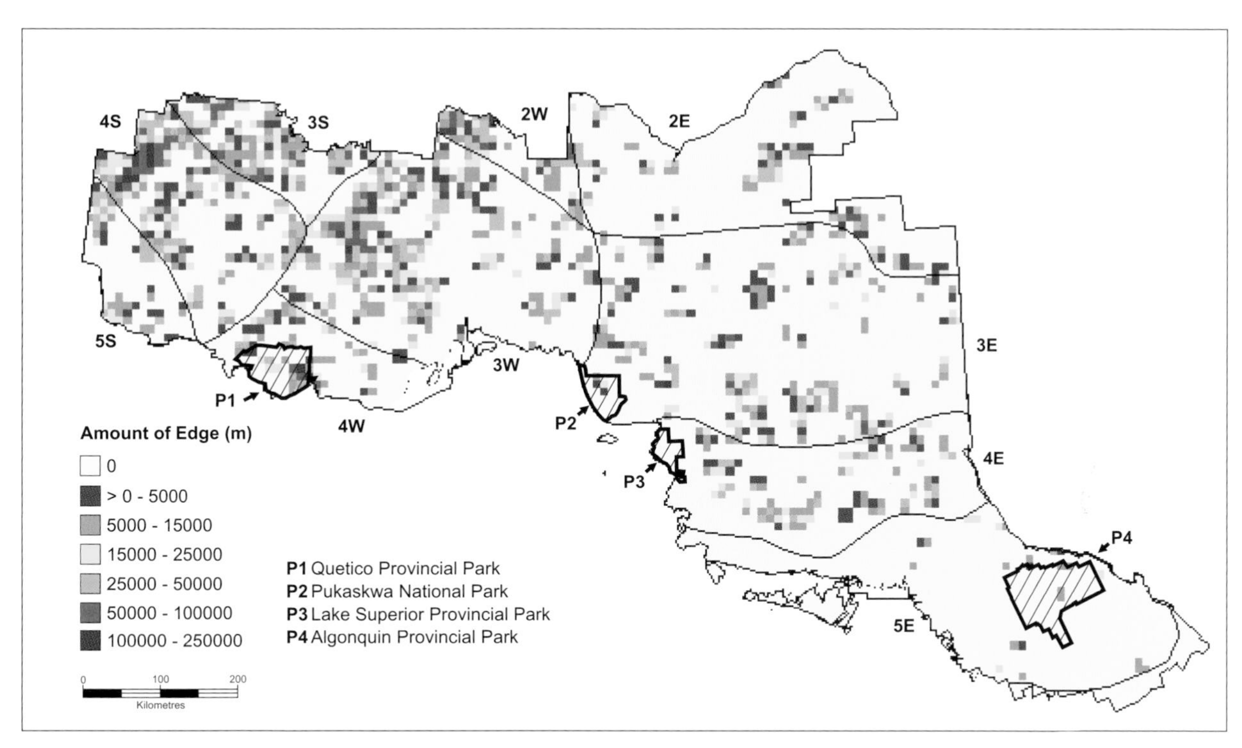

Figure 5.20a Amount of edge created by forest fires in the forest management zone from 1951 to 1995, calculated on a 10-km by 10-km grid.

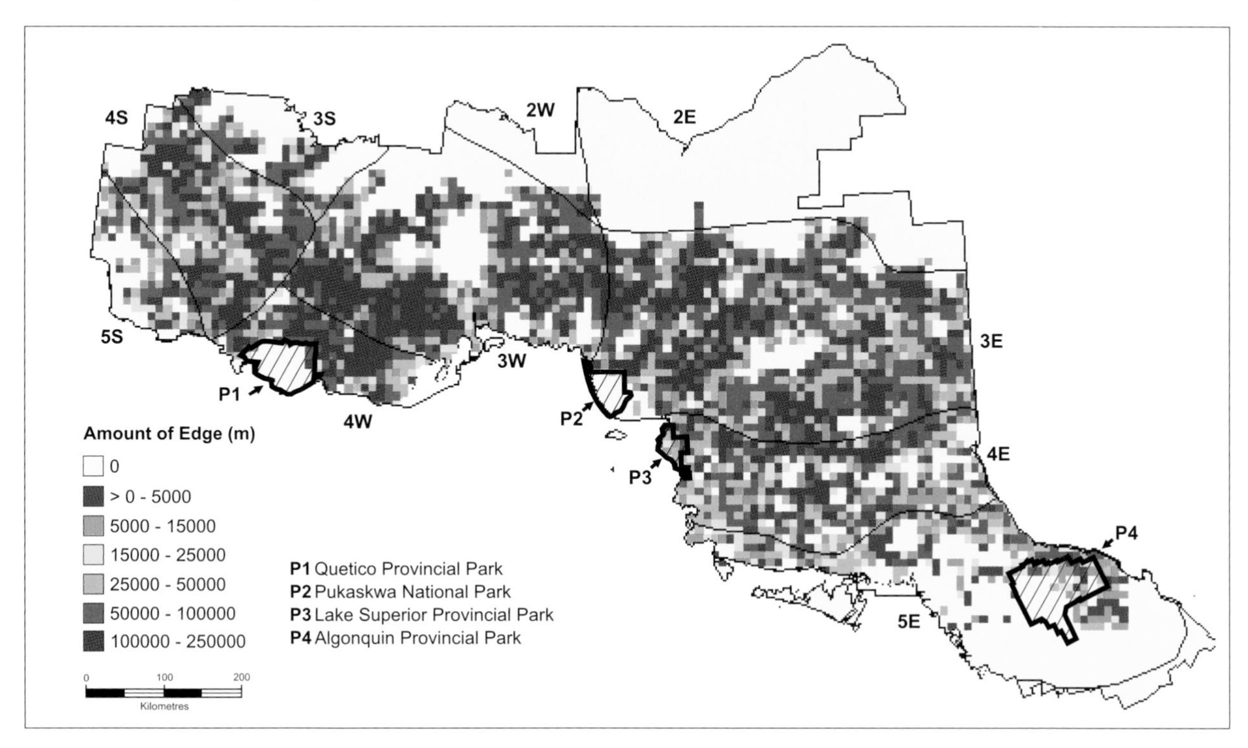

Figure 5.20b Amount of edge created by forest harvest in the forest management zone from 1951 to 1995, calculated on a 10-km by 10-km grid.

Lacking direct recourse to experience from other jurisdictions, and recognizing that efforts to interpret the ecological meaning of forest spatial patterns in Ontario have barely begun, we can only suggest some possible implications of the patterns we have described. The trend toward progressively simpler forest patch shapes, smaller patches, and larger amounts of cut edge may have ecological implications for habitat structure, at least temporarily. One direct consequence is to increase the habitat of species that favour edge and open-canopy conditions. Perhaps a parallel can be drawn with the Oregon Cascades, where Ripple et al. (1991) found that increased forest edge was advantageous to some ungulates, but adverse to interior-dependent species such as the spotted owl and pileated woodpecker. The magnitude and significance of the structural changes depend on the degree to which the organisms adapt to the changes, and the degree to which the changes are "natural," or correspond to what would occur without human intervention. In any case, the relationship between structure and species distribution may be more complex. Rempel et al. (1997) showed that species response to changed habitat structure at a regional scale may be confounded by many other factors, such as hunting. We refer the reader to Voigt et al. (2000, this volume) and Carleton (2000, this volume) for extensive discussions of habitat and population changes in the context of forest management, from a faunistic and floristic perspective, respectively.

The inevitable increase in road density and accessibility with increased forest harvest may have direct ecological implications, in the sense of changes from "nature," because the linear corridors of roads are definitely not natural, especially if they are not temporary. The effects of roads range from the obvious interruption of previously undisturbed forest patches (e.g., Miller et al. 1996), the alteration of faunistic habitat (e.g., Thiel 1985), and the creation of corridors for potential invasion by exotic species. In addition, roads and associated harvest areas decrease the degree of "naturalness," in terms of wilderness and ecotourism values (e.g., Boyd et al. 1994).

The implications of the proportion of the forest management area represented by forest less than 45 years old cannot be determined in the absence of knowledge of what constitutes a "natural" disturbance regime. Likewise, the preponderance of young forest in some ecoregions (for example, 45 percent in ecoregion 4S) cannot be judged "unnatural." Whether it corresponds to or differs from "natural" conditions, however, the abundance of young forest in some regions does have a direct bearing on the sustainability of the regional timber supply.

We ask the reader to keep the scope of these inferences in mind. The results we have discussed here were derived from one data set at one point in time. They were also based on large scales of observation and analysis, relatively low-resolution source data, and a specific ecoregional framework. Using a different unit of analysis (for example, a 10,000-km^2 cell instead of 100-km^2 cell) or a different scale of source data (aerial photographs, for example, instead of Landsat TM images) may yield different results and inferences. Our inability to predict ecological processes from our present knowledge of Ontario's forest cover patterns, and our seeming hesitation to extrapolate, does not derive from any lack of confidence in the results described here. Instead, it is due to the lack of spatially explicit null hypotheses and models of ecological patterns and processes in Ontario's forests.

Future Directions for Landscape Ecological Studies in Ontario

The monitoring of spatial land cover patterns, wildlife populations, primary productivity, and many other forest variables, has been made mandatory in Ontario by the *Forest Management Planning Manual for Ontario's Crown Forests* (OMNR 1996). This work will be supported by the increasing availability of large-scale land cover data and the development of analytical techniques.

To derive the fullest benefit from this spatial information, the focus of Ontario landscape ecological study needs to shift. First, it may be necessary to go beyond concepts and hypotheses adopted from classical landscape ecology (Forman 1995b). The structure and function of a patch-matrix landscape may be inadequate or even inappropriate for identifying ecological effects and implications in a continuous forest landscape such as that of Ontario. Contrary to the classical scenarios of island biogeography and habitat fragmentation, in managed forest landscapes, both the formation of "islands" and their impact are likely to be temporary. This may be especially true in disturbance-based boreal forest landscapes, where the return of non-forest islands to forested land creates an ever-shifting mosaic of disturbed patches.

The use of neutral models (Caswell 1976) has already been explored in landscape ecology (e.g., Gardner et al. 1987; Pearson and Gardner 1997), and these may provide a more robust and appropriate approach to investigating continuous forest landscapes. This null-model approach has been recognized as a very powerful means of understanding "nature" in community ecology (e.g., Gotelli and Graves 1996). The same principles can be extended to an understanding of landscape processes, in the absence of any other influences.

One set of neutral models, or null hypotheses, for Ontario would define intrinsic spatial patterns that exist as a function of underlying geoclimatic variables, patterns which would permit the prediction of "natural" vegetation cover, for example. Null hypotheses are essential to the assessment of deviations originating from any source, and to the formation of any judgment on management-derived spatial patterns. For example, if spatial patterns such as juxtaposition, contiguity, or even spatial correlation should change, the change could not be detected without having spatially explicit knowledge of the underlying geoclimatic and autogenic causes of the land cover patterns.

Another essential set of null hypotheses concerns "natural disturbances." In a landscape system that is shaped by disturbances, such as the boreal forest, the effects of management measures can only be assessed in the context of large-scale disturbances from fire and insect infestation. For example, inferences from the forest composition mosaic, from the distribution of age classes, from measurements of edge and interior, or from the temporality of post-disturbance effects, cannot be conclusive unless they are made in the context of a natural disturbance model. One of the most important and immediate research requirements in the management of forest landscapes for objectives of sustainability is, therefore, to develop methods of spatially predicting the patterns of natural vegetation and natural disturbance regimes at large scales (Perera 1998).

The final set of null hypotheses required concerns the causal effects of land cover patterns on ecological processes. The majority of studies on land cover processes, which were conducted in the context of heavily settled or farmed landscapes (Forman 1995b), may be useful in deriving null hypotheses for certain ecological processes; for example, for the influence of habitat extent or quality on the response of a specific wildlife species. One shortcut to understanding the effects of management measures on ecological processes, in a relative sense, would be carefully planned adaptive management practices and studies, as defined by Walters (1986).

SUMMARY

Almost all of Ontario's forest cover lies within the forest management zone. This managed forest cover is not patchy or fragmented, in the sense that a heavily farmed or settled landscape is fragmented. The forest patches are generally small and have complex shapes, so they also have a lower amount of interior. The patches are not isolated from each other, however, and border mostly on other forest cover types rather than non-forest land cover.

The Hills ecoregions of Ontario do not differ greatly in land cover composition and diversity, but differences do exist among regions in the dominant land cover type and the spatial pattern of land cover patches. One distinct difference among the ecoregions consists of the relative proportion of the area covered by young forest. Nearly 20 percent of the forest cover of the forest management zone is less than 45 years old, although values range among the ecoregions from 5 percent to 45 percent. The main source of disturbance in this young forest is harvest, although regional differences again occur. The total area harvested has increased, doubling every decade since the 1950s, while the area burned, under a regime of fire suppression, has remained largely the same. In general, harvest activities are broadly dispersed across the managed forest, while forest fires are relatively concentrated. Under a heavy fire management regime, forest fires have been restricted almost exclusively to northwestern Ontario, while forest harvest is evenly distributed throughout the forest management zone.

It is clear that forest harvest has progressively created a vast amount of edge, in comparison to forest fires, and has done so at a rate disproportionately higher than the ratio of total burned area to total harvest area. This difference is caused primarily by the shift during the last two decades away from the harvest pattern of larger and fewer clear-cut blocks and toward the harvest pattern of many small clear-cut blocks, interconnected by roads and other clearings. Almost all of this edge is likely to be temporary, however, because it lies between different forest types rather than against non-forest types created by human activity.

There are several reasons for the lack of specific conclusions in this chapter with respect to the ecological consequences of very widespread forest harvest disturbances. First and foremost, knowledge of ecological cause and effect for Ontario, at large scales and based on sound experimentation and monitoring, is still very scarce. Our understanding of natural processes at large scales is still in its infancy. Both generalizing from samples in Ontario and extrapolating from findings in other jurisdictions are a challenge, given the uncommonly large scale of Ontario's forest management. In the short term, particular care and restraint must be exercised in interpreting the forest land cover patterns of the province, except in the form of well-constructed and testable hypotheses. The long-term solution we propose is a concerted effort to develop null models for landscape patterns and processes, as a means of understanding the present situation, identifying the trajectory of change, and setting goals for land cover management.

REFERENCES

Baldwin D.J.B., L.E. Band, and A.H. Perera. In press. A Quantitative Basis for Using Ontario's Existing Eco-regionalization Systems. Forest Research Report No. 151. Sault Ste. Marie, Ontario: Ontario Forest Research Institute, Ontario Ministry of Natural Resources.

Baldwin, D.J.B., J.R. Desloges, and L.E. Band. 2000. Physical geography of Ontario. In: A.H. Perera, D.L. Euler, and I.D. Thompson (editors). Ecology of a Managed Terrestrial Landscape: Patterns and Processes in Forest Landscapes of Ontario. Vancouver, British Columbia: University of British Columbia Press. 12-29.

Band, L.E. 2000. Forest productivity in Ontario. In: A.H. Perera, D.L. Euler, and I.D. Thompson (editors). Ecology of a Managed Terrestrial Landscape: Patterns and Processes in Forest Landscapes of Ontario. Vancouver, British Columbia: University of British Columbia Press. 163-177.

Band, L.E., F. Csillag, A. H. Perera, and J.A. Baker. 1999. Deriving an Eco-regional Framework for Ontario through Large-Scale Estimates of Net Primary Productivity. Forest Research Report No. 149. Sault Ste. Marie, Ontario: Ontario Forest Research Institute, Ontario Ministry of Natural Resources. In press.

Boyd, S.W., R.W. Butler, W. Haider, and A.H. Perera. 1994. Identifying areas for ecotourism in northern Ontario: application of a geographical information system methodology. Journal of Applied Recreation Research 19(1): 41-66.

Canadian Forest Service. 1997. The state of Canada's forests 1996-1997. Publication Fo-6/1997E. Ottawa, Ontario: Canadian Forest Service. 123 p.

Carleton, T.J. 2000. Vegetation responses to the managed forest landscape of central and northern Ontario. In: A.H. Perera, D.L. Euler, and I.D. Thompson (editors). Ecology of a Managed Terrestrial Landscape: Patterns and Processes in Forest Landscapes of Ontario. Vancouver, British Columbia: University of British Columbia Press. 179-197.

Caswell, H. 1976. Community structure: a neutral model analysis. Ecological Monographs 46: 327-354.

Ecological Stratification Working Group. 1996. A National Ecological Framework for Canada. Report and map at 1:750 000 scale. Ottawa, Ontario: Agriculture and Agri-food Canada, Research Branch, Centre for Land and Biological Resources Research and Environment Canada; State of the Environment Directorate, Ecozone Analysis Branch. 125 p.

Eng, M. 1998. Spatial patterns in forested landscapes: implications for biology and forestry. In: J. Voller and S. Harrison (editors). Conservation Biology Principles for Forested Landscapes. Vancouver, British Columbia: University of British Columbia Press. 42-75.

Epp, A.E. 2000. Ontario forests and forest policy before the era of sustainable forestry. In: A.H. Perera, D.L. Euler, and I.D. Thompson (editors). Ecology of a Managed Terrestrial Landscape: Patterns and Processes in Forest Landscapes of Ontario. Vancouver, British Columbia: University of British Columbia Press. 237-275.

Euler, D. 1981. A moose habitat strategy for Ontario. *Alces* 17: 180-192.

Flannigan, M.D. and M.G. Weber. 2000. Influences of climate on Ontario forests. In: A.H. Perera, D.L. Euler, and I.D. Thompson (editors). Ecology of a Managed Terrestrial Landscape: Patterns and Processes in Forest Landscapes of Ontario. Vancouver, British Columbia: University of British Columbia Press. 103-114.

Fleming, R.A., A.A. Hopkin, and J.-N. Candau. 2000. Insect and disease disturbance regimes in Ontario's forests. In: A.H. Perera, D.L. Euler, and I.D. Thompson (editors). Ecology of a Managed Terrestrial Landscape: Patterns and Processes in Forest Landscapes of Ontario. Vancouver, British Columbia: University of British Columbia Press. 141-162.

Forest Landscape Ecology Program, Ontario Ministry of Natural Resources. 1998. Forest Landscape Ecology Program. 1998. Ontario's Forest Fire History: An Interactive Digital Atlas [CD-ROM]. Sault Ste. Marie, Ontario: Ontario Forest Research Institute, Ontario Ministry of Natural Resources.

Forman, R.T.T. 1995a. Land Mosaics: The Ecology of Landscapes and Regions. New York, New York: Cambridge University Press. 632 p.

Forman, R.T.T. 1995b. Some general principles of landscape and regional ecology. Landscape Ecology 10(3): 133-142.

Forman, R.T.T. and M. Godron. 1986. Landscape Ecology. New York, New York: John Wiley. 619 p.

Franklin J.F. and R.T.T. Forman. 1987. Creating landscape patterns by forest cutting: ecological consequences and principles. Landscape Ecology 1(1): 5-18.

Gardner, R.H, B.T. Milne, M.G. Turner, and R.V. O'Neil. 1987. Neutral models for the analysis of broad-scale landscape pattern. Landscape Ecology 1(1): 19-28.

Gluck, M.J. and R.S. Rempel. 1996. Structural characteristics of post-wildfire and clearcut landscapes. Environmental Monitoring and Assessment 39: 435-450.

Gotelli, N.J. and G.R. Graves. 1996. Null Models in Ecology. Washington, DC: Smithsonian Institution Press. 368 p.

Gustafson, E. J. 1998. Quantifying landscape spatial patterns: what is the state-of-art? Ecosystems 1(2): 143-156.

Haila, Y., I.K. Hanski, J. Niemela, P. Puttila, S. Raivio, and H. Tukia. 1994. Forestry and the boreal forest: Matching management with natural forest dynamics. *Annales Zoologici Fennici* 31: 187-202.

Hills, G. A. 1959. A Ready Reference to the Description of the Land of Ontario and its Productivity. Preliminary Report. Toronto, Ontario: Ontario Department of Lands and Forests. 142 p.

Li, C. 2000. Fire regimes and their simulation with reference to Ontario. In: A.H. Perera, D.L. Euler, and I.D. Thompson (editors). Ecology of a Managed Terrestrial Landscape: Patterns and Processes in Forest Landscapes of Ontario. Vancouver, British Columbia: University of British Columbia Press. 115-140.

Li, H. and J.F. Reynolds. 1993. A new contagion index to quantify spatial patterns of landscapes. Landscape Ecology 8(3): 155-162.

McGarigal, K. and B. Marks. 1995. FRAGSTATS: Spatial Pattern Analysis Program for Quantifying Landscape Structure. General Technical Report PNW-GTR-351. Portland, Oregon: USDA Forest Service. 122 p.

Miller, J.R., L.A. Joyce, R.L. Knight, and R.M. King. 1996. Forest roads and landscape structure in the southern Rocky Mountains. Landscape Ecology 11(2): 115-127.

Mladenoff, D.J., M.A. White, J. Pastor, and T.R. Crow. 1993. Comparing spatial pattern in unaltered old-growth and disturbed forest landscapes. Ecological Applications 3: 294-306.

O'Neill, R.V., J.R. Krummel, R.H. Gardner, G. Sugihara, B. Jackson, D.L. DeAngelis, B.T. Milne, M.G. Turner, B. Zygmunt, S.W. Christensen, V.H. Dale, and R.L. Graham. 1988. Indices of landscape pattern. Landscape Ecology 1: 153-162.

Ontario Ministry of Natural Resources. 1996. Forest Management Planning Manual for Ontario's Crown Forests. Toronto, Ontario: Queen's Printer for Ontario. 452 p.

Patton, D.R. 1975. A diversity index for quantifying habitat "edge." Wildlife Society Bulletin 3(4): 171-173.

Pearson, S.M. and R.H. Gardner. 1997. Neutral models: useful tools for understanding landscape patterns. In: J. A. Bissonette (editor). Wildlife and Landscape Ecology: Effects of Pattern and Scale. New York, New York: Springer-Verlag. 215-230.

Perera, A.H. 1998. An integrated spatial toolbox for forest landscape planning and management. In: Proceedings of GIS'98/RT'98 Conference, April 6-9, 1998, Toronto, Ontario. Fort Collins, Colorado: GIS World Inc. 43-46.

Perera, A.H. and D.J.B. Baldwin. 1993. Spatial Characteristics of Eastern White Pine and Red Pine Forests in Ontario. Forest Fragmentation and Biodiversity Project Report No. 9. Sault Ste. Marie, Ontario: Ontario Forest Research Institute, Ontario Ministry of Natural Resources. 82 p.

Perera, A.H. and D.J.B. Baldwin. In press. Spatial Patterns of Ontario's Forest Cover: A Multi-scale Assessment of Landscape Metrics. Forest Research Report No. 150. Sault Ste. Marie, Ontario: Ontario Forest Research Institute, Ontario Ministry of Natural Resources.

Perera, A.H., J. Baker, L.E. Band, and D.J.B. Baldwin. 1996. A strategic framework to eco-regionalize Ontario. Environmental Monitoring and Assessment 39: 85-96.

Perera, A.H., D.J.B. Baldwin, F. Schnekenburger, J.E. Osborne, and R.E. Bay. In press. Forest Fires in Ontario: A Spatio-temporal Perspective. Forest Research Report No. 147. Sault Ste. Marie, Ontario: Ontario Forest Research Institute, Ontario Ministry of Natural Resources.

Perera. A.H. and D.L. Euler. 2000. Landscape ecology in forest management: an introduction. In: A.H. Perera, D.L. Euler, and I.D. Thompson (editors). Ecology of a Managed Terrestrial Landscape: Patterns and Processes of Forest Landscapes in Ontario. Vancouver, British Columbia: University of British Columbia Press. 3-11.

Rempel, R.S., P.C. Elkie, A.R. Rodgers, and M.J. Gluck. 1997. Timber-management and natural-disturbance effects on moose habitat: landscape evaluation. Journal of Wildlife Management 61(2): 517-524.

Ripple, W.J., G.A. Bradshaw, and T.A. Spies. 1991. Measuring forest landscape patterns in the Cascade Range of Oregon, USA. Biological Conservation 57: 73-88.

Rowe, J.S. 1972. Forest Regions of Canada. Publication No. 1300. Ottawa, Ontario: Canadian Forest Service. 172 p.

Spectranalysis Inc. 1994. Development of a spatial forest data base for the eastern boreal forest region of Ontario. Forest Fragmentation and Biodiversity Project Report No. 14. Sault Ste. Marie, Ontario: Ontario Forest Research Institute, Ontario Ministry of Natural Resources. 22 p.

Spies, T.A., W.J. Ripple, and G.A. Bradshaw. 1994. Dynamics and pattern of a managed coniferous landscape in Oregon. Ecological Applications 4: 555-568.

Suffling, R. 1988. Catastrophic disturbance and landscape diversity: the implications of fire control and climate changes in subarctic forests. In: Proceedings of the First Symposium of the Canadian Society of Landscape Ecology and Management, November 17-20, 1986, University of Guelph, Guelph, Ontario. Ottawa, Ontario: Polyscience. 111-120.

Tang, S.M., J.F. Franklin, and D.R. Montgomery. 1997. Forest harvest patterns and landscape disturbances. Landscape Ecology 12(6): 349-363.

Thiel, R. 1985. Relationship between road densities and wolf habitat suitability in Wisconsin. American Midland Naturalist 113: 404-407.

Turner, M.G. 1989. Landscape ecology: The effect of pattern on process. Annual Review of Ecological Systematics 20: 171-197.

Turner, M.G. and R.H. Gardner (editors). 1991. Quantitative Methods in Landscape Ecology. New York, New York: Springer-Verlag. 536 p.

Voigt, D.R., J.A. Baker, R.S. Rempel, and I.D. Thompson. 2000. Forest vertebrate responses to landscape-level changes in Ontario. In: A.H. Perera, D.L. Euler, and I.D. Thompson (editors). Ecology of a Managed Terrestrial Landscape: Patterns and Processes in Forest Landscapes of Ontario. Vancouver, British Columbia: University of British Columbia Press. 198-233.

Walters, C.J. 1986. Adaptive Management of Renewable Resources. New York, New York: McGraw Hill. 374 p.

Wickware, G.M. and C.D.A. Rubec. 1989. Ecoregions of Ontario. Ecological Land Classification Series No. 26. Ottawa, Ontario: Environment Canada. 37 p.

Wiken, E.B. and G. Ironside. 1977. The development of ecological (biophysical) land classification in Canada. Landscape Planning 4: 273-275.

SECTION II:

Forest Landscape Ecological Processes

Section II comprises six chapters that examine how climate and biological processes, coupled with forest management, have affected development of Ontario's forest landscapes and the distribution patterns of the plants and vertebrates that live there.

• Chapter 6 discusses the expected climate changes that will have a profound effect on forests in Ontario. • Chapter 7 examines the dynamics of Ontario's forest fire regimes by comparing the effects of three levels of fire suppression, and discusses objectives and methods of fire simulation. These chapters are closely linked to the descriptions of current climate and historical fire disturbance patterns provided in Chapters 3 and 4, respectively.

• Chapter 8 describes the insect and disease disturbance regimes that have large-scale effects on Ontario forests and provides a detailed discussion of the dynamics of spruce budworm infestation. • Chapter 9 provides a provincial overview of quantitative primary productivity in Ontario's forest ecosystems and discusses methods and issues involved in obtaining these data.

The two final chapters address the response of Ontario's forest vegetation and wildlife to change from natural processes and disturbance and from human activity. • Chapter 10 provides a synthesis of the successional response of Ontario's forest vegetation to large-scale human intervention in forest management through forest harvest and the suppression of forest fire. • Chapter 11 discusses the response of wildlife to various causes of change in forest landscapes, and examines reasons for change observed in the ranges of particular forest vertebrate species.

The purpose of Section II is to provide a synthesis of processes affecting the development of forest communities, both plant and animal, and to illustrate to the reader the large-scale nature of these processes. As foresters and biologists, we are often concerned with sites and stands, but important processes operate at spatial and temporal scales far beyond what managers may influence on the ground over the course of a few years.

6 Influences of Climate on Ontario Forests

MICHAEL D. FLANNIGAN* AND MICHAEL G. WEBER**

INTRODUCTION

Climate and vegetation are intimately linked (Woodward 1987). This linkage is dynamic, because climate is always changing. Climate and its associated weather influence the structure and functioning of vegetation directly through such elements as temperature and precipitation, and indirectly through disturbance and permafrost. Climate is the total of all statistical weather information that describes the variation in weather at a given place for a specific interval of time (Greer 1996). In common usage, climate is the synthesis of weather; that is, the weather at some location averaged over a specified time period, typically 30 years, plus information on the variability and extremes of weather recorded during the same period.

The factors which control the climate at any one location include variations in solar radiation due to latitude, the distribution of continents and oceans, atmospheric pressure and wind systems, ocean currents, major terrain features, proximity to waterbodies, and local features (see Trewartha and Horn [1980] for more detail). As climate changes, the corresponding weather variables change. Temperature is a good example. Traditionally, in studies and in documentation of climate, much of the focus has been on changes in the mean temperature. In terms of the impact of temperature on vegetation, however, the variability of temperature might be even more important. Specifically, extreme minimum temperatures that drop below -40°C are lethal to many tree species. In addition, unusually late frosts in spring or early summer can severely damage seedlings. Similar principles apply to other weather variables, such as precipitation and wind: extreme drought and extreme wind speeds are capable of exerting a significant impact on vegetation.

The distribution of vegetation results from the interaction of many factors, such as climate, physical geography (topography, soil nutrients, and soil drainage), the sum total of past history, disturbance (natural and anthropogenic), and competition among plants and among animals. Climate is a key determinant of species presence or absence. The objective of this chapter is to examine the influence of climate and its associated weather on the vegetation of the boreal forest and the Great Lakes-St. Lawrence forest regions (Rowe 1972), the biomes which comprise most of the commercial forest area in Ontario. We outline how climate influenced the development of Ontario's forest vegetation in the past and describe how climate accounts for present-day patterns of vegetation distribution. We discuss predictions for future vegetation change based on the use of global climate models and an assumption that the atmospheric carbon dioxide will double. We then provide a detailed description of certain direct and indirect processes by which climate affects vegetation. The direct influences described include temperature and precipitation; the indirect influences include forest pests and diseases, and the presence of permafrost in the soil. Throughout the chapter, we discuss the interaction of climate and other causes of forest change, but we conclude by considering the influence which vegetation itself exerts on climate.

* Canadian Forest Service, Northern Forest Research Centre, 5320 - 122nd Street, Edmonton, Alberta T6H 3S5

** Canadian Forest Service, Great Lakes Forest Research Centre, 1219 Queen Street East, Sault Ste. Marie, Ontario P6A 5M7

AN OVERVIEW OF PAST, PRESENT, AND FUTURE CLIMATIC EFFECTS ON VEGETATION

Ontario is a large, floristically diverse geographic region. The province is characterized by a striking south-north gradient in vegetation cover, from the Carolinian forest in the south, through the Great Lakes-St Lawrence forest and the boreal forest, to the forested barrenland and the tundra in the north. This pattern is caused, in part, by a north-south gradient in temperature, but there is also a northwest-southeast gradient in moisture (see Figure 2.5 in Baldwin et al. [2000, this volume]). The climate of Ontario is diverse, as one might expect given the size of the region. The Great Lakes have a significant influence (Hare and Thomas 1974). Influences are exerted on the vegetation of Ontario at a number of different scales in space and time. Woodward (1987) provides an excellent overview of the time scales involved and the impact of effects at these different scales on vegetation.

In this chapter, we will discuss changes in climate and vegetation during the Holocene period, which is the most recent geologic epoch of the Quaternary period, extending from the end of the Pleistocene, approximately 10,800 years ago, to the present. This interval represents the current interglacial period. We also address the effects of climate on vegetation at spatial scales ranging from the individual forest stand, to the landscape, to the forest biome. When interpreting the influence of climate on vegetation, it is important to consider the climate and weather in the context of the life cycle characteristics of individual species. For example, a late spring frost that is not lethal to mature trees of a particular species may be lethal to its seedlings. Such a frost might be harmful to the production of viable seeds and thus might limit the distribution or expansion of the species (e.g., Pigott and Huntley 1978; Black and Bliss 1980). The impact of climate on vegetation must, therefore, be examined for all stages of the life cycle, including germination, seedling establishment, growth to sexual maturity, and production of viable seed. Sensitivities to climate vary by species and also with the developmental stage.

Past Climate and the Establishment of Ontario Forest Vegetation

Climate changes periodically, owing in part to a number of changes in the earth's orbit. The eccentricity of the earth's revolution around the sun has a 105,000-year cycle; there is a 41,000-year cycle in the obliquity of the earth's axis, and there is a 21,000-year cycle in the precession of the earth's axis about the pole of the ecliptic (that is, the precession of the equinoxes). Milankovitch (1941) stated that the periodic or cyclic warming and cooling of the earth's surface is caused by these orbital changes. Other factors play a role in the natural variation of the climate as well (see Webb 1992). Discussion in this section is restricted to changes in climate and vegetation during the last 10,000 years. Ten thousand years ago, Ontario was still greatly influenced by the continental ice sheet, which covered much of northern Ontario. The climate warmed to a point where it was warmer than the present day for the period from 7000 to 3000 years BP. A general cooling trend has been experienced in the last 3000 years, in which there have been relatively short periods of warming such as the recent warming period since the end of the Little Ice Age (about 1850 AD).

The vegetation in Ontario has changed dramatically during the Holocene. Paleoecological evidence suggests that boreal tree species such as white spruce (*Picea glauca*) and jack pine (*Pinus banksiana*) were among the first to appear following the retreat of the glaciers. These pioneer species were quickly followed by black spruce (*Picea mariana*) and white birch (*Betula papyrifera*), and then by the poplars (*Populus* spp.). After the invasion of the boreal species, the warming climate favoured the development of mixed forests of conifers and deciduous species. The predominant species in these mixed forests included white pine (*Pinus strobus*), hemlock (*Tsuga canadensis*), sugar maple (*Acer saccharum*) and beech (*Fagus* spp.) (Ritchie 1987; Liu 1990). These mixed forests spread farther north than the present day Great Lakes-St. Lawrence forest limit during the warm period 3000 to 7000 years ago, before retreating to the present-day limits during the general cooling trend which has taken place over the last 3000 years. The abundance of some key species has changed considerably during this time. For example, hemlock showed a marked decline around 4000 years ago, and has never regained its former stature. White pine has also decreased significantly over the last 1000 years, possibly because of the prevalence of cooler and moister conditions, which favour spruce. Naturally, there is a great deal of regional variation according to site-specific conditions.

Modelling the Effects of Climate Change

The present climate of Ontario can be described as humid continental, except for those areas close to Hudson Bay that have a more maritime climate. A more detailed description of Ontario's climate is provided by Baldwin et al. (2000, this volume). The present vegetation of Ontario is discussed by Thompson (2000, this volume). Hills (1959, 1960) divided the province of Ontario into 13 site regions or ecoregions (see Figure 5.1 in Perera and Baldwin [2000, this volume]), based on a qualitative description of climate, soils, topography, and vegetation communities. Rowe (1972) provides a general description of the forest geography of Canada in terms of forest regions and forest sections. An overview of ecoregionalization of Ontario is provided by Perera and Baldwin (2000, this volume). The present climate of Ontario is warming (Gullett and Skinner 1992), and indications are that the warming will continue in the next century (Intergovernmental Panel on Climate Change [IPCC] 1996). There is consensus in the scientific community that human activities are responsible for recent changes in the climate (IPCC 1996). Specifically, increases in radiatively active gases, such as carbon dioxide, methane, and the chlorofluorocarbons in the atmosphere are causing a significant warming of the earth's surface. Significant increases in temperature are anticipated in the next century, more rapid increases than have occurred in the last 10,000 years. Other climatic elements are also expected to change, including precipitation, wind, and cloudiness. More importantly, the variability of the climate appears to be increasing; therefore, more extreme events such as droughts, floods, major freezing-rain storms, heat waves, and cold snaps might be in store for the next century. All of these may do serious harm to vegetation.

The use of general circulation models (GCMs) enables researchers to simulate the future climate. There are a number of shortcomings associated with GCMs; nevertheless, most models are in agreement in predicting that the greatest warming will occur at high latitudes and in winter. Significant warming is expected to occur by the middle of the next century, but temperatures are expected to continue rising beyond 2100, even if the atmospheric concentrations of greenhouse gases are stabilized by that time (IPCC 1996). The confidence is lower for estimates of precipitation, but many models suggest an increase in water stress on vegetation, particularly in the centre of continents.

Many researchers have addressed the topic of climate in relation to vegetation using different types of modelling approaches. Most use a biome approach, which relates the current areal extent of biomes to current climate and uses those relationships to predict where the vegetation might be in the future, or at least to identify the region most climatically suitable for that biome. Examples of this type of model are provided in Figure 6.1, which shows the equilibrium potential of natural vegetation under climate change already in progress, and Figure 6.2, which shows the potential distribution of major biomes under predicted climate change, defined by the Mapped Atmosphere-Plant-Soil System (MAPSS) model (Neilson 1993). The present climate is provided by the climate database of the International Institute for Applied Systems Analysis (IIASA) (Leemans and Cramer 1991), while the future climate is derived from the difference between the control run and a scenario of carbon dioxide doubling from the GCM of the Geophysical Fluid Dynamics Laboratory, termed the GFDL model (Weatherald and Manabe 1986), with aerosols included. The projected shifts in the boundaries of vegetation classes are generated by a model that simulates steady-state leaf-area index, calculated from a sub-model of site water and heat balance (Neilson 1993). Figure 6.1 is similar to Figure 3.1 in Thompson 2000 (this volume), which shows the present vegetation in Ontario (see also Olson et al. 1983), and also to Figure 5.1 in Perera and Baldwin (2000, this volume), except that the MAPSS model does not reflect the northern Ontario wetlands. One main difference between Figure 6.1 and Figure 3.1 (Thompson 2000, this volume) is that the Carolinian and Great Lakes-St. Lawrence forests are combined in Figure 6.1.

Striking differences are obvious between Figures 6.1 and 6.2, which show the equilibrium potential of natural vegetation now and in the future. Figure 6.2 depicts the savanna-woodland forest type as extending over most of southern and eastern Ontario and depicts the temperate mixed forest as moving north to James Bay, or approximately 500 km north of its present-day limit. Many other models exist, based on a variety of GCMs , so many potential outcomes have been derived. For example, Warrick et al. (1986) use a Holdridge life-zone classification (Holdridge 1947) with the GFDL model, and suggest that the potential vegetation would be temperate forest over all of Ontario, except for a narrow band of boreal forest along

Hudson Bay. Box (1981) relates vegetation to a number of meteorological variables and uses these relationships to determine new patterns of vegetation under the climate regime resulting from a doubling of atmospheric carbon dioxide. Rizzo and Wiken (1992) apply a classification model derived from the current ecological setting to simulate the effects of climate change from carbon dioxide doubling on Canada's ecosystems. For additional information on simulated changes in vegetation distribution under global warming see Appendix C in IPCC (1998).

There are numerous caveats to the use of models of this kind, in addition to the caveats associated with the GCMs themselves. Most models use biomes and move the vegetation as a community. We know that this result cannot be accurate, because vegetation is an assemblage of different species in which each species is distributed according to its own physiological requirements, as constrained by competitive interactions (Gleason 1926). Species of vegetation move as individuals, not as a community (Whitney 1986; Davis 1989). The issue is further complicated by disturbance, which plays a major role in determining the abundance and distribution of individual species (Flannigan 1993; Suffling 1995; Bergeron et al. 1997) and is not fully incorporated into these models. Caution is also advised when interpreting results from physiological models because of the inherent problems involved in scaling up from a leaf or a tree to a stand, and eventually to a continental scale (Coleman et al. 1992). Finally, these models display regions where the climate is instantaneously suitable for the various vegetation types; however, the time required for the vegetation to come into equilibrium with the projection could take centuries, as determined by migration rates, competition, and altered disturbance regimes.

The Impact of Climate Change on Ontario's Vegetation

As we have already seen in Figures 6.1 and 6.2, models suggest that the climate suitable for the major biomes in Ontario will shift northwards by 500 km or more by the end of the next century. Paleoecological studies have shown, however, that maximum rates of migration are much less than would be required for the vegetation to keep pace with projected climate change (Prentice et al. 1991; Webb and Bartlein 1992). These maximum rates are, if anything, greater than can be expected in the future, as they represent migration over a recently deglaciated landscape. The existing forests in the transition zone between forest and grassland will not necessarily be rapidly replaced by grassland.

Another factor which might slow down the anticipated vegetation transition is a decrease in disturbance regimes that might be associated with climate warming. For example, Bergeron and Archambault (1993) have shown for a region near Lake Abitibi in Quebec that the fire frequency has decreased since the end of the Little Ice Age despite temperature increases of more than 1°C over the same period, because of increased precipitation frequency. Modelling results from Flannigan et al. (1998) suggest that fire weather severity will decrease in portions of eastern Canada with a doubling of atmospheric carbon dioxide, because increased precipitation in the warmer climate will more than compensate for the increase in temperature. Decreased disturbance in the Claybelt region of Ontario might lead to an increased abundance of balsam fir (*Abies balsamea*) and cedar (*Thuja occidentalis*) because of their shade tolerance. These species would be difficult to replace with southern competitors, not only because of their shade tolerance, but also because decreased disturbance rates would mean smaller and fewer areas for the southern competitors to exploit. In regions where disturbances from fire, insect pests, and disease increase, the transition of the vegetation assemblages to the adjacent types may be accelerated (Suffling 1995). The vegetation changes associated with the new climate may lead, moreover, to new assemblages of species (Martin 1993). Competition may be a key factor in defining the vegetation composition. Bonan and Sirois (1992) have suggested that the southern limit of black spruce is dictated by competition rather than climate, as black spruce is at its optimum climate for growth at its present-day southern limit. Thompson et al. (1998) present an overview of possible changes to Ontario's forested landscapes as a result of climate change.

Increases in climate variability under a new climate could have major impacts on the vegetation of Ontario (Mearns et al. 1989; Solomon and Leemans 1997). Models have suggested that synoptic storm frequency would decrease in the long term, but that there would be an increase in the overall intensity of disturbances (Lambert 1995). In the next century, there may thus be fewer storms, but more extreme

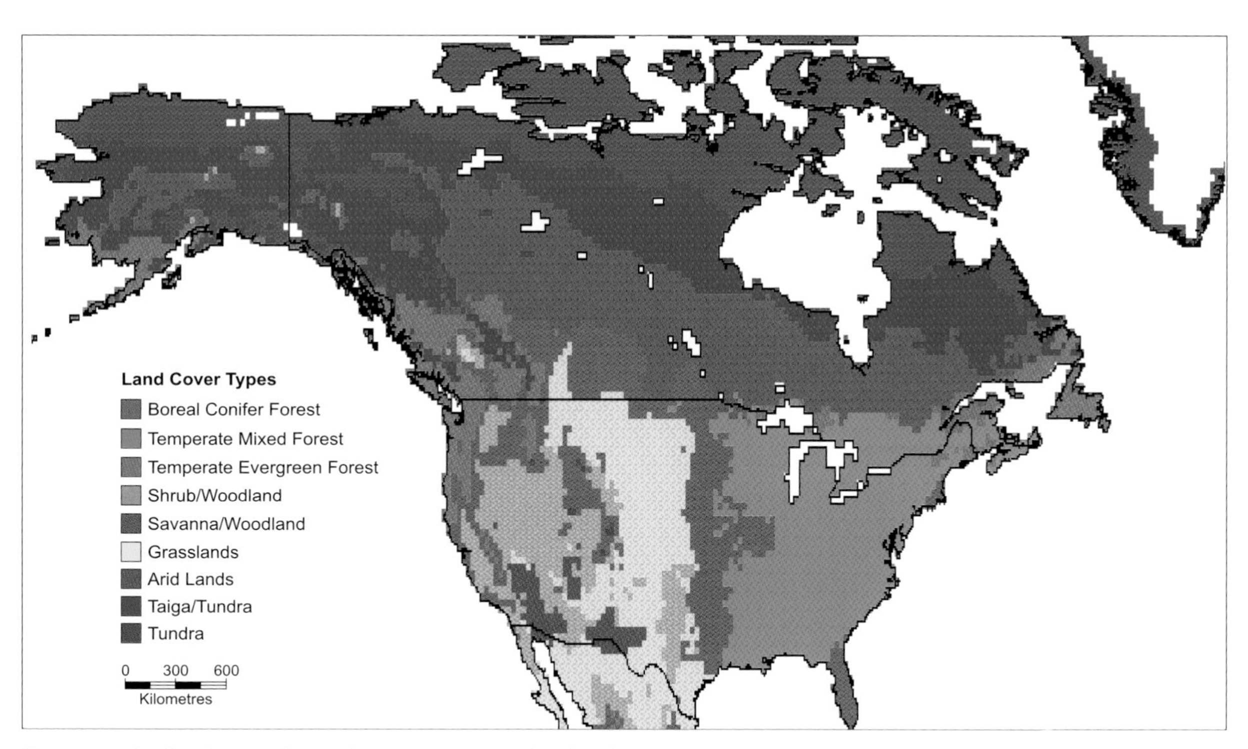

Figure 6.1 The distribution of major biome types as simulated under current climate change by the Mapped Atmosphere-Plant-Soil System (MAPSS) model. (Adapted from Neilson and Drapek 1998)

weather (for example, extreme wind speeds or very heavy precipitation causing flooding). Research has also suggested that the persistence of blocking ridges in the upper atmosphere will increase in a climate scenario of doubled carbon dioxide (Lupo et al. 1997). This factor could have significant impact on forest fires, as these upper ridges are associated with dry and warm conditions at the earth's surface that are conducive to forest fires. Extreme environmental conditions caused by prolonged drought, floods, extreme heat, extreme cold, and the increased occurrence of severe winds, can be expected to have a negative influence on forest health. These environmental stresses predispose individual plants, species, and ecosystems to secondary stressors, such as outbreaks of insect infestation and disease. Research has shown that resistance to drought increases with increased carbon dioxide (Townend 1993). Recent research has also suggested that increased carbon dioxide may lead to increased tolerance of cold temperatures (Boese et al. 1997).

The anticipated changes in climate will have significant impacts on physiological processes and the cycling of nutrients. The global atmospheric concentration of carbon dioxide has risen from pre-industrial levels of 280 parts per million by volume (ppmv) to 360 ppmv in 1994 (Amthor 1995). Plants and ecosystems are closely coupled with nitrogen and carbon cycles, which might be altered by the elevated carbon dioxide and by climate change. The nitrogen and carbon cycles are closely linked (Reynolds et al. 1996) through decomposition and litter quality. Temperature increases will greatly influence decomposition and nutrient cycling (Anderson 1992). Historically, the boreal forest has been presumed to be a carbon sink in the global carbon budget. This carbon sink likely will be reduced under climate change (Kurz and Apps 1993; Kurz et al. 1995), or may even become a carbon source. Increased temperatures will lead to an increase in soil temperature and an associated increase in the active layer over permafrost. Improved soil drainage as a result of soil warming, especially at northern latitudes, is an important consideration, because of the implications for organic layer drying, and hence fire severity (Anderson 1992).

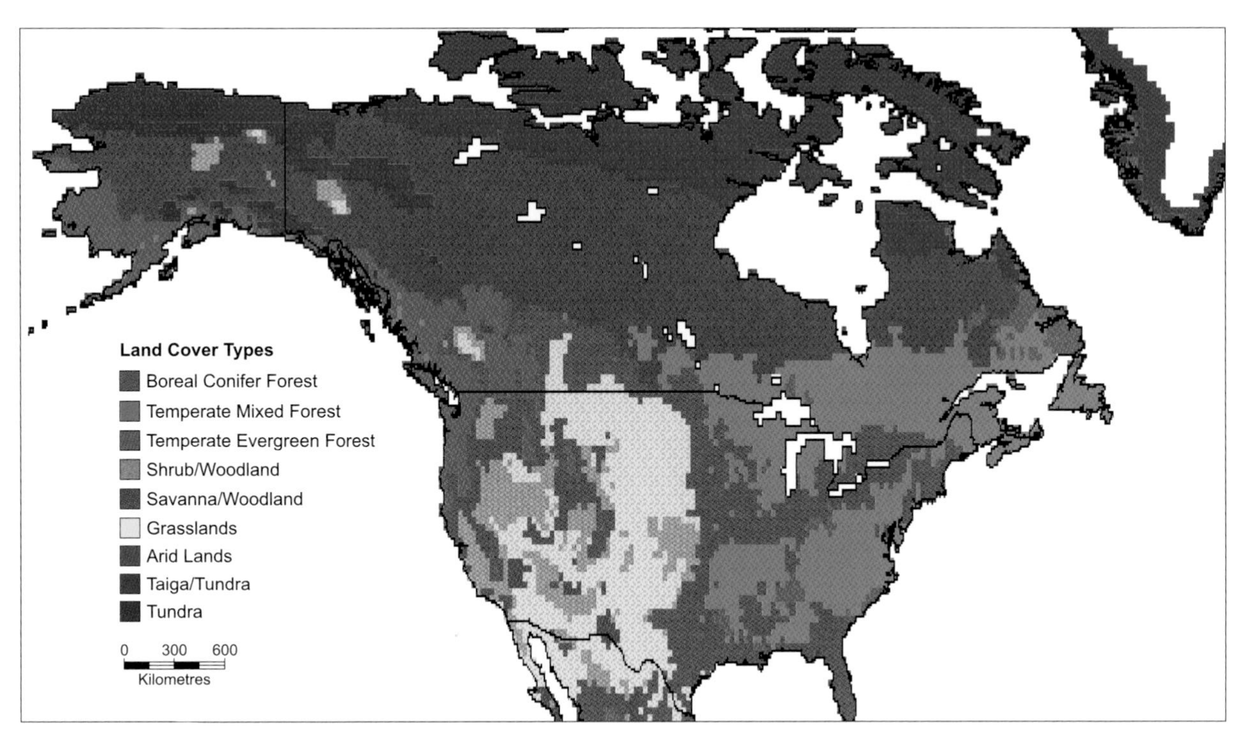

Figure 6.2 The potential distribution of major biomes as simulated under the Geophysical Fluid Dynamics Laboratory (GFDL) Global Climate Model, with aerosols included, by MAPSS. (Adapted from Neilson and Drapek 1998)

PROCESSES OF CLIMATE INFLUENCE ON VEGETATION

Weather variables such as temperature, precipitation, and wind have a direct influence on vegetation in terms of growth, mortality, species abundance, and composition. Weather also exerts an indirect influence on vegetation through such factors as forest fires, pest and disease outbreaks, and the presence or absence of permafrost.

Direct Effects of Climate

Influences of Temperature on Vegetation

Various aspects of temperature can have a significant impact on vegetation. These include winter minimum temperature, frost during the growing season, and warmth during the growing season.

Winter minimum temperatures are important in determining the distribution of tree species. Many studies suggest that the poleward limit of a tree species is controlled by the minimum winter temperature that is regularly experienced (Sakai and Weiser 1973; George et al. 1974; Sakai 1978; Larcher and Bauer 1981; Woodward 1987; and Arris and Eagleson 1989). In Ontario, this is probably true for most, if not all, of the non-boreal, deciduous tree species. Most deciduous species cannot tolerate temperatures below -30°C to -40°C, the limit of the strategy which they use to survive freezing temperatures. There are three standard strategies that plants use to survive freezing temperatures: deep supercooling, extracellular freezing, and extraorgan freezing (Sakai and Larcher 1987; Woodward and Williams 1987). Deep supercooling allows water in the plant cells to remain liquid despite temperatures well below 0°C, owing to a lack of ice nucleation sites. As long as ice does not form within the cell, there is no mechanical damage. Typically, the coldest temperature that plants can survive using deep supercooling for pure water is about -40°C. Survival at lower temperatures (to about -55°C) using deep supercooling is possible only in the presence of high concentrations of solutes in the cell water (Gusta et al. 1983). Most deciduous tree species, except the birches (*Betula* spp.) and poplars, use deep supercooling and typically cannot tolerate temperatures below -40°C.

Extracellular and extraorgan freezing occurs after the migration of water out of the plant cells or organs and into intercellular spaces, where freezing can usually occur without damage. The intercellular spaces are usually large enough to accommodate the influx of water and the expansion associated with the phase change from liquid to solid, without damage to the surrounding cells and organs. The survival of plants using extracellular and extraorgan strategies is limited by the extent to which the plant can withstand extreme dehydration of the cell or organ caused by the outward migration of water from cells (Sakai 1979), which results in desiccation. The boreal conifers use extracellular or extraorgan strategies and can survive temperatures of -70° to -80°C, which is colder than anything experienced in Ontario, and these conifers are not limited by extreme minimum temperatures.

During the growing season, tree species, and in particular seedlings, are not frost-hardy, so that temperatures of -2°C to -5°C can be lethal. Growing-season frost can also damage reproductive structures. Female conifer flowers and conelets are particularly susceptible to frost damage in early spring, which can limit seed production (Schooley et al. 1986). Frost can also damage other parts of the tree, including the stem, bud, and root collar, and can cause leaf and needle damage. If the initial damage from temperatures below freezing is not lethal, then the damaged areas often become sites of infection by canker and other diseases, or become susceptible to insect attack (Hiratsuka and Zalasky 1993). Growing-season frost can be critical in plantations, especially where topography creates low-lying areas (Stathers 1989). Studies of frost hardiness have been conducted on many coniferous species found in Ontario (Glerum et al. 1966; Glerum 1973; Joyce 1987). Results suggest that there is little difference in frost-hardiness between those conifer species (Glerum 1973).

Growing-season warmth can also be an important determinant in vegetation distribution. For example, Black and Bliss (1980) found that the northern limit of black spruce was determined by the summer warmth required for seed germination. Pigott and Huntley (1978) found that insufficient warmth at the northern limit of small-leaf linden (*Tilia cordata*) during the flowering period resulted in non-viable seed. For most tree species, there is a critical temperature that needs to be exceeded for growth to begin. The growing-degree concept was developed from the fact that many grasses require temperatures of 5°C or higher for growth to occur. For some trees, such as red pine (*Pinus resinosa*), the critical temperature for initiating and maintaining growth is 10°C; therefore, if summer mean annual temperatures did not exceed 10°C, it would be unlikely that red pine could remain established in such a climate. Summer warmth is critical in plantations, where site treatments such as mounding have serious micro-meteorological implications on the local thermal regime (Spittlehouse and Stathers 1990). McCaughey et al. (1997) provide a good overview of the weather and climate associated with Canadian forests.

Influences of Precipitation on Vegetation

The lack of precipitation, if prolonged, results in drought that can damage or kill trees. Drought can be restricted to one growing season or may persist for several growing seasons. If drought is severe enough, leaf abscission will occur. Summer drought is different from winter desiccation. Drought is caused by inadequate soil moisture; whereas desiccation occurs when soil moisture is unavailable because the ground is frozen. Drought-stressed trees are prone to attacks from insects and diseases. On the other hand, if the precipitation is too heavy, flooding can occur and cause extensive damage in low-lying areas.

Freezing rain and heavy snow can accumulate on the vegetation to such an extent that the added weight on the foliage and branches causes physical damage. The build-up of snow and ice is influenced by stand density and the shape of the crown. This is a common cause of damage in plantations (Powers and Oliver 1970). The amount of damage can be significant; there are reports of more than 20 percent of stems broken in a stand (Van Cleve and Zasada 1970). When trees are laden with a coating of ice, they are more prone to windthrow. A severe freezing rain event in January 1998 damaged millions of trees in eastern Ontario and southern Quebec. Hail also can cause extensive damage to vegetation (Riley 1953; Laut and Elliot 1966). Seedlings and saplings are especially prone to damage; whereas mature trees typically sustain only minor damage. As with other types of physical damage, the parts of the trees damaged are potential sites of infection by pathogens.

Indirect Effects of Climate

Climatic Aspects of the Influence of Insects and Disease on Vegetation

Climate and weather play a major role in the life cycle of many forest insects, some of which have a major influence on forest productivity (Fleming and Volney 1995). Additionally, climate and weather can be important in disease contraction and spread. If climate changes, as the GCMs suggest, the greatest impact of climate change on the structure and function of the boreal forest will be mediated through changes in disturbance regimes such as insect outbreaks and fire. Discussion of a large number of insect defoliators is beyond the scope of this chapter, so the spruce budworm (*Choristoneura fumiferana*) is chosen as a representative species. Fleming et al. (2000, this volume) provide a detailed description of the effects of various insect pests and forest diseases on Ontario's forest landscapes.

Fleming (1996) reviewed the possible influences of climate change on defoliating insects in North America's boreal forests and outlined the interrelationships among climate, vegetation, and insect populations. The direct influence of climate on vegetation may have a secondary impact on insect populations. Climate influences the synchrony of host plant phenology with spruce budworm development as well as the synchrony with natural invertebrate enemies. Finally, weather elements such as drought and late-spring frost may have a direct impact on spruce budworm populations; in fact, Cerezke and Volney (1995) suggest that late-spring frosts coincide with the collapse of the spruce budworm outbreak. Spruce budworm is only one example of the many types of insects that influence the forest, but that work does highlight the complex interactions and feedbacks among vegetation host, climate/weather, and natural enemies. As the climate and weather change, non-linear and perhaps unexpected interactions may have devastating effects, allowing insects to become an additional agent of accelerated change in the forests.

Climate directly influences vegetation, its pathogens, and its insects, including pathogen vectors. The relationships between weather and tree diseases have been studied for many years (Hepting 1963). So-called "forest declines" (Manion 1981) may be a result of an interaction between climate and disease. For example, red spruce (*Picea rubens*) decline consists of an interaction between winter injury and air pollution, which allows pathogenic fungi such as *Cytospora* sp., *Fomes* sp., *Armillaria* sp., needle-cast diseases, rust diseases, and several other butt-rot and stem-rot fungi (Johnson 1992) to injure or kill the tree. Coakley (1988) suggests that a change in climatic conditions or a change in climatic variability may alter plant disease development by affecting the following factors: (1) the speed of pathogen development; (2) the geographical range of the host, pathogen, or vector, especially at the boundaries of their respective distributions; and (3) control of the disease. Predicting the impact of climate change on forest diseases is made more complicated by the need to take into account the interactions among climate, pathogens, and insect vectors of the pathogens, but it is clear that, with warming, the potential for rapid outbreaks of forest disease across Ontario is a real threat.

Influences of Permafrost on Vegetation

In some northern Ontario forested landscapes, permafrost is an important agent, exerting control over forest ecosystem structure and function. Although of concern only locally, permafrost is a terrain feature that may be of concern to ecosystem managers charged with maintaining the integrity of Ontario's northernmost areas. The terrain sensitivity of landscapes underlain by permafrost must be considered in planning both commercial and non-commercial northern development activities, such as the construction of roads, settlements, or fire-guards. According to Brown (1973), continuous permafrost underlies only a narrow, treeless band along the Hudson Bay coast of northern Ontario. Discontinuous permafrost, consisting of scattered islands of permanently frozen ground, each a few square metres to several hectares in size, occurs mainly in peatlands. Other areas where discontinuous permafrost may be encountered are on north-facing slopes of east-west oriented valleys, or along isolated patches of forested stream-banks, where increased shading reduces summer thaw and winter snow cover (Brown 1973).

The southern limit of discontinuous permafrost in Ontario lies at about latitude 51°N, to 52°N around James Bay and coincides with the mean annual air temperature isotherm of -10°C. The area occupied by discontinuous permafrost, also known as the Hudson Bay Lowland physiographic region, contains the northern limits of all boreal forest tree species in Ontario and is characterized by a fire-dominated disturbance regime.

The impact of potential climate change on the northern Ontario boreal forest of the Hudson Bay Lowland may be envisaged from simulation studies carried out for other parts of the North American boreal forest, where permafrost and fire interact to dominate forest ecosystem structure and function. An example has been provided by Bonan et al. (1990) for interior Alaska. Their simulations assumed climate change scenarios of warming by 1°C, 3°C, and 5°C, factorially coupled with increases of 120 percent, 140 percent, and 160 percent in monthly precipitation values. To emphasize the importance of site conditions in response to expected climate change, the simulations were performed for two contrasting forest types: a black spruce (*Picea mariana*) forest growing on a permafrost-dominated, poorly drained, north-facing slope, and a forest of white spruce, paper birch, and aspen located on a well-drained, permafrost-free, south-facing slope. According to these simulations, the effects of climatic warming on ecosystem structure and function in the northern boreal forest may not be so much a direct response to increased air temperature as to increased potential evapotranspiration demands. Analysis of their simulation results also revealed the importance of the forest floor organic layers in controlling ecosystem response to climatic warming. For example, the thick forest floor layer of 20 cm to 30 cm typical of many black spruce forests in interior Alaska and elsewhere is the major factor responsible for cold, wet soil conditions which restrict nutrient availability and tree growth (Weber and Van Cleve 1981, 1984).

In the absence of fire, the short-term response of these permafrost-dominated sites to climate warming was a decrease in the depth of the active soil layer (that is, the layer of soil lying above the permafrost that thaws out annually in response to summer warming). This decrease occurred from a drying of the forest floor, which impeded the conduction of heat into deeper soil layers. In the long term, however, with recurrent forest fires, the drier organic layers were conducive to increased fire severity, and thus to the removal of greater amounts of forest floor material. As a result, the depth of the active layer increased, and soil drainage further improved (Bonan 1989; Bonan et al. 1990). The complete elimination of shallow, discontinuous permafrost would be a possible scenario under these conditions. The final outcome of this simulation run was the fire-caused conversion of the low-productivity black spruce forests to mixed forests of spruce and hardwood growing on warmer soils. In contrast, on the well-drained, south-facing spruce and hardwood forest sites, increased potential water loss in the warmer climate reduced soil moisture and resulted in the site-conversion of these stands to dry aspen forests. The greatest simulated reduction in soil moisture resulted in steppe-like vegetation and an elimination of the tree overstory on these sites. Bonan et al. (1990) thus highlighted the sensitivity of divergent forest ecosystems to water balance and to its interaction with the fire regime under climate change (Weber and Flannigan 1997).

The Influence of Vegetation on Climate

The link between climate and vegetation is well known, but the reverse link is not as well known. The link between vegetation and climate is found at all scales, from microscales to the global scale. At smaller spatial scales, differences in temperature, wind, and relative humidity would be expected to exist between an agricultural field and an adjacent forest stand, because of differences in the energy budget between the two areas. At larger scales, for example, across the entire boreal forest biome, the influence of vegetation on the climate can be significant. From using the GCM of the United Kingdom Meteorological Office (UKMO), Thomas and Rowntree (1992) suggest that, in the absence of boreal forests, northern hemisphere temperatures would be 2.8°C cooler and precipitation would decrease. These changes would result from the difference in albedo between the forest and non-forest vegetation, especially in winter, as the albedo of snow is particularly high. (Albedo is the amount of electromagnetic radiation reflected by a body relative to the amount incident upon it, and is commonly expressed as a percentage [Greer 1996].) Also using a GCM, Bonan et al. (1992) suggest that, if tundra or bare ground replaced the boreal forest, the climate of the entire northern hemisphere would be significantly cooler, and that latent heat flux and atmospheric moisture would increase. The warming effect of the boreal forest consists of masking the high reflectance of snow over vast areas of the northern hemisphere. Other researchers (Otterman et al. 1984; Crowley and Baum 1997) confirm that vegetation does play a significant role in regional to global temperature and precipitation patterns. Foley et al. (1994) argue that the interaction of vegetation with climate was operating during the Holocene and gave rise to large positive feedback between the climate and the boreal forests, which resulted in warmer temperatures in the northern hemisphere.

SUMMARY

Climate and vegetation interact across the range of spatial and temporal scales in a complex fashion. Climate determines the suite of species that is available to colonize the landscape. The actual vegetation present over the landscape is the result of many factors among which climate is of primary importance. Climate exerts direct control over vegetation through either beneficial or deleterious effects of temperature, precipitation, and wind, and indirect control through climatic influences on fire and insect disturbances, disease, and soil properties such as permafrost, which, in turn, influence vegetation.

Across the province of Ontario, there are large north-south and northwest-southeast climatic gradients in temperature and precipitation, respectively, which give rise to a great diversity of vegetation types. As climate changes, so does the vegetation, although at a slower pace. Should the climate continue to warm, dramatic change in the forests of Ontario can be expected, especially if the climate changes as rapidly as the global climate models suggest. The interaction between climate change and disturbance regimes has the potential to overshadow the importance of the direct effects of global warming on species distribution, migration, substitution, and extinction. Disturbance could thus be the most effective agent of change, and the rate and magnitude of disturbance-induced changes to the forested landscape of Ontario could greatly exceed anything caused by atmospheric warming alone.

ACKNOWLEDGEMENTS

We thank Ron Neilson and Ray Drapek for providing Figures 6.1 and 6.2. Thanks also go to Mike Wotton, who helped prepare those figures for this chapter.

REFERENCES

Amthor, J.S. 1995. Terrestrial higher-plant response to increasing atmospheric [CO_2] in relation to the global carbon cycle. Global Change Biology 1: 243-274.

Anderson, J.M. 1992. Response of soils to climate change. Advances in Ecological Research 22: 163-210.

Arris, L.L. and P.S. Eagleson. 1989. Evidence of a physiological basis for the boreal-deciduous forest ecotone in North America. *Vegetatio* 82: 55-58.

Baldwin, D.J.B., J.R. Desloges and L.E. Band. 2000. Physical geography of Ontario. In: A.H. Perera, D.L. Euler, and I.D. Thompson (editors). Ecology of a Managed Terrestrial Landscape: Patterns and Processes of Forest Landscapes in Ontario. Vancouver, British Columbia: University of British Columbia Press. 12-29.

Bergeron, Y. and S. Archambault. 1993. Decreasing frequency of forest fires in the southern boreal zone of Québec and its relation to global warming since the end of the "Little Ice Age." Holocene 3: 255-259.

Bergeron, Y., A. Leduc, and T.-X. Li. 1997. Explaining the distribution of *Pinus* spp. in a Canadian boreal insular landscape. Journal of Vegetation Science 8: 37-44.

Black, R.A. and L.C. Bliss. 1980. Reproductive ecology of *Picea mariana* (Mill.) at the tree line near Inuvik, Northwest Territories, Canada. Ecological Monographs 50: 331-354.

Boese, S.R., D.W. Wolfe, and J.J. Melkonian. 1997. Elevated CO_2 mitigates chilling-induced water stress and photosynthetic reduction during chilling. Plant Cell and Environment 20: 625-632.

Bonan, G.B. 1989. A computer model of the solar radiation, soil moisture, and soil thermal regimes in boreal forests. Ecological Modelling 45: 275-306.

Bonan, G.B., D. Pollard, and S.L. Thompson. 1992. Effect of boreal forest vegetation on global climate. Nature 359: 716-718.

Bonan, G.B., H.H. Shugart, and D.L. Urban. 1990. The sensitivity of some high-latitude boreal forests to climatic parameters. Climatic Change 16: 9-29.

Bonan, G.B. and L. Sirois. 1992. Air temperature, tree growth, and the northern and southern range limits to *Picea mariana*. Journal of Vegetation Science 3: 495-506.

Box, E.O. 1981. Macroclimate and Plant Forms: An Introduction to Predictive Modelling in Phyto-Geography. The Hague, Netherlands: Dr. W. Junk Publishers. 258 p.

Brown, R.J.E. 1973. Permafrost in Canada. Toronto, Ontario: University of Toronto Press. 234 p.

Cerezke, H.F. and W.J.A. Volney. 1995. Forest insect pests in the northwest region. In: J.A. Armstrong and W.G.H. Ives (editors). Forest Insect Pests in Canada. Ottawa, Ontario: Canadian Forest Service. 59-72.

Coakley, S.M. 1988. Variation in climate and prediction of disease in plants. Annual Review Phytopathology 26: 163-181.

Coleman, D.C., E.P. Odum, and D.A. Crossley, Jr. 1992. Soil biology, soil ecology, and global change. Biology and Fertility of Soils 14: 104-111.

Crowley, T.J. and S.K. Baum. 1997. Effect of vegetation on an ice-age climate model simulation. Journal of Geophysical Research 102: 16,463-16,480.

Davis, M.B. 1989. Lags in vegetation response to greenhouse warming. Climatic Change 15: 75-82.

Flannigan, M.D. 1993. Fire regime and the abundance of red pine. International Journal of Wildland Fire 3:241-247.

Flannigan, M.D, Y. Bergeron, O. Engelmark, and B.M. Wotton. 1998. Future wildfire in circumboreal forest in relation to global warming. Journal of Vegetation Science 9: 469-476.

Fleming, R.A. 1996. A mechanistic perspective of possible influences of climate change on defoliating insects in North America's boreal forests. *Silva Fennica* 30: 281-294.

Fleming, R.A., A.A. Hopkin, and J.-N. Candau. 2000. Insect and disease disturbance regimes in Ontario's forests. In: A.H. Perera, D.L.

Euler, and I.D. Thompson (editors). Ecology of a Managed Terrestrial Landscape: Patterns and Processes of Forest Landscapes in Ontario. Vancouver, British Columbia: University of British Columbia Press. 141-162.

Fleming, R.A. and W.J.A. Volney. 1995. Effects of climate change on insect defoliator population processes in Canada's boreal forest: some plausible scenarios. Water, Air and Soil Pollution 82: 445-454.

Foley, J.A., J.E. Kutzbach, M.T. Coe, and S. Levis. 1994. Feedbacks between climate and boreal forests during the holocene epoch. Nature 371: 52-54.

George, M.F., M.J. Burke, H.M. Pellet, and A.G. Johnson. 1974. Low temperature exotherm and woody plant distribution. Horticultural Science 9: 519-522.

Gleason, H.A. 1926. The individualistic concept of the plant association. Bulletin of the Torrey Botanical Club 53: 7-26.

Glerum, C. 1973. Annual trends in frost hardiness and electrical impedance for seven coniferous species. Canadian Journal of Plant Science 53: 881-889.

Glerum, C., J.L. Farrar, and R.L. McLure. 1966. A frost hardiness study of six coniferous species. Forestry Chronicle 42: 69-75.

Greer, I.W. (editor). 1996. Glossary of Weather and Climate with Related Oceanic and Hydrologic Terms. Boston, Maine: American Meteorological Society. 272 p.

Gullett, D.W. and W.R. Skinner. 1992. The state of Canada's climate: temperature change in Canada 1895-1991. State of the Environment Report No. 92-2. Ottawa, Ontario: Environment Canada. 36 p.

Gusta, L.V., N.J. Tyler, and T.H.H. Chen. 1983. Deep undercooling in woody plants north of the -40°C isotherm. Plant Physiology 72: 122-128.

Hare, F.K. and M.K. Thomas. 1974. Climate Canada. Toronto, Ontario: John Wiley. 256 p.

Hepting, G.H. 1963. Climate and forest diseases. Annual Review Phytopathology 1: 31-50.

Hills, G.A. 1959. A Ready Reference to the Description of the Land of Ontario and its Productivity. Preliminary Report. Maple, Ontario: Ontario Department of Lands and Forests. 142 p.

Hills, G.A. 1960. Regional site research. Forestry Chronicle 36: 401-423.

Hiratsuka, Y. and H. Zalasky. 1993. Frost and Other Climate-Related Damage of Forest Trees in the Prairie Provinces. Information Report NOR-X-331. Edmonton, Alberta: Canadian Forest Service. 25 p.

Holdridge, L.R. 1947. Determination of world plant formations from simple climate data. Science 105: 367-368.

Intergovernmental Panel on Climate Change (IPCC). 1996. Climate Change 1995. Impacts, adaptions and mitigation of climate change: scientific-technical analyses. Cambridge, UK: Cambridge University Press. 878 p.

Intergovernmental Panel on Climate Change (IPCC). 1998. The Regional Impacts of Climate Change: An Assessment of Vulnerability. Cambridge, UK: Cambridge University Press. 517 p.

Johnson, A.H. 1992. The role of abiotic stresses in the decline of red spruce in high elevation forest of the eastern United States. Annual Review of Phytopathology 30: 349-367.

Joyce, D.G. 1987. Adaptive variation in cold hardiness of eastern larch, *Larix laricina*, in Northern Ontario. Canadian Journal of Forest Research 18: 85-89.

Kurz, W.A. and M.J. Apps. 1993. Contribution of northern forests to the global C cycle: Canada as a case study. Water, Air and Soil Pollution 70: 163-176.

Kurz, W.A., M.J. Apps, B.J. Stocks, and J.A. Volney. 1995. Global climate change: disturbance regimes and biospheric feedbacks of temperate and boreal forests. In: G.M. Woodwell and F.T. Mackenzie (editors). Biotic Feedbacks in the Global Climatic System. Will the Warming Feed the Warming? New York, New York: Oxford University Press. 119-133.

Lambert, S.J. 1995. The effect of enhanced greenhouse warming on winter cyclone frequencies and strengths. Journal of Climate 8: 1447-1452.

Larcher, W. and H. Bauer. 1981. Ecological significance of resistance to low temperature. In: O.L. Lange, P.S. Nobel, C.B. Osmond, and H. Ziegler (editors). Physiological Plant Ecology I. Encyclopedia of Plant Physiology, Volume 12A. Berlin, Germany: Springer-Verlag. 403-437.

Laut, J.G. and K.R. Elliot. 1966. Extensive hail damage in northern Manitoba. Forestry Chronicle 42: 198.

Leemans, R. and W.P. Cramer. 1991. The IIASA Database for Mean Monthly Values of Temperature, Precipitation and Cloudiness on a Global Terrestrial Grid. Research Report RR-91-18. Laxenburg, Austria: International Institute of Applied Systems Analysis (IIASA). 62 p.

Liu, K.-B. 1990. Holocene paleoecology of the Boreal Forest and Great Lakes -St. Lawrence Forest in northern Ontario. Ecological Monographs 60: 179-212.

Lupo, A.R., R.J. Oglesby, and I.I. Mokhov. 1997. Climatological features of blocking anticyclones: a study of Northern Hemisphere CCM1 model blocking events in present-day and double CO_2 concentration atmospheres. Climate Dynamics 13: 181-195.

Manion, P.D. 1981 Tree Disease Concepts. Englewood Cliffs, New Jersey: Prentice Hall. 399 p.

Martin, P. 1993. Vegetation responses and feedbacks to climate: A review of models and processes. Climate Dynamics 8: 201-210.

McCaughey, J.H., B.D. Amiro, A.W. Robertson, and D.L. Spittlehouse. 1997. Forest environments. In: T.R. Oke and W.R. Rouse (editors). The Surface Climates of Canada. Montreal, Quebec: McGill-Queen's University Press. 247-276.

Mearns, L.O., S.H. Schneider, S.L. Thompson, and L.R. McDaniel. 1989. Climate variability statistics from General Circulation Models as applied to climate change analysis. In: G.P. Malanson (editor). Natural Areas Facing Climate Change. The Hague, Netherlands: SPB Academic Publishing. 51-73.

Milankovitch, M. 1941. Kanon der Erdbestrahlung und seine Anwendung auf des Eiszeitproblem. Belgrade, Yugoslavia: Royal Serbian Academt. Translated to English in 1969. 484 p.

Neilson, R.P. 1993. Vegetation redistribution: A possible biosphere source of CO_2 during climate change. Water, Air, and Soil Pollution 70: 659-673.

Neilson, R.P. and Drapek, R.J. 1998. Potentially complex biosphere responses to transient global warming. Global Change Biology 4:505-521.

Olson, J.S., J.A. Watts, and L.J. Allison. 1983. Carbon in live vegetation of major world ecosystems. Report ORNL-5862. Oak Ridge, Tennessee: Oak Ridge National Laboratory. 152 p.

Otterman, J., M.D. Chou, and A. Arking. 1984. Effects of non-tropical forest cover on climate. Journal of Climatology and Applied Meteorology 23: 1626-1634.

Perera, A.H. and D.J.B. Baldwin. 2000. Spatial patterns in the managed forest landscapes of Ontario. In: A.H. Perera, D.L. Euler, and I.D. Thompson (editors). Ecology of a Managed Terrestrial Landscape: Patterns and Processes of Forest Landscapes in Ontario. Vancouver, British Columbia: University of British Columbia Press. 74-99.

Pigott, C.D. and J.P. Huntley. 1978. Factors controlling the distribution of *Tilia cordata* at the northern limits of its geographical range.

1. Distribution in north-west England. New Phytologist 81: 429-441.
Powers, R.F. and W.W. Oliver. 1970. Snow Breakage in a Pole-Sized Ponderosa Pine....More Damage at High Stand Densities. Research Note PSW-218. Berkeley, California: USDA Forest Service. 3 p.
Prentice, I.C., M.T. Sykes, and W. Cramer. 1991. The possible dynamic response of northern forests to global warming. Global Ecology and Biogeography Letters 1: 129-135.
Reynolds, J.F., P.R. Kemp, B. Acock, J.-L. Chen, and D.L. Moorhead. 1996. Progress, limitations, and challenges in modeling the effects of elevated CO_2 on plants and ecosystems. In: H.A. Mooney (editor). Carbon Dioxide and Terrestrial Ecosystems. San Diego, California: Academic Press. 347-380.
Riley, C.G. 1953. Hail damage in forest stands. Forestry Chronicle 29: 139-143.
Ritchie, J.C. 1987. Postglacial Vegetation of Canada. Cambridge, UK: Cambridge University Press. 198 p.
Rizzo, B. and E. Wiken. 1992. Assessing the sensitivity of Canada's ecosytems to climatic change. Climatic Change 21: 37-55.
Rowe, J.S. 1972. Forest Regions of Canada. Publication No. 1300. Ottawa, Ontario: Canadian Forest Service. 172 p.
Sakai, A. 1978. Freezing tolerance of evergreen and deciduous broad-leaved trees in Japan with reference to tree regions. Low Temperature Science Series Biological Sciences 36: 1-19.
Sakai, A. 1979. Freezing avoidance mechanism of primordial shoots of conifer buds. Plant and Cell Physiology 20: 1381-1390.
Sakai, A. and W. Larcher. 1987. Frost Survival of Plants: Responses and Adaption to Freezing Stress. Berlin, Germany: Springer-Verlag. 321 p.
Sakai, A. and C.J. Weiser. 1973. Freezing resistance of trees in North American with reference to tree regions. Ecology 54: 118-126.
Schooley, H.O., D.A. Winston, R.L. McNaughton, and M.L. Anderson. 1986. Frost killing red pine female flowers. Forestry Chronicle 62: 140-142.
Solomon, A.M. and R. Leemans. 1997. Boreal forest carbon stocks and wood supply: past, present and future responses to changing climate, agriculture and species availability. Agricultural and Forest Meteorology 84: 137-151.
Spittlehouse, D.L. and R.J. Stathers. 1990. Seedling Microclimate. Land Management Report No. 65. Victoria, British Columbia: British Columbia Ministry of Forests. 28 p.
Stathers, R.J. 1989. Summer Frost in Young Forest Plantations. Report No. 73. Victoria, British Columbia: British Columbia Forest Resource Development Agreement. 24 p.
Suffling, R. 1995. Can disturbance determine vegetation distribution during climate warming? A boreal test. Journal of Biogeography 22: 501-508.
Thomas, G. and P.R. Rowntree. 1992. The boreal forests and climate. Quarterly Journal of the Royal Meteorological Society 118: 469-497.
Thompson, I.D. 2000. Forest vegetation of Ontario: factors influencing landscape change. In: A.H. Perera, D.L. Euler, and I.D. Thompson (editors). Ecology of a Managed Terrestrial Landscape: Patterns and Processes of Forest Landscapes in Ontario. Vancouver, British Columbia: University of British Columbia Press. 30-53.
Thompson, I.D., M.D. Flannigan, B.M. Wotton, R. Suffling, and R.E. Munn. 1998. The effects of climate change on landscape diversity: an example in Ontario forest. Environmental Monitoring and Assessment 49: 213-233.
Townend, J. 1993. Effects of elevated carbon dioxide and drought on the growth and physiology of clonal Sitka spruce plants (*Picea sitchensis* (Bong.) Carr.). Tree Physiology 13: 389-399.
Trewartha, G.T. and L.H. Horn. 1980. An Introduction to Climate, 5th edition. New York, New York: McGraw-Hill. 416 p.
Van Cleve, K. and J. Zasada. 1970. Snow breakage in black and white spruce stands in interior Alaska. Journal of Forestry 68: 82-83.
Warrick, R.A., H.H. Shugart, M.J. Antonovsky, J.R. Tarrant, and C.J. Tucker. 1986. The effects of increased CO_2 and climatic change on terrestrial ecosystems. In: B. Bolin, R.B. Döös, J. Jager, and R.A. Warrick (editors). SCOPE 29: The Greenhouse Effect, Climatic Change, and Ecosystems. Chichester, UK: John Wiley. 363-392.
Weatherald, R.T. and S. Manabe. 1986. An investigation of cloud cover change in response to thermal forcing. Climatic Change 8: 5-23.
Webb, T. 1992. Past changes in vegetation and climate: lessons for the future. In: R.L. Peters and T.E. Lovejoy (editors). Global Warming and Biological Diversity. New Haven, Connecticut: Yale University Press. 59-75.
Webb, T. and P.J. Bartlein. 1992. Global changes during the last 3 million years: climatic controls and biotic response. Annual Review of Ecology and Systematics 23: 141-173.
Weber, M.G. and M.D. Flannigan. 1997. Canadian boreal forest ecosystem structure and function in a changing climate – impacts on fire regimes. Environmental Reviews 5: 145-166.
Weber, M.G. and K. Van Cleve. 1981. Nitrogen dynamics in the forest floor of interior Alaska black spruce ecosystems. Canadian Journal of Forest Research 11: 743-751.
Weber, M.G. and K. Van Cleve. 1984. Nitrogen transformations in feather moss and forest floor layers of interior Alaska black spruce ecosystems. Canadian Journal of Forest Research 14: 278-290.
Whitney, C.G. 1986. Relation of Michigan's presettlement pine forests to substrate and disturbance history. Ecology 67: 1548-1559.
Woodward, F.I. 1987. Climate and Plant Distribution. Cambridge, UK: Cambridge University Press. 174 p.
Woodward, F.I. and B.G. Williams. 1987. Climate and plant distribution at global and local scales. *Vegetatio* 69: 189-197.

7 Fire Regimes and their Simulation with Reference to Ontario

CHAO LI*

INTRODUCTION

Forest dynamics, the processes and causes of change in forests, is one of the central issues in ecology and forest science. Various disturbances affect forest growth rates and mortality. By removing trees from the canopy, these disturbances create gaps that alter the structure of the forest landscape. In Ontario, forest fire has been one of the major natural disturbances shaping forest landscape structure. Studies of fire disturbance patterns are being conducted to advance our understanding of Ontario forest dynamics and to improve the planning of forest resource management. Detailed studies are also conducted on the behaviour of forest fires to improve the accuracy of predictions of fire start and spread and to make fire protection efforts more effective (e.g., Forestry Canada Fire Danger Group 1992; Hirsch 1996).

With the goals of sustaining forest resources and maintaining biodiversity, forest resource managers in Ontario have adopted a policy that forest management should emulate natural disturbance patterns (Euler and Epp 2000, this volume). Their rationale is that long-term forest stability may depend on how closely human practices of forest harvest and regeneration correspond to the effects of natural disturbance (Engelmark et al. 1993). To evaluate and implement this forest policy, it is essential to understand natural forest fire disturbance patterns. The fires in issue, and discussed in this chapter, are primarily stand-replacing fires, those which consume the forest by burning either very rapidly through the crowns or very intensely at the ground surface. Stand-replacing fires cause abrupt changes in the pattern of forest patches across the landscape. Clear-cut harvesting produces similarly abrupt changes in the forest mosaic, so both types of disturbance may have a similar ecological impact on the forest landscape.

My first objective in this chapter is to provide an overview of the nature of fire disturbance and the role of fire in the development of forest landscapes. The second is to present an ecological modelling approach to the exploration of forest fire dynamics at the landscape scale. Thus, the chapter begins with definitions of a number of parameters used to describe fire regimes and to assess their ecological consequences. An overview is then provided of the fire regimes in Ontario, a subject discussed in greater depth by Thompson (2000, this volume). The interaction of various factors influencing fire regime dynamics is described. The role of landscape-level simulation models in integrating existing knowledge and information on fire regime dynamics is explored, and several models are examined. Finally, the ability of such models to provide information important for the development of forest policy is demonstrated through a recent case study conducted in northwestern Ontario.

THE DESCRIPTION OF FIRE REGIMES

Research into fire disturbance is concerned with both short-term fire behaviour and long-term fire dynamics. Fire behaviour studies include the processes of fuel ignition, flame development, fire spread, and other

* Canadian Forest Service, Northern Forest Research Centre, 5320 - 122nd Street, Edmonton, Alberta T6H 3S5

related phenomena that are determined by the fire's interaction with weather, fuels, and topography. Studies of fire behaviour are conducted on a time scale ranging from a whole fire season, to a month, a day, an hour, or even a minute. The objective of such research is to predict how fires will behave under various physical conditions. Numerous investigations have been conducted on the ignition of fires by lightning, the processes of fire combustion, and the spread of fires to the surrounding forest (e.g., Rothermel 1972, 1983; Turner and Lawson 1978; Albini and Stocks 1986; Van Wagner 1987; Todd and Kourtz 1991; Weber 1991; Johnson 1992; Kourtz and Todd 1992).

Results from fire behaviour studies in Canada have been incorporated into the Canadian Forest Fire Behaviour Prediction System, or FBP (Forestry Canada Fire Danger Group 1992; Hirsch 1996). Both the FBP and the Canadian Forest Fire Weather Index system, or FWI (Van Wagner 1987), provide equations for calculating indices important in fire management operations. Fire scientists at the Canadian Forest Service have conducted numerous studies on the prediction of precise fire events based on very detailed information, including the simulation of weather conditions.

The study of long-term fire dynamics is concerned with the history of fire disturbances in a given area, or the "fire regime" of that area (Stokes and Dieterich 1980). A fire regime consists of a general pattern of fire activity over periods beyond one fire season, usually on the scale of decades or centuries. The methods used to reconstruct fire history include compiling historical records, examining the present age structure of forest stands, gathering evidence from old fire scars on trees, excavating the remains of burned forests from old layers of lake mud, and mapping forest age classes and fire boundaries (Heinselman 1973). Once fire years are determined, then fire frequency can be calculated.

Natural Fire Regimes

The ultimate goal of studying fire history is to understand the natural fire regime under which the present biota have developed (Heinselman 1971). The 1997 National Workshop on Wildland Fire Activity in Canada defined a natural fire regime as the end result of fires occurring in wildlands over broad areas and long periods; that is, as the climatology of fire behaviour (Simard 1997). A natural fire is any fire of natural origin, one caused by lightning, spontaneous combustion, or volcanic activity (Food and Agriculture Organization of the United Nations 1986).

The natural fire regimes discussed herein consist mainly of fires that have burned unchecked. As discussed by Euler and Epp (2000, this volume), Ontario's *Crown Forest Sustainability Act* (1994) specifies "natural disturbance" and the "bounds of natural variation" as broad guidelines for forest management. A natural fire regime might be adequately interpreted, therefore, as the best estimate available of fire dynamics as they occur under a minimum of human influence. The major challenge faced by researchers in helping forest managers interpret and implement this policy lies in reconstructing natural disturbance regimes so as to include not only forest fires, but also forest insect pests and other disturbances that predispose forests to fire.

As pure natural conditions no longer exist, true natural fire regimes may never be known. Pure natural forests have not existed in North America since the beginning of human activity, at least 14,500 years ago (e.g., Butzer 1971). Since that time, forest conditions have been modified to various degrees by human influence. The use of fire by the Aboriginal peoples of North America prior to European settlement, however, is thought to have had only local impact (Johnson 1992; Weber and Stocks 1997). The greatest influence of human activity on fire regimes has occurred in the twentieth century, especially from the 1920s onward, with increasing improvements in access to forested lands, in fire detection, and in fire suppression technology. Empirical data on fire dynamics are increasingly available, a fact that enhances our basis for a comprehensive understanding of fire regimes. Most of the empirical data have been obtained from point-based sampling designs, however, so it is still difficult to draw general conclusions about spatial fire dynamics.

The Parameters of Fire Regimes

Disturbance regimes are usually characterized by the size, intensity, and severity of the disturbance events, by the times and frequency of occurrences, by the fire rotation period (the period of time within which an area of a specified size can be expected to experience fire over all of its extent), and by the synergism with other types of disturbance (White and Pickett 1985). Fire regimes are no exception. Heinselman (1981) described fire regimes according to fire type, intensity, size, and frequency or cycle, but placed particular emphasis on fire frequency and intensity. Malanson (1987) focused on the parameters of size, intensity, season of occurrence, and frequency. Simard (1997) included the parameters of cycle, type, and severity.

Fire Frequency

Fire frequency, or the number of fires per unit time, is essentially a point-based parameter, although it can be assessed over a specific area (e.g., Johnson and Van Wagner 1985; Food and Agriculture Organization of the United Nations 1986). Fire frequency, on its own, indicates merely that that fires have occurred a certain number of times within the period measured and says nothing about the nature of the fires. A related concept is fire-return interval, or simply "fire interval," which is the reciprocal of frequency. Fire interval signifies the average number of years between fires at a given location (e.g., Merrill and Alexander 1987). In practice, of course, the intervals between fires are not equal, so what is actually calculated is the mean fire interval.

Both fire frequency and fire interval can be obtained from empirical data collected from point-based samples. One possible source of such data lies in the chronology of fire scars on tree boles. If there are five fire scars spanning 200 years, for example, then the mean fire return interval is 40 years and the fire frequency is 0.025 fire per year. The use of fire scars to determine fire interval is only appropriate for studying fires of less than stand-replacing severity, fires that were not intense enough to kill the scarred trees.

Fire Cycle or Rotation

A fire cycle is defined as the average number of years needed to burn an area equal in size to a particular area under study (e.g., Heinselman 1973; Johnson and Van Wagner 1985; Turner and Romme 1994). This concept takes into account that some sites in the area may burn more than once and others not at all. Heinselman (1973) used the term "natural fire rotation" for the concept. He thought that the average time required for a natural fire regime to burn over an area equivalent to the total area of a given ecosystem was analogous to the forester's "rotation," the period required to replace a forest stand. Although fire is a semi-random process in many ecosystems, each vegetation type and physiographic site tends to have its own characteristic return interval.

As fire cycle is an area-based fire regime parameter, data on fire size are used to calculate it. Fire cycle, FC, is calculated as follows from the mean annual area burned, MYAB. TA is the total study area, TY the total number of years for the investigation, and TB the total area burned.

$$FC = TA/MYAB \qquad [7.1]$$

and

$$MYAB = TB/TY \qquad [7.2]$$

Total area burned (TB) is calculated as follows:

$$TB = \sum fire_i \quad i = 1, \ldots, n \qquad [7.3]$$

where n is the total number of fires.

Summaries of fire cycles in various regions can be found in Heinselman (1973), Turner and Romme (1994), and Baker (1995). Duchesne and Hinrichs (1996) have compiled a summary of fire frequencies in Canadian northern ecosystems before and after European settlement.

Fire Size and Size Distribution

Two fire size statistics, mean fire size and fire size distribution, are usually used to describe a fire regime. Mean fire size denotes the average size, in units of area, of the individual fires that occur across a landscape of interest within a given period. When the fires are classified by size range, the frequency with which fires in each class have occurred signifies fire size distribution. Fire size distribution is a complementary description of fire size variation in a fire regime.

Mean fire size differs from one forest type to another. For example, fires in the boreal forest landscapes of northern Ontario, which are composed largely of continuous coniferous stands, are generally larger than fires in the mixedwood and deciduous forest landscapes of the Great Lakes-St. Lawrence region (e.g., Thompson 2000, this volume). One explanation lies in the potential of various fuel types identified in the FBP system for different fire behaviour. Coniferous fuel types generally show higher rates of fire spread than deciduous or mixedwood fuel types, given similar weather and topography (Forestry Canada Fire Danger Group 1992).

Several probability distribution functions have been used to describe fire size distribution. These include the negative exponential probability distribution employed by Johnson (1992), Johnson and Van Wagner

(1985), and Suffling (1991), among many others. This function is used more frequently than either the Weibull distribution (used, for example, by Johnson and Van Wagner 1985) or geometric distributions (used, for example, by Boychuk et al. 1997).

Fire Intensity and Severity

Fire intensity is the heat energy released per unit time, per unit length of fire front. It is the product of the available heat of combustion per unit area and rate of spread (e.g., Merrill and Alexander 1987). Intensity is a major determinant of the impact of a fire. High-intensity fire kills trees easily and is usually associated with stand-replacing crown or surface fires. As reported by Van Wagner (1977a), for example, fires in boreal forests usually fall into this category. Low-intensity fire does not usually kill trees and is generally associated with non-stand-replacing surface fires. Fire severity is measured by the quantity of fuel consumed. Fire intensity and severity influence both the immediate effects of fire on the forest and the ecological consequences.

Types of Fire Regimes

Fire regimes are classified by their frequency and intensity. Heinselman (1978, 1981) and Kilgore (1981) distinguished the following six types of fire regimes:

Type 1: Frequent, light surface fires occurring in a cycle of 1 to 25 years;
Type 2: Infrequent, light surface fires occurring in a cycle of more than 25 years;
Type 3: Infrequent, severe surface fires occurring in a cycle of more than 25 years;
Type 4: A combination of crown fires with a short fire cycle and severe surface fires, together occurring in a cycle of 25 to 100 years;
Type 5: A combination of crown fires with a long fire cycle and severe surface fires, together occurring in a cycle of 100 to 300 years.
Type 6: A combination of crown fires with a very long fire cycle and severe surface fires, together occurring in a cycle longer than 300 years.

Each fire regime is associated with a particular type of forest. The fire regimes observed in Ontario include both surface and crown fires with various frequencies and intensities.

Ecological Roles of Forest Fire

Forest fire has been an integral component of vegetation dynamics in boreal forests for millions of years and has played an important role in the development of the tree species assemblages that exist in present-day forests (Hopkins 1967; Weber and Stocks 1997). Fire is the key environmental factor initiating succession. Fire controls the species composition and age structure of forests, and produces the vegetation patterns upon which the animal components of the ecosystem depend (Heinselman 1971). As will be discussed in a later section, the influence which fire exerts on forest development is significantly modified by human activity to suppress forest fires.

Forest Fire and Species Adaptation

Most tree species in the boreal forest are fire-dependent. They are generally adapted to intense crown fires occurring on a short cycle and over large areas. Jack pine (*Pinus banksiana*), for example, requires fire to open serotinous cones that accumulate in the canopy and to prepare the mineral bed for seed germination (e.g., Cayford et al. 1967; Cayford 1970).

Red pine (*Pinus resinosa*) is a fire-dependent species that occurs in both boreal coniferous forests and Great Lakes-St. Lawrence mixedwood and deciduous forests. This shade-intolerant species requires fire to expose a mineral-soil seedbed and also to remove the canopy so that seedlings receive enough light to survive (e.g., Bergeron and Brisson 1990; Engstrøm and Mann 1991).

Aspen occurs as a pioneer species on many burns and generally succeeds to white or black spruce (e.g., Wilton 1964). Suckering in aspen is evidence that the species evolved in a fire-rich environment. The thick, fire-resistant bark of red pine and ponderosa pine (*Pinus ponderosa*) gives evidence of a similar origin. Other instances of species adaptation to fire are described by Thompson (2000, this volume). Although the process by which these adaptations occurred may never be known, it is believed that the adaptation process and subsequent successional pathways were fully developed by the end of the last glaciation, 15,000 years ago (Weber and Stocks 1997).

Forest Fire and Species Composition

In a fire environment, tree species composition is determined by the frequency of fires. From a long-term perspective, a species cannot survive where the mean fire-return interval is shorter than its time to maturity. Too long a fire-return interval, however, favours shade-tolerant species. Flannigan (1993) provided evidence that species abundance and distribution are related to fire regimes. In particular, he found a strong link between fire intensity and the abundance of red pine, as expressed by an inverse relationship between the percentage of conifers in the forest and the volume of red pine.

Forest Fire and Soil Chemistry

Fire plays an important role in recycling nutrients and hence may influence long-term site productivity and ecosystem function. Paré et al. (1993) reported that, at sites in the southern boreal forest in eastern Canada, the forest floor showed a steady accumulation of organic matter and total nutrients with the passage of time since the last fire, and a 50 percent decrease in the concentrations of available phosphorus and potassium. The availability of nitrogen and calcium was more strongly affected by tree species and gaps than by time since fire. Fire increases soil pH and the rate of nitrogen mineralization by converting biomass to ash and causing soil temperature to rise (e.g., Viro 1974; Woodmansee and Wallach 1981).

In the boreal mixedwood forest near Thunder Bay, Ontario, Johnston and Elliott (1998) reported significant differences in the chemistry of the forest floor and in woody fuel consumption between areas of burned forest and areas where the trees had been cut before the fire but remained on the site. These researchers found that the pH level was higher and phosphorus concentration lower in the ash of unharvested blocks. They found the highest phosphorus concentrations in whole trees cut before the fire. They also found that the concentrations of nitrogen, phosphorus, and magnesium in the foliage were higher in the unharvested, burned plots than in the harvested, burned plots, and that these concentrations were negatively correlated with the depth of forest floor materials consumed by the fire.

Forest Fire and the Pattern of Vegetation Patches

Another ecological role of forest fire is the production of vegetation patterns across landscapes, including patterns of age structure. The mosaic of vegetation types is composed of forest stands representing various age classes or successional stages. The distribution of age classes provides basic information for forest resource management planning, and an understanding of fire disturbance regimes is essential for predicting future age-class distribution.

Vegetation mosaic patterns suitable to the needs of particular species are essential for supporting the animal life of ecosystems (Voigt et al. 2000, this volume). Herbivores in forest ecosystems are adapted to the fire environment; for example, white-tailed deer (*Odocoileus virginianus*) and moose (*Alces alces*) are adapted to recent burns and to the open areas that occur in the early successional stages of the forest. The distribution of these animals changes when the forest matures and the open areas disappear. Correspondingly, the populations of predators depending on the herbivores fluctuate and their distribution changes. Forest birds are also distributed according to the forest conditions. For example, in a study carried out in Missouri on the effects of local habitat and landscape variables on the abundance of 22 bird species, Latta et al. (1998) demonstrated that 5 of the 22 bird species examined were sensitive to local vegetation variables, and 17 species responded most strongly to one of four landscape variables (mean patch size, core area, edge density, or forest cover). The diversity of the animal community, therefore, depends on the characteristics of vegetation patterns (Holling 1992a). Fire disturbance thus plays a crucial role in maintaining biodiversity.

Hypotheses of Forest Succession

Various hypotheses of forest succession have been proposed to explain the patterns of forest dynamics and the development of present-day forests. These hypotheses feature some of the ecological roles of fire. The concept of the climax forest (Clements 1916) contains a forest succession hypothesis consisting of two functions: exploitation and conservation. Exploitation refers to the rapid colonization of recently disturbed areas by a pioneer tree species, while conservation refers to the slow accumulation and storage of energy and material after the pioneer tree species is established. True climax forests are probably uncommon and can occur only on sites where fires are rare. In boreal forests, fires periodically initiate new successional processes before the forest achieves its climax stage.

Holling (1986, 1992b) revised the hypothesis of Clements to include two additional functions: release and reorganization. Release, or creative destruction, refers to a process by which tightly-bound accumulations of biomass and nutrients become increasingly fragile, until they are suddenly released by a disturbance such as forest fire. Organization, or mobilization, refers to the soil processes that minimize nutrient loss and reorganize the nutrients, so that they become available for the next phase of exploitation. Under this hypothesis, the renewal cycle of forest succession is composed of four stages. The cycle begins with exploitation and proceeds slowly to conservation, which is then followed by a very rapid release of nutrients, by a rapid stage of reorganization, and finally by rapid resumption of the exploitation stage. The hypothesis containing these four stages is termed the "four-box model." As one of the agents of release, forest fire disrupts the renewal cycle of forest succession and resets the timing of succession processes.

Holling's hypothesis explicitly incorporates several of the ecological roles of forest fire. These include the transformation of organic standing biomass into its original components (water, carbon dioxide, and minerals) and the initiation of new successions. The mineral soil exposed by fire and the dense ash covering the forest floor create a favourable environment for the growth of young trees. The seed sources of some coniferous species, if previously stored in the canopy, are generally sufficient to seed a new forest.

The Influence of Human Activities

The development of fire management organizations, of powerful technology for fire detection, and of effective initial measures for attacking fire, have greatly increased the ability of present-day forest managers to manipulate fire disturbances. The events leading to today's fire-fighting capabilities in Ontario cover a span of more than 150 years. In 1849, the provincial government formed a Royal Commission to seek ways of protecting the forests from unnecessary destruction (Aviation and Fire Management Centre 1986). Succeeding commissions studied the problem, but little action was taken. Destructive fires continued, and remained uncontrolled. In fact, little credence was placed in the possibility of controlling forest fire. During this period, fire regimes were approximately natural, as human influence was limited.

The province enacted its first forest fire prevention legislation in 1878. Fire districts were established, and the use of fire in the forest was restricted during periods of particular danger. In 1886, fire rangers were appointed. By 1897, 191 fire rangers had been assigned to Crown lands, and a fire control system had begun to develop (Hughes 1980; Aviation and Fire Management Centre 1986).

In 1917, the Ontario Government passed the *Forest Fires Prevention Act* (1917). This step was taken in reaction to a disastrous fire in July 1916 which swept through areas around the north-central settlements of Matheson and Cochrane. This fire killed a large number of people and burned everything in its path, including culverts, corduroy roads, bridges, homes, crops, and part of the nearby town of Iroquois Falls. Under this statute and amendments that followed, a fire-detection system was established which included lookout towers at key points on high ground. Fire ranger stations and supply warehouses for fire-fighting were also built across the north of the province. By 1924, 20 World War I surplus "flying boats" were in use for detecting fire and carrying fire crews and equipment to the fire scene. As the fire control system grew, Ontario developed a policy of total suppression, with the ultimate goal of suppressing all fires in the province, at all cost (Aviation and Fire Management Centre 1986).

The 1970s and early 1980s were a time of important organizational changes in Ontario's fire management system. The province was divided into eight regions, which were subdivided into 49 districts. A two-level field fire organization was established in 1973. By 1983, command and control of forest fire management took place at a regional level, and a provincial fire centre (the Aviation and Fire Management Centre) had been established in Sault Ste. Marie.

After a severe fire season in 1980, the Ontario Ministry of Natural Resources (OMNR) launched an initiative known as the Fire Management Improvement Project (FMIP) to develop and implement an improved provincial fire management system. The outcome included changes in fire policy and the division of fire management zones. The new OMNR fire policy had the following three objectives: (1) to prevent loss of human life and personal injury and to minimize social disruption; (2) to ensure that fires had no more than a minimal effect on public works, private property, and natural resources; and (3) to use the natural benefits of fire to achieve other objectives of the Ministry for land and resource management.

At the present time, all of Ontario, except for segments in the extreme north and south, is divided into three fire management zones, a zone of intensive management, a zone of measured management, and a zone of extensive management. The intensive zone, located in central and southern Ontario, includes areas where population densities are high. Fire suppression is performed here to protect settlements, industry, recreational areas, and natural resources. The extensive zone is located in the far north of the province. Here, fires are allowed to burn unchecked, except in the rare case when values are threatened. In the zone of measured protection, located between the intensive and extensive zones, fires are monitored, but usually no action is taken to suppress them. The boundaries of the three fire management zones have gradually shifted northward as the ability to control fire has been enhanced, so that the area of human influence over fire dynamics has steadily increased.

Changing Fire Regimes in Ontario

Ontario's fire regimes are diverse. Ward and Tithecott (1993) summarized fire history studies and concluded that Ontario's fire cycle in the pre-suppression era ranged from 20 to 135 years, depending on location, although the average fire cycle was about 65 years. Thompson (2000, this volume) describes the various fire disturbance patterns which occur across the province.

The characteristics of natural fire regimes are determined by the interaction of fire events, weather patterns, fuel conditions, and landscape topography. Human manipulation of fire disturbance is believed to alter the natural fire regimes. This premise can be tested by comparing fire regime statistics from the three fire-management zones of the province. Under the Ontario fire policy, fire regimes in the southernmost, intensive-management zone come under the

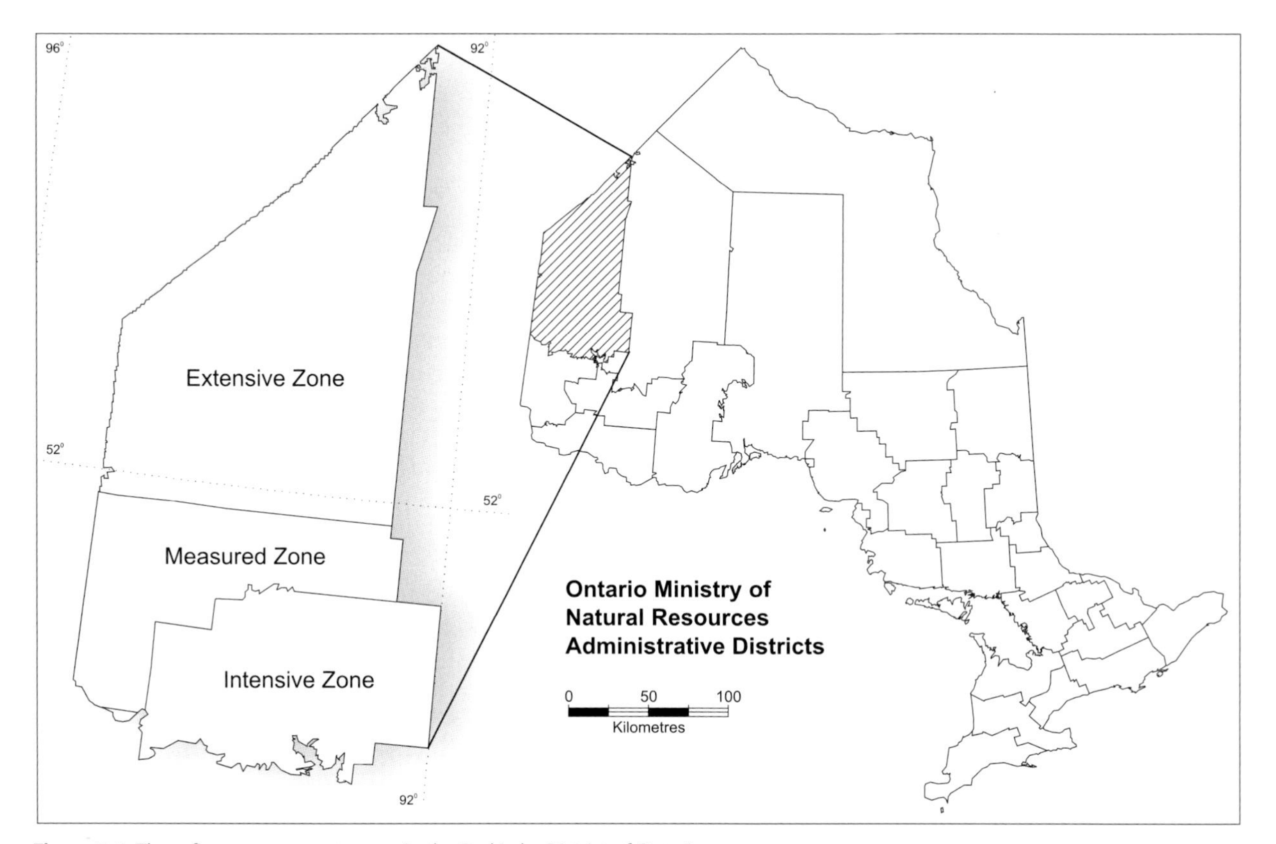

Figure 7.1 Three fire management zones in the Red Lake District of Ontario.

influence of intensive human activity. Those in the extensive zone receive a minimum of human influence and thus remain approximately natural fire regimes. The fire regimes of the measured zone represent an intermediate state.

The Red Lake District in the northwestern part of Ontario's boreal region (Figure 7.1) provides an appropriate site for making this comparison, as all three fire management zones are represented in the district. Fire records from 1976 to 1993 provided by the Aviation, Flood and Fire Management Branch of OMNR have been used in this analysis.

Table 7.1 summarizes fire occurrences in the three fire management zones of Red Lake District from 1976 to 1993. During this period, human activities in the intensive-management zone caused more frequent fires than in either of the other zones, where fire was caused mainly by lightning strikes. This finding agrees with other observations that fire occurrence increases with population growth (Stocks 1991). Fire suppression in the intensive-management zone, however, significantly reduced both the mean fire size and the percentage of the total area burned. It also extended the fire cycle, calculated on the basis of total forested land, from 50.67 years in the extensive-management zone to 134.66 years in the intensive-management zone.

These results are qualitatively consistent with Ward and Tithecott (1993). Those authors indicated that the total land area of both the intensive and measured zones of Ontario is about 47 million ha. Calculating with an average natural fire cycle of 65 years, derived from a number of natural fire history studies, they found that an average of about 1.54 percent of the total land area would have been burned each year. Under the fire management scenario, however, the observed average annual area burned was only about 0.17 percent from 1978 to 1992, yielding a fire cycle of 580 years. The fire cycle under the fire management scenario is thus significantly longer than that without fire management.

The values in Table 7.1 can be interpreted as indicating qualitative changes in fire regimes among the three fire management zones. It is often difficult, however, to make an accurate comparison between the fire regimes present in various forest types and geographic regions, even if the overall areas are of similar size (Kilgore 1981). Detailed analyses of the effect of topography, fuel type, and weather patterns in each area under comparison are required to obtain quantitative conclusions. Figure 7.2 shows the mean fire size by each of these factors in the three zones.

Table 7.1

Historical fire occurrences in the three fire management zones of the Red Lake District of Ontario, from 1976 to 1993 (forested land only).

Fire management zone	*Intensive*	*Measured*	*Extensive*
Total no. of fires (10^3km^2)	73.29	58.01	17.82
Mean fire size (ha)	182.40	352.56	1993.39
Total area burned (%)	13.37	20.45	35.52
Mean annual burn (%)	0.74	1.14	1.97
Fire cycle (year)	134.66	88.02	50.67

In the extensive-management zone, fires caused by lightning and human activity have a similar mean size, probably because they were allowed to spread freely. In the measured and intensive zones, however, fires caused by human activity were much smaller than those caused by lightning. This difference can be explained by three factors that exist in the intensive-management zone: better access to forested areas, more effective fire detection, and fire suppression efforts (Ward and Tithecott 1993; Weber and Stocks 1997).

The differences among the fire regimes depicted by Table 7.1 may, indeed, result from the application of different fire-management policies. They are unlikely to result from such factors as differences in the length of fire season or from any intrinsic difference in fire cause (i.e., human activity or lightning). If length of fire season were a dominant factor in determining fire regime, the longer fire season in the intensive zone,

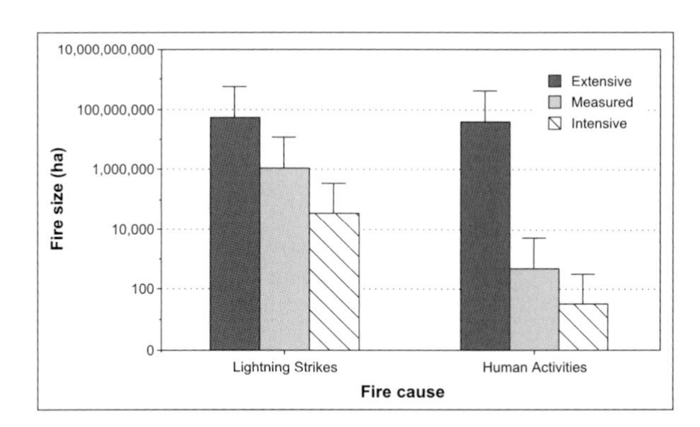

Figure 7.2 Comparison of mean fire size between fires caused by lightning and fires caused by human activities in the Red Lake District of Ontario from 1976 to 1993.

with its accompanying higher risk of fire occurrence, should give rise to a fire cycle shorter than in the extensive and measured zones. If fire regime were influenced purely by the fact that fires are of human origin, then the intensive zone should, once again, have a shorter fire cycle, because of its higher fire risk.

Finally, if lightning activity is a key factor in and of itself, there should be significant differences in lightning activity among the three fire management zones. An analysis of fire cause shows that the fires caused by lightning detected in the intensive-management zone (56.35 fires per 1000 km^2) were, in fact, more numerous than in either the measured-management zone (47.57 fires per 1000 km^2) or the extensive zone (13.55 fires per 1000 km^2). These statistics suggest that the intensive-management zone may experience a higher risk of lightning fire than the measured-management zone or the extensive-management zone. The only other explanation for the greater number of fires caused by lightning in the intensive zone is that, as one proceeds northward, the number of days on which thunderstorms occur become fewer (Johnson 1992). All of these factors indicate that a shorter fire cycle should be expected in the intensive management zone – just the opposite of the results shown in Table 7.1.

Another finding of the analysis carried out for Red Lake District is that the lightning-caused fires were responsible for a very large percentage of the total area burned. Fires caused by lightning represent 78.25 percent of all the fires in the district from 1976 to 1993 and are responsible for 83.28 percent of the total area burned. Fires of human origin were the cause of fire over only 3.4 percent of the area burned, and the cause of the remainder of the fires is unknown. This result is consistent with Johnson (1992), who reported that lightning is the most significant cause of fires in the boreal forest, accounting for about 90 percent of the area burned.

The results from Red Lake District also indicate that 5.33 percent of the fires larger than 1000 ha in size are responsible for 96.42 percent of the total area burned from 1976 to 1993. This result is consistent with the report by Stocks (1991) that the 2 or 3 percent of fires that grow larger than 200 ha are responsible for about 98 percent of the total burned area in Canada. The Red Lake results are also compatible with a theoretical investigation by Strauss et al. (1989). From a concept of fire size inequality, those authors developed an extreme proportion function and then applied this function to the analysis of empirical fire size distributions. The results showed that only 1 percent of the largest fires could burn 80 to 96 percent of the total area burned in the western United States.

Although no data on pure, natural fire regimes exist, examining fire regimes that occur under various fire-management policies may provide one means of understanding the natural fire regimes. By assuming that the natural fire regime is more similar to a regime in which fire management does not occur than to a regime in which it does occur, one can estimate the influence exerted on fire regimes by fire management. The picture of a natural fire regime achievable by this means may need to be further evaluated by studying the interactions among natural variables of fire events, weather, forest fuel, and topography. Forest ecosystems are complex systems and thus resilient to certain disturbances (e.g., Holling 1973, 1986). Fire disturbance, however, may act as a force of self-organization within the forest ecosystems (e.g., Holling et al. 1996; Müller 1997).

The Interaction of Fire, Weather, Fuel, and Topography

A natural fire regime can be interpreted as the result of interaction among fire events, weather patterns, fuel conditions, and topography, in the area under investigation. In most fire regimes that can be observed, however, it is necessary to take into account the effect of human activities. To understand how all of these factors interact, one must integrate the available information on both fire processes and landscape responses.

The Fire Disturbance Event

A natural fire event starts with a lightning ignition. Once the fire is ignited, its ability to spread to the surrounding area is determined by topography, weather, and fuel conditions. The process of spread will stop when these conditions are no longer supportive.

Fire spread processes start horizontally with the ignition of a series of fuel particles at or near the edge of the ignition point. The flames begin to burn at the source point, then move outward and accumulate enough heat to continue burning. As the flames move out from the source point of ignition, their perimeter can be elongated into an elliptical shape by either wind or slope. The head of the fire is defined as the portion of the fire edge having the greatest rate of spread and frontal fire intensity. The head generally lies in the downwind or upslope part of the fire (Food and Agriculture Organization of the United Nations 1986; Merrill and Alexander 1987). If there is wind blowing, the head of the fire advances faster than other portions of the fire perimeter and burns elliptical patches (Van Wagner 1969; Alexander 1985; Johnson 1992). In the absence of wind, the head of the fire lies in the upslope direction. Over a flat landscape, the head of the fire extends more or less equally in all directions and burns circular patches.

Fuel quantity and quality (that is, dryness) are critical factors as well. Once a fire is started, the accumulation of fuel around the ignition point determines whether enough horizontal heat flux can be generated to ignite unburned fuel in the neighbouring forest stands. The moisture content and quantity of the fuel in those adjacent stands will determine whether the fire spreads. During this process, wind conditions modify the direction and speed of the spread.

Fire will stop spreading when it encounters a barrier, especially a non-flammable feature such as a lake or river. Some forest stands, such as wet sites or lower slopes, have a lower probability of sustaining the fire. Newly burned forests may also have a lower probability of burning again; in fact, the majority of the historical burns in Ontario over the last seven decades do not overlap each other. This latter finding is one of the results obtained from an analysis of the historical fire disturbance map of fires larger than 200 ha from 1921 to 1992 compiled by the Forest Landscape Ecology Program of the Ontario Forest Research Institute (Li, unpub.). This conclusion regarding the overlap of fires is consistent with findings reported by such authors as Clark (1989, 1990). The fire spread stage ends when the fire ceases to spread in any of the possible directions.

As stated above, the pattern of non-overlapping historical burns in Ontario is derived from records of the occurrence and extent of fires over a span of 72 years. Further investigation may be required to interpret this finding, because the forest and weather conditions at each burn are not known. Some studies have already suggested that fire spread processes may not be significantly influenced either by vegetation conditions or by the length of time that has passed since the last burn. For example, Johnson and Larsen (1991) found a negative exponential distribution of fire frequencies in the Kananaskis River Watershed of western Canada. This result suggests a constant fire hazard function and implies that fire spread processes are not sensitive to the age of forest stands.

From the perspective of landscape ecology, forest fire is assessed in terms of a minimum unit of area, the actual size of which depends on the issue of interest. Within that minimum area, forest and fire dynamics are assumed to be homogeneous. Once that assumption is made, fire disturbance can be studied in two separate stages, fire initiation and fire spread (Li and Apps 1995, 1996).

The only difference in describing a fire from a landscape perspective and describing it as a continuous event lies in distinguishing the initial stage from the remainder of the fire process. The initial stage consists of lightning ignition of the fire and the burning of an area equal to the minimum area of interest set for describing the fire. (Since the focus of this chapter is stand-replacing fires, we assume that most of the trees in the burned area are killed.) If the fire burns an area that is significantly smaller than the minimum unit of area, then that fire will be ignored because of its trivial contribution to the fire regime at the landscape level. Whether lightning initiates a fire when it strikes, however, depends on the conditions of the site and certain characteristics of the lightning itself.

The Number of Fire Ignition Sources

A general scenario of fire activity over a particular study area can be obtained by averaging the number of annual ignitions over a given period. Fluctuations around this mean number are caused by weather conditions and by the intensity of human activity. For example, dry storm lightning occurring in a dry season is likely to increase the number of fires in that season, while lightning strikes during a wet season have less chance of igniting fires.

Table 7.2 shows the mean number of fires larger than 200 ha that occurred each year over a span of 72 years within the Hills site regions of Ontario. (Figure 5.1 in Perera and Baldwin [2000, this volume] shows the locations of the Hills site regions.) The annual mean values for each site region are expressed both by the total number of fires and by the number of fires occurring per square kilometre.

The temporal pattern of fire ignition sources can be obtained by analyzing historical fire occurrence (Li et al 1997). For example, from 1976 to 1994, 11,295

Table 7.2

Historical fire occurrence in Ontario (1921-1992) by Hills site regions (Hills 1959). Non-forested areas are included.

Site Region	*Total Fires*	*Fires/Year/10⁴km²*
1E	–	–
2E	29	0.0399
2W	43	0.0531
3E	315	0.5198
3S	157	0.4440
3W	319	0.7024
4E	210	0.7573
4S	231	0.9191
4W	127	1.1052
5E	349	0.8893
5S	90	1.0865
6E	7	0.0219
7E	2	0.0152
Total/Average	1879	0.3681

recorded fire ignitions occurred across a total land area of 144,960 km² in northwestern Ontario. A grid of 32 squares, each measuring 40 km by 40 km, was laid on part of this area, and the fires which occurred in each grid cell were classified by cause, mainly lightning or human activity. To identify cyclical fluctuations, a time series analysis was applied the following three data sets: the fires caused by lightning, the fires caused by human activity, and the total number of fires. The results showed that the most significant fluctuation in all three data sets occurred on a 3.6-year cycle. Although the process by which this cycle is produced remains unclear, a similar pattern of fluctuation has been observed in the province of Alberta, Canada (Cumming et al. 1995), and in Sweden (Granstrøm 1993). In Alberta, ten peak years of fire incidence between 1960 and 1994 yielded an average cycle of 3.5 years. In Sweden, a time series of the number of fires started by lightning per square kilometre per year from 1944 to 1975 indicated a four-year cycle in southeastern Sweden and a three-year cycle in northern Sweden.

Fuel Condition

Fuel condition determines whether the potential of lightning strikes or human activity to start fires is realized. The critical characteristics of fuel condition are fuel type, moisture, and quantity, as affected by climate and weather.

Fuel Type

Fuel type classification is a component of fire behaviour prediction in both the Canadian and US systems. Merrill and Alexander (1987, p. 24) report that, in the Canadian system, fuel type is defined as "an identifiable association of fuel elements of distinctive species, form, size, arrangement, and continuity that will exhibit characteristic fire behaviour under defined burning conditions." More specifically, fuel type consists of "a fuel complex of sufficient homogeneity, extending over an area of sufficient size that equilibrium fire behaviour can be maintained over a considerable time period" (Forestry Canada Fire Danger Group 1992, p. 11).

The fuel type classification in the Canadian FBP system is based on factors present at the forest floor, in the understory, and in the canopy. The amount and distribution of fire-carrying fuel on the forest floor is determined by the depth of the organic, or "duff," layer, and the distribution and load of downed, woody fuel. In the understory, the most important factor is the presence of surface and ladder fuels. The main factors in the overstory are stocking and crown closure. The FBP system identifies a total of 16 discrete fuel types and organizes them into the following five major groups: coniferous, deciduous, mixedwood, slash, and open. For a detailed description of this classification system, the reader may refer to Forestry Canada Fire Danger Group (1992) and Johnson (1992).

In the FBP system, fuel type provides the basis for calculating the rate of fire spread, which is a key factor in fire intensity. Where many different fuel types are distributed generally across a landscape, they may give rise to a heterogeneous mixture of fire intensity levels and, in turn, to a complex spatial mosaic of burn patterns.

Fuel Moisture Content

Fuel moisture content, or degree of dryness, has a direct effect on fire behaviour potential. The Canadian FWI System uses three codes to reflect the moisture content of components of a fuel complex (Van Wagner 1987). The first code is the Fine Fuel Moisture Code (FFMC), which represents the moisture content of small fuels in a forest stand having a dry weight of about 0.25 kg·m⁻². FFMC values indicate the potential for fire ignition, if an ignition source is present, and are determined by rainfall, relative humidity, wind speed, and temperature. Because of the short time required by this fine fuel component to reach equilibrium moisture content (a point at which the fuel neither gains nor loses moisture, as long as conditions remain constant), its dynamics are thought to be quick fluctuations around a mean (e.g., Harrington et al. 1983).

The second code is the Duff Moisture Code (DMC), which represents the moisture content of organic matter having a dry weight of about 5 $kg \cdot m^{-2}$. DMC is a function of rainfall, relative humidity, and temperature. It is a slow-changing variable that indicates a period of drought that is substantial but of less than one month in duration, during which time fire could become established over a large area (Harrington et al. 1983).

The third code is the Drought Code (DC), which represents the presence of a deep layer of compact organic matter having a dry weight of about 25 $kg \cdot m^{-2}$. The DC values are calculated from rainfall and temperature. The DC code is an indicator of potential long-term drought. In accordance with the cumulative effect of precipitation during a fire season, DC values increase with time from the start of the fire season (McAlpine 1990).

The concept of "timelag" is used to represent the rate of change in the moisture content of a fuel, and relates to the size of the elements comprising the fuel. Timelag has been defined as the drying time required for dead fuels to lose 63 percent of the difference between their initial moisture content and their equilibrium moisture content, under standard conditions of dry-bulb temperature, relative humidity, wind speed, and time of year (Food and Agriculture Organization of the United Nations 1986; Merrill and Alexander 1987). Dead forest fuels exhibit timelag values ranging from minutes to months. The fuels represented by the FFMC, DMC, and DC codes in the Canadian FWI System have timelag values of two-thirds of a day, 12 days, and 52 days, respectively, in average weather conditions (Van Wagner 1987; Johnson 1992).

The three moisture-content codes are used in predicting short-term fire behaviour. For the study of long-term fire regimes, it appears reasonable that attention should be paid to the average changes in these codes over longer time periods, or to fuels with a longer timelag. Fluctuations in longer timelag values are an indication of average weather conditions among fire seasons. The Thousand-Hour Timelag Fuel Moisture (THRFM) Code is a method of measuring long-term moisture conditions in the US system of fire behaviour prediction. Renkin and Despain (1992) investigated the relative variation in THRFM values over fire seasons from 1965 to 1988 in Yellowstone National Park. The study resulted in an approximate normal probability distribution; in other words, it demonstrated that dry years could, indeed, bring about an increase in overall fire probability, with more fire ignitions and easier fire spread, and that wet years could bring about a corresponding decrease in fire probability.

Since Canada's DC code is defined by a timelag of 1248 hours, or almost one-quarter again the length of the US 1000-hour index, one may speculate that long-term DC code dynamics could be described by the same probability distribution, or at least a very similar one. The longer the timelag is, the more slowly the moisture change takes place, and the more stable the result should be.

Fuel Quantity

Fuel quantity depends on the balance between accumulation and decomposition and is a limiting criterion for all fires spreading in vegetation (Van Wagner 1983). Measured as above-ground biomass, fuel quantity has been observed to follow a distinct temporal pattern. First, there is a slow increase in fuel quantity, then an accelerated increase. A less rapid increase follows, and finally maximum stand biomass is achieved, after which point fuel quantity either remains constant or declines slightly. This temporal pattern has been found in a wide range of forest ecosystems after a large-scale disturbance (e.g., Bormann and Likens 1979; Sprugel 1985).

Different temporal patterns, however, have been observed in fuel components of different sizes. For example, Romme (1983) described temporal patterns in various fuels in the subalpine forests of Yellowstone National Park. The small fuels examined in this study included needle litter and other fuels with a one-hour timelag that were capable of supporting fire initiation and initial fire spread. The availability of small fuels is low in very young stands, reaches a maximum after 150 to 200 years, then remains constant or declines slightly in older stands of mature even-aged lodgepole pine (*Pinus contorta*) forest in the Yellowstone area. Dead, woody fuel is important in determining potential fire intensity and flame height. The availability of that fuel is high immediately after a fire, declines to a minimum when the new stand reaches an age of between 70 and 200 years, and then increases again to reach a peak at about 350 years of age. The vertical continuity of fuels controls the initiation of a crown fire. Vertical continuity is high in very young stands, but then a gap develops between the live crown base and the ground. Continuity reaches its maximum at about 200 years.

Sturtevant et al. (1997) found a similar temporal pattern in coarse woody debris (CWD) in the boreal forest of Newfoundland. The CWD in these forests consists mostly of decaying residual material until the new stand reaches 50 or 60 years of age; after that, it is composed of the woody debris from the new forest itself. This study indicated that CWD volume followed the general "U-shaped" temporal trend of accumulation that has been observed in other northern forest systems composed of balsam fir (*Abies balsamea*), lodgepole pine, Douglas-fir (*Pseudo-tsuga canadensis*), and Douglas-fir with western hemlock (*Tsuga canadensis*).

While this temporal pattern in the increase of above-ground biomass is supported by studies carried out in northern hardwood forests and in certain other forest ecosystems, it is not supported by field observations made in lodgepole pine forest (Ryan and Waring 1992) or in the boreal forest (Paré and Bergeron 1995). In the southern portion of the Canadian boreal forest, for example, Paré and Bergeron (1995) found that the living, above-ground biomass increased linearly to 17.3 $kg \cdot m^{-2}$ by year 75 after fire and then declined strongly to 7.7 $kg \cdot m^{-2}$ by year 97.

A complete description of fuel dynamics should thus combine surface, crown, and ladder fuels. Despite the number of existing dynamic fuel models for surface fuel (e.g., Rothermel and Philpot 1973) and crown fuel (e.g., Bilgili and Methven 1994), no single model has been able to characterize the simultaneous dynamics of different kinds of fuels for a given forest type. Without such a fuel model, any estimate of fire probability as a function of stand age can only be an approximation from qualitative knowledge. Such approximations have been used in a number of studies of fire dynamics, such as those of Holling (1981), Antonovski et al. (1992), Peterson (1994), Ratz (1995), and Holling et al. (1996).

Fire behaviour studies have also suggested the possible importance of fuel quantity in fire initiation. The initiation of crown fires, which usually cause stand replacement, depends heavily on the presence of ladder fuel between the ground surface and the canopy. Most crown fires result from high-intensity surface fires, and to generate enough vertical heat flux to start a crown fire, the surface fire must reach a certain threshold of intensity. Surface fire intensity is determined by the accumulation of surface fuel. For example, the average maximum temperatures of grassland headfires in western Texas were found to be a linear, increasing function of the amount of fine fuel per acre (Wright and Bailey 1982). When the base of the crown is high off the ground, a crown fire is difficult to start without enough bridge fuel (Van Wagner 1977a).

Notwithstanding the variety of patterns in fuel quantity dynamics reported, it remains in issue whether fuel quantity significantly affects the probability of fire occurrence or fire spread. This question has been extensively discussed among fire researchers. Heinselman (1973) assumed that increased fuel loads lead to increased fire behaviour potential. This hypothesis was supported by observations made in a mosaic of young and old stands in the chaparral of the United States and Mexico (Minnich 1983; Riggan et al. 1988). Van Wagner (1983) argued that fire behaviour in the boreal forest is not a simple function of age, and that potential fire intensity is not proportional to fuel weight per unit area. He pointed out that "the rate of fire spread depends more on the quantity and arrangement of fine fuel than on the accumulation of downed logs or deep organic matter" (p. 77). He suggested that the intensity of surface fires in boreal forests increases with age until canopy closure, then gradually decreases; and that the potential for crown fire rises until the age of about 60 years, and declines thereafter. When stands become overmature, some of the downed trees are always in an ideal state for combustion, so fire intensity increases once again.

It has been hypothesized that fire frequency in Yellowstone National Park is controlled by the slow development of a fuel complex capable of supporting an intense crown fire (Despain and Sellers 1977). Forest managers and researchers have observed that wildfires burned readily in late-successional forests, but often stopped spreading when they reached stands representing an early or intermediate stage of succession (Turner et al. 1994). These observations suggest that fire-spread processes are largely constrained by the spatial patterns of forest successional stages; that is, by the amount of fuel accumulated. That hypothesis did not hold true, however, for the high-intensity fires of 1988 that took place under extreme weather favourable to fire spread. In response to this anomaly, the hypothesis was developed that, while the spatial configuration of the fuel complex may control fire spread in most years, this constraint is overridden by extreme weather that favours fire processes (Turner et al. 1994).

The latter hypothesis, in turn, raises the question of the comparative importance of fuels and weather on fire intensity and crown fire occurrence. Bessie and Johnson (1995) investigated the relative roles played by weather (that is, fuel moisture content and wind speed) and fuel (that is, fuel loads, fuel depth, mass density, heat of combustion, and the ratio of surface area to volume) in forest fire behaviour. They used two well-known models, the surface fire intensity model of Rothermel (1972) and the crown fire initiation model of Van Wagner (1977a), to compare the contribution made by fuel load and crown structure to variations in surface fire intensity and crown fire initiation, with the contribution made by weather. They found that the years in which large areas were burned had higher fire-intensity predictions than the years in which smaller areas burned, and that the frequency distribution of weather variables shifted toward more extreme values in years when large areas had burned. This quantitative examination supports the concept that forest fire behaviour is determined primarily by weather variation among years rather than by fuel variation associated with stand age.

The results achieved by Bessie and Johnson (1995) suggest that fuel accumulations may not be as important in determining fire behaviour in boreal forests as they are in other types of forest. Canadian fire researchers have found no significant changes in fire probability among boreal stands burned from 20 to 30 years ago (M.G. Weber, K.G. Hirsch, M.E. Alexander, and M.D. Flannigan, pers. comm., 1997). These observations appear to differ from the observations made on fuel accumulation in the subalpine forests of Yellowstone National Park, and they support the hypothesis that forest flammability is independent of age. In short, although fuel load increases with the length of time since the last fire, fire probability may not be significantly affected by dynamic patterns in the fuel complex, especially in the scenario of the Canadian boreal forest.

Climate and Weather Conditions

Climate and weather conditions affect both the number of fire ignitions and the spread of fires. Weather is a short-term phenomenon resulting from atmospheric processes. Weather conditions include temperature, precipitation, relative humidity, and wind. As has been shown, fuel moisture is a function of all of these factors. Wind speed, in particular, influences the rate of fire spread. Wind direction apparently affects directional rate of spread and results in fires of various shapes. Climate is the average weather condition over a given period of time. Climate may be relatively stable and may change only slowly across a particular landscape, although human activities increase the rate of climate change (Flannigan and Weber 2000, this volume). Fire history studies have shown that climate change in various periods in the past was accompanied by corresponding changes in fire dynamics. Our understanding of these historical processes provides a foundation for predicting future change in fire dynamics in relation to projected changes in climate.

Several studies have documented that the dynamics of fire disturbances are a function of climate conditions. In a study of fire scars from giant sequoia (*Sequoiadendron giganteum*), Swetnam (1993) found that the frequency of long-term fire occurrence at the regional level was mainly determined by climatic conditions. During a warm period from about AD 1000 to 1300, small fires occurred frequently; whereas, during cooler periods from about AD 500 to 1000 and after AD 1300, less frequent but more widespread fires occurred. These changes indicate how the projected global climate change could potentially change current fire regimes. Clark (1988) presumed that climate warming would cause an increase in the frequency of fires in Minnesota. Price and Rind (1994) predicted increases of 44 percent in fire ignitions by lightning and a 78 percent increase in area burned with a doubling of carbon dioxide in the atmosphere. Flannigan and Van Wagner (1991) predicted a potential 46 percent increase in seasonal fire severity ratings in Canada. Wotton and Flannigan (1993) predicted an increase in the length of the fire season in Canada by 28 to 29 days.

The empirical evidence, however, has not always supported predictions of increased fire disturbance under climate warming. On islands in Lac Duparquet in the southern boreal forest of Quebec, for example, Bergeron (1991) found a 34 percent decrease in fire frequency during the period from 1870 to 1989, as compared with the preceding 74 years. This change is thought to result from a reduction in the frequency of drought. From running the Canadian General Circulaton Model (GCM), Bergeron and Flannigan (1995) concluded that, with a doubling of carbon dioxide in the atmosphere, the average FWI decreased over eastern Canada and increased dramatically over western Canada. These results support the hypothesis that warming may lead to a decrease in the drought periods that are conducive to fires in the southeastern boreal forest. Such contradictory findings suggest that general predictions on how fire dynamics will change with climate conditions are difficult to make. To be accurate, the predictions may have to take many other factors into account, and may thus become location-dependent.

Within a given fire season, however, it is certain that fire processes are greatly influenced by weather conditions. Dry fire seasons mean lower fuel moisture content and an increased probability of fire across the landscape; similarly, wet fire seasons mean higher fuel moisture content and a reduced probability of fire. As a result, a larger area is generally burned in dry seasons than in wet ones. Fast-changing weather conditions determine the final size and shape of a particular fire. Individual fires, in turn, determine forest stand mosaics and age distributions.

Landscape Structure

The components of a forest landscape are characterized by different rates of change. For example, topography, which includes elevation, slope, and aspect, changes so slowly as to be considered stable. Changes in soil components caused by the decomposition of dead materials also proceed slowly. Changes in vegetation cover type or tree species composition result from interactions among various factors, including seed availability, species competition, soil variation, and climatic conditions. Forest age is the fastest-changing variable in a forest landscape. It determines the amount of fuel available within a forest stand and the connectivity of the fuel bed (or the "landscape fragmentation," in the terminology of landscape ecology), and thus essentially constitutes the spatial configuration of forest fuel.

Land Cover Type

Forest cover type expresses the dominant tree species which occur in a land unit. For example, sixteen classes of land cover types resulted from a supervised classification of Thematic Mapper (TM) satellite image data of Ontario (Spectranalysis Inc. 1992; Perera and Baldwin 2000, this volume). Twelve of these classes consist of forest-related vegetation. These classes represent types of forest canopy differentiated by the major tree species occurring within them, sampled by a minimum unit of land area. The different forest types may have different probabilities of fire disturbance. Throughout the discussion that follows, the probability of fire disturbance is defined in a relative sense as the probability that one forest stand will burn more readily and with greater intensity than another within the same landscape (Li et al. 1996).

The forest cover types usually correspond to different fuel types, or combinations of fuel types, defined in the Canadian FBP system. Fire behaviour studies have shown that rates of fire spread have been associated with fuel type and that the rate of spread is a function of the initial spread index, which is, in turn, a function of weather conditions. The relative probability of fire disturbance can, therefore, be estimated from relative rates of fire spread for different fuel types (Table 7.3), calculated under similar weather conditions. Table 7.3 indicates that forests with coniferous canopy cover types generally have higher probabilities of fire disturbance than forests with deciduous

Table 7.3

The relative rates of fire spread of forest land cover types for use in simulations (Li et al. 1997).

Land Cover Type	*Relative Fire Probability*
Dense jack pine	0.9665
Dense black spruce	0.9165
Sparse coniferous	0.5649
Mixed coniferous	0.9415
White pine	0.3034
Red pine	0.3034
White and red pine	0.3034
Mixed deciduous	0.2202
Dense deciduous	0.2202
Sparse deciduous	0.1321
Poorly vegetated area	0.0793
Recently cut and burned	0.0793

Table 7.4

Differences in fire disturbance pattern between the boreal and Great Lakes-St. Lawrence forest regions of Ontario.

	Forest Type	
	Boreal	*Great Lakes-St. Lawrence*
Fire type	Crown fires	Frequent surface fires and infrequent crown fires
Fire cycle	50-100 years	100-350 years
Fire size	Usually large	Usually small

canopy cover types. If this is a reasonable assumption, then fire disturbances should be more frequent in the boreal forests than in the Great Lakes-St. Lawrence forests. This inference is supported by the observed fire disturbance patterns described by Thompson (2000, this volume) and summarized in Table 7.4.

Topography

Topography has a significant effect on the process of fire spread. In a flat landscape, fire has a more or less even probability of spreading to neighbouring forest stands. In a hilly landscape, however, there is a greater probability of spread uphill than downhill. Consequently, some parts of the landscape may be burned more frequently than others.

Van Wagner (1977b, 1988) investigated the effect of slope, both uphill and downhill, on fire spread processes in Canadian forest landscapes. With respect to uphill slope, Van Wagner (1977b) collected data from five studies to obtain the following expression of the relationship between relative fire spread rate and the tangent of the slope angle:

$$SF = \exp[3.533(\tan S)^{1.2}] \quad [7.4]$$

where SF is the spread factor relative to 1 for level surface, and S is the radian angle between slope and horizontal. With respect to downhill slope, Van Wagner (1988) carried out a laboratory experiment with beds of pine needles. His results are summarized in the following equation:

$$SF = (1 - 0.033A + 0.000749\,A^2) \times \cos(S) \quad [7.5]$$

where A is the angle of the slope to horizontal in degrees.

Soil Moisture

The soil moisture regime may have an indirect influence on fire disturbance events. In land units with drier soil moisture regimes, the moisture content of fuel also tends to be low, so that less heat energy is required to preheat the fuel and dry it. Consequently, the fire probabilities on these land units might be higher than those with wetter soil moisture regimes.

RECONSTRUCTION OF NATURAL FIRE REGIMES

Having examined factors by which fire regimes can be described and reviewed the principal characteristics of the existing fire regime in Ontario, we will now consider the means by which the natural fire regimes of the province may be discovered, or reconstructed. Natural fire regimes are those that would have existed in the absence of fire suppression activities. The ability to discount human influences on landscapes is important for implementing Ontario's current forest management policy of emulating the effects of natural disturbance regimes. Two different approaches to reconstructing natural fire regimes have been developed, one based on empirical data and the other on modelling.

Reconstruction Based on Empirical Data

This approach attempts to estimate natural fire regimes by constructing maps of time since fire or stand origin (Heinselman 1973; Johnson and Gutsell 1994). Time-since-fire maps record the date of the most recent fire in each unit of the landscape. Alternatively, historical records of fire disturbance are compiled.

The construction of a time-since-fire or stand-origin map requires a complete inventory of the entire landscape or a random (that is, unbiased) sampling of the area, under the assumption that all the forest stands are of fire origin (Johnson and Gutsell 1994). Once samples are taken, the distribution of the time since fire is determined, and the results are superimposed on a map of the landscape. This method has been used in fire history studies, such as the famous Boundary Water Canoe Area Study in western Minnesota (Heinselman 1973). The method for constructing a time-since-fire map was described in detail by Johnson (1992) and Johnson and Gutsell (1994). Time since fire can be reconstructed from tree-ring data, lake sediment data (e.g., MacDonald et al. 1991), or vegetation inventory data. The resulting maps provide empirical evidence for conclusions about forest age structure, for the study of fire frequency, and for the calculation of total area burned in different years.

They are also useful in estimating parameter values and testing a negative exponential or Weibull probability distribution. The estimated probability distribution is also used as an approximation of fire size distribution in a number of fire regime models, as will be discussed later in the chapter.

Another method of understanding natural fire regime on the basis of empirical data is to use historical fire disturbance records directly. In OMNR, historical fire disturbance records available for Ontario were analyzed for the purpose of determining natural fire size distributions in various Hills site regions (J. McNicol, pers. comm., 1996). This work was undertaken to provide scientific support and guidance for determining plot sizes for forest harvest planning in each site region.

Two sets of historical fire disturbance records were used in McNicol's analysis. One was the historical fire disturbance map, which contains historical burns larger than 200 ha from 1921 to 1993. Computerized fire report records kept by the Aviation, Flood and Fire Management Branch of OMNR were also used. From these two sets of data, McNicol was able to calculate the sizes of recorded fires and to group the fires into six predetermined size classes ranging from 11 ha to 10,000 ha. From this classification, McNicol and his research team obtained fire size distributions for the site regions, which provide insight into the fire regime of each region. They found that the fire size distribution in most site regions had a negative exponential shape.

Some difficulties arise in explaining the resulting fire size distributions. Ontario's fire control policy of total suppression can be traced back to the early 1920s. Since that time, the capability to suppress fires efficiently has greatly increased. Most of these historical fire records were obtained under the intensive fire suppression scenario. Consequently, the data may not necessarily represent natural fire disturbance patterns, but rather the patterns produced by various levels of fire suppression.

Reconstruction Based on Models

The use of models to assist the reconstruction of historical fire regimes is a theoretical approach intended to complement the use of incomplete empirical data. This approach translates or integrates available knowledge and information about a fire process into mathematical expressions and their logical linkages in order to investigate a particular question. The models have the potential to provide logical consequences under conditions set by the assumptions employed in them. The results are useful in identifying possible gaps in existing knowledge and in suggesting parameters important for future measurement.

Theoretical Models

Theoretical fire regime models are mathematical models based on theoretical assumptions under ideal conditions. These can be either analytical models or simulation models. Two analytical models well known to fire researchers are the Weibull and negative exponential fire history models (Johnson and Van Wagner 1985; Johnson 1992; Johnson and Gutsell 1994). Both of these models apply to homogeneous, stochastic processes and need to approximate the following two stability criteria for their use: that all the forest stands in the study have the same fire regime, and that each of the stands has, on average, a constant fire regime during the time period of study.

The negative exponential model is referred to as the "random selection" model (Johnson and Van Wagner 1985). This model assumes that the probabilities that forest stands in a given study area will burn follow a Poisson probability distribution, and that they are independent of stand age. The same proportion of the forest is assumed to burn during each time interval. Therefore, the instantaneous rate of burn is always the same for every stand, no matter how long the time since fire. The Weibull model is characterized as an "age selection" model. This model assumes that the probabilities that forest stands in a study area will burn follow a power function of time since fire. Through the use of different parameter values, the Weibull model is capable of describing a wide range of fire ecology scenarios, ranging from age selection to the random selection of the negative exponential model.

FLAP-X (Boychuk et al. 1997) is a model developed for investigating the influence of fire disturbance on forest age distribution. This model simulates fire dynamics within a landscape (represented by a grid of rectangular cells with equal numbers of rows and columns) which is subjected to random fires assumed to have an elliptical shape. The number of fires per year is determined by a random variable following a Poisson probability distribution. Fires are randomly located within the landscape, and wind direction is randomly determined on a yearly time step. The sizes of the simulated fires are determined by a negative exponential distribution. Consequently, the simulated changes in forest age structure are the result of user-defined fire-disturbance patterns. FLAP-X is thus useful in the investigation of forest age distribution if one assumes that no interaction takes place between landscape structure and fire process.

The fire models described above focus on simulating fire scenarios based on assumptions taken from actual observations. The negative exponential model has been widely used to guide forest resource management practices in Ontario.

System models have also been developed to investigate whether the dynamics of an ecosystem might display a pattern of self-organization. A theoretical fire model developed by Peterson (1994) is one example. This model was based on abstract scenarios set in the boreal forests of Manitoba, Canada. The model assumes that a fire is able to spread to all neighbouring cells according to probability, and probability is a function of time since fire. A fixed number of fire ignitions is assumed for each yearly time step of the simulations, and the final sizes of fires is determined by the cell age mosaic of the hypothetical forest landscape. Peterson's model was used to test two hypotheses, the interaction hypothesis and the lump hypothesis. Simulation results from the model supported the interaction hypothesis, which states that fire produces and is influenced by forest patterns, but only weakly supported the lump hypothesis, which states that the interaction of forest and fire processes produces spatial and temporal discontinuities in forest landscapes.

Theoretical models cannot be linked to real forest landscapes characterized by data sets in geographic information systems (GIS). Although these models are all based on hypothetical landscapes, they are useful (as Bessie and Johnson [1995] have suggested) in exploring realistic dynamics for landscapes across which there is little variation in the influence of fuels on fire probability.

GIS-Linked Models

GIS-linked fire models attempt to describe fire dynamics within a real forest landscape through linkage to a GIS data set characterizing the landscape. This type of simulation is intended to provide more relevant information on the forest landscape than can be produced by theoretical models alone. Three models will be discussed in order to demonstrate how GIS-linked models can provide useful information on natural fire regimes.

DISturbance PATCH (DISPATCH)

DISPATCH was developed for studying the effects of climatic change on the structure of landscapes subject to large disturbances (Baker et al. 1991). This model consists of the following five major components: (1) the climatic regime, which provides a probabilistic way of modelling the occurrence of a variety of weather conditions on a seasonal time scale; (2) the disturbance regime, which assumes a negative exponential distribution of disturbance size; (3) GIS map layers which include vegetation type, patch age, elevation, slope, and aspect; (4) a disturbance probability map produced by a user-defined combination of the five GIS map layers; and (5) a structural analysis program for the output of quantitative indices and measures of the landscape.

In the initial DISPATCH model, the landscape is simulated on a grid of 200 by 200 cells. Landscape structure is initialized by randomly assigning to each cell a stand age between 0 and 250 years; then weekly changes in the landscape are simulated. The model checks whether antecedent conditions favour a disturbance. If they do, then a disturbance size is generated from one of four negative exponential distributions, corresponding to each of the four seasons. A disturbance probability map is then formulated. The location of disturbance initiation is determined either as the cell with the highest disturbance probability, or as a cell randomly chosen among cells containing disturbance probabilities above some user-defined minimum value. After fire initiation, the disturbance may spread to one of the eight neighbouring cells, whichever has the highest disturbance probability, or to a randomly-chosen neighbour having a disturbance probability value above a user-defined minimum. The algorithm continues until either the predetermined disturbance size is reached, or until there are no neighbouring cells with the minimum probability of being disturbed.

DISPATCH runs within the GRASS GIS package. The size of each of the grid cells comprising the study area can be set to suit the question under study. This is the means by which the system simulates the impact of a user-specified fire regime on landscape dynamics.

Ecological Model for Burning the Yellowstone Region (EMBYR)

EMBYR is a modified percolation model developed to simulate the causes and consequences of the large-scale fires that burned the Yellowstone National Park during 1988 (Gardner et al. 1996). A percolation model is a theoretical model of the spread of disturbance across a landscape represented by a raster grid. This type of model has been used to study how random features occurring at a local scale may determine the overall behaviour of a system, on the basis of one critical probability required to make a fire self-sustaining. In fire models that use the percolation theory, either of two identities (0 or 1), representing susceptibility to fire disturbance, is randomly assigned to each cell in a square matrix representing the landscape. The proportion of susceptible cells is determined by a user-determined probability. A starting point for the fire disturbance is chosen. The fire then spreads to any of the four neighbouring cells sharing a common edge with the initial cell, in accordance with a predetermined critical probability. This process of spread continues until the simulated fire stops moving in all directions or reaches the edges of the matrix. A detailed review of percolation models and related fire models can be found in Li and Perera (1997). One research focus in the development of percolation models was to determine critical values by which a disturbance will spread across a landscape (O'Neill et al. 1992).

The EMBYR model represents two major departures from the other percolation models: the ability to link with a GIS database, and an improvement in the fire spread algorithm. The modified fire spread algorithm allows fire to spread to all of its neighboring cells, both those sharing a common edge with the initial cell and those sharing a common corner. The fire spread probability is formulated as a table, in which each element is a function of successional stage. Fire is ignited when a particular weather condition that is stochastically simulated exceeds the minimum value set by the user for fire initiation. Fire ignition sources are assumed to be present all the time, awaiting only the favourable weather condition for fire to begin. The locations of fire ignition are randomly selected from a data set containing the actual origins of historical fires; the model assumes that fire starts nowhere else. The EMBYR model has been used in long-term simulations to investigate how climate change will influence the dynamics of landscapes that are subject to fire disturbance (Gardner et al. 1996).

ONtario FIre REgime model (ON-FIRE)

ON-FIRE was developed for simulating natural fire regimes in Ontario, especially in the boreal region of northwestern Ontario. Initially, the DISPATCH model seemed more appropriate than other models for use in this research, because it was spatially explicit and linked to real landscape data sets; however, the input-output structure of the model was ultimately unsuitable. DISPATCH requires input on fire frequency, fire size distribution, and weather, as well as initial landscape structure. Fire frequency and size distribution, however, were part of the final results expected from the research.

ON-FIRE is designed to eliminate a hidden problem in the algorithm commonly used in spatial fire models: when fire frequency and fire size distribution are represented by two independent, random input variables, the fire regime output may not correspond to user expectations. This discrepancy occurs because very large fire sizes may be randomly assigned in years of high fire occurrence. Under these extreme conditions, the study area may even be burned more than once in a single year, or burned repeatedly in successive years. One way to avoid this unrealistic situation is to seek more controllable ways of generating model behaviour (e.g., Andison 1996). Another way is to develop a relationship between the two variables, such as is implemented in the LANDIS model (He and Mladenoff 1999). The ON-FIRE algorithm avoids this problem by allowing a fire to grow freely according to rules of how fuel, topography, and weather conditions influence fire spread. There is, therefore, no need for the user to specify a particular fire size before the simulation is conducted. This solution is probably realistic, because of the difficulties involved in predicting final fire size before a fire is initiated or at the beginning of the fire ignition process.

The need to represent issues currently debated in fire research is a critical element in the development of any spatial fire model. It is desirable for the model to be capable of exploring the consequences of various assumptions or hypotheses. Efforts were made to achieve this goal in the development of ON-FIRE. For example, the model takes into account the issue of whether various fuel accumulations have a different effect on fire probability or potential fire behaviour by assuming different shapes for the baseline fire probability function, described by the following logistic equation:

$$P_{fire} = c / [1 + \exp(a - b \times age)] \quad [7.6]$$

where a, b, and c are parameters. Various hypotheses can be formulated in the model by using different combinations of parameter values. In a case study on the temporal patterns of fires over a specific landscape in northwestern Ontario, ON-FIRE simulations suggested that different hypotheses provided similar predictions of temporal disturbance pattern for small or intermediate-size fires, but different predictions for large fires. Li et al. (1997) provide a detailed description of this study.

The rules in ON-FIRE governing fire initiation and spread depend on the relative fire probabilities of various vegetation cover types. These probabilities represent a best estimate based on the rates of spread of various fuel types in the FBP system, and assume that canopy cover types correspond to fuel types. In addition to the rules for fire disturbance, ON-FIRE also contains simplified rules for forest growth during fire-free periods and for regeneration following fire. These rules provide logical linkages among the various components of forest dynamics. A detailed description of the ON-FIRE model is provided by Li and Perera (1997) and by Li et al (1995, 1996, 1997).

A sensitivity analysis of ON-FIRE has revealed that the impact of weather conditions on fire dynamics is more significant in this model than other factors (Li et al. 1997). Consequently, ON-FIRE is capable of simulating the scenario observed in Yellowstone National Park, that the spatial configuration of the fuel complex may control fire spread in most years, but that extreme weather that favours fire processes overrides this constraint (Turner et al. 1994). These results suggested that weather conditions were represented in a reasonable way in this model.

ON-FIRE does not attempt to provide every detail of a fire process, as the FIREMAP model does (Vasconcelos and Guertin 1992). Nevertheless, information valuable for understanding and reconstructing natural fire regimes can be obtained by running ON-FIRE to explore the natural fire size distribution produced under a given fire cycle over a given forest landscape.

A Case Study of Natural Fire Regime Reconstruction Using ON-FIRE

Description of the Study Area

A study area measuring 130 km by 130 km was chosen from Hills site region 3S in northwestern Ontario. Waterbodies represent approximately 20 percent of this area, conifer forests 47 percent, deciduous forests 22 percent, poorly vegetated sites 6 percent, and non-forest land use the remainder. GIS data sets available for the study area, at a spatial resolution of one hectare, provided model input for the variables of land cover type, elevation, and soil moisture.

Description of the Experiment

Fire and forest dynamics over 1200 years were simulated using the ON-FIRE model. Seven replications were conducted for the experiment, but only the results of the final 1000-year simulation were used for analysis. The first 200 years of simulation results were discarded in order to eliminate possible undue effects of the initial conditions. The potential influence of climate change scenarios was not incorporated into this simulation experiment, but will be addressed elsewhere (Li et al., unpub.). The parameter values for equation 7.6 used in this simulation experiment were 0.6, 2.5, and 0.05 for c, a, and b, respectively. These parameter values result from a model calibration study performed using empirical data available for a period of 19 years over a 10,000-ha landscape in northwestern Ontario (Li et al. 1997). The model was run on a DEC Alpha UNIX workstation. Each replication took four to five days of CPU time.

Simulated fires were grouped according to the predefined fire size classes, in hectares, used by McNicol: 11-130, 131-260, 261-520, 521-1040, 1041-5000, and 5001-10,000. Two additional classes were identified, 1-10 and over 10,000. For each replication, the number of fires and the total area burned in each fire size class were summarized. The arithmetic averages and their 95-percent confidence limits were then calculated. The range between lower and upper limits at the 95-percent confidence level for the different fire size classes served to approximate the variability of fires occurring under natural conditions.

The simulation results were compared with the results from an empirical data analysis conducted by OMNR (J. McNicol, pers. comm., 1996). In the strict sense, this comparison may not indicate whether the simulated fire size distribution is consistent with McNicol's results, because of the difficulties involved in comparing fire regimes in different fire history studies, conducted across various landscapes. The objective of this comparison, therefore, was to provide additional information on whether the fire size distributions obtained from the OMNR analysis could be used for approximating natural fire disturbance patterns.

The rationale behind this simulation experiment is the exploration of potential system behaviour using a calibrated model (Rykiel 1996). The method employed in the experiment is not unlike the method used to map ecosystems through the analysis of remote sensing spatial data over large areas. First, the image analysis system is calibrated (or "trained") using representative samples, then the trained system is used to classify unsampled areas, as a first approximation. The simulation results produced using the ON-FIRE model are not accurate predictions, but they serve to indicate what the logical consequences will be if the rules used in the model are reasonable.

Simulated Natural Fire Size Distribution

The average simulated fire cycle found for the study area was 121.67 years. The simulated natural fire size distributions obtained are summarized in Figure 7.3. Both the number of fires and the total area burned in each size class are expressed as a percentage of the study area totals.

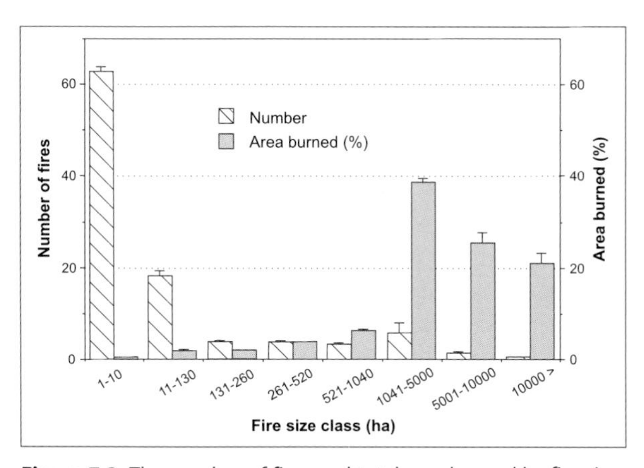

Figure 7.3 The number of fires and total area burned by fire size class. Area burned is expressed as a percentage of the total area burned for all fire size classes.

Figure 7.3 shows that the frequency of fires decreases as fire size increases. The simulated fire size distributions, therefore, have a negative exponential shape, a result which appears compatible with the reports of Baker (1989) and Van Wagtendonk (1986). This negative exponential fire size distribution suggests that the natural fire regimes might consist of a large number of small fires and a few large fires. The simulation results show that 62.89 percent of the simulated fires fall within the smallest size class (1-10 ha) and only 0.55 percent of the fires fall within the largest size class (larger than 10,000 ha). The simulation results also suggest that the frequency distributions of fires in size classes ranging from 131 ha to 1040 ha may be similar, in this case ranging between 3.32 percent and 3.88 percent. The resulting fire size distribution may deviate from the theoretical negative exponential distribution, but the basic shape remains.

The simulation results in Figure 7.3 also suggest that a large percentage of the total area burned might result from only a few large fires. In this simulation experiment, 85.32 percent of the total area burned resulted from fires having a final size greater than 1040 ha, and only 0.42 percent of the total area burned resulted from fires having a final size less than 10 ha. This result also qualitatively supports the conclusion obtained from historical data for Canada on fires larger than 200 ha, that about 5 percent of the fires are responsible for 95 percent of the area burned (e.g., Stocks 1991).

Comparison with Empirical Results

For the 11-130 ha and 521-1040 ha size classes, the simulated fire results were consistent with the empirical data (Figure 7.4a). For fires larger than 1041 ha, however, and for the 261-520 ha size class, the results from the empirical data were lower than the simulated results. For the 131-260 ha size class, the empirical results were higher than the simulated results. When a comparison was made of the percentage of area burned, the empirical results for fires smaller than 1040 ha were higher than the simulated results for the same size range (Figure 7.4b). For fires larger than 1040 ha, however, the empirical results were lower than the simulated results. One possible reason for the difference in fire frequency in different size classes is the influence of fire suppression. Fire

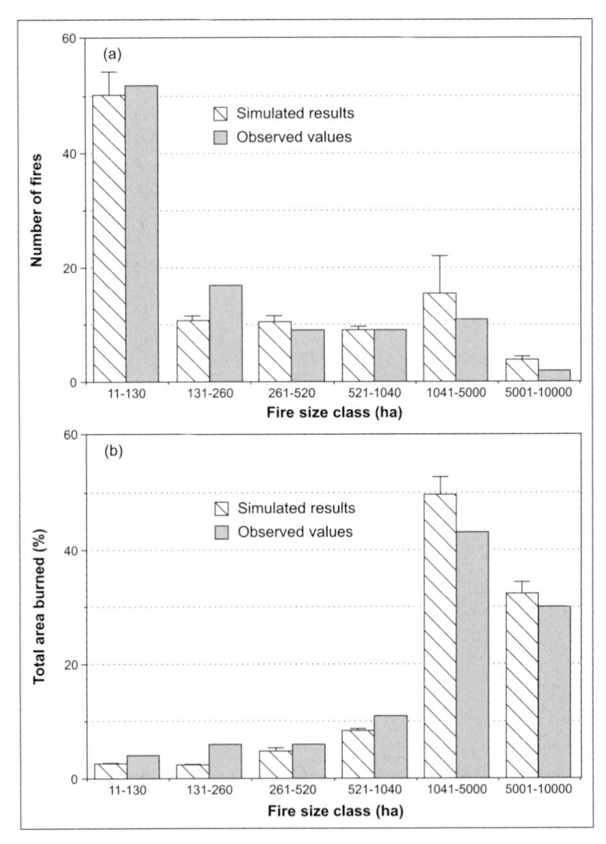

Figure 7.4 A comparison by fire size class between simulated and observed results for (a) number of fires and (b) burned areas. Striped bars indicate simulated results and shaded bars indicate observed values. The area burned by size class is expressed as a percentage of the total area burned for all size classes.

suppression would be expected to reduce the final size of fires, and many fires recorded in the smaller size classes might have been recorded in the larger size classes, if fire suppression had not been practised.

This case study demonstrates that the ON-FIRE model is capable of providing useful information on fire size distributions and their ranges of variability under a scenario of no fire suppression. It should be noted that the information provided by ON-FIRE simulations is not intended for use in predicting detailed fire events, but rather to provide insight into fire dynamics to assist strategic-level planning.

SUMMARY

The analysis of forest management policy options in Ontario requires a conceptual understanding of the fire disturbance regimes of the province at the landscape scale. Fire suppression influences the forest fire disturbance patterns across the province; however, the extensive fire-management zone appears to provide the closest approximation of a natural fire regime.

Predictions of long-term forest dynamics are deeply influenced by the spatial and temporal patterns of fire disturbance. The incompleteness of current knowledge and the need for further study of these patterns is evident from the range of hypotheses advanced and opinions held on a number of fundamental questions. One way to support decision making by forest managers planning processes is to integrate the available data, thus making the best use of existing information resources.

Ecological modelling provides a useful tool when it is either impossible or too costly to observe processes or events of interest in the real world. Modelling usually represents one step beyond a purely conceptual understanding of ecosystems and provides a way of linking available knowledge and information in logical relationships. To the extent that the models developed to date are capable of indicating the behaviour of real ecosystems, exploring the logical consequences of model system behaviour provides useful insight into real-world dynamics.

The ON-FIRE model was developed in an effort to gain a better understanding of Ontario's landscape-level forest dynamics, in response to current requirements for a strategic-level perspective. This is a coarse-grained model with a time step of one year and a spatial resolution of one hectare. Even at this early stage in its development, however, ON-FIRE has shown various meaningful and encouraging results.

Interest in the use of coarse-grained modelling to investigate fire effects at landscape scales is growing all over the world. The Global Change Terrestrial Ecosystems section of the International Geosphere-Biosphere Program, for example, includes a special task for the development of landscape fire models. The explanation and prediction of fire disturbance patterns and associated forest dynamics at the landscape scale will certainly benefit from advanced theories, hypotheses, and methods developed by other fields of research, such as complex systems theory. A multidisciplinary approach to the study of fire dynamics should thus be promoted.

ACKNOWLEDGEMENTS
I wish to acknowledge the Ontario Forest Research Institute and the Ontario Ministry of Natural Resources for their support toward the development of the ON-FIRE model. This manuscript has benefited from discussions with scientists in the Fire and Landscape Management Networks of the Canadian Forest Service.

REFERENCES

Albini, F.A. and B.J. Stocks. 1986. Predicted and observed rates of spread of crown fires in immature jack pine. Combustion Science and Technology 48: 65-76.

Alexander, M.E. 1985. Estimating the length-to-breadth ratio of elliptical forest fire patterns. In: Proceedings of the 8th Conference on Fire and Forest Meteorology, April 29 to May 2, 1995, Detroit, Michigan. Bethesda, Maryland: Society of American Foresters. 287-304.

Andison, D.W. 1996. Managing for Landscape Patterns in the Sub-Boreal Forests of British Columbia [Ph.D. dissertation]. Vancouver: University of British Columbia. 197 p.

Antonovski, M.Y., M.T. Ter-Mikaelian, and V.V. Furyaev. 1992. A spatial model of long-term forest fire dynamics and its applications to forests in western Siberia. In H.H. Shugart, R. Leemans and G.B. Bonan (editors). A Systems Analysis of the Global Boreal Forest. Cambridge, UK: Cambridge University Press. 373-403.

Aviation and Fire Management Centre. 1986. Forest Fire Management in Ontario. Toronto: Ontario Ministry of Natural Resources. 28 p.

Baker, W.L. 1989. Landscape ecology and nature reserve design in the Boundary Waters Canoe Area, Minnesota. Ecology 70: 23-35.

Baker, W.L. 1995. Long-term response of disturbance landscapes to human intervention and global change. Landscape Ecology 10: 143-159.

Baker, W.L., S.L. Egbert, and G.F. Frazier. 1991. A spatial model for studying the effects of climatic change on the structure of landscapes subject to large disturbances. Ecological Modelling 56: 109-125.

Bergeron, Y. 1991. The influence of island and mainland lakeshore landscapes on boreal forest fire regimes. Ecology 72: 1980-1992.

Bergeron, Y. and J. Brisson. 1990. Fire regime in red pine stands at the northern limit of the species range. Ecology 71: 1352-1364.

Bergeron, Y. and M.D. Flannigan. 1995. Predicting the effects of climate change on fire frequency in the southeastern Canadian boreal forest. Water, Air and Soil Pollution 82: 437-444.

Bessie, W.C. and E.A. Johnson. 1995. The relative importance of fuels and weather on fire behaviour in subalpine forests. Ecology 76: 747-762.

Bilgili, E. and I.R. Methven. 1994. A dynamic fuel model for use in managed even-aged stands. International Journal of Wildland Fire 4: 177-185.

Bormann, F.H. and G.E. Likens. 1979. Pattern and Process in a Forested Ecosystem: Disturbance, Development, and the Steady State. New York: Springer-Verlag. 253 p.

Boychuk, D., A.H. Perera, M.T. Ter-Mikaelian, D.L. Martell, and C. Li. 1997. Modelling the effect of spatial scale and correlated fire disturbances on forest age distribution. Ecological Modelling 95: 145-164.

Butzer, K.W. 1971. Environment and Archeology. Chicago: Aldine-Atherton. 492-515.

Cayford, J.H. 1970. The role of fire in the ecology and silviculture of jack pine. In: Proceedings of the 10th Annual Tall Timbers Fire Ecology Conference. August 20-21, 1970, Fredericton, New Brunswick. Tallahassee, Florida: Tall Timbers Research Station. 221-244.

Cayford, J.H., Z. Chrosciewicz, and H.P. Sims. 1967. A Review of Silvicultural Research in Jack Pine. Forestry Branch Publication No. 1173. Ottawa: Canadian Department of Forestry and Rural Development. 255 p.

Clements, F.E. 1916. Plant succession: An analysis of the development of vegetation. Publication No. 242. Washington, DC: Carnegie Institute. 512 p.

Clark, J.D. 1989. Ecological disturbance as a renewal process: theory and application to fire history. *Oikos* 56: 17-30.

Clark, J.D. 1990. Fire and climate change during the last 750 years in northwestern Minnesota. Ecological Monographs 60: 135-159.

Clark, J.S. 1988. Effect of climate change on fire regimes in northwestern Minnesota. Nature 334: 233-235.

Cumming, S.G., P.J. Burton, M. Joy, B. Klinkenberg, F.K.A. Schmiegelow, and J.N.M. Smith. 1995. Experimental habitat fragmentation and simulation of landscape dynamics in the boreal mixedwood: a pilot study. Publication Fo42-91/137-1995E. Ottawa, Ontario: Canadian Forestry Service. 75 p.

Despain, D.G. and R.E. Sellers. 1977. Natural fire in Yellowstone National Park. Western Wildlands 4: 20-24.

Duchesne, L.C. and T. Hinrichs. 1996. The ecological impact of fire protection and its role in forest ecosystem management. In: G.S. Ramsey (compiler). Proceedings: International Wildland Fire Foam Symposium and Workshop, May 3-5, 1994, Thunder Bay, Ontario. Information Report PI-X-123. Chalk River, Ontario: Petawawa National Forestry Institute, Canadian Forestry Service. 102-105.

Engelmark, O., R. Bradshaw, and Y. Bergeron. 1993. Disturbance dynamics in boreal forests: introduction. Journal of Vegetation Science 4: 730-732.

Engstrøm, F.B. and D.H. Mann. 1991. Fire ecology of red pine (*Pinus resinosa*) in northern Vermont, U.S.A. Canadian Journal of Forest Research 21: 882-889.

Euler, D.L. and A.E. Epp. 2000. A new foundation for Ontario forest policy for the 21st century. In: A.H. Perera, D.L. Euler, and I.D. Thompson (editors). Ecology of a Managed Terrestrial Landscape: Patterns and Processes of Forest Landscapes in Ontario. Vancouver: University of British Columbia Press. 276-294.

Flannigan, M.D. 1993. Fire regime and the abundance of red pine. International Journal of Wildland Fire 3: 241-247.

Flannigan, M.D. and C.E. Van Wagner. 1991. Climate change and wildfire in Canada. Canadian Journal of Forest Research 21: 66-72.

Flannigan, M.D. and M.G. Weber. 2000. Influences of climate on Ontario forests. In: A.H. Perera, D.L. Euler, and I.D. Thompson (editors). Ecology of a Managed Terrestrial Landscape: Patterns and Processes of Forest Landscapes in Ontario. Vancouver, University of British Columbia Press. 103-114.

Food and Agriculture Organization of the United Nations 1986. Wildland fire management terminology. Forestry Paper No. 70. Rome, Italy: Food and Agriculture Organization of the United Nations. 257 p.

Forestry Canada Fire Danger Group. 1992. Development and Structure of the Canadian Forest Fire Behaviour Prediction System. Information Report ST-X-3. Ottawa: Forestry Canada, Science and Sustainable Development Directorate. 63 p.

Gardner, R H., W.W. Hargrove, M.G. Turner, and W. H. Romme. 1996. Climate change, disturbances and landscape dynamics. In: B.H. Walker and W.L. Steffen (editors). Global Change and Terrestrial

Ecosystems. IGBP Book Series No. 2. Cambridge, UK: Cambridge University Press. 289-307.

Granstrøm, A. 1993. Spatial and temporal variation in lightning ignitions in Sweden. Journal of Vegetation Science 4: 737-744.

Harrington, J.B., M.D. Flannigan, and C.E. Van Wagner. 1983. A Study of the Relation of Components of the Fire Weather Index to Monthly Provincial Area Burned by Wildfire in Canada (1953-80). Information Report PI-X-25. Chalk River, Ontario: Petawawa National Forestry Institute, Canadian Forestry Service. 65 p.

He, S.H. and D.J. Mladenoff. 1999. Spatially explicit and stochastic simulation of forest-landscape fire disturbance and succession. Ecology 80: 81-99.

Heinselman, M.L. 1971. The natural role of fire in northern conifer forest. In C.W. Slaughter, R.J. Barney, and G.M. Hansen (editors). Proceedings, Fire in the Northern Environment: A Symposium, University of Alaska College, April 13-14, 1971, Fairbanks, Alaska. Portland, Oregon: USDA Forest Service. 61-72.

Heinselman, M.L. 1973. Fire in the virgin forest of the Boundary Waters Canoe Area, Minnesota. Quaternary Research 3: 329-382.

Heinselman, M.L. 1978. Fire in wilderness ecosystems. In: J.C. Hendee, G.H. Stankey, and R.C. Lucas (editors). Wilderness Management. Miscellaneous Publication No. 1365. Washington, DC: USDA Forest Service. 248-278.

Heinselman, M. L. 1981. Fire intensity and frequency as factors in the distribution and structure of northern ecosystems. In: H.A. Mooney, T.M. Bonnicksen, N.L. Christensen, J.E. Lotan, and W.A. Reiners (technical coordinators). Proceedings of the Conference on Fire Regimes and Ecosystem Properties, December 11-15, 1978, Honolulu, Hawaii. General Technical Report WO-26. Washington, DC: USDA Forest Service. 7-57.

Hills, G.A. 1959. A Ready Reference to the Description of the Land of Ontario and its Productivity. Preliminary Report. Toronto: Ontario Department of Lands and Forests. 140 p.

Hirsch, K.G. 1996. Canadian Forest Fire Behaviour Prediction (FBP) System: User's Guide. Special Report No. 7. Edmonton, Alberta: Canadian Forest Service. 121 p.

Holling, C.S. 1973. Resilience and stability of ecological systems. Annual Review of Ecology and Systematics 4: 1-23.

Holling, C.S. 1981. Forest insects, forest fires, and resilience. In: H.A. Mooney, T.M. Bonnicksen, N.L. Christensen, J.E. Lotan, and W.A. Reiners (technical coordinators). Proceedings of the Conference on Fire Regimes and Ecosystem Properties, December 11-15, 1978, Honolulu, Hawaii. General Technical Report WO-26. Washington, DC: USDA Forest Service. 445-464.

Holling, C.S. 1986. The resilience of terrestrial ecosystems: local surprise and global change. In: W.C. Clark and R.E. Munn (editors). Sustainable Development of the Biosphere. Cambridge, UK: Cambridge University Press. 292-317.

Holling, C.S. 1992a. Cross-scale morphology, geometry, and dynamics of ecosystems. Ecological Monographs 62: 447-502.

Holling, C.S. 1992b. The role of forest insects in structuring the boreal landscape. In: H.H. Shugart, R. Leemans, and G.B. Bonan (editors). A Systems Analysis of the Global Boreal Forest. Cambridge, UK: Cambridge University Press. 170-191.

Holling, C.S., G. Peterson, P. Marples, J. Sendzimir, K. Redford, L. Gunderson, and D. Lambert. 1996. Self-organization in ecosystems: lumpy geometries, periodicities, and morphologies. In: B.H. Walker and W.L. Steffen (editors). Global Change and Terrestrial Ecosystems. Cambridge, UK: Cambridge University Press. 346-384.

Hopkins, D.M. 1967. The Bering Land Bridge. Stanford, California: Stanford University Press. 495 p.

Hughes, S.R. 1980. A Background Report on Forest Fire Management Policies and Operations in the Province of Ontario. Internal Report. Sault Ste. Marie, Ontario: Ontario Ministry of Natural Resources. 91 p.

Johnson, E A. 1992. Fire and Vegetation Dynamics: Studies from the North American Boreal Forest. Cambridge, UK: Cambridge University Press. 129 p.

Johnson, E.A. and S.L. Gutsell. 1994. Fire frequency models, methods and interpretations. Advances in Ecological Research 25: 239-287.

Johnson, E.A. and C.P.S. Larsen. 1991. Climatically induced change in fire frequency in the southern Canadian Rockies. Ecology 72: 194-201.

Johnson, E.A. and C.E. Van Wagner. 1985. The theory and use of two fire history models. Canadian Journal of Forest Research 15: 214-220.

Johnston, M. and J. Elliott. 1998. The effect of fire severity on ash, and plant and soil nutrient levels following experimental burning in a boreal mixed wood stand. Canadian Journal of Soil Science 78: 35-44.

Kilgore, B.M. 1981. Fire in ecosystem distribution and structure: western forests and scrublands. In: H.A. Mooney, T.M. Bonnicksen, N.L. Christensen, J.E. Lotan, and W.A. Reiners (technical coordinators). Proceedings of the Conference on Fire Regimes and Ecosystem Properties, December 11-15, 1978, Honolulu, Hawaii. General Technical Report WO-26. Washington, DC: USDA Forest Service. 58-89.

Kourtz, P. and B. Todd. 1992. Predicting the Daily Occurrence of Lightning-Caused Forest Fires. Information Report PI-X-112. Chalk River, Ontario: Petawawa National Forestry Institute, Forestry Canada. 18 p.

Latta, S.C., C.A. Howell, K. Reynard, G.R. Parks, P. Porneluzi, and J.R. Faaborg. 1998. Landscape-sensitivity and population trends in Missouri breeding birds, 1991-1996. In: Proceedings of the 13th Annual Conference of the United States Regional Association of the International Association for Landscape Ecology, March 17-21, 1998, East Lansing, Michigan. East Lansing, Michigan: US-IALE. 112.

Li, C. and M.J. Apps. 1995. Disturbance impact on forest temporal dynamics. Water, Air and Soil Pollution 82: 429-436.

Li, C. and M.J. Apps. 1996. Effects of contagious disturbance on forest temporal dynamics. Ecological Modelling 87(1-3): 143-151.

Li, C. and A. Perera. 1997. ON-FIRE: A landscape model for simulating the fire regime of northwest Ontario. In: X. Chen, X. Dai, and T. Hu (editors). Ecological Research and Sustainable Development. Beijing, China: China Environmental Science Press. 369-393.

Li, C., M. Ter-Mikaelian, and A. Perera. 1995. Modelling interactions between fire regime and forest landscape. In: Proceedings of the 7th International Conference on Geomatics, June 11-15, 1995, Ottawa, Ontario [CD-ROM].

Li, C., M. Ter-Mikaelian, and A. Perera. 1996. Ontario Fire Regime Model: Its Background, Rationale, Development and Use. Forest Fragmentation and Biodiversity Project Report No. 25. Sault Ste. Marie, Ontario: Ontario Forest Research Institute, Ontario Ministry of Natural Resources. 42 p.

Li, C., M. Ter-Mikaelian, and A. Perera. 1997. Temporal fire disturbance patterns on a forest landscape. Ecological Modelling 99: 137-150.

MacDonald, G.M., C.P.S. Larsen, J.M. Szeicz, and K.A. Moser. 1991. The reconstruction of boreal forest fire history from lake sediments:

a comparison of charcoal, pollen, sedimentological, and geochemical indices. Quaternary Science Reviews 10: 53-71.

Malanson, G.P. 1987. Diversity, stability and resilience: effects of fire regime. In: L. Trabaud (editor). The Role of Fire in Ecological Systems. The Hague, Netherlands: SPB Academic Publishing. 157 p.

McAlpine, R.S. 1990. Seasonal Trends in the Drought Code Component of the Canadian Forest Fire Weather Index System. Information Report PI-X-97E/F. Chalk River, Ontario: Petawawa National Forestry Institute, Forestry Canada. 37 p.

Merrill, D.F. and M.E. Alexander (editors). 1987. Glossary of Forest Fire Management Terms. Fourth Edition. Publication NRCC No. 26516. Ottawa: National Research Council of Canada, Canadian Committee on Forest Fire Management. 91 p.

Minnich, R.A. 1983. Fire mosaics in Southern California and Northern Baja California. Science 219: 1287-1294.

Müller, F. 1997. State-of-the-art in ecosystem theory. Ecological Modelling 100: 135-161.

O'Neill, R.V., R.H. Gardner, M.G. Turner, and W.H. Romme. 1992. Epidemiology theory and disturbance spread on landscapes. Landscape Ecology 7: 19-26.

Paré, D. and Y. Bergeron. 1995. Above-ground biomass accumulation along a 230-year chronosequence in the southern portion of the Canadian boreal forest. Journal of Ecology 83: 1001-1007.

Paré, D, Y. Bergeron, and C. Camiré. 1993. Changes in the forest floor of Canadian southern boreal forest after disturbance. Journal of Vegetation Science 4: 811-818.

Perera, A.H. and D.J.B. Baldwin. 2000. Spatial patterns in the managed forest landscape of Ontario. In: A.H. Perera, D.L. Euler, and I.D. Thompson (editors). Ecology of a Managed Terrestrial Landscape: Patterns and Processes of Forest Landscapes in Ontario. Vancouver: University of British Columbia Press. 74-99.

Peterson, G.D. 1994. Spatial modelling of fire dynamics in the Manitoba boreal forest [M.Sc. thesis]. Gainsville, Florida: University of Florida. 112 p.

Price, C. and D. Rind. 1994. The impact of a 2xCO_2 climate on lightning-caused fires. Journal of Climate 7: 1484-1494.

Ratz, A. 1995. Long-term spatial patterns created by fire: a model oriented towards boreal forests. International Journal of Wildland Fire 5: 25-34.

Renkin, R.A. and D.G. Despain. 1992. Fuel moisture, forest type, and lightning-caused fire in Yellowstone National Park. Canadian Journal of Forest Research 22: 37-45.

Riggan, P.J., S. Goods, P.M. Jacks, and R.N. Lockwood. 1988. Interaction of fire and community development in chaparral of southern California. Ecological Monographs 58: 155-176.

Romme, W.H. 1983. Fire frequency in subalpine forests of Yellowstone National Park. In: Proceedings of the Fire History Workshop, October 20-24, 1980, Tucson, Arizona. General Technical Report RM-81. Fort Collins, Colorado: USDA Forest Service. 27-30.

Rothermel, R.C. 1972. A Mathematical Model for Predicting Fire Spread in Wildland Fuels. Research Paper INT-115. Ogden, Utah: USDA Forest Service. 40 p.

Rothermel, R.C. 1983. How to predict the spread and intensity of forest and range fires. General Technical Report INT-143. Ogden, Utah: USDA Forest Service. 161 p.

Rothermel, R.C. and C.W. Philpot. 1973. Predicting changes in chaparral flammability. Journal of Forestry. 71: 640-643.

Ryan, M.G. and R.H. Waring. 1992. Maintenance respiration and stand development in a subalpine lodgepole pine forest. Ecology 73: 2100-2108.

Rykiel, E.J., Jr. 1996. Testing ecological models: the meaning of validation. Ecological Modelling 90: 229-244.

Simard, A. J. 1997. National Workshop on Wildland Fire Activity in Canada, Edmonton, Alberta: Workshop Report. Information Report ST-X-13. Ottawa: Canadian Forest Services, Science Branch. 38 p.

Spectranalysis Inc. 1992. Development of a spatial database of red and white pine old-growth forest in Ontario - west. Forest Fragmentation and Biodiversity Project, Technical Report No. 5. Sault Ste. Marie, Ontario: Forest Landscape Ecology Program, Ontario Ministry of Natural Resources. 41 p.

Sprugel, D.G. 1985. Natural disturbance and ecosystem energetics. In: S.T.A. Pickett and P.S. White (editors). The Ecology of Natural Disturbance and Patch Dynamics. Orlando, Florida: Academic Press. 335-352.

Statutes of Ontario. 1994. *Crown Forest Sustainability Act*. S.O. 1994, c. 25.

Statutes of Ontario. 1917. *Forest Fires Prevention Act* . S.O. 1917, c. 54.

Stocks, B.J. 1991. The extent and impact of forest fires in northern circumpolar countries. In: J.S. Levine (editor). Global Biomass Burning: Atmospheric, Climatic, and Biospheric Implications. Cambridge, Massachusetts: Massachusetts Institute of Technology Press. 197-202.

Stokes, M.A. and J.H. Dieterich (editors). 1980. Proceedings of the Fire History Workshop, October 20-24, 1980, Tucson, Arizona. General Technical Report RM-81. Fort Collins Colorado: USDA Forest Service. 142 p.

Strauss, D., L. Bednar, and R. Mees. 1989. Do one percent of forest fires cause ninety-nine percent of the damage? Forest Science 35: 319-328.

Sturtevant, B.R., J.A. Bissonette, J.N. Long, and D.W. Roberts. 1997. Coarse woody debris as a function of age, stand structure, and disturbance in boreal Newfoundland. Ecological Applications 7: 702-712.

Suffling, R. 1991. Wildland fire management and landscape diversity in the boreal forest of northwestern Ontario during an era of climatic warming. In: S.C. Nodvin and T.A. Waldrop (editors). Fire and the Environment: Ecological and Cultural Perspectives: Proceedings of an International Symposium, March 20-24, 1990, Knoxville, Tennessee. General Technical Report SE-GTR-69. Asheville, North Carolina: USDA Forest Service. 97-106.

Swetnam, T.W. 1993. Fire history and climate change in giant sequoia groves. Science 262: 885-889.

Thompson, I.D. 2000. Forest vegetation of Ontario: factors influencing landscape change. In: A.H. Perera, D.L. Euler, and I.D. Thompson (editors). Ecology of a Managed Terrestrial Landscape: Patterns and Processes of Forest Landscapes in Ontario. Vancouver: University of British Columbia Press. 30-53.

Todd, B. and P. Kourtz. 1991. Predicting the daily occurrence of people-caused forest fires. Information Report PI-X-103. Chalk River, Ontario: Petawawa National Forestry Institute, Forestry Canada. 16 pages.

Turner, M.G., W.W. Hargrove, R.H. Gardner, and W.H. Romme. 1994. Effects of fire on landscape heterogeneity in Yellowstone National Park, Wyoming. Journal of Vegetation Science 5: 731-742.

Turner, J.A. and B.D. Lawson. 1978. Weather in the Canadian Forest Fire Danger Rating System: A User's Guide to National Standards and Practices. Information Report BC-X-177. Victoria, British Columbia: Canadian Forestry Service. 40 p.

Turner, M.G. and W.H. Romme. 1994. Landscape dynamics in crown fire ecosystems. Landscape Ecology 9: 59-77.

Van Wagtendonk, J.W. 1986. The role of fire in the Yosemite Wilderness. In: R.C. Lucas (editor). Proceedings, National Wilderness Research Conference: Current Research, July 23-26, 1985, Fort Collins, Colorado. General Technical Report INT-212. Ogden, Utah: USDA Forest Service. 2-9.

Van Wagner, C.E. 1969. A simple fire-growth model. Forestry Chronicle 45: 103-104.

Van Wagner, C.E. 1977a. Conditions for the start and spread of crown fire. Canadian Journal of Forest Research 7: 23-34.

Van Wagner, C.E. 1977b. Effects of slope on fire spread rate. Canadian Forest Service, Bi-Monthly Research Notes 33: 7-8.

Van Wagner, C.E. 1983. Fire behaviour in northern conifer forests and shrublands. In: R.W. Wein and D.A. MacLean (editors).The Role of Fire in Northern Circumpolar Ecosystems. New York: John Wiley. 65-80.

Van Wagner C.E. 1987. Development and structure of the Canadian Forest Fire Weather Index System. Forestry Technical Report No. 35. Ottawa: Canadian Forestry Service. 36 p.

Van Wagner, C.E. 1988. Effect of slope on fires spreading downhill. Canadian Journal of Forest Research 18: 818-820.

Vasconcelos, M.J. and D.P. Guertin. 1992. FIREMAP - Simulation of fire growth with a geographic information system. International Journal of Wildland Fire 2: 87-96.

Viro, P.J. 1974. Effects of fire on soil. In: T.T. Kozlowski and C.E. Ahlgren (editors). Fire and Ecosystems. New York: Academic Press. 7-45.

Voigt, D.R., J.A. Baker, R.S. Rempel, and I.D. Thompson. 2000. Forest vertebrate responses to landscape-level changes in Ontario. In: A.H. Perera, D.L. Euler, and I.D. Thompson (editors). Ecology of a Managed Terrestrial Landscape: Patterns and Processes of Forest Landscapes in Ontario. Vancouver: University of British Columbia Press. 198-233.

Ward, P.D. and A.G. Tithecott. 1993. The Impact of Fire Management on the Boreal Landscape of Ontario. Publication No. 305. Sault Ste. Marie, Ontario: Aviation, Flood and Fire Management Branch, Ontario Ministry of Natural Resources. 12 p.

Weber, M.G. and B.J. Stocks. 1997. Forest fires in the boreal forests of Canada. In: J.S. Moreno (editor). Large Forest Fires. Leiden, Netherlands: Backbuys Publishers. 215-233.

Weber, M.G. and S.W. Taylor. 1992. The use of prescribed fire in the management of Canada's forested lands. Forestry Chronicle 68: 324-334.

Weber, R.O. 1991. Toward a comprehensive wildfire spread model. International Journal of Wildland Fire 1: 245-248.

White, P.S. and S.T.A. Pickett. 1985. Natural disturbance and patch dynamics: an introduction. In: S.T.A. Pickett and P.S. White (editors). The Ecology of Natural Disturbance and Patch Dynamics. Orlando, Florida: Academic Press. 3-13.

Wilton, W.C. 1964. The Forests of Labrador. Publication No. 1066. Ottawa, Ontario: Canada Department of Forestry, Forest Research Branch. 72 p.

Woodmansee, R.G. and L.S. Wallach. 1981. Effects of fire regimes on biogeochemical cycles. In: H.A. Mooney, T.M. Bonnicksen, N.L. Christensen, J.E. Lotan, and W.E. Reiners (editors). Fire Regimes and Ecosystem Properties. General Technical Report WO-26. Washington, DC: USDA Forest Service. 379-400.

Wotton, B.M. and M.D. Flannigan. 1993. Length of the fire season in a changing climate. Forestry Chronicle 69: 187-192.

Wright, H.A. and A.W. Bailey. 1982. Fire Ecology: United States and Southern Canada. Toronto: John Wiley. 501 p.

8 Insect and Disease Disturbance Regimes in Ontario's Forests

RICHARD A. FLEMING,* ANTHONY A. HOPKIN,* and JEAN-NOËL CANDAU**

INTRODUCTION

The forests of Ontario can be divided into three main regions (Rowe 1972), the boreal forest, the Great Lakes-St. Lawrence forest, and the deciduous forest. The boreal forest region is the largest area and contains the majority of commercial forests in the province (see Thompson 2000, this volume). The forests in this region are composed largely of white spruce (*Picea glauca*), black spruce (*Picea mariana*), balsam fir (*Abies balsamea*), and tamarack (*Larix laricina*). The forests of the more southern regions of the ecozone contain white pine (*Pinus strobus*), red pine (*Pinus resinosa*), and jack pine (*Pinus banksiana*), trembling aspen (*Populus tremuloides*), balsam poplar (*Populus balsamifera*), and white birch (*Betula papyrifera*). The Great Lakes-St. Lawrence forest is characterized by white and red pine and a significant hardwood component, primarily of red maple (*Acer rubrum*) and sugar maple (*Acer saccharum*). The deciduous forest region is unique to Ontario and is the most southerly forested area in Canada. As the name of the region indicates, this is a region dominated by hardwoods. Forested areas, although they represent a smaller proportion of the land cover than in the other two zones, are extremely diverse. The common species include sugar maple, red oak (*Quercus rubra*), white oak (*Q. alba*), and beech (*Fagus grandifolia*). Butternut (*Juglans cinerea*), black walnut (*J. nigra*), and elm (*Ulmus americana*) are also widely distributed through this region.

Numerous insects and diseases cause damage and localized mortality in Ontario's forests, and under certain circumstances many of these pests damage large areas. For example, the larch sawfly (*Pristiphora erichsonii*) was believed to be responsible for the destruction of most mature stands of larch in the 1800s in Canada (Jardon et al. 1994). Similarly, in Ontario, this insect was believed responsible for extensive mortality in larch stands between 1910 and 1926 (Howse 1983). At present, larch sawfly is not considered a major pest, primarily due to the reduced commercial importance of larch, but also because extensive damage from this insect has not been recorded in the last 50 years.

Only a few pests cause damage over large areas on a routine basis and thereby cause changes to the forest landscape. The major insect pests of trees in Ontario include the spruce budworm (*Choristoneura fumiferana*) on spruce and balsam fir; jack pine budworm (*Choristoneura pinus pinus*) on pine species; and forest tent caterpillar (*Malacosoma disstria*) on poplar and other hardwood species such as sugar maple. The gypsy moth (*Lymantria dispar*), an introduced insect, has been associated with large-scale defoliation in southern Ontario's hardwood forests since 1981.

Major disturbances caused by insects are similar to fire-caused disturbances, in that extensive areas of forest are affected on an annual basis. Also like fire (Van Wagner 1978), insect outbreaks are usually cyclical

* Canadian Forest Service, Great Lakes Forest Research Centre, 1219 Queen Street East, Sault Ste. Marie, Ontario P6A 5M7

** Forest Landscape Ecology, Ontario Forest Research Institute, Ontario Ministry of Natural Resources, 1235 Queen Street East, Sault Ste. Marie, Ontario P6A 2E5

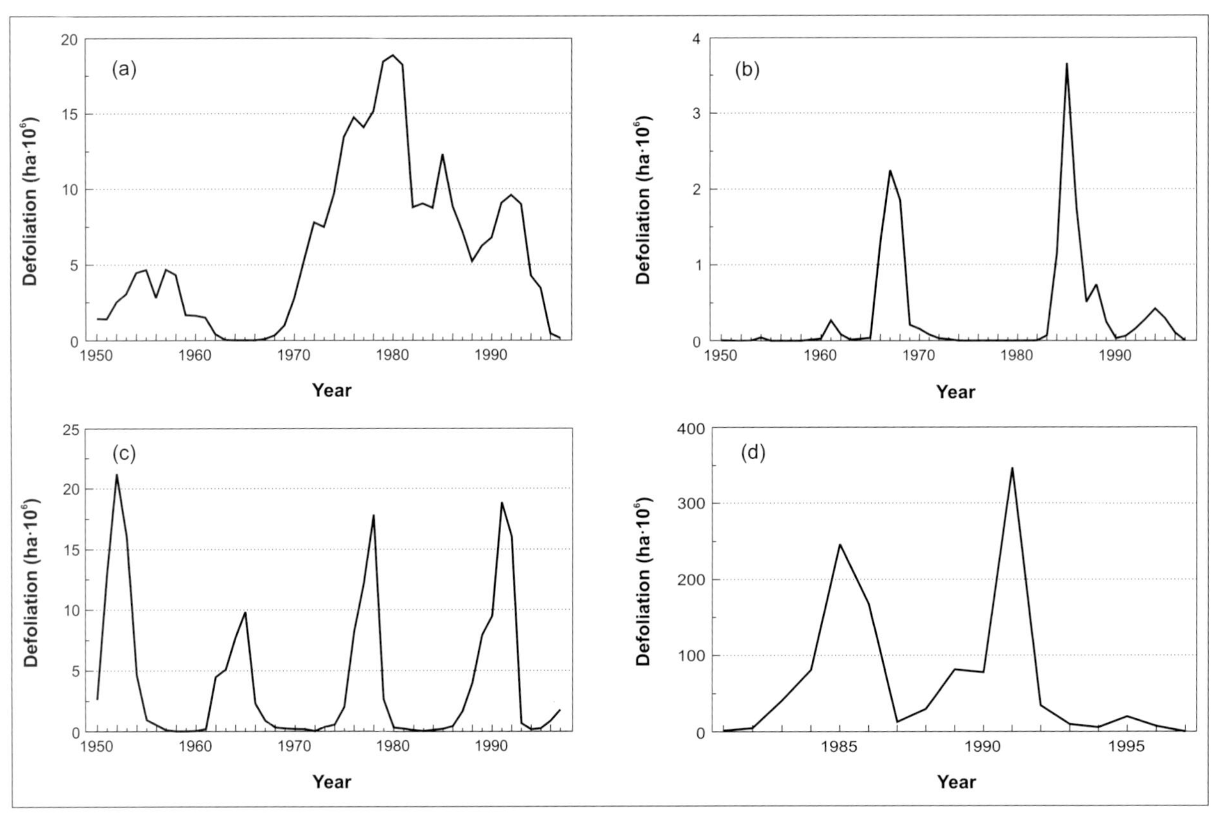

Figure 8.1 The gross area of forest in Ontario subjected to moderate to severe defoliation by: (a) spruce budworm; (b) jack pine budworm; (c) forest tent caterpillar; and (d) gypsy moth.

and play a role in forest succession. Unlike fire, however, insects are host-specific, and their outbreaks typically result in continuous damage over a period of years rather than the catastrophic damage over days or weeks characteristic of fire.

Although insect defoliators are generally regarded as the most important group of biotic disturbance factors in Ontario's northern forests, diseases also play a role. Diseases usually have an impact at the stand level, but occasionally they affect forests at broader scales as well. Diseases caused by root rots such as *Armillaria* species and *Inonotus tomentosus* are chronic problems in older stands of both conifers and hardwoods in the boreal forest. These diseases do not cause extensive damage, but kill individual trees, or pockets of trees, and cause gaps in the canopy. Such diseases are often most active after prolonged insect defoliation or drought has weakened the trees. Although diseases are less evident than fire or insects as causes of major forest disturbances, they influence the forest landscape by contributing to tree mortality in unfit individuals and regenerating material in natural ecosystems (Castello et al. 1995). Diseases such as white pine blister rust (*Cronartium ribicola*), Dutch elm disease (*Ophiotoma ulmi*), beech bark disease (*Nectria coccinea*), and chestnut blight (*Cryphonectria parasitica*) have changed the forest landscape by affecting the distributions and abundances of their respective hosts. More recently, an introduced disease, butternut canker (*Sirococcus clavigignenti-juglandacearum*), has threatened the existence of butternut in the deciduous forest region of Ontario (Davis et al. 1992).

The mosaic patterns and diversity of forests in all forest regions are a result of climate and major forest disturbances, particularly harvesting, fire, and insect activity. All tree species are vulnerable to attack by numerous insects and diseases. Insect activity and disease account for a large portion of the wood volume lost each year in Ontario's forests. Over the period from 1982 to 1987, the losses attributed to damage

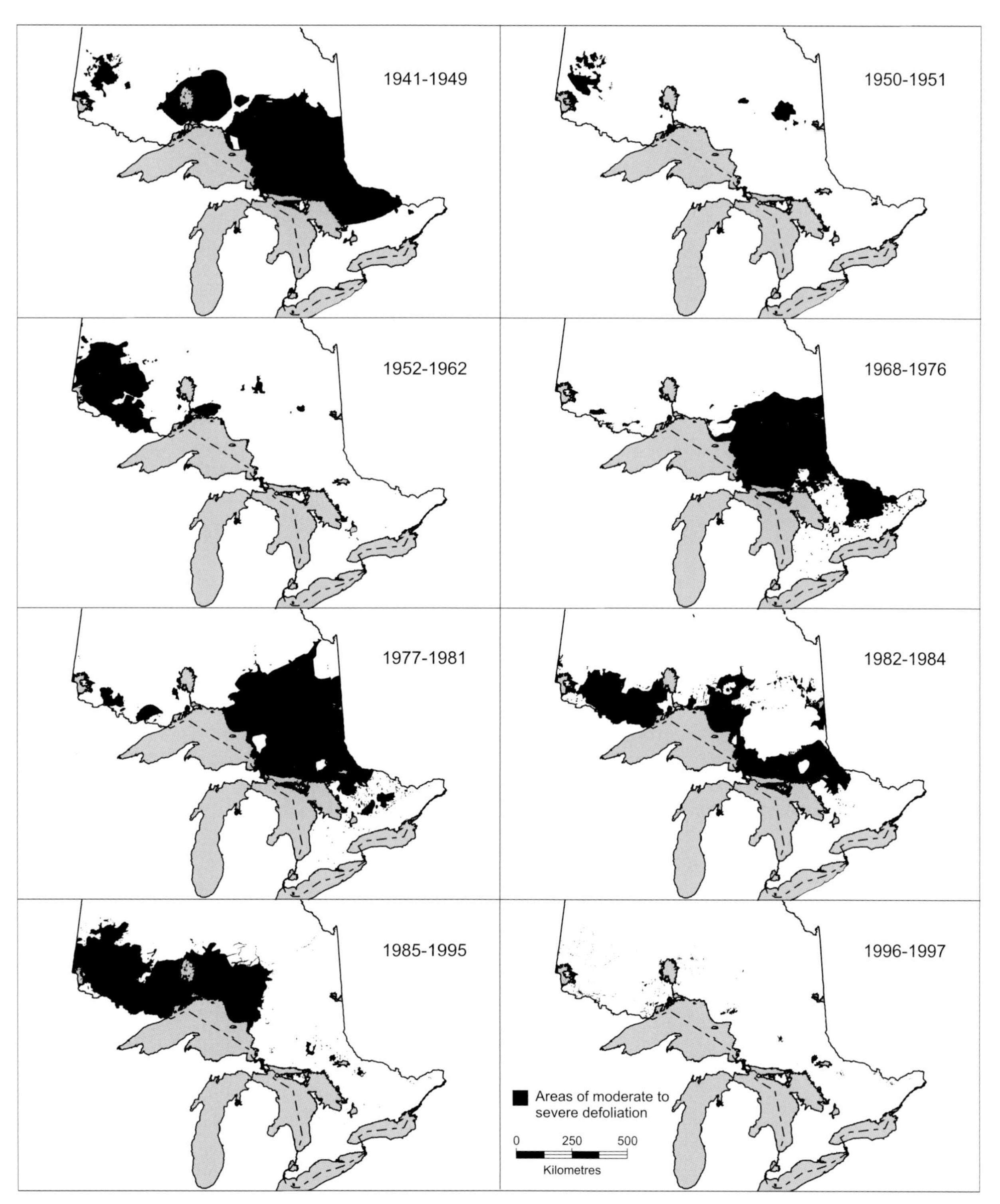

Figure 8.2 Areas of Ontario showing moderate to severe defoliation during outbreaks of spruce budworm. The dates on each map refer to the period during which current defoliation was observed at various locations within the delimited areas.
(Canadian Forest Service, unpublished data)

by insects and diseases, including decay, were estimated at 31 million m^3 annually (Gross et al. 1992). Insect and disease pests are considered to be problems because of the economic importance of the forests. As noted by Aber and Melillo (1991), however, major forest disturbances such as fire and defoliation are a necessary part of most terrestrial ecosystems and a mechanism for reversing declining rates of nutrient cycling and stand stagnation. In an extensive review of the literature, Attiwill (1994) illustrates how natural disturbances provide the impetus for forest diversity, structure, and function, and that, although they are at times devastating, natural disturbances are necessary for the sustainability of forests. In addition, insects contribute to biomass decomposition, to nutrient cycling (Schowalter et al. 1986), and to carbon cycling (Haack and Byler 1993) by digesting foliage and releasing nutrients through feces (Kimmins 1972; Fogal and Slansky 1985).

ONTARIO'S MAJOR INSECT AND DISEASE DISTURBANCES

Softwood Defoliators

The spruce budworm is considered the most important of the biotic disturbances in the boreal forest of Ontario. The immediate effect of spruce budworm outbreaks is to reduce the proportion of balsam fir in the upper canopy. In addition, during long outbreaks, white spruce is usually badly damaged, and often killed. Although white and black spruce are less vulnerable than balsam fir to budworm-induced mortality (Howse 1981), their seed production is reduced. Spruce budworm larvae attack the female reproductive structures and thus reduce the number of cones (Blais 1985).

Although spruce budworm outbreaks are a natural phenomenon and an integral part of the boreal forest ecosystem, some believe that the outbreaks have increased in size and duration in recent history. For instance, Blais (1983) suggested that harvesting practices and fire suppression have inadvertently favoured balsam fir, a preferred host tree of the spruce budworm, at the expense of less suitable host trees. From 1966 to 1996, spruce budworm outbreaks were more prolonged and much more extensive in Ontario than during the preceding episode of outbreaks that ended in 1963 (Figure 8.1a). At its peak in 1980, approximately 18 million ha were recorded as defoliated to either a moderate or severe level (Sterner and Davidson 1981). The population declined slowly after this point, but remained at high levels until 1996, 30 years after the episode of outbreaks began (Figure 8.1a). By 1996, the gross area within which balsam fir and white spruce mortality was recorded stood at 8,319,210 ha in Ontario (Howse 1997).

This 30-year episode was actually a series of outbreaks that occurred within the provincial boundaries and extended over much of the range of spruce and fir forest. After the previous outbreak in the northwestern portion of the province ended in 1961, an episode of outbreaks began in northeastern Ontario

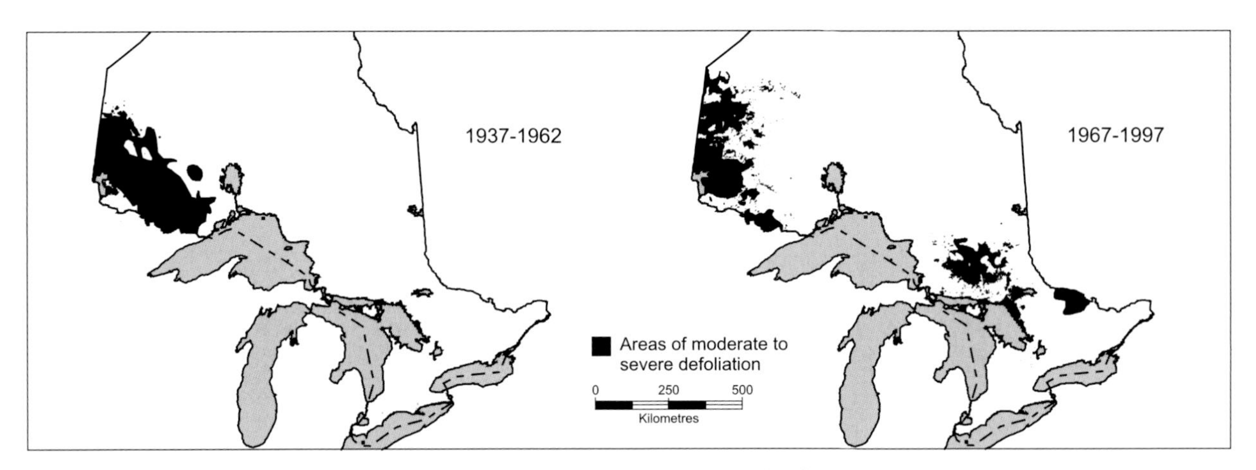

Figure 8.3 Areas of Ontario showing moderate to severe defoliation during outbreaks of jack pine budworm. The dates on each map refer to the period during which current defoliation was observed at various locations within the delimited areas.
(Canadian Forest Service, unpublished data)

in 1965 (Figure 8.2) and continued until 1995. Some new outbreaks also began at approximately the same time in the northwest, but the population of the spruce budworm built up much more slowly there, possibly because of an aggressive aerial spray program carried out until 1976 (Howse 1995; Howse et al. 1995).

Jack pine budworm is the most important insect pest of jack pine in Ontario. Unlike the closely related spruce budworm, this pest does not usually cause much mortality to its host over extensive areas, although pockets of mortality are common in defoliated stands. Severe defoliation is most likely to occur in semi-mature to mature stands, and losses to this insect mostly take the form of reduced tree growth and mortality of understory trees (Gross 1992; Hopkin and Howse 1995). Jack pine budworm outbreaks occur at about ten-year intervals (Figure 8.1b) and last only two to four years (Howse 1986; Volney and McCullough 1994). Although this insect occurs across the range of jack pine and is found as far east as New Brunswick, there are historical records of large-scale outbreaks only in Saskatchewan, Manitoba, and northwestern Ontario. Small, scattered outbreaks have also occurred in Quebec, New Brunswick, and in southern and central Ontario (Cerezke 1986; Howse 1986; Magasi 1995). After 1967, outbreaks of this insect extended eastward (Figure 8.3) and have since been detected over large areas of central and eastern Ontario, in the Great Lakes-St. Lawrence Forest region (Howse 1986; Howse and Meating 1995).

Hardwood Defoliators

Many insects are defoliators of the hardwood trees in the boreal forest (Rose and Linquist 1982; Ives and Wong 1988; Howse 1995). In Ontario, the most important of these are the bruce spanworm (*Operophtera bruceata*), which feeds on maples, beech, and aspen; the large aspen tortrix (*Choristoneura conflictana*), a pest of aspen; and the forest tent caterpillar, whose principal host species in the boreal forest is aspen (and, in Southern Ontario, sugar maple and oaks). In the Great Lakes-St. Lawrence region, the oak leaf shredder (*Croesia semipurpurana*) and the gypsy moth are major defoliators of oaks.

Of all the hardwood-feeding insects in Ontario, the forest tent caterpillar affects the largest area. Outbreaks of this insect occur over huge areas (Figure 8.4). For instance, when the recent outbreak in Ontario peaked in 1992, 19 million ha of boreal forest had been defoliated at moderate to severe levels (Figure 8.1c). Although the forest tent caterpillar causes heavy defoliation over large areas, it is not usually responsible for much tree mortality (Howse 1981; Moody and Amirault 1992). Defoliation by this insect might, in any case, be beneficial in releasing spruce from the understory in mixedwood stands (Fitzgerald 1995). On the other hand, Gross (1991) reported forest tent caterpillar defoliation to be a key factor in the general decline of maples that occurred in southwestern Ontario in 1977 and 1978. Gross found high dieback in defoliated stands at the end of the defoliation period, from 1974 to 1977, with subsequent mortality of more than 25 percent in some stands. Forest tent caterpillar outbreaks are extremely regular and dramatic: the affected area can increase from hundreds of hectares to millions in two to three years (Figure 8.1c).

Recent work has shown that the population dynamics of this insect may be strongly affected by the spatial structure of the forest. In an analysis of data concerning historical outbreaks in the boreal forest of Ontario, Roland (1993) observed that the amount of forest edge affected the duration of the outbreak. He suggested that the increased fragmentation of aspen forest caused by both harvesting and the conversion of land from forest to other uses was exacerbating outbreaks of the forest tent caterpillar. Subsequent work (Roland and Taylor 1997) revealed the cause: the impact of three important natural enemies (parasitic insects) is substantially reduced in fragmented forest, as compared to continuous forest. (This particular example also illustrates the dangers of generalization: it shows that insect herbivores do not necessarily thrive on large monocultures of their host plants). A recent audit of Ontario's forests (Hearnden et al. 1992) indicates that hardwood is becoming more widespread in northern Ontario's forests, suggesting that future outbreaks of the forest tent caterpillar might become more extensive. At present, however, large-scale outbreak patterns seem remarkably similar to those of previous events (Figures 8.4 and 8.1c).

The gypsy moth is an introduced insect in both Ontario and North America. Though a problem in eastern North America since 1889, this insect was not discovered in Ontario until 1969, near Kingston. Two outbreaks have occurred in Ontario since then, a fact which indicates a steady northward migration of the

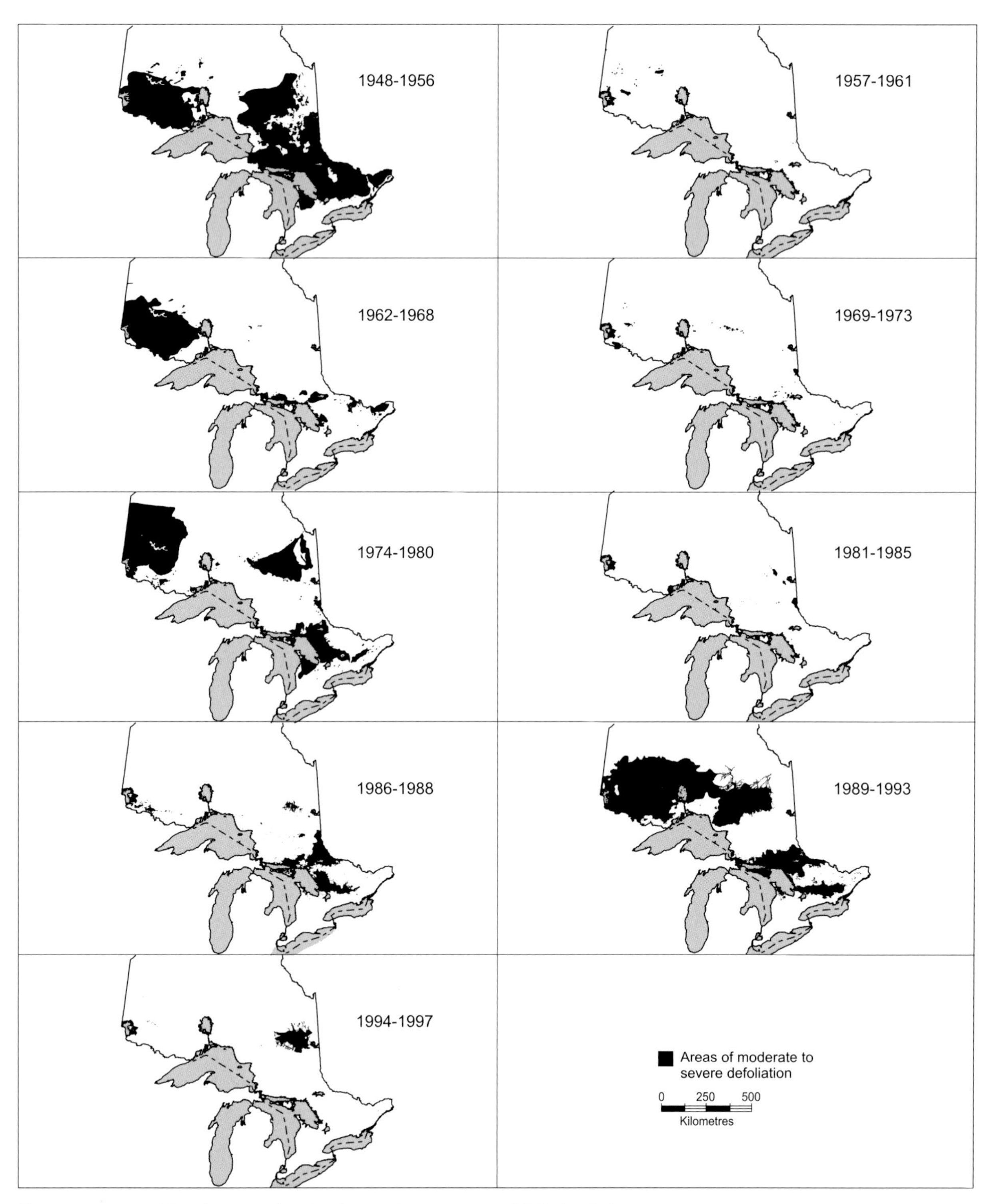

Figure 8.4 Areas of Ontario mapped as showing moderate to severe defoliation during outbreaks of the forest tent caterpillar. The dates on each map refer to the period during which current defoliation was observed at various locations within the delimited areas. (Canadian Forest Service, unpublished data)

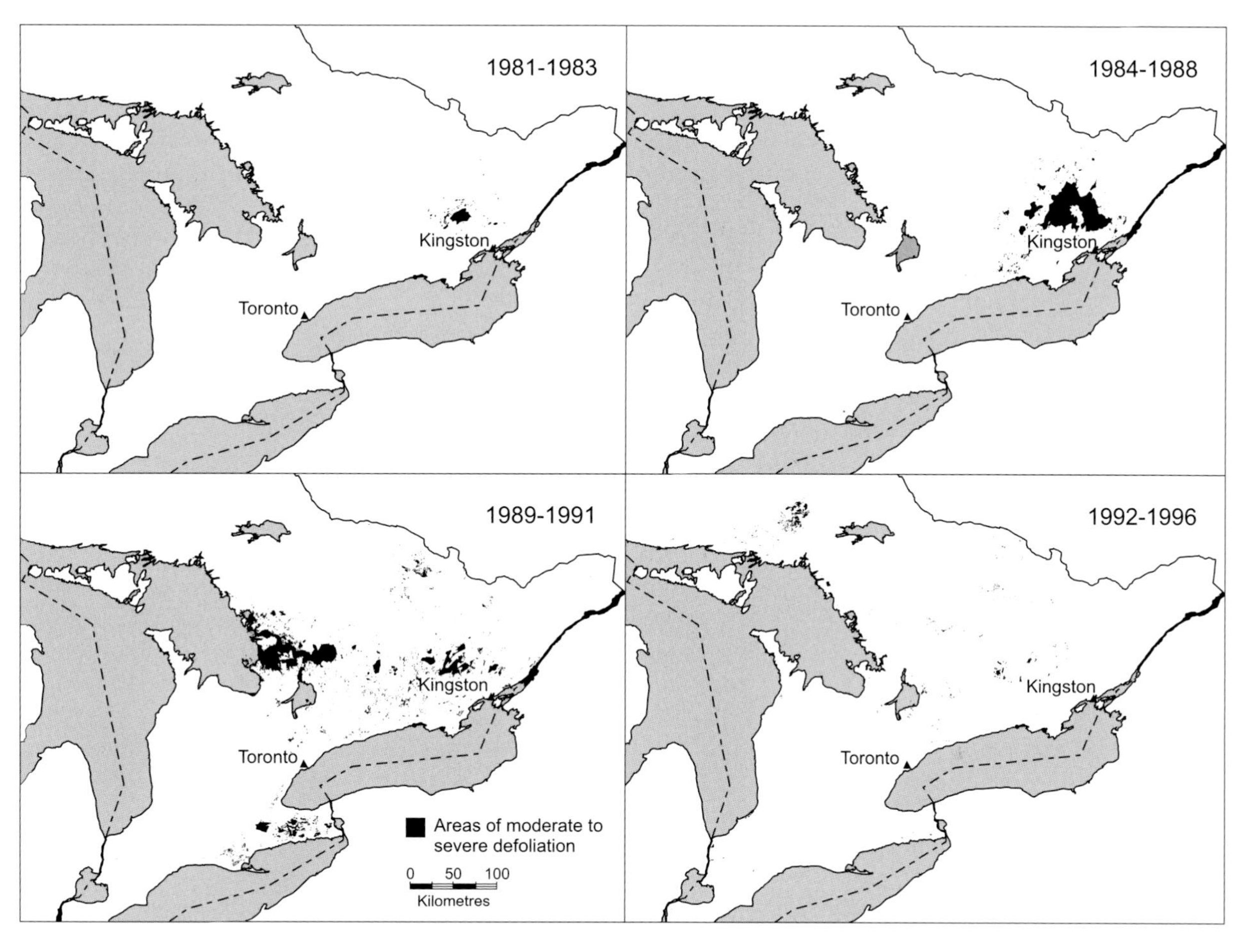

Figure 8.5 Areas of Ontario showing moderate to severe defoliation during outbreaks of the gypsy moth. The dates on each map refer to the period during which current defoliation was observed at various locations within the delimited areas. (Canadian Forest Service, unpublished data)

insect (Figure 8.5). Defoliation first became evident in 1981, and peaked at 246,000 ha in 1984. This first outbreak was followed by a second, which peaked in 1991 (Figure 8.1d), and then by a total collapse in 1997. Gypsy moth defoliation patterns in Ontario have been smaller in their spatial extent than those of the insect pests mentioned above. Although the gypsy moth will attack a wide range of host trees (Lechowicz and Jobin 1983; Nealis and Erb 1993), it is widely accepted that oak is its primary host, and that oak must be common in a forest stand before severe damage will occur.

The four major insects considered to be disruptive at the landscape level differ greatly in their host ranges, ecology, population dynamics, and impact on the forest. In Ontario, episodes of damaging spruce budworm outbreaks have recurred at 20- to 70-year intervals (Blais 1983). Moderate to severe defoliation by this insect can occur over large areas for periods of 10 to 15 years and cause whole-tree mortality over large areas. There is much debate about the population ecology of the spruce budworm and the degrees to which resource depletion, predation, parasitism, and disease contribute to outbreak collapse (e.g., Royama 1984; Blais 1985).

The forest tent caterpillar similarly affects extensive areas, although it causes generally low levels of tree mortality. Outbreaks of this insect occur at regular intervals. The duration of outbreak episodes is generally short, as can be seen in Figure 8.1c, although local outbreaks sometimes last as long as nine years (Sippell 1962). The collapse of the populations is usually associated with mortality caused by virus epizootics (Myers 1993) and by parasites (Sippell 1962). The food source of the forest tent caterpillar is fairly constant from year to year, so, resource depletion is considered, at best, a minor contributor to population collapses.

Similarly, gypsy moth populations are largely controlled by virus and, more recently, by a fungal pathogen, *Entomophaga maimaiga* (Reardon and Hajek 1993; Nealis et al. 1999). The outbreaks of jack pine budworm are both short and surprisingly small in area, given the range of this insect's host tree species. Preliminary work suggests that past outbreaks have occurred under a fairly narrow range of climatic conditions, conditions which are narrower than those typically associated with jack pine (McKenney et al. 1995). Climate may be a limiting factor, not so much in the distribution of the insects or the host tree, but in the likelihood of an outbreak. Nealis (1995) found that outbreak collapse is largely related to natural enemies, although other factors, such as host tree phenology, are thought to contribute to insect survival (Nealis and Lomic 1994).

ONTARIO'S MAJOR FOREST DISEASES

The disturbance patterns caused by disease contrast greatly with those of the major insects. Disease has an important effect on stand development, primarily by eliminating unfit individuals from the overstory and thus creating openings where new species can gain a foothold. This process is usually less dramatic than the disturbances caused by the major insect defoliators, so it receives relatively little attention. A few diseases, most of them exotics, have had a dramatic impact on the forest landscape in Ontario.

The best known of these exotic diseases are white pine blister rust, chestnut blight, and Dutch elm disease. Chestnut blight was discovered in the eastern United States in the early 1900s, and spread across the eastern hardwood forest and into Ontario over the following 30 years (Anagnostakis 1987). The disease eliminated the American chestnut (*Castanea dentata*) as a dominant or co-dominant tree species. Prior to the introduction of the disease, chestnut was a large component of the southernmost forest in Ontario. This tree species has been replaced by a number of tree species, but primarily by maples and oaks. White pine blister rust (*Cronartium ribicola*) has changed the forest landscape by affecting the distribution of mature eastern white pine (Castello et al. 1995). This is particularly the case in northeastern Ontario, in the southern portion of the boreal region where the disease has the greatest impact. This area is considered a high-hazard zone because of the regional climate (Gross 1985). Here, the disease kills younger white pine and makes regeneration difficult after harvest, reducing the white pine component of Ontario's forests.

Dutch elm disease was first discovered in Ontario in 1946. The spread of the disease from both the southwest and southeast suggests multiple entry points. The disease moved progressively northward through the range of elm in the southern deciduous forest and into the Great Lakes-St. Lawrence and southern boreal forest regions. In these northerly areas, the host grows in scattered pockets and as planted trees in urban centres (Figure 8.6). The disease also spread into northwestern Ontario in 1976 from Minnesota (Sterner and Davidson 1981). Dutch elm disease provides an excellent example of a relationship between a disease and insect infestation. Because this disease is vectored by beetles, by the native elm beetle (*Hylurgopinus rufipes*) and the European elm bark beetle (*Scolytus multistriatus*), it has spread rapidly.

Butternut canker has caused serious damage to the butternut population of the northeastern United States (Ostry et al. 1994). The disease was not reported in Ontario until 1991 (Davis et al. 1992), although it had likely been present for some time. Subsequent surveys by the Forest Insect and Disease Survey group (FIDS) of the Canadian Forest Service found the disease across the range of butternut in Ontario. It threatens the entire butternut component in southern Ontario.

Beech bark disease, while not yet present in Ontario, has altered the forest landscape in northeastern North America (Houston 1994), including the state of New York and the adjacent province of Quebec, by causing mortality to semi-mature and mature beech. The fungus causing the disease is not native to North America, and is carried from tree to tree by an introduced insect, *Cryptococcus fagisuga*. This insect was not officially reported in Ontario until 1981 (Bisessar et al. 1985), although surveys conducted by FIDS have recorded the presence of the insect across the range of beech in Ontario since 1966. These findings suggest that beech in the province is at future risk.

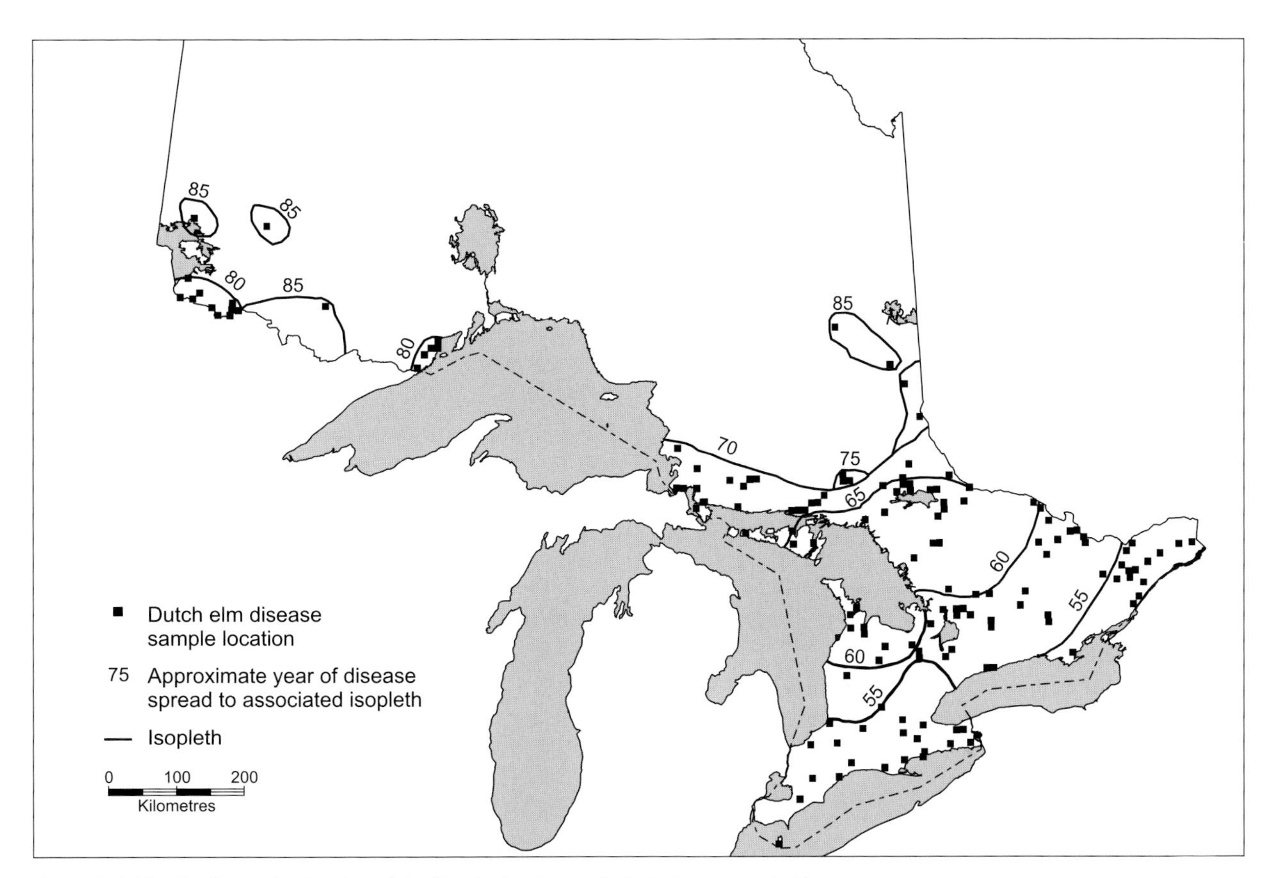

Figure 8.6 Distribution and migration of the Dutch elm disease in Ontario, as recorded in Forest Insect and Disease Surveys between 1955 and 1985. Numbers refer to the approximate year in which the disease had extended its range to the associated isopleth.
(Canadian Forest Service, unpublished data)

ABIOTIC DISTURBANCES

Abiotic disturbances often interact with insect and disease disturbances; therefore, some background knowledge of the former is required for a full understanding of the dynamics of the latter. The major abiotic disturbances include fire, drought, and blowdown. Drought has been frequently recorded across the boreal forest, where the damage is most noticeable on shallow sites. Drought stress is directly harmful to forest vegetation but more often acts in concert with secondary organisms to cause mortality. In 1993, mortality was recorded over 31,000 ha of drought-stricken jack pine in the boreal forest of Ontario (Howse and Applejohn 1993). Bark beetle damage and root rots were found in the dead trees. Drought conditions can also increase the level of stress to which defoliated trees are subjected during insect outbreaks and can result in higher than normal tree mortality. While drought does not directly predispose jack pine to insect defoliation, Riemenschneider (1985) reported that drought conditions stimulate flowering on jack pine. This, in turn, increases the survival of jack pine budworm (Nealis and Lomic 1994) and hence the likelihood of damage during an outbreak (Volney 1988). There is also evidence suggesting that spruce budworm reproductivity (Sanders et al. 1978) and survival (Lucuik 1984) increase directly as a result of the warmer and drier conditions (Mattson and Haack 1987) of drought. Indirect effects are also important for the spruce budworm. At high temperatures the vulnerable larval stages can escape many natural enemies, because (a) the larvae develop faster than at normal temperatures (Lysyk 1989) and thus reduce the duration of their exposure, and (b) they develop more quickly at high

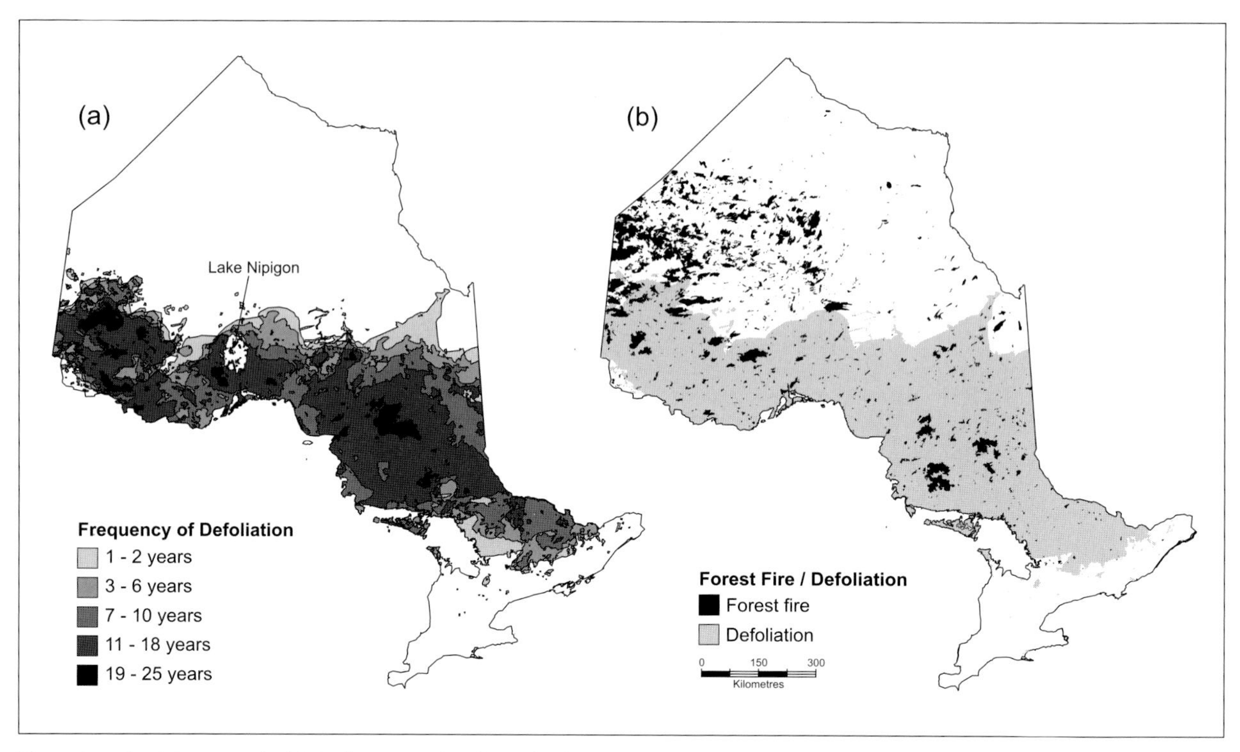

Figure 8.7 Spatial patterns in (a) the frequency (see legend) of moderate to severe spruce budworm defoliation as recorded by aerial survey from 1941 to 1996; and (b) the occurrence of forest fire (black) overlaid on the occurrence of defoliation (grey) in the same period.
(Canadian Forest Service, unpublished data)

temperatures than do many of their natural enemies (Fleming and Candau 1998).

Blowdown, caused by strong surface winds which can uproot extensive areas of forest, is a common feature of boreal forests and plays an important role in determining the structure of the forest (Flannigan et al. 1989). Blowdown is often reported in various parts of the boreal and differs greatly in extent. In 1991, extensive damage caused by blowdown was observed over 207,700 ha of forest in northwestern Ontario (Sajan and Brodersen 1992). As the sudden appearance of many uprooted trees greatly accelerates fuel accumulation, fire hazard often increases after blowdown occurs (Flannigan et al. 1989). In 1973, severe wind damage and blowdown occurred in an area of 40,000 ha in northwestern Ontario. In the following year, 32,000 ha of that area burned (Stocks 1975).

LARGE-SCALE DYNAMICS OF INSECT AND DISEASE DISTURBANCES

So far, we have provided aggregated descriptions of the temporal dynamics (see Figure 8.1) and spatial dynamics (see Figure 8.2) of insect and disease disturbances; however, understanding the landscape-level processes underlying disturbance dynamics requires a more analytical approach. Such an approach is best illustrated by a case study of particular species, because, although there are commonalities among species in terms of the dynamics of their disturbance regimes, there are also some substantial differences. These differences (for example, in length of outbreak cycle) probably result from the unique suite of ecological factors associated with each species, factors such as host tree species, natural enemies, and climatic "preferences." It is thus important to consider each insect or disease individually, at least to some

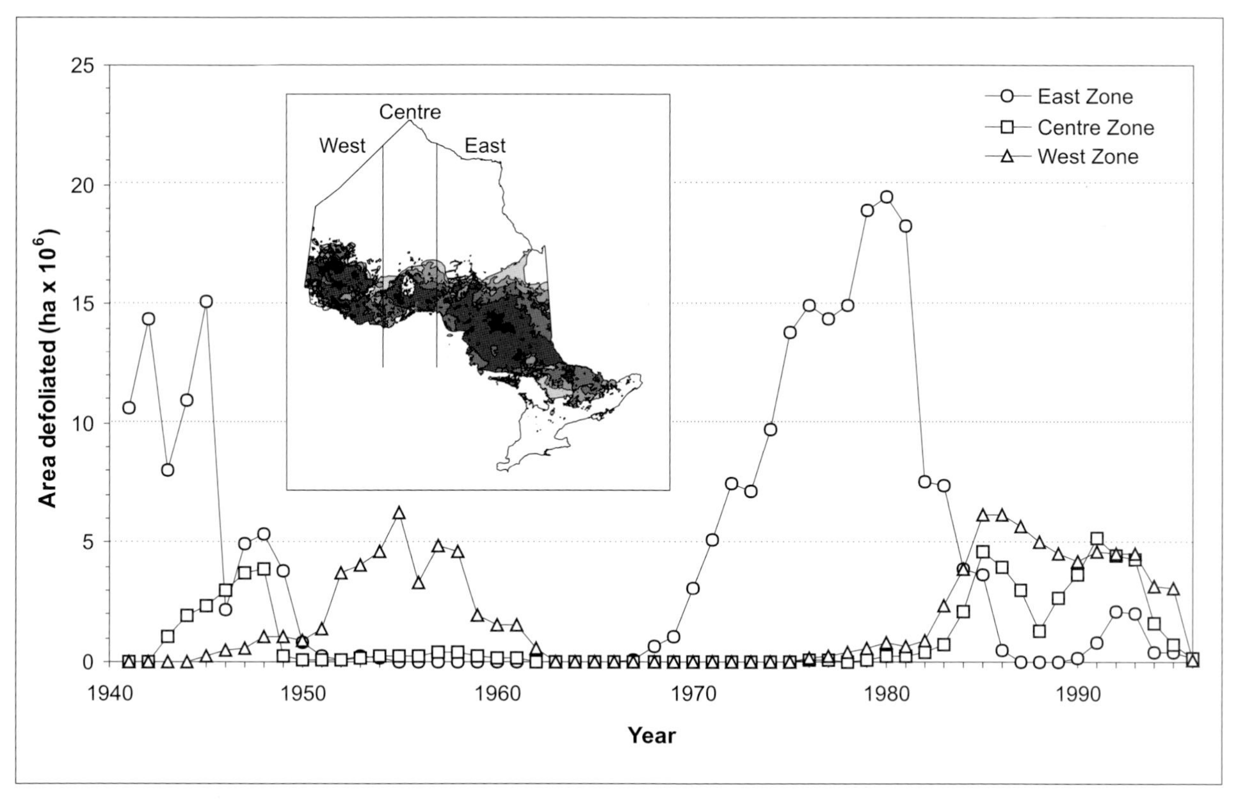

Figure 8.8 Moderate to severe spruce budworm defoliation from 1941 to 1996 in the eastern, central, and western defoliation zones, shown on the map of defoliation frequency (Figure 8.7a).

extent. The spruce budworm is a logical choice of species for such a case study in Ontario, because it is more economically important, and has been studied longer and in greater detail, than any other forest insect or disease in the province.

Large-Scale Dynamics of Spruce Budworm Disturbances

The best information for studying the large-scale dynamics of spruce budworm disturbances comes from survey data collected by FIDS. Since 1941, that group has been conducting aerial reconnaissance of large-scale defoliation events in Ontario's productive and exploitable forest area. Ideally, the reconnaissance flights begin as soon as the current season's defoliation is completed, generally in mid- to late-July. Survey planes usually fly at about 170 km/hr. at altitudes between 360 m and 600 m. Survey lines are typically 6 km to 10 km apart. An observer in the aircraft delineates defoliation on 1:125,000- or 1:250,000-scale maps with a maximum error of about 500 m, and this information is later transferred to basemaps at smaller scales (for example, 1:500,000). Sanders (1980), Dorais and Kettela (1982), and Allen et al. (1984), provide additional technical details about aerial sketchmapping and describe some of the uncertainties associated with this data collection technique.

Figure 8.7a shows how many times the annual FIDS surveys between 1941 and 1996 mapped moderate to severe defoliation from the spruce budworm at a given location. For purposes of illustration, the frequencies of defoliation have been grouped into five classes defined by the number of years of defoliation: 1 to 2, 3 to 6, 7 to 10, 11 to 18, and 19 to 25 years. The defoliation follows a belt located between 45°N and 52°N latitude. The northern and southern fringes of this belt have rarely been defoliated. In the northeast, defoliation was recorded along the banks of the Moose River just once, in 1979. That fact helps to explain the unusual northeastern boundary of the class containing one to two years of defoliation.

Within the belt of defoliation, there appear to be spatial patterns in the cumulative frequencies. Relatively small areas of very frequent defoliation (that is, areas defoliated 19 to 25 times over the 55 years) are surrounded by regions where frequencies decline approximately with distance from these foci. Two narrow corridors about 100 km wide were less frequently defoliated. These corridors run approximately from north to south and separate a central zone of more frequent defoliation centred near Lake Nipigon from eastern and western zones, where more frequent defoliation also occurred. The defoliation belt can thus be split into three zones, according to biological factors reflected in the frequency of defoliation by the spruce budworm (Figure 8.8, inset).

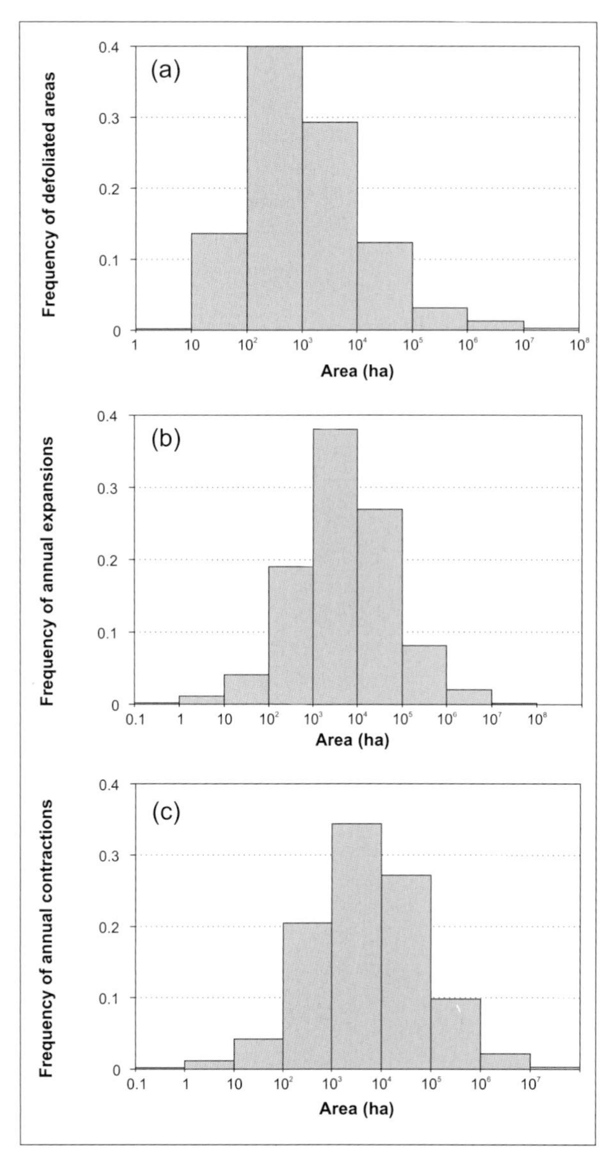

Figure 8.9 Histograms illustrating the annual occurrence frequency of size classes of (a) total extent of contiguous areas having moderate to severe defoliation by spruce budworm from 1941 to 1996; (b) expansion of these areas of defoliation; and (c) contraction of these areas of defoliation.

The three zones of spruce budworm defoliation shown in the inset of Figure 8.8 vary greatly in area. The eastern zone (25.5 million ha) encompasses over 2.5 times the area of the western zone (9.6 million ha) and almost four times that of the central zone (6.6 million ha). Figure 8.8 shows the historical record of defoliation from 1941 to 1996 and suggests the presence of an outbreak cycle in each area. Statistical analyses (Candau et al. 1998) have shown that the outbreak cycles in the eastern and western zones each have periods of 32 to 36 years, and that the eastern cycle has run five or six years ahead of the western cycle since 1941. The outbreak in the 1940s in the central zone overlapped with the end of the outbreak in the eastern zone and the start of the outbreak in the western zone. The last outbreaks in the central and western zones occurred almost simultaneously. When the area of defoliation in each zone is considered as a proportion of the zone's total area, the defoliation cycles are also remarkably similar in amplitude.

The sum of the defoliated areas in the three zones shown in Figure 8.8 corresponds to the historical fluctuation shown in Figure 8.1a, except that Figure 8.8 starts nine years earlier. This difference in starting point is important, however, because it demonstrates how Figure 8.1a could be misinterpreted as showing a dramatic increase in the extent of outbreak episodes since 1950. In fact, as Figure 8.8 shows, Figure 8.1 captures none of the outbreak episodes that occurred in the eastern and western zones in the 1940s.

Spruce budworm outbreaks sometimes occur over vast areas. Figure 8.9a shows how often contiguous areas of annual defoliation from spruce budworm attained certain orders of size, over the years 1941 to 1996. Most of these areas are between 10 ha and 10,000 ha, but the distribution is skewed and has a maximum of over 10 million ha. This area of contiguous defoliation of over 10 million ha is equivalent to 40 percent of the eastern zone of defoliation (see Figure 8.8) and dwarfs the size of even the largest fires (Figure 8.7b).

The area of spruce budworm outbreaks can also expand and contract quickly. The size distributions of the contiguous areas of outbreak expansion and contraction from one year to the next are shown in Figure 8.9b and 8.9c, respectively. Most of these areas are between 10 ha and 10,000 ha, but the distributions are skewed and have maximums of over 1 million ha. The similarity between the distributions in Figures 8.9b and 8.9c suggests that the absolute rates of expansion during outbreak buildup are similar to those of contraction during outbreak collapse.

Figure 8.9 focuses on the large-scale spatial dynamics of spruce budworm outbreaks; Figure 8.10 deals with the corresponding temporal dynamics. In considering Figure 8.10, one should be aware that it represents the overall averages throughout the defoliation belt illustrated in Figure 8.7, but does not pretend to deal with the spatial variation in defoliation frequencies shown there. Figure 8.10a shows that often spruce budworm defoliation does not reach moderate or severe intensities for two consecutive years in the same area. The frequencies of occurrence of consecutive years of detectable defoliation gradually decline with the number of years, up to a maximum of 17 years. These intervals can be compared with the four or five consecutive years of 100-percent current defoliation that it typically takes to kill balsam fir (MacLean 1985). White and black spruce are less vulnerable (Morris 1963) than balsam fir, so the longer periods of continuous defoliation may be concentrated in stands of these species. Alternatively, spruce budworm densities may be very uneven from tree to tree in stands of balsam fir, and trees which initially escape may not be attacked until after the first trees have started to die.

The distribution of continuous time intervals in the FIDS records during which no spruce budworm defoliation was detected appears to have many peaks (see Figure 8.10b. The shortest intervals range from one to six consecutive years and may be caused by differences in the duration and intensity of defoliation in the course of an outbreak. The longer group of time intervals without detectable defoliation ranges from 23 to 44 years, and may correspond to the periods of time separating outbreaks (see Figure 8.8). The variation in these longer intervals probably reflects spatial differences in local forest vulnerability to noticeable levels of defoliation (see Figure 8.7a).

Insect Interactions with Fire

Insect defoliation and fire are the major disturbances affecting the succession of northern forests. That both factors are important is particularly apparent in the spruce and fir boreal forest, where a relationship seems to exist between fire and outbreaks of the spruce budworm (e.g., Bergeron and Dansereau 1993), but the underlying cause of this relationship is uncertain. It is often argued (e.g., Furyaev et al. 1983) that this relationship must be considered in order to understand the role of either kind of disturbance in the boreal

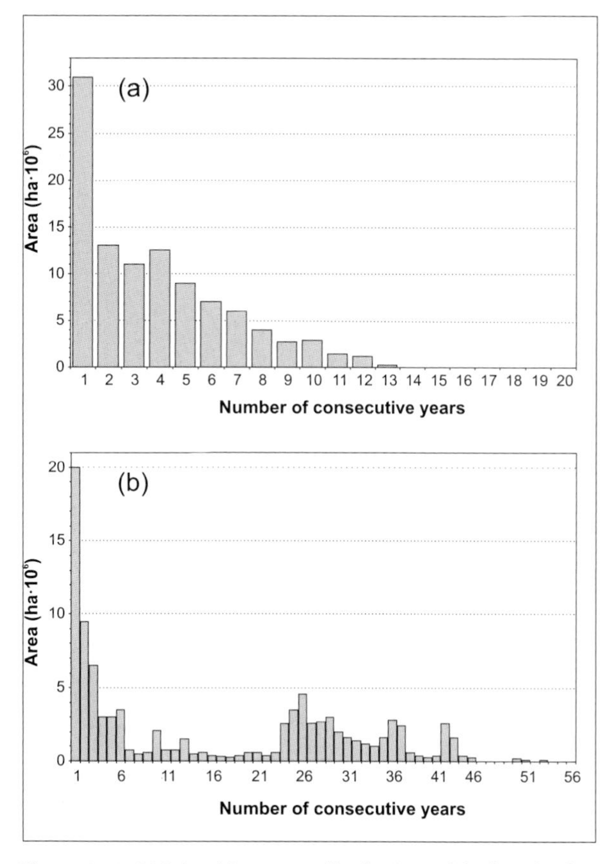

Figure 8.10 Weighted frequency distributions of the length of the time intervals in the Forest Insect and Disease Survey records, showing: (a) continuous occurrence of spruce budworm defoliation, and (b) continuous absence of spruce budworm defoliation. The frequencies have been weighted by area to account for differences in extent.

forest. It is a common notion that heavy defoliation by the spruce budworm can alter successional patterns in mature stands by making the stands more susceptible to crown fires. For instance, Wein (1990) reasoned that increased stress on host trees (for example, from chronic defoliation or greater insect outbreak frequencies) accelerates host tree mortality. He suggested that greater tree mortality adds to the fuel load and ultimately leads to more frequent and more intense forest fires.

By contrast, in a study of balsam fir killed by spruce budworm, Stocks (1987) argued that the forest fire potential of such stands increases for just a few years after an outbreak. He explained that spruce budworm defoliation opens up the canopy and lets understory vegetation proliferate. The moist, green layer of understory vegetation effectively prevents fire spread by isolating the dry surface fuels from the crown fuels. After about five years of crown breakage and windthrow, however, so much dead and downed material has accumulated on the ground that the surface fuel can overcome the dampening effect of the understory vegetation. The potential for summer fires peaks between five and eight years after stand mortality. After that, the potential for forest fires begins to gradually decline as the understory vegetation continues to proliferate and the dead and downed surface fuel begins to decompose and absorb moisture.

Johnston (1996) summarizes the general effects of spruce budworm and fire on the boreal forests of Ontario. He notes that after severe crown fires, post-fire stands can be dominated by birch, because that tree species has the ability to colonize extensive areas by wind-disseminated seed. Where extensive crown fires have not occurred, shade-tolerant species, particularly balsam fir and white spruce, are likely to dominate, creating a forest composed largely of tree species susceptible to spruce budworm. Where fire does not occur after a budworm outbreak, the tree mortality resulting directly from the budworm or secondary insects, such as beetles, and diseases, such as root rots, produces gaps in the canopy. Deciduous (non-host) species such as birch and aspen, which require ample light, rarely become established in the smaller openings. Instead, host species of the spruce budworm such as balsam fir and white spruce, which are more tolerant of shade than birch and aspen, eventually occupy the gaps and thus maintain a forest susceptible to spruce budworm.

Other disturbance-causing insects also interact with wildfire (McCullough et al. 1998). For instance, jack pine is a tree species greatly influenced by both fire and a major disturbance-causing insect, the jack pine budworm. Volney (1988) noted that outbreaks of the jack pine budworm in the prairie provinces of Canada have increased in areal extent over time. He suggested that this is the result of improved control of forest fire, which allows more jack pine stands to reach the older ages at which they are most susceptible to the insect. A similar situation probably exists in Ontario. In the absence of fire for extended periods of time, jack pine is often replaced in the forest by more shade-tolerant species (Fowells 1965). Defoliation by the jack pine budworm may, however, reduce the likelihood of such extended absences of fire. Nealis (1995) suggested that the tree mortality resulting from defoliation contributes to fire in affected jack pine stands by producing gaps in the canopy that expose the forest floor and its litter layer to increased direct sunlight and air circulation. The consequent drying probably promotes ground fires, which cause the serotinous cones of jack pine to open and deposit their seed on an exposed mineral bed, while simultaneously burning back competing tree species (which are less tolerant of ground fire). This scenario concludes with jack pine's successful regeneration and retention of the site.

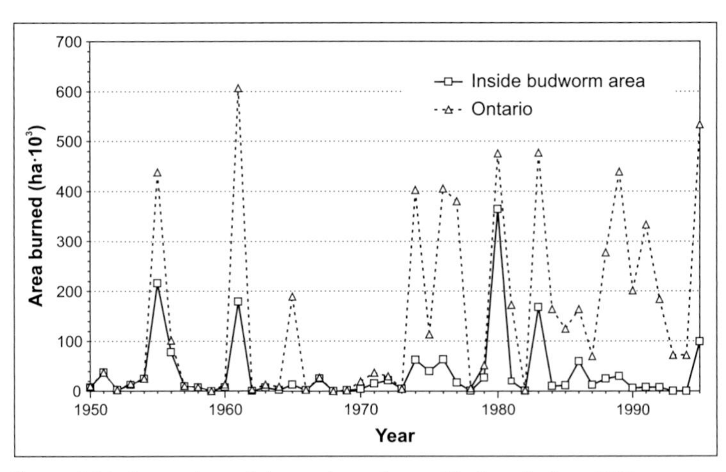

Figure 8.11 Comparison of the total area burned in Ontario from 1950 to 1995 with the area burned within the belt defoliated by spruce budworm over the same period (see Figure 8.7b).

So far, we have discussed possible causal relationships between insect outbreaks and wildfire. These relationships have generally been revealed through small-scale experimental studies. It is also useful to search historical records for large-scale patterns in the interaction between insect and fire disturbance regimes. For this, we focus on spruce budworm and fire, because this interaction offers the most extensive set of historical records currently available for such a search. Figure 8.11 illustrates the historical records of total area burned in the province and the area burned inside the budworm defoliation belt (see Figure 8.7b). Comparing this graph with Figures 8.1 and 8.8 suggests that fire has a cycle more erratic than that of the major insect disturbance agents. In addition, fire tends to be a less extensive disturbance agent in the province. For instance, the largest total area burned since 1950 was just over 600,000 ha (in 1961). At their peaks, spruce budworm, jack pine budworm, and forest tent caterpillar outbreak episodes cover millions of hectares (see Figure 8.1). The largest total area burned in the spruce budworm defoliation belt (360,000 ha) occurred in 1980, at the end of the peak defoliation years for spruce budworm since 1950, and just after a forest tent caterpillar outbreak. The degree to which fire suppression affects these relationships is uncertain. (Fire suppression efforts can be quite successful, and are often targeted to protect particular threatened areas rather than all of the forest). Matching the spatial locations of fire and spruce budworm defoliation adds another dimension to interpretations of the interaction depicted in Figures 8.7b and 8.11.

Figure 8.12 quantitatively illustrates the relationship between fire and spruce budworm defoliation from a spatial perspective. The few areas which have been burnt twice (or more) have been counted twice (or more) in the figure, so mathematically it is possible to have percentage burns exceeding 100 percent. The figure suggests that the interaction between fire and spruce budworm outbreaks continues during post-disturbance succession. The plotted data suggest that the probability of fire increased slowly with the frequency of spruce budworm defoliation up to 12 to 14 occurrences of moderate to severe defoliation in the 56 years between 1941 and 1996. (Statistical analysis is needed to show that the plot does imply an interaction between spruce budworm and fire and does not merely reflect variation in the total areas experiencing different frequencies of spruce budworm defoliation.) It may be that the climate, forest composition, and other environmental conditions on these areas are such as to support Wein's (1990) reasoning that increased stress on host trees results in more intense and more frequent fires.

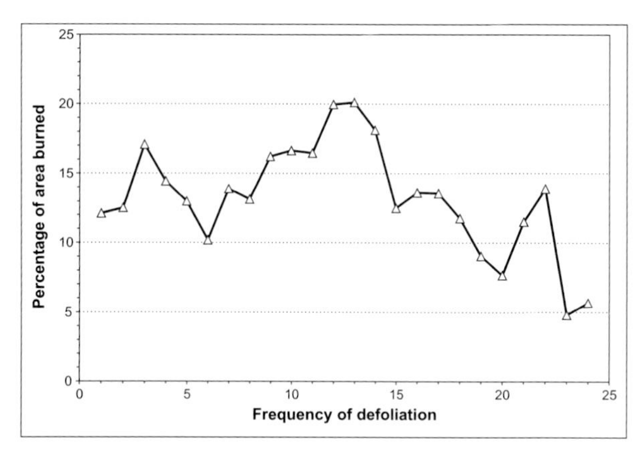

Figure 8.12 The spatial association between the proportion of area burned and the frequency of defoliation by spruce budworm. The plot shows the percentage of those areas experiencing various frequencies of spruce budworm defoliation from 1941 to 1996 (see Figure 8.7a) that also burned between 1921 and 1995.

The plotted data, however, suggests that the probability of fire declined as defoliation frequency increased beyond 12 to 14 occurrences from 1941 to 1996. Johnston (1996) argues that, where extensive crown fires rarely occur (possibly as a result of a wet climate or of fire control), shade-tolerant species, particularly balsam fir and white spruce (both favoured spruce budworm hosts) are likely to dominate. The result is a forest susceptible to spruce budworm outbreaks.

Forest fire may also affect the probability or severity of future spruce budworm attack in the same area (Bergeron et al. 1995). After severe crown fires, birch (which is not a host for spruce budworm) can dominate post-fire stands due to the ability of that tree species to colonize extensive areas by wind-disseminated seed. Within the belt of spruce budworm defoliation, the area burnt is so small (see Figure 8.7) that its relative impact on large-scale spruce budworm defoliation is likely negligible. In conclusion, the available information suggests that the interaction between forest fire activity and spruce budworm outbreaks is complex, that it is important to understanding either disturbance regime, but especially fire, and that it needs to be quantified.

POPULATION BIOLOGY OF THE SPRUCE BUDWORM OUTBREAK CYCLE

The biological processes behind the rise and fall in time, and the spread and contraction over space, of spruce budworm outbreaks are profitably discussed from the perspective of spruce budworm ecology. Such a perspective begins with a more intensive examination of the insect and its life-style.

The spruce budworm is the most damaging defoliator in North America's boreal forests. Although it is Ontario's most important disturbance-causing insect (see Figure 8.1), it has been studied most intensively in New Brunswick (Morris 1963), where it causes greater economic concerns. The spruce budworm is a native, naturally outbreaking insect which feeds on balsam fir and spruces. It often kills most of the trees in dense, mature balsam fir stands, unless control measures are applied (MacLean 1985), and accounts for an average of just over 50 percent of all insect-caused losses to forest productivity in Canada each year (Hall and Moody 1994). It is not widely recognized that insects as a group have a 50-percent greater impact than fire, and that average annual losses to the spruce budworm alone amount to about 75 percent of the total losses to fire in Canada (Hall and Moody 1994).

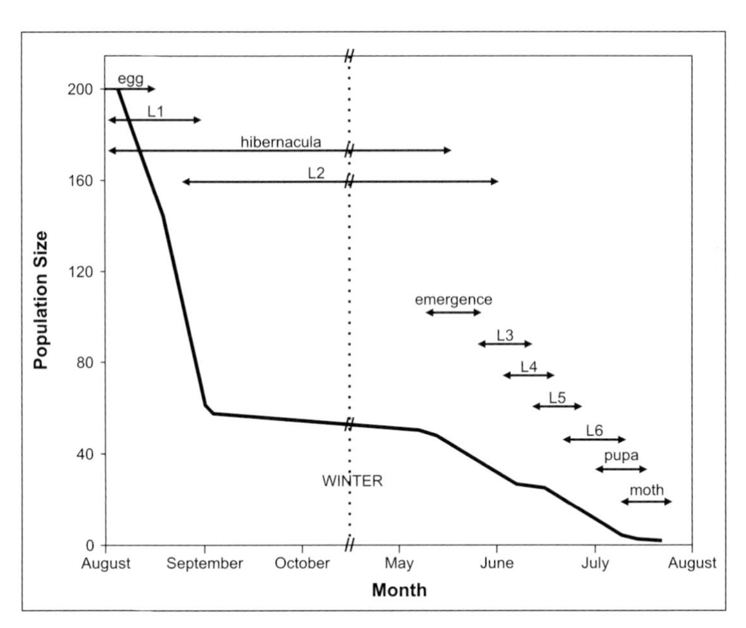

Figure 8.13 A generalized survivorship curve for the spruce budworm. Larval instars are denoted "LN", where "N" is the instar number (Morris 1963).

The geographical distribution of spruce budworm extends well beyond Ontario and follows the distribution of white spruce throughout Canada's boreal forest and southward into the cool temperate forests of the Great Lakes-St. Lawrence region. The northern boundary is less certain, but reports exist of heavy defoliation within 150 km of the Arctic circle in the Northwest Territories (Volney and Cerezke 1992). In outbreaks, which usually last from 5 to 15 years, populations can reach 10^8 fourth-instar larvae/ha, while between outbreaks, the budworm can remain relatively rare (10^5 fourth-instar larvae/ha) for 20 to 60 years, as the forest recovers (Fleming and Candau 1998).

Like many insect species which outbreak periodically (Wallner 1987), the spruce budworm feeds on a range of host plants and proliferates in a variety of habitats. Moths disperse well but are short-lived. Possibly because of the resulting evolutionary and ecological constraints, they lay their eggs in masses, which produce very aggregated larval populations. In its life cycle, the spruce budworm winters as a minute second-instar larva (L2 in Figure 8.13), becoming active in early May, and going through four feeding instars before pupation in late June. Mating, moth dispersal, and oviposition typically take place during late July. The first-instar larvae hatch in mid-August and immediately crawl to wintering sites on the branches of their host trees.

As climate affects almost every part of the spruce budworm's life cycle, climate must be considered when describing spruce budworm population dynamics. For instance, climate can directly influence spruce budworm survival (Lucuik 1984) and fecundity (Harvey 1983). Climate also has more subtle influences through its effects on processes such as herbivory, predation, and parasitism (Fleming and Candau 1998). The net result of both these direct and indirect influences has been to increase per-capita growth rates in warm, dry years (Wellington 1948).

The first evidence of the potential importance of natural enemies in the spruce budworm outbreak cycle was provided by the detailed life-table studies conducted at the Green River watershed in New Brunswick (Morris 1963). This work was carried out mainly from 1947 to 1959, and remains the best data set of its kind on spruce budworm ever published. Analysis of those data using various approaches (e.g., Morris 1963; Royama 1984) has consistently shown that it is the rate of survival from fourth instar to pupation (see Figure 8.13) that is most closely associated with changes in population trend. The nature of this association is consistent with theories of regulation by natural enemies (e.g., Morris 1963; Royama 1984). A complication, however, in adopting such theories is the fact that the species composition of the complex of natural enemies differs among regions and among phases of the spruce budworm's outbreak cycle (Miller and Renault 1976).

Processes Behind the Spruce Budworm Outbreak Cycle

Socioeconomic concerns have motivated the development of models for forecasting the ecological responses of the forest ecosystems subject to spruce budworm to alternative options in management policy. Such models are key requirements for adaptive resource management, and the current ones essentially encapsulate one of four general theories (Table 8.1) about the fundamental processes behind the spruce budworm's outbreak cycle (Fleming and Shoemaker 1992).

Blais's (1983, 1985) "tree mortality" theory is the simplest of the four. Blais argues that extensive areas of mature stands of balsam fir are a prerequisite for outbreak development. He concludes that outbreaks collapse as a result of stand destruction after five years or more of severe defoliation, and that the next outbreak cannot begin until the regenerating stand has matured to the point where it is again vulnerable to attack.

Natural enemies play a role in the other three theories. In his "fast-enemies" hypothesis, Royama (1984) reasons that populations of certain natural enemies increase (or decrease) quickly in response to corresponding changes in spruce budworm populations, their food source. Because of the resulting changes in the rate of mortality in spruce budworm, he argues that rising spruce budworm populations are sometimes driven back to low densities before destroying the stand. The collapse of the spruce budworm populations deprives the natural enemies of their principal food, so that natural enemy populations are expected to decline soon afterwards. According to this theory, therefore, oscillations in the size of the populations of natural enemies are thought to lag slightly behind those of the spruce budworm, as in a typical predator-prey relationship (e.g., Krebs 1972).

Both the "moth-invasion" theory (Stedinger 1984) and the "slow-enemies" theory (Holling 1992) assume that the key natural enemies cannot increase in numbers quickly enough to prevent outbreaking populations of spruce budworm from destroying the forest. These theories also assume that it is forest destruction that causes outbreak collapse. Factors other than budworm availability, such as the accessibility of alternative hosts, prey, and overwintering habitat; or territoriality; or relatively low potential rates of population increase, are thought to prevent populations of natural enemies, such as predatory

Table 8.1

Features of four theories of the basic processes underlying the outbreak cycle of spruce budworm.

	Prerequisites for Local Outbreak		
Theory	*Development*	*Collapse*	*Role of Natural Enemies*
Tree Mortality (Blais 1983)	Mature forest	Forest destruction	Unspecified
Fast Enemies (Royama 1984)	Low natural enemy populations	High natural enemy populations	Respond numerically to N and drive the outbreak cycle
Moth Invasion (Stedinger 1984)	Mature forest + moth invasion	Forest destruction	Background mortality, saturated at high N
Slow Enemies (e.g., Holling 1992)	Mature forest (a) by itself (b) + moth invasion (c) + "good" climate	Forest destruction	Backgound mortality, saturated at high N

"N" represents spruce budworm population density.

birds, from responding numerically in any major way to increases in budworm densities. Hence, the budworm has little impact on the natural enemies in these theories, and the impact of these "slow" or "background" natural enemies is presumed to occur between outbreaks, when spruce budworm populations are small. Under such conditions, the mortality inflicted by the natural enemies is assumed to be enough to keep budworm populations suppressed, even after the forest has first matured enough to support another outbreak. In the "slow-enemies" theory, outbreaks eventually become inevitable as the forest matures further.

The "moth-invasion" theory (Stedinger 1984) suggests that at a local level, "slow" natural enemies extinguish low-density budworm populations, and that outbreaks occur only when enough moths invade from other stands to "swamp" the effects of the local natural enemies. In this scenario, local spruce budworm populations then continue to increase until outbreak levels are reached, even without more moth invasions. This theoretical spruce budworm forest system thus relies on the presence of outbreaking stands to provide the immigrants to infect other mature stands, where local, low-density populations of the spruce budworm are being slowly driven to extinction.

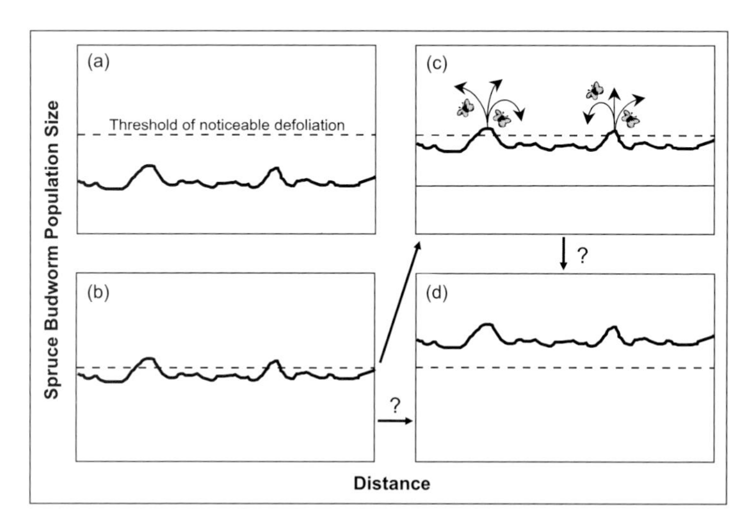

Figure 8.14 Schematic contrasting two opposing theories of spruce budworm outbreak development: (a) population sizes vary over distances of several hundred kilometres, but everywhere are too low to cause defoliation heavy enough for detection by aerial survey; (b) population increase throughout the area produces pockets of detectable damage. Do moths dispersing from these pockets, or epicentres (c), lead to an outbreak (d), or is (d) the result of continued population increase from (b)? (Sanders 1983)

Figure 8.14 uses a spatial perspective to distinguish the "moth-invasion" or "epicentre" theory, which requires panel (c) for outbreak development, from the "fast-enemies" theory, for which panel (c) does not apply. The "moth-invasion" theory (e.g., Stedinger 1984) suggests that outbreaks begin as epicentres or localized patches of high densities of spruce budworm. Moths dispersing from these epicentres are thought to spread the outbreak into neighbouring forests. On the other hand, proponents of the "fast-enemies" theory (e.g., Royama 1984) have maintained that apparent epicentres are merely the first places where budworm population increase is detected in regions throughout which budworm populations are rising simultaneously. The tendency for outbreak episodes to begin in the eastern part of the province and, contrary to the prevailing winds, finish in the west (see Figure 8.8) suggests that the "moth-invasion" theory is not relevant at large scales.

The "slow-enemies" theory (e.g., Holling 1992) is similar in some respects to the "moth-invasion" theory, except for the assumption that, after outbreak collapse, small local populations of the spruce budworm are driven down, not to extinction, but to a low, "equilibrium" density. From this low, "equilibrium" density, the theoretical budworm population cannot reach outbreak levels again until the forest has matured. After that point, local populations of spruce budworm can outbreak by saturating the natural enemies (either through moth invasion, or as a consequence of increased survival brought about by favourable weather or a maturing forest). Thus the "slow-enemies" theory differs from the "moth-invasion" theory, in that the former does not require moth invasion to initiate local outbreaks.

Questions have been raised about each of these theories. While the nature of these questions is important, it is often rather technical and is beyond the scope of this chapter. More detail can be found in Fleming and Shoemaker (1992) and Regniere and Lysyk (1995).

SUMMARY

Forest insect and disease disturbances are a key cause of forest change in Ontario, and they have important effects on biodiversity, forest succession, and ecosystem sustainability. All tree species are vulnerable to attack; annual losses in Ontario's forests have been estimated at 31 million m^3. Although many insect and disease species cause damage and localized mortality, and will damage large areas of forest under unusual circumstances, only four routinely affect Ontario's forest at landscape scales. Three are indigenous insects: the spruce budworm on spruce and balsam fir, the jack pine budworm on pine species, and the forest tent caterpillar on poplar and other hardwood species. The fourth, the gypsy moth, was first discovered in Ontario in 1969, near Kingston. This introduced insect has been associated with large-scale defoliation in southern Ontario's hardwood forests since 1981. The disturbances caused by these four insects last longer than, and typically cover a much greater area than, those caused by wildfire. Diseases are less spectacular in their impacts, although some exotics, such as white pine blister rust and Dutch elm disease, have severely restricted the distribution of their host species.

These four insects differ greatly in their host ranges, ecology, population dynamics, and impact on Ontario's forest. Episodes of spruce budworm outbreaks have recurred at 20- to 70-year intervals and resulted in heavy defoliation for 10 to 15 years and whole-tree mortality over large areas (8,319,210 ha in 1996). Four theories about the population ecology of this insect and the degrees to which resource depletion, dispersal, climate, and natural enemies influence the outbreak cycle, have been discussed.

Outbreaks of jack pine budworm occur at about 10-year intervals, typically last two to four years, and cover only a small part of the host range. In contrast to the closely related spruce budworm, this insect rarely causes much mortality over extensive areas, although pockets of mortality are common in defoliated stands. Severe defoliation is most likely in semi-mature to mature stands, and losses to this insect are primarily in the form of reduced tree growth and mortality of intermediate and suppressed trees.

Of Ontario's hardwood-feeding insects, the forest tent caterpillar affects the largest area (19 million ha in 1992), but usually causes little tree mortality. Its populations oscillate at a regular, approximately 10-year, frequency, and the duration of outbreak episodes is generally two to four years. The collapse of populations of forest tent caterpillar is usually associated with mortality caused by virus epizootics and parasites.

Gypsy moth populations are also largely controlled by virus, but have also been more recently controlled by a fungal pathogen. Gypsy moth defoliation patterns in Ontario have been smaller in their spatial extent than those of the other three major insects, even though the gypsy moth has a wide range of host trees. Oak is the primary host and must be common in a forest stand before severe damage occurs.

An important feature of disturbances is that they interact with one another. Dutch elm disease and beech bark disease are both spread from tree to tree by insects. Defoliation by spruce budworm and jack pine budworm increases the probability of subsequent fire. Drought provides opportunities for bark beetles, and spruce and jack pine budworm, and increases the likelihood of fire. Blowdown can accelerate fuel build-up for fires and provide habitat for bark beetles. In the presence of such interactions, it is doubtful that any disturbance regime can be understood completely in isolation from the others.

An analytical approach is needed to develop both a thorough understanding of the processes driving disturbance regimes and a predictive capacity to support management decision-making. Survey data on spruce budworm defoliation were analyzed as an illustrative example, and basic relationships underlying the landscape-level dynamics and interactions with fire were characterized for this insect. The more intensive, but smaller-scale, research needed to develop detailed life-tables was also discussed. Because the relationships between processes operating at large and small scales is often uncertain, working directly at the landscape level, if possible, offers a promising approach for developing the ability to forecast the dynamics of disturbance regimes.

ACKNOWLEDGEMENTS

We thank Tim Burns and Terry Dumond of the Canadian Forest Service for their help in producing the figures. A Collaborative Research Agreement (LA#105817) with the Ontario Ministry of Natural Resources provided J-N Candau with financial support.

REFERENCES

Aber, J.D. and J.M. Melillo. 1991. Terrestrial Ecosystems. Orlando, Florida: Saunders College Publishing, Holt Rinehart and Winston. 429 p.

Allen, D.C., L. Dorais, and E.G. Kettela. 1984. Survey and detection. In: D.M. Schmitt, D.G. Grimble, and J.L. Searcy (Technical Coordinators). Managing the Spruce Budworm in Eastern North America. Agriculture Handbook No. 620. Washington, D.C.: USDA Forest Service. 21-36.

Anagnostakis, S.L. 1987. Chestnut blight: the classic problem of an introduced pathogen. Mycologia 79: 23-37.

Attiwill, P.M. 1994. The disturbance of forest ecosystems: the ecological basis for conservative management. Forest Ecology and Management 63: 247-300.

Bergeron, Y. and P. Dansereau. 1993. Predicting the composition of Canadian southern boreal forest in different fire cycles. Journal of Vegetation Science 3: 827-832.

Bergeron, Y., H. Morin, A. Leduc, and C. Joyal. 1995. Balsam fir mortality following the last spruce budworm outbreak in northwestern Quebec. Canadian Journal of Forest Research 25: 1375-1384.

Bisessar, S., D.L. McLaughlin, and S.N. Linzon. 1985. The first occurrence of the beech scale insect on American beech trees in Ontario. Journal of Arboriculture 11: 13-14.

Blais, J.R. 1983. Trends in the frequency, extent and severity of spruce budworm outbreaks in eastern Canada. Canadian Journal of Forest Research 13: 539-547.

Blais, J.R. 1985. The ecology of the eastern spruce budworm: a review and discussion. In: C.J. Sanders, R.W. Stark, E.J. Mullins, and J. Murphy (editors). Recent Advances in Spruce Budworm Research: Proceedings of a Symposium, September 16-20, 1984, Bangor, Maine. Ottawa, Ontario: Canadian Forest Service. 49-59.

Candau, J-N., R.A. Fleming, and A.A. Hopkin. 1998. Spatio-temporal patterns of large-scale defoliation caused by the spruce budworm in Ontario since 1941. Canadian Journal of Forest Research 28: 1-9.

Castello, J.D., D.J. Leopold, and P.J. Smallidge. 1995. Pathogens, patterns, and processes in forest ecosystems. Bioscience 45: 16-24.

Cerezke, H.F. 1986. Impact studies of the jack pine budworm, *Choristoneura pinus pinus*, in Nesbit Provincial Forest, Saskatchewan. In: Jack Pine Budworm Information Exchange, January 14-15, 1986, Winnipeg, Manitoba. Winnipeg, Manitoba: Department of Natural Resources, Forest Protection Branch. 25-38.

Davis, C.N., D.T. Myren, and E.J. Czerwinski. 1992. First report of butternut canker in Ontario. Plant Disease 76: 972.

Dorais, L. and E.G. Kettela 1982. A Review of Entomological Survey and Assessment Techniques Used in Regional Spruce Budworm, *Choristoneura fumiferana* (Clem.), Surveys and in the Assessment of Operational Spray Programs. Report of the Committee for Standardization of Survey and Assessment Techniques, Eastern Spruce Budworm Council. Quebec, Quebec: Quebec Department of Energy and Resources. 43 p.

Fitzgerald T.D. 1995. The Tent Caterpillars. Cornell Series in Arthropod Biology. Ithaca, New York: Cornell University Press. 303 p.

Flannigan, M.D., T.J. Lynham, and P.C. Ward. 1989. An extensive blowdown occurrence in northwestern Ontario. In: D.C. MacIver, H. Auld and R. Whitewood. Proceedings of the 10th Conference on Fire and Meteorology, April 17-21, 1989, Ottawa, Ontario. Chalk River, Ontario: Petawawa National Forestry Institute, Forestry Canada. 65-71.

Fleming, R. A. and J.-N. Candau. 1998. Influences of climatic change on some ecological processes of an insect outbreak system in Canada's boreal forests and the implications for biodiversity. Environmental Monitoring and Assessment 49: 235-249.

Fleming, R.A. and C.A.S. Shoemaker. 1992. Evaluating models for spruce budworm - forest management: comparing output with regional field data. Ecological Applications 2: 460-477.

Fogal, W.H. and F. Slansky, Jr. 1985. Contribution of feeding by European pine sawfly larvae to litter production and element flux in Scots pine plantations. Canadian Journal of Forest Research 15: 484-487.

Fowells, H.A. 1965. Silvics of Forest Trees of the United States. Agriculture Handbook No. 271. Washington, DC: USDA Forest Service. 762 p.

Furyaev, V.V., R.W. Wein, and D.A. MacLean. 1983. Fire influences in *Abies*-dominated forests. In: R.W. Wein and D.A. MacLean (editors). The Role of Fire in Northern Circumpolar Ecosystems. New York: John Wiley. 221-234.

Gross, H.L. 1985. White pine blister rust: A discussion of the disease and hazard zones for Ontario. Proceedings of the Entomological Society of Ontario 116, Supplement: 73-79.

Gross, H.L. 1991. Dieback and growth loss of sugar maple associated with defoliation by the forest tent caterpillar. Forestry Chronicle 67: 33-42.

Gross, H.L. 1992. Impact analysis for a jack pine budworm infestation in Ontario. Canadian Journal of Forest Research 22: 818-831.

Gross, H.L., D.B. Roden, J.J. Churcher, G.M. Howse, and D. Gertridge. 1992. Pest Caused Depletions to the Forest Resource of Ontario, 1982-1987. Joint Report No. 17. Sault Ste. Marie, Ontario: Forestry Canada. 23 p.

Haack, R.A. and J.W. Byler. 1993. Insects and pathogens. Regulators of forest ecosystems. Journal of Forestry 91(9): 32-37.

Hall, J.P. and B.H. Moody. 1994. Forest Depletions Caused by Insects and Diseases in Canada, 1982-1987. Information Report ST-X-8. Ottawa, Ontario: Canadian Forest Service. 14 p.

Harvey, G.T. 1983. Environmental and genetic effects on mean egg weight in spruce budworm (Lepidoptera: Tortricidae). Canadian Entomologist 115: 1109-1117.

Hearnden, K.W., S.V. Millson, and W.C. Wilson. 1992. A Report on the Status of Forest Regeneration. Toronto, Ontario: Ontario Ministry of Natural Resources, Independent Forest Audit Committee. 116 p.

Holling, C.S. 1992. The role of forest insects in structuring the boreal landscape. In: A Systems Analysis of the Global Boreal Forest. H.H. Shugart, R. Leemans, and G.B. Bonan (editors). Cambridge, UK: Cambridge University Press. 170-191.

Hopkin, A.A. and G.M. Howse. 1995. Impact of the jack pine budworm in Ontario: a review. In: W.J.A. Volney, V.G. Nealis, G.M. Howse, A.R. Westwood, D.R. McCullough, and B.L. Laishley (editors). Jack Pine Budworm Biology and Management. Proceedings, Jack Pine Budworm Symposium, January 24-26, 1995, Winnipeg, Manitoba. Information Report NOR-X-342. Edmonton, Alberta: Canadian Forest Service. 111-119.

Houston, D.R. 1994. Major new tree disease epidemics: beech bark disease. Annual Review of Phytopathology 32: 75-87.

Howse, G.M. 1981. Losses from and control of spruce budworm and other insects in the boreal mixedwood forest. In: R.D. Whitney and K.M. McClain (co-chairs). COJFRC Symposium Proceedings O-P-9. Boreal Mixedwood Symposium, September 16-18, 1980, 1980, Thunder Bay, Ontario. Sault Ste. Marie, Ontario: Canadian Forest Service. 239-251.

Howse, G.M. 1983. Pests of larch: biology, damage and control. In: C.M. Graham, H.L. Farintosh, and B.J. Graham (editors). Larch Symposium, Potential for the Future, Ontario Ministry of Natural Resources, University of Toronto. November 9, 1982. Toronto, Ontario: Ontario Ministry of Natural Resources, Toronto, Ontario. 35-45.

Howse, G.M. 1986. Jack pine budworm in Ontario. In: Jack Pine Budworm Information Exchange, January 14-15, 1986, Winnipeg, Manitoba. Winnipeg, Manitoba: Manitoba Department of Natural Resources, Forest Protection Section. 47-50.

Howse, G.M. 1995. Forest insect pests in the Ontario Region. In: J.A. Armstrong and W.G.H. Ives (editors). Forest Insect Pests in Canada. Ottawa, Ontario: Canadian Forest Service. 41-57.

Howse, G.M. 1997. Forest Health Conditions in Ontario. Online Bulletin (www.glfc.forestry.ca). Sault Ste. Marie, Ontario: Canadian Forest Service.

Howse, G.M. and M.J. Applejohn. 1993. Forest Insect and Disease Conditions in Ontario, Summer 1993. Survey Bulletin. Sault Ste. Marie, Ontario: Forestry Canada. 14 p.

Howse, G.M. and J.H. Meating. 1995. Jack pine budworm situation in Ontario 1981-1994. In: W.J.A. Volney, V.G. Nealis, G.M. Howse, A.R. Westwood, D.R. McCullough, and B.L. Laishley (editors). Jack Pine Budworm Biology and Management. Proceedings, Jack Pine Budworm Symposium, January 24-26, 1995, Winnipeg, Manitoba. Information Report NOR-X-342. Edmonton, Alberta: Canadian Forest Service. 31-34.

Howse, G.M., J.H. Meating, and J.J. Churcher. 1995. Insect control in Ontario, 1974-1987. In: J.A. Armstrong and W.G.H. Ives (editors). Forest Insect Pests in Canada. Ottawa, Ontario: Canadian Forest Service. 679-700.

Ives, W.G.H and H.R. Wong. 1988. Tree and shrub insects of the prairie provinces. Information Report NOR-X-292. Edmonton, Alberta: Canadian Forest Service. 327 p.

Jardon, Y., L. Filion, and C. Cloutier. 1994. Tree ring evidence for endemicity of the larch sawfly in North America. Canadian Journal of Forest Research 24: 742-747.

Johnston, M. 1996. The role of disturbance in boreal mixedwood forests of Ontario. In: C.R. Smith and G.W. Crook (compilers). Advancing Boreal Mixedwood Management in Ontario. Proceedings, Workshop, October 17-19, 1995, Sault Ste. Marie, Ontario. Sault Ste. Marie, Ontario: Canadian Forest Service. 33-40.

Kimmins, J.P. 1972. Relative contributions of leaching, litter-fall, and defoliation by *Neodiprion sertifer* (Hymenoptera) to the removal of cesium-134 from red pine. *Oikos* 23: 226-232.

Krebs, C.J. 1972. Ecology: The Experimental Analysis of Distribution and Abundance. New York: Harper and Row. 694 p.

Lechowicz, M.J. and L. Jobin. 1983. Estimating the susceptibility of tree species to attack by the gypsy moth, *Lymantria dispar*. Ecological Entomology 8: 171-183.

Lucuik, G.S. 1984. Effect of climatic factors on post-diapause emergence and survival of spruce budworm (*Choristoneura fumiferana*) larvae (Lepidoptera: Tortricidae). Canadian Entomologist 116: 1077-1084.

Lysyk, T.J. 1989. Stochastic model of eastern spruce budworm (Lepidoptera: Tortricidae) phenology on white spruce and balsam fir. Journal of Economic Entomology 82: 1161-1168.

MacLean, D.A. 1985. Effects of spruce budworm outbreaks on forest growth and yield. In: C.J. Sanders, R.W. Stark, E.J. Mullins, and J. Murphy. (editors). Recent Advances in Spruce Budworm Research. Ottawa, Ontario: Canadian Forest Service. 148-174.

Magasi, L.P. 1995. Forest insect pests in the Maritimes Region. In: J.A. Armstrong and W.G.H. Ives (editors). Forest Insect Pests in Canada. Ottawa, Ontario: Canadian Forest Service. 13-25.

Mattson, W.J. and R.A. Haack. 1987. The role of drought in outbreaks of plant-eating insects. Bioscience 37: 110-118.

McCullough, D.B., R.A. Werner, and D. Neumann. 1998. Fire and insects in northern and boreal forest ecosystems of North America. Annual Review of Entomology. 43: 107-127.

McKenney, D.W., B.G. Mackey, J.E. McKee, V. Nealis, and A.A. Hopkin. 1995. Towards Environmental Stratifications for Optimizing Forest Plot Locations. Sault Ste. Marie, Ontario: Canadian Forest Service. NODA Note No. 15. 8 p.

Miller, C.A. and T.R. Renault. 1976. Incidence of parasitoids attacking endemic spruce budworm (Lepidoptera: Tortricidae) populations in New Brunswick. Canadian Entomologist 108: 1045-1052.

Moody, B.H. and P.A. Amirault. 1992. Impacts of Major Pests on Forest Growth and Yield in the Prairie Provinces and the Northwest Territories: A Literature Review. Information Report NOR-X-324. Edmonton, Alberta: Forestry Canada. 35 p.

Morris, R.F. (editor). 1963. The dynamics of epidemic spruce budworm populations. Memoirs of the Entomological Society of Canada 31: 1-332.

Myers, J.H. 1993. Population outbreaks in forest lepidoptera. American Scientist 81: 240-251.

Nealis, V.G. 1995. Population biology of the jack pine budworm. In: W.J.A Volney, V.G. Nealis, G.M. Howse, A.R. Westwood, D.R. McCullough, and B L. Laishley (editors). Jack Pine Budworm Biology and Management. Information Report NOR-X-342. Edmonton, Alberta: Canadian Forest Service. 55-72.

Nealis, V.G. and S. Erb. 1993. A Sourcebook for Management of the Gypsy Moth. Joint Report, Forestry Canada-Ontario Ministry of Natural Resources. Sault Ste. Marie, Ontario: Forestry Canada, Ontario Region. 48 p.

Nealis, V. and Lomic, P.V. 1994. Host-plant influence on the population ecology of the jack pine budworm, *Choristoneura pinus* (Lepidoptera Tortricidae) Ecological Entomology 19: 367-373.

Nealis, V.G., P.M. Roden, and D.A. Ortiz. 1999. Natural mortality of the gypsy moth along a gradient of infestation. Canadian Entomologist 131. In press.

Ostry, M.E., M.E. Mielke, and D.D. Skilling. 1994. Butternut - Strategies for Managing a Threatened Tree. General Technical Report NC-165. St. Paul, Minnesota: USDA Forest Service. 8 p.

Reardon, R. and A. Hajek. 1993. Entomophaga maimaiga in North America: A Review. Forest Health and Protection Report NA-TP-15. Morgantown, West Virginia: USDA Forest Service. 22 p.

Riemenschneider, D.E. 1985. Water Stress Promotes Early Flowering in Jack Pine. Note NC-331. St. Paul, Minnesota: USDA Forest Service. 3 p.

Regniere, J. and T.J. Lysyk. 1995. Forest insect pests in the Ontario Region. In: J.A. Armstrong and W.G.H. Ives (editors). Forest Insect Pests in Canada. Ottawa, Ontario: Canadian Forest Service. 95-105.

Roland, J. 1993. Large scale forest fragmentation increases the duration of tent caterpillar outbreak. *Oecologia* 93: 25-30.

Roland, J. and P.D. Taylor. 1997. Insect parasitoid species respond to forest structure at different spatial scales. Nature 386: 710-713.

Rose, A.H. and O.H. Linquist. 1982. Insects of Eastern Hardwood Trees. Technical Report No. 29. Ottawa, Ontario: Canadian Forest Service. 304 p.

Rowe, J.S. 1972. Forest Regions of Canada. Publication No. 1300. Ottawa, Ontario: Canadian Forest Service. 172 p.

Royama, T. 1984. Population dynamics of the spruce budworm, *Choristoneura fumiferana*. Ecological Monographs 54: 429-462.

Sajan, R.J. and H. Brodersen. 1992. Results of Forest Insect and Disease Surveys in the Northwestern Region of Ontario 1991. Miscellaneous Report No. 116. Sault Ste. Marie, Ontario: Forestry Canada. 32 p.

Sanders, C.J. 1980. A Summary of Current Techniques Used for Sampling Spruce Budworm Populations and Estimating Defoliation in Eastern Canada. Information Report O-X-306. Sault Ste. Marie, Ontario: Canadian Forest Service. 33 p.

Sanders, C.J. 1983. Research on the spruce budworm - understanding population dynamics. In: The Spruce Budworm Problem in Ontario - Real or Imaginary? Sault Ste. Marie, Ontario: Canadian Forest Service. 53-56.

Sanders, C.J., D.R. Wallace, and G.S. Lucuik. 1978. Flight activity of female eastern spruce budworm (Lepidoptera: Tortricidae) at constant temperatures in the laboratory. Canadian Entomologist 107: 1289-1299.

Schowalter, T.D., W.W. Hargrove, and D.A. Crossley, Jr. 1986. Herbivory in forested ecosystems. Annual Review of Entomology 31: 177-196.

Sippell, W.L. 1962. Outbreaks of the forest tent caterpillar, *Malacosoma disstria* Hbn., a periodic defoliator of broad-leaved trees in Ontario. Canadian Entomology 94: 408-416.

Stedinger J.R. 1984. A spruce budworm-forest model and its implications for suppression programs. Forest Science 30: 597-615.

Sterner, T.E. and A.G. Davidson. 1981. Forest Insect and Disease Conditions in Canada 1980. Ottawa, Ontario: Canadian Forest Service. 43 p.

Stocks, B.J. 1975. The 1974 Wildfire Situation in Northwestern Ontario. Information Report O-X-232. Sault Ste. Marie, Ontario: Canadian Forest Service. 27 p.

Stocks, B.J. 1987. Fire potential in the spruce budworm-damaged forests of Ontario. Forestry Chronicle 63: 8-14.

Thompson, I.D. 2000. Forest vegetation of Ontario: factors influencing landscape change. In: A.H. Perera, D.L. Euler, and I.D. Thompson (editors). Ecology of a Managed Terrestrial Landscape: Patterns and Processes of Forest Landscapes in Ontario. Vancouver, British Columbia: University of British Columbia Press. 30-53.

Van Wagner, C.E. 1978. Age class distribution and the fire cycle. Canadian Journal of Forest Research 8: 220-227.

Volney, W.J.A. 1988. Analysis of historic jack pine budworm outbreaks in the Prairie provinces of Canada. Canadian Journal of Forest Research 18: 1152-1158.

Volney, W.J.A. and H.F. Cerezke. 1992. The phenology of white spruce and the spruce budworm in northern Alberta. Canadian Journal of Forest Research 22: 198-205.

Volney, W.J.A. and D.G. McCullough. 1994. Jack pine budworm population behavior in northwestern Wisconsin. Canadian Journal of Forest Research 24: 502-510.

Wallner, W.E. 1987. Factors affecting insect population dynamics: differences between outbreak and non-outbreak species. Annual Review of Entomology 32: 317-340.

Wein, R.W. 1990. The importance of wildfire to climate change - hypotheses for the taiga. In: J.G. Goldammer and M.L. Jenkins (editors). Fire in Ecosystem Dynamics. The Hague, Netherlands: Academic Press. 185-190.

Wellington, W.G. 1948. The light reactions of the spruce budworm, *Choristoneura fumiferana* Clemens (Lepidoptera: Tortricidae). Canadian Entomology 80: 56-82.

9 Forest Ecosystem Productivity in Ontario

LAWRENCE E. BAND*

INTRODUCTION

Forest ecosystem productivity is a central and basic ecosystem variable. It is directly linked with forest growth and renewal, the forest trophic chain, and the atmospheric carbon budget. Productivity is indirectly linked with the formation of the landscape mosaic, and with the hydrologic processes of evapotranspiration, runoff, and sedimentation. Forest ecosystem productivity is defined in a number of different ways, depending on whether it is used to describe production levels of merchantable timber, the net change in standing biomass, or the net carbon exchange between the full surface ecosystem and the atmosphere. The various processes controlling both the growth of forests and carbon exchange are controlled, in turn, by a range of climate, soil, drainage, canopy, and disturbance conditions, the latter including fire and herbivory.

This chapter will first briefly discuss various definitions of forest ecosystem productivity and the processes contributing to productivity. A fuller treatment of these processes, with reference to analysis and prediction over extensive regions, can be found in the recent texts of Waring and Running (1998) and Landsberg and Gower (1997). The focus of this chapter is the dominant characteristics and processes of northern forests, with emphasis on the boreal and subboreal biomes that comprise most of Ontario. As used here, forest productivity is defined in terms of net biomass accumulation or net carbon exchange, rather than production of timber. Methods of estimating productivity based on plot-based and stand-based structural measurements and/or gas exchange at local levels are described. Methods that have been developed over the last decade for estimating productivity over large spatial extents using remote sensing and GIS technology are discussed. Finally, the limited information available on large-scale trends of ecosystem productivity across Ontario is presented, together with the major ecological limitations, including climate, soils, and terrain, that define its patterns.

DEFINITIONS AND PROCESSES CONTRIBUTING TO ECOSYSTEM PRODUCTIVITY

Forest net primary production (NPP) is defined as

$$NPP = GPP - R_m - R_s \quad [9.1]$$

where GPP is gross photosynthetic production, and R_m and R_s are the autotrophic respiration losses of carbon associated, respectively, with the physiological processes of maintenance and synthesis of living tissue (Waring and Running 1998). NPP is considered a source to the surface ecosystem when GPP is greater than R_a (the combined autotrophic respiration). GPP is primarily related to the leaf-area index (LAI),or the surface area available for carbon exchange (although certain species also have photosynthetic bark). R_a is primarily dependent on the live, respiring biomass and the allocation of carbon (C) to structural components.

As the canopy ages, foliar biomass accumulates quickly and then asymptotes toward a limiting value, so that the LAI peaks relatively early in a stand's development. However, respiring stemwood takes much

* Department of Geography, University of North Carolina, CB# 3220 Chapel Hill, North Carolina, USA 27599

longer to accumulate as it is allocated carbon only after the demands of root and leaf are met. As a result, stand GPP can greatly exceed R_a early in stand development, while the two quantities will eventually converge over time, such that NPP tends toward zero and the stand reaches equilibrium. Further reductions in GPP occur as stands tend toward old growth stages and the conducting tissue becomes less efficient (Yoder et al. 1994). A portion of NPP is shed as detritus, through foliage, branch, and root turnover, and contributes to the carbon pool of the forest floor. Finally, the reduction in NPP with canopy age leads to a reduced ability to withstand and repair damage from a variety of stresses, so that mortality and a reduction in standing biomass become more likely.

NPP contributes to the production of merchantable timber volume by providing the net photosynthate that can be allocated to root, leaf, or stem biomass. The processes by which this allocation is accomplished and the conditions that determine the percentage of biomass contributing to stem growth are not fully understood, although a number of hypotheses and models have been proposed. Often it is the above-ground NPP (ANPP) that is measured and reported, because of the difficulty of measuring the stock, production, and turnover of below-ground biomass. Translating stem biomass increment into volume of growth, furthermore, depends on the density of the woody biomass, which varies substantially among species. Losses from herbivory, wind, and ice damage, and other types of disturbance, decrease net biomass accumulation. Wildfire plays a major role in determining landscape patterns and regional forest productivity in the boreal and sub-boreal forest biomes. For all the foregoing reasons, small-scale geographic trends in NPP do not necessarily correspond to trends in timber volume production, although at large scales they are similar.

Finally, net ecosystem productivity (NEP) is the carbon residual of the overstory and understory NPP and the microbial, heterotrophic respiration (R_h). NEP is what is measured by sampling the net flux of carbon dioxide, methane, and other carbon-based gases toward and away from the surface, or the net accumulation of carbon *in situ*. Understandably, few full and accurate measurements of NEP have been made in forest ecosystems. Measurements made from towers of net carbon dioxide exchange resulting from the processes of assimilation, respiration, and decomposition have only rarely been complemented by measurements of changing carbon stores in standing biomass and soils. In terms of the productivity of standing volume, NPP and ANPP are probably more useful concepts.

The distribution of NPP and ANPP in Ontario's terrestrial ecosystems follows an expected north-south trend in temperature and, to a lesser extent, an east-west trend in precipitation. Hills (1961) described a hierarchy of controls on the distribution of forest communities and on forest productivity across the province. At large scales, the primary controls are the long-term average and range of regional climate variables. Embedded in these large-scale trends are a set of nested variations in community structure and productivity that respond to differences in mesoscale climate (lake effects, for example); to geologic controls on topography, soils, and drainage; to dominant lifeform; to age-class distribution; and to disturbance. At the microscale level (that is, at the scale of individual landforms), differences in exposure, drainage, and disturbance provide a finer level of heterogeneity. Each of these scales of variation interact, so that a hierarchical approach is required to explain and understand the spatial and temporal dynamics of ecosystem productivity.

The Hills site classification system (Hills 1959; Burger 1993) was meant to stratify the province into successively finer and more uniform regions, in terms of both community composition and productivity. When that classification system was designed, no methods of estimating large-scale ecosystem productivity were available, so the system was designed to provide an ecoregionalization method based on the dominant controls of productivity. For that reason, site boundaries often correspond to the positions of mean annual isotherms, isohyets, or major geologic and topographic boundaries. The system has been used for planning purposes by the Ontario Ministry of Natural Resources for decades. Other systems of ecoregionalization based on nested systems of climate, geology, terrain, and soils which are used in Ontario include that of Wickware and Rubec (1989) which made extensive use of remote sensing imagery to outline ecoregional boundaries. Band et al. (1999) developed a hierarchical ecoregionalization by first estimating province-wide productivity using simulation based on remote sensing data, and then spatially partitioning a set of abiotic controls of the estimated forest productivity. This system of ecoregionalization is discussed further below.

MEASUREMENT AND ESTIMATION OF FOREST ECOSYSTEM PRODUCTIVITY

Two basic approaches are taken to obtaining estimates of nested patterns of productivity. The "bottom-up" approach is based on either empirical measurements from plots and stands or on site-specific simulation. The second is a "top-down" approach, based on simulation models using larger-scale remote sensing data and GIS analyses. As it is based on specific inventory and mensuration information, on the direct measurement of gas exchange, or on detailed canopy-level process models, the bottom-up method has the potential to provide more accurate estimates of productivity at local levels; however, as this approach is very labour intensive, it is usually data-limited.

The top-down methods that have recently been developed use information collected by earth-resource or meteorological satellites to estimate canopy leaf area and/or radiation absorption. The estimates, in turn, are used to estimate photosynthetic activity. Top-down approaches are difficult to validate, as they generally provide information on productivity at a coarse spatial resolution and over much larger units than can feasibly be measured with plot-scale methods. The distinction between the bottom-up and top-down approaches is actually gradational: plot-based measurement and methods based on large-scale remote sensing represent the opposite ends of a gradient. The most promising approach for resource agencies is to combine techniques drawn from different portions of this gradient, taking advantage of the respective strengths of each.

Bottom-up methods, based on mensuration, measure the net accumulation of stand volume or biomass at intervals of at least one year, but typically five years, or over the full period since stand establishment. The information gained from standard forest inventory records or from plot data is generally limited to merchantable timber, and contains only measurements taken for stems exceeding a given diameter at breast height (DBH). While these sources are appropriate for estimating timber reserves and harvest yield potential, they may provide a biased estimate of ecosystem productivity. Determining full ecosystem NPP requires estimates of understory productivity, as well as some assessment of the contribution of below-ground dynamics. The latter is particularly difficult to measure directly (e.g., Steele et al. 1997). In certain northern environments, the NPP of the understory, including bryophytes, is a significant component of the ecosystem NPP, and may exceed that of the overstory (e.g. Van Cleve et al. 1983).

Stand-Level Measurement of NPP and ANPP

To date, only limited work has been done to synthesize plot and stand measurements into province-wide spatial patterns of forest ecosystem productivity. Despite an active program of plot mensuration carried out by the Province of Ontario, interpolation or aggregation of these data into large regional trends has not been done in a comprehensive manner. One reason is the difficulty of fully determining NPP or ANPP at the plot level. A second is the difficulty of adequately weighting plot information on the basis of relative areal representation, a process which must make use of a stratification system to account for the major variations and gradients controlling productivity. As stated above, these include climate, soils, landforms, disturbance, dominant life-form (or species composition), stand age, and other factors. Detailed determinations of plot or stand growth at the level of NPP or ANPP, which require measurement of both biomass increment and detritus production, are much less available than volume increment measurements. The same is true of NEP, which requires the full net exchange between surface and atmosphere. For the purpose of establishing the limits of the range of detailed productivity estimates, a portion of the available literature on plot productivity measurements is reviewed below. The major controls on and variations in these estimates are discussed. It is beyond the scope of this chapter to provide an exhaustive survey of the productivity measurements reported for all of Ontario, so certain areas are reported for which NPP or ANPP have been determined, or where alternative techniques of measurement (of gas exchange, for example) have been applied.

Few comparative investigations have been conducted that determine productivity using uniform methodologies across different stand types and across regions. Significant differences among investigations in the productivity values reported may be caused by differences in the methods of measurement and estimation used, so comparison among studies must be done with care. In addition, the literature shows that there is as much variation in the different forms of productivity (in either NPP or NEP, for example) within small areas as there is over much larger areas, as a result of variations in canopy and environmental conditions. Gower et al. (1997) developed ANPP estimates for a set of plots within the study areas of the Boreal Ecosystem Atmosphere Study (BOREAS) in northern Manitoba and central Saskatchewan, while Fassnacht and Gower (1997) developed ANPP estimates in northern Wisconsin. The Manitoba and Wisconsin sites are in ecosystems similar to Ontario's, and have the advantage of sharing a uniform methodology. The sites in northern Manitoba and the sites in southern Saskatchewan showed significant differences in ANPP among species. Aspen (*Populus tremuloides*) had the highest values, followed by black spruce (*Picea mariana*) and jack pine (*Pinus banksiana*).

A comparison among the values reported shows that there was more variance between species than between sites in the BOREAS areas. The 1994 values for ANPP and NPP, respectively, in kilograms of carbon per hectare per year ($kg{\cdot}C{\cdot}ha{\cdot}yr^{-1}$), were determined to be roughly 3500 and 3900 for the southern aspen site, 3500 and 3800 for the northern aspen site, 1660 and 2600 for the southern black spruce site, 1360 and 2200 for the northern black spruce site, 1170 and 2100 for the southern jack pine site, and 1220 and 2100 for the northern jack pine site. ANPP for the eight jack pine sites sampled by Fassnacht and Gower (1997) averaged 2050 $kg{\cdot}C{\cdot}ha^{-1}{\cdot}yr^{-1}$. While a general decrease in ANPP was evident from south to north within species, it is clear that local conditions, species physiology, and associated site conditions account for much of the variation. Steele et al. (1997) estimated belowground productivity (BNPP) for the BOREAS stands, which showed that the deciduous stands had a much lower allocation of NPP to root growth, in addition to a higher photosynthetic capacity. These differences are dependent on both genetic factors and on characteristic differences in site micro-environments, in terms of soil temperature, soil frost persistence, and nutrient status.

Morrison (1990, 1991) estimated carbon budgets for stands of old-growth sugar maple (*Acer saccharum*) in the Algoma Highlands of central Ontario. These measurements recorded five-year growth increments and biomass increments approaching 12 $mg{\cdot}ha^{-1}$ of dry biomass. Taking carbon content to be roughly 45 percent of dry biomass and converting to an average annual production gives values much lower than the deciduous (aspen) sites sampled in the BOREAS project (approximately 1200 $kg{\cdot}C{\cdot}ha^{-1}{\cdot}yr^{-1}$), although it should be kept in mind that the Algoma sites were old-growth stands in which ages ranged from 150 to 300 years and included increased effects of mortality. In addition, the difference in method of estimating biomass increments may have increased the difference between the results of the studies. This is an example of some of the difficulties involved in making comparative or interpolative estimates of forest productivity over large areas or different biomes.

In the biome to the southeast of Ontario, ANPP has been determined by mensuration in the Harvard Forest in Massachusetts, in conjunction with tower measurements of net ecosystem exchange (NEE). Aber and Federer (1992) report ANPP values of about 4200 $kg{\cdot}C{\cdot}ha^{-1}{\cdot}yr^{-1}$ (taking C content to be approximately 45 percent biomass), substantially higher than the values reported above, although the stands in the Harvard Forest are substantially younger than the stands at the Algoma site. These values are comparable to the ANPP for young aspen stands (roughly 4000 $kg{\cdot}C{\cdot}ha^{-1}{\cdot}yr^{-1}$), and significantly higher than the ANPP for mature aspen stands (roughly 1000 $kg{\cdot}C{\cdot}ha^{-1}{\cdot}yr^{-1}$) near Chalk River, Ontario, cited by Cannell (1982), and near Dorset, Ontario (roughly 1400 $kg{\cdot}C{\cdot}ha^{-1}{\cdot}yr^{-1}$), also cited by Cannell (1982). These observations illustrate that any comparative study of plot-level productivity over a large area can only be synthesized by sampling similar species of similar ages growing in similar edaphic conditions. Likewise, any attempt to synthesize large-scale patterns from plot data needs to carefully stratify the samples and areally weight the results on the basis of terrain, soils, and canopy attributes.

Measurement of Net Ecosystem Productivity

NEP can be estimated by a combination of comparatively short-term sampling of gas exchange with fast-response sensors, or by longer-term estimates of accumulated stand carbon. Direct measurement of the net ecosystem exchange (NEE) of carbon dioxide has greatly improved over the past decade, although there are still very few studies that provide sufficiently long records to estimate seasonal or annual NEP. The measurement can be performed in small chambers, from towers, or from aircraft. Chamber methods have limited application for estimating NEP, as sufficient sampling of all components of the ecosystem (that is, the overstory, understory, and forest floor) would need to be carried out. Measurement from towers or aircraft employs the methodology of eddy flux correlation, in which the covariances of gas concentration and the vertical velocity time series are used to estimate net gas exchange. Measurement from aircraft has the advantage of providing spatial "snapshots" of net gas exchange (within the time lag of the flight path) over landscape extents, while measurement from towers has the advantage of providing longer time series over smaller areas, on the order of 1 km^2, depending on tower height, wind velocity, and surface roughness. An explanation of the details of tower and aircraft eddy flux methods in northern forest ecosystems is beyond the scope of this chapter, and the reader is referred to papers in the special issue of the Journal of Geophysical Research on the BOREAS project (Sellers et al. 1998).

Continuous measurements of net carbon exchange from towers carried out in the Harvard Forest in Massachusetts and in the BOREAS northern old black spruce site (NOBS) near Thompson, Manitoba, probably bracket Ontario conditions best, at least in terms of climate and major biome, although, once again, the large variations that can occur from stand to stand within the same region should be kept in mind. Goulden et al. (1996) estimated NEP for the Harvard Forest at approximately 210 $g \cdot C \cdot m^{-2} \cdot yr^{-1}$, about half the value of ANPP reported above.

Tower flux estimates of NEP for the BOREAS sites showed consistency with the ANPP estimates: the southern aspen stand appears to be a significant carbon sink, while the conifer sites appear to vary between sink and source from year to year (Sellers et al. 1998). Goulden et al. (1997) estimated NEP for the NOBS site for 1994 and 1995, respectively, at near zero and potentially negative. Time-series sampling of NEE at tower sites can be augmented by longer-term estimates of NEP based on sampling of carbon accumulation, if initial carbon storage values can be estimated from a previous time. Harden et al. (1997) has used this methodology to estimate long-term NEP at BOREAS tower flux sites, by estimating the net carbon accumulation since stand-clearing fires that can be dated. These authors estimated a rate of 100 kg to 300 $kg \cdot C \cdot ha^{-1} \cdot yr^{-1}$ for the NOBS site, as a decadal average, based on estimated carbon accumulation rates in soil and moss following stand-clearing fires.

The very low current rates are reasonable, considering the location of the ecosystem (near the northern limit of the boreal biome) and the age of the stand (approximately 120 years). As expected, NEP values are lower than ANPP, because they account for the effects of both below-ground NPP and heterotrophic respiration. Low NEP values for the conifer stands may be partially explained by stand age, which for the black spruce is well beyond the time of maximum yield for stands growing on similar sites in the Claybelt of Ontario, according to volume curves generated by Whynot and Penner (1990). The questionable representativeness of these sites thus needs to be considered in drawing conclusions about NEP for conifer sites in Ontario or for the boreal forest in general. Finally, although direct estimation of net carbon exchange or net carbon accumulation is the most accurate method of estimating productivity, either as NEP or NPP, these methods are necessarily spatially sparse because of sampling difficulty. In terms of the regional estimation of forest productivity, therefore, these methods are most useful in providing either calibration or validation sites for larger-scale approaches.

Stand-Level to Landscape-Level Models

Empirical Approaches

Operationally, estimates of stand productivity can be predicted by the construction of species-specific and environment-specific growth curves which are calibrated from permanent plot data (typically on volume increment, not NPP). Given current above-ground merchantable volume, these curves are used to predict volume increment, potential yield, and the time to maximum yield. The curves can be stratified for different ranges of a site index, so as to reflect the optimality of the local growth environment. In Ontario, province-wide yield tables linked to site classes were developed by Plonski (1981) for these purposes, but are recognized to be both overly broad and unrepresentative of specific regions. Consequently, effort has been put into creating more regionally specific growth and yield curves which are related to available forest classification systems, such as the Forest Ecosystem Classification (FEC) (e.g., Whynot and Penner 1990). Results have been restricted by the limited usefulness of the FEC as a stratification for productivity. McKenney et al. (1994) have shown, moreover, that the FEC data set may not include adequate representation from key strata in Ontario. This finding illustrates the comment made earlier that the bottom-up approach is data-limited; nevertheless, further efforts to develop volume yield curves specific to different environments and site class conditions may, if coupled with extrapolation techniques based on remote sensing and GIS technologies, produce stand and regional estimates of sufficient quality for the estimation of potential yield.

Long-term regional forest productivity, estimated from knowledge of above-ground biomass or timber volume at maturity, can be used in combination with stand age to compute mean annual increment (MAI) in either biomass or volume. The MAI typically considers only merchantable timber, and does not separate out either the components of ecosystem productivity or potential losses during stand development from such processes as competition, herbivory, and disease (Lowe et al. 1991). MAI may thus be considered as the minimum average net growth rate over the lifespan of the stand, although there may be an advantage to incorporating the potential losses listed above.

Aggregation of stand-based MAI to areally-based MAI requires stratification and area weighting by species or life-form, site quality, and land cover, and applies only to the specific landscape strata sampled; for example, to forest lands either available for or protected from timber harvest. Gray (1995) compiled forest inventory data from provincial sources for all of Canada, and presented estimated volumetric MAI, aggregated to the level of forest sections. The resulting estimated levels of productivity for Ontario are presented in Figure 9.1 for all species, broadleaf and conifer. As expressed by wood-volume MAI, productivity for all species decreases along a northern gradient, with the exception of a belt of lower MAI running through central Ontario along the shore of Lake Superior and through the Muskoka and Algonquin region. It is difficult to ascribe a reason, but this trend may correspond to a level of disturbance that has selectively removed better-growth stands, or a transition from primarily broadleaf dominance to conifer dominance in the mixedwood communities. Shifts in dominant species effectively prevent the comparison of volume MAI and biomass production, as different wood densities would need to be incorporated.

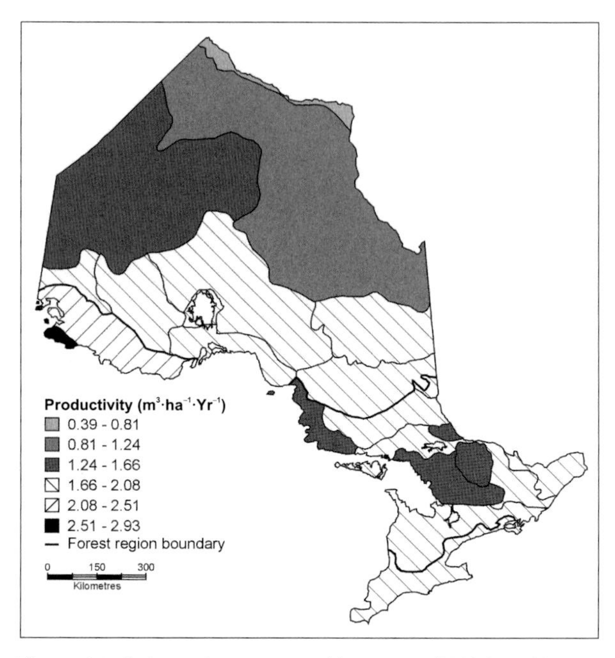

Figure 9.1 Estimated mean annual increment (MAI), in cubic metres per hectare per year, for forest sections in Ontario. (Data from Gray 1995)

Mechanistic Forest Ecosystem Models

Process models of stand productivity, based on the estimation of carbon flux in response to environmental and canopy variables, have the potential to be linked to GIS systems and used to predict landscape-level NPP or NEP. Landsberg and Gower (1997) and Battaglia and Sands (1998) have reviewed a set of physiologically based forest ecosystem models that simulate carbon, water, and nutrient cycles, and have evaluated the suitability and promise of these models for forest management. Detailed description of the models lies beyond the scope of this chapter, and the reader is referred to the citations provided above, or to the original model references. The models reviewed by these studies range from very detailed models requiring extensive parameterization at the plot level (e.g., the Maestro model [Wang and Jarvis 1990]), to models that are purposely simplified to allow their extrapolation and parameterization over larger regions. The latter type includes BIOMASS (McMurtrie et al. 1990), CENTURY (Parton et al. 1988), FOREST-BGC (Running and Coughlan 1988), and PnET (Aber and Federer 1992). The trade-offs of all these models revolve around the detail and rigour of the process representation, weighed against the ability to parameterize and to operate the models in the more data-sparse environments typical of any larger-scale scenario.

The FOREST-BGC model (Running and Coughlan 1988) uses a simplified description of canopy structure and biophysical processes to derive the major carbon fluxes of GPP, R_a and R_s, and so gain estimates of NPP. A later version of this model, named BIOME-BGC, updated a number of physiological processes and added R_h, in order to simulate NEP (Running and Hunt 1993). Simulations of NEP at the BOREAS tower flux sites by Kimball et al. (1997) using BIOME-BGC largely agreed with recorded NEP at annual levels, with some departures at specific times. This model computes a daily carbon and water budget, given general stand and soil conditions and daily meteorological inputs. Band (1994) used another version of this model to estimate NPP and stand growth over different regions of Ontario. Comparison of the model simulations with measured above-ground production and standing biomass in the Turkey Lakes Watershed of the Algoma Highlands (estimated by Morrison [1990, 1991]) and with volume increment growth for stands in Petawawa (estimated by Bonner and Magnussen [1986]) showed reasonable agreement. Simulated above-ground biomass after 150 years and the (last) five-year growth increment for stands in the Turkey Lakes Watershed (TLW) are approximately 350 and 10 mg·ha^{-1}, respectively, while observed values given by Morrison (1990) and Jeffries et al. (1988) for the same variables are 200 mg to 345 mg·ha^{-1} and just over 12 mg·ha^{-1}. The fact that the simulated biomass reported by Band (1994) is at the high end of the range of values reported from the observed stands, but that the last five-year increment is much closer to the observed values, is consistent with the lack of explicit disturbance as a factor in long-term biomass accumulation in the model. Similar comparisons in the Petawawa site by Band (1994) showed reasonable agreement between observed and simulated volume incremental growth for conifer, hardwood, and mixed stands.

At landscape levels, high-resolution satellite data can be used to obtain spatial information on canopy cover, information including life-form, LAI, and potentially standing biomass, with which to parametrize the models described above. Thematic Mapper (TM) data has largely been used for these purposes (e.g., Spanner et al. 1990; Nemani et al. 1993). The TM data can also be combined with digital terrain data and soil information to form a full landscape description. A set of ecosystem models can then be driven with a meteorological time series to obtain direct estimates of canopy absorbed radiation, photosynthesis, respiration, evapotranspiration, and runoff. The ecosystem models have the advantage of providing estimates of net biomass accumulation, runoff production, snowmelt, and soil moisture patterns, which can be directly compared with spatially distributed empirical information. Initial work with the Regional HydroEcological Simulation System (RHESSys) model over Ontario has successfully predicted watershed hydroecological processes, including the distribution of forest growth and hydrology, in a set of experimental forests and watersheds (Band 1993, 1994; Band et al. 1998; Creed et al. 1996; Mackay and Band 1997). An important feature of this approach is a hierarchical representation of the landscape, built up from forest patches derived from remote sensing data classification, embedded within a watershed landform template consisting of nested watersheds, sub-watersheds, hill slopes, and bottomlands. This feature permits a telescoping of model results from large watersheds, to sub-catchments, down to individual patches. Further comparisons of snowmelt depletion curves, daily runoff hydrographs from small watersheds, and spatial distribution of canopy cover over topography, can be found in Band (1994) and Mackay and Band (1997).

Landscape to Regional Scales

In theory, the same approach can be extended to much larger regions, such as a full province, using information from coarse-resolution satellites. The primary data source on surface ecosystem distribution and function at larger scales has been the Normalized Difference Vegetation Index (NDVI) generated from data recorded by the AVHRR satellite sensor. The NDVI is computed as

$$\mathrm{NDVI} = (\mathrm{IR}-\mathrm{R})/(\mathrm{IR}+\mathrm{R}) \quad [9.2]$$

where IR and R are the reflectance values in the near infrared and red bands, respectively. The AVHRR sensor is operated by the National Oceanic and Atmosphere Administration (NOAA), which has archived NDVI since 1981, at best resolutions of 1.1 km. The data set is now being standardized and distributed as part of the Pathfinder project (Agbu and James 1994). AVHRR information can also be used to define canopy conditions with which to operate the range of forest ecosystem models described above.

Running et al. (1994) took this approach to mapping carbon and water exchange over regions of western Montana, using a version of FOREST-BGC at a 1-km resolution.

One limitation to this approach is the lack of available meteorological information for locations away from population centres, particularly in remote areas such as northern Ontario. In addition, there is a general lack of high-quality soils information at landscape levels. Extrapolating these landscape-level models to large areas, such as a whole province, thus requires the use of additional methods to gain reasonable input data. Liu et al (1997) have developed a model modified from FOREST-BGC, referred to as BEPS (Boreal Ecosystem Process Simulator), which includes more detailed canopy radiation terms. BEPS is meant to be driven entirely from coarse-resolution estimates of stand conditions derived from remote sensing data, and from interpolated or satellite-derived daily meteorological surfaces. Liu et al. (1997) have demonstrated the operation of BEPS by simulating carbon and water budgets at a 1-km resolution using information drawn from AVHRR data to characterize the forest canopy across the full province of Quebec. BEPS essentially drives a plot-based model at a resolution of 1 km. Much of the effort required to use this model is thus involved in generating 1-km fields of required meteorological information, including daily time steps of temperature, shortwave radiation, and precipitation. This meteorological information is some of the most problematic, as it is generated from a combination of samples from satellite data taken at one-degree intervals and data interpolated from widely-scattered meteorological stations. The information, therefore, typically has a much lower spatial resolution than the underlying land surface data. Attempts to test model predictions of ANPP with plot-generated values suffer from the mismatch between the effective AVHRR resolution and the typical forest plot size: the forest plot data shows a much greater range than the BEPS predictions. The development of BEPS is, however, a promising approach that should be continued and tested against productivity estimates gained over target areas, perhaps estimates generated from more detailed landscape-level simulations.

Ollinger et al. (1998) used a version of the Pnet II model with a monthly time-step to simulate NPP and ANPP over New England at a 1-km resolution. The use of monthly meteorological information eases some of the concerns over interpolating daily values; furthermore, New England has a much denser meteorological station network than northern Ontario and Quebec. This model showed reasonable agreement with plot-based estimates of stand ANPP, which ranged from approximately 350 g to 750 g $C \cdot m^{-2} \cdot yr^{-1}$ for ANPP, reported by life-form (for both hardwood and conifer forest types), and also with regional differences in runoff production. The range of ANPP reported (750 to 1260 g $C \cdot m^{-2} \cdot yr^{-1}$) is high, compared to the limited estimates of these quantities reported for Ontario and surrounding regions.

Energy Conversion Efficiency Methods

Extending the regional methodology to the full globe increases the difficulty and lack of precision with which meteorological and surface information can be defined. The Pathfinder data sets (Agbu and James 1994) contain AVHRR information at resolutions of 1 km and 8 km for the full globe and can yield land cover and canopy information. While the modelling approach outlined above can be used to simulate global-scale carbon and water budgets at lower resolutions, a number of models have been developed with significantly reduced mechanistic complexity, commensurate with the much sparser information at the global scale. NDVI derived from satellite data has been used to estimate canopy LAI, or the fractional, absorbed, photosynthetically active radiation (FPAR), both of which are key ecosystem variables. FPAR has been shown to scale approximately linearly with the NDVI. Combining estimates of incident, photosynthetically active radiation (PAR) with FPAR produces the absorbed PAR, or APAR. APAR has been shown to correlate linearly with net primary productivity (NPP) (Monteith 1977), as

$$NPP = \varepsilon \cdot APAR \qquad [9.3]$$

where ε is the constant of proportionality, or a radiation conversion efficiency, expressed here as grams of carbon assimilated (both above and below ground), minus the respiration losses per megajoule APAR.

Some confusion exists in the literature regarding proper values of ε, because of the difference between the unit used to measure production (i.e., grams of carbon or grams of dry biomass), and that used to measure absorbed radiation (i.e., total absorbed solar radiation, or total PAR). The values of ε also differ if the constant is used to compute gross or net productivity. Landsberg et al. (1997) have stated that ε should be estimated taking all environmental limitations and physiological feedbacks into consideration, rather than relying on limited experimental methods.

Global estimates of NPP have been produced by several groups through the use of satellite data (e.g., Prince 1991; Potter et al. 1993). These are typically limited to best resolutions of 0.5° by 0.5° cells. Intercomparison of a set of these methods showed some differences, but mainly a remarkable amount of uniformity, in estimates of NPP within the Ontario region. Average values ranged from 500 g or 600 g $C{\cdot}m^{-2}{\cdot}yr^{-1}$ in southern Ontario, to less than 200 g $C{\cdot}m^{-2}{\cdot}yr^{-1}$ in the north (Moore et al. 1995). It is not surprising that these models yield broadly similar NPP values, as they either use the same NDVI data sets, and often the same meteorological time series, or are based on similar energy conversion efficiency approaches. Some of the models, but not all, appear to reflect the band of decreased productivity extending along the shore of Lake Superior and through the Algonquin region. It is difficult, moreover, to compare the model estimates of annual biomass increment for one year with average net volume increments. Using approximate wood-density conversion factors, the MAI estimates appear to be on the order of 50 percent lower than the satellite-derived estimates, but this difference is consistent with the difference in what is measured in each case.

Remote Sensing Estimation of NPP Over Ontario

Top-down methods have been developed in Ontario to produce estimates of ecosystem productivity at a landscape level (that is, at a level at which comparison, calibration, and validation of model output can be carried out using empirical plot or stand data), and to extend these methods to the full province (Band et al. 1998). An important attribute of the present approach has been the development of a full hierarchy of resolutions, from individual forest patches which can be compared to field data, up through resolutions measured in kilometres. The RHESSys model has been used at local scales to calibrate lower-resolution, simpler models that emphasized the energy conversion efficiency approach, through the use of AVHRR Pathfinder data. Given the 8-km resolution of the Pathfinder NDVI data, a method must be used to estimate ε at landscape levels, levels commensurate with the sampling resolution of the Pathfinder data set, and subject to the caveat of Landsberg et al. (1997) noted above.

One important observation made from this work is that no homogeneous landscapes exist within an 8-km cell; therefore, landscape-level estimates of productivity area are required, rather than estimates for individual species or life-forms. RHESSys was used to simulate, among other processes, the absorption of PAR and the net assimilation of carbon over a set of representative landscapes. Averaging these processes over areas large enough to be discriminated by AVHRR data provides landscape average ε values at scales over an ecosystem gradient in Ontario that are also commensurate with the satellite resolution incorporating the combination of major ecosystem limitations on productivity.

Sites across Ontario used as part of this gradient are the Petawawa National Forestry Institute, a site in the Temagami District of Central Ontario, the Turkey Lakes Experimental Watershed in the Algoma Highlands, and the Rinker Lake study site near Lake Nipigon (Figure 9.2). While additional sites are being added to this transect to provide a more representative sample of Ontario ecosystems, the transect provides a reasonable (albeit simple) gradient across the Great Lakes and Laurentian regions into the southern boreal biome of the province. The landscape-level simulations described above produce landscape-averaged estimates of the canopy APAR, as well as the net canopy productivity. These estimates provide a basis for the estimation of average landscape ε for each of the sites along the Ontario transect.

Figure 9.3 shows the variation in the radiation conversion efficiency, ε, over the 1987 growing season for three sites along the transect. The curves in this figure represent areally averaged values for the full areas of each study site. There is substantial variation in ε within each site at certain times of the growing season because of different local edaphic and physiologic conditions at the stand level, but these local effects cannot be resolved at Pathfinder resolution or for the full provincial area. They are, however, very important in forming the areal average ε values, as the controlling factors can be very nonlinear (Band

1993). These local variations are also important in terms of the landscape-level patterns of productivity that can be resolved at the eco-element level. A rough progression of phenologies can be seen, based on climatic and biome conditions. Colder, more boreal sites show a delayed increase in ε in the spring and an earlier reduction in the late summer and early autumn. This phenomenon reflects colder temperatures and also, to some extent, a greater proportion of conifer life-forms: conifers cannot make use of the absorbed PAR while temperatures are too low, while the broadleaf life-forms have lower absorbed PAR before leafout. The higher values in the summer for the southern boreal site may reflect a more favourable water balance, particularly a greater abundance of surface water. This apparent increase in peak ε with a decrease in mean annual temperature has also been noted by Landsberg et al. (1997).

Figure 9.2 East-west transect of four sites across Ontario used to operate the RHESSys model and calibrate AVHRR-based productivity estimates.

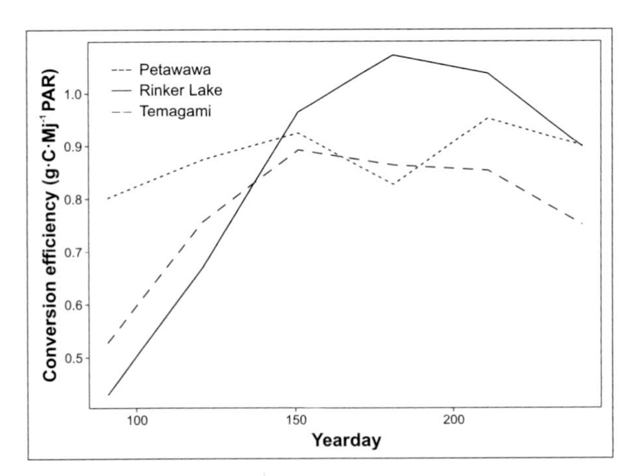

Figure 9.3 Monthly composited trajectories of the energy conversion (grams of carbon per megajoule of photosynthetically active radiation, PAR) for three of the four transect sites identified in Figure 9.2.

Band et al. (1999) have shown that a simple linear regression explains roughly 50 percent of the variation in monthly ε over a set of landscapes, when mean monthly temperature, precipitation, and an annual degree-day summation are used as independent variables. The first two variables address some of the limitations on NPP from temperature and water availability, while the degree-day variable acts as a biome indicator and accounts for the trend in ε with mean annual temperature. We have used this simple explanation of ε (that is, ε generated from monthly aggregated values simulated over representative landscapes) in combination with long-term average climate variables, to produce estimates of long-term patterns of the efficiency factor at monthly levels. Extending the number of sites across a greater range of conditions, and adding supplementary variables may improve the statistical explanation of ε. This possibility is under investigation.

Calibration of Productivity Using Coarse-Resolution Satellite Data

Surfaces of ε were produced for monthly time periods through 1987 using the regression with monthly temperature, precipitation, and cumulative degree days. Estimation of APAR was done with methods similar to those used by the global-scale NPP models described above. The advantage to this approach is that the space-time averaging of meteorological conditions at a one-month period is more appropriate than the daily data used by Liu et al. (1997); nevertheless, the effects of the daily distribution of precipitation and temperature on productivity are retained through the calibration of ε. Figure 9.4 shows the monthly composited NDVI for the months of May through September of 1987. Figure 9.5 shows the NPP for the

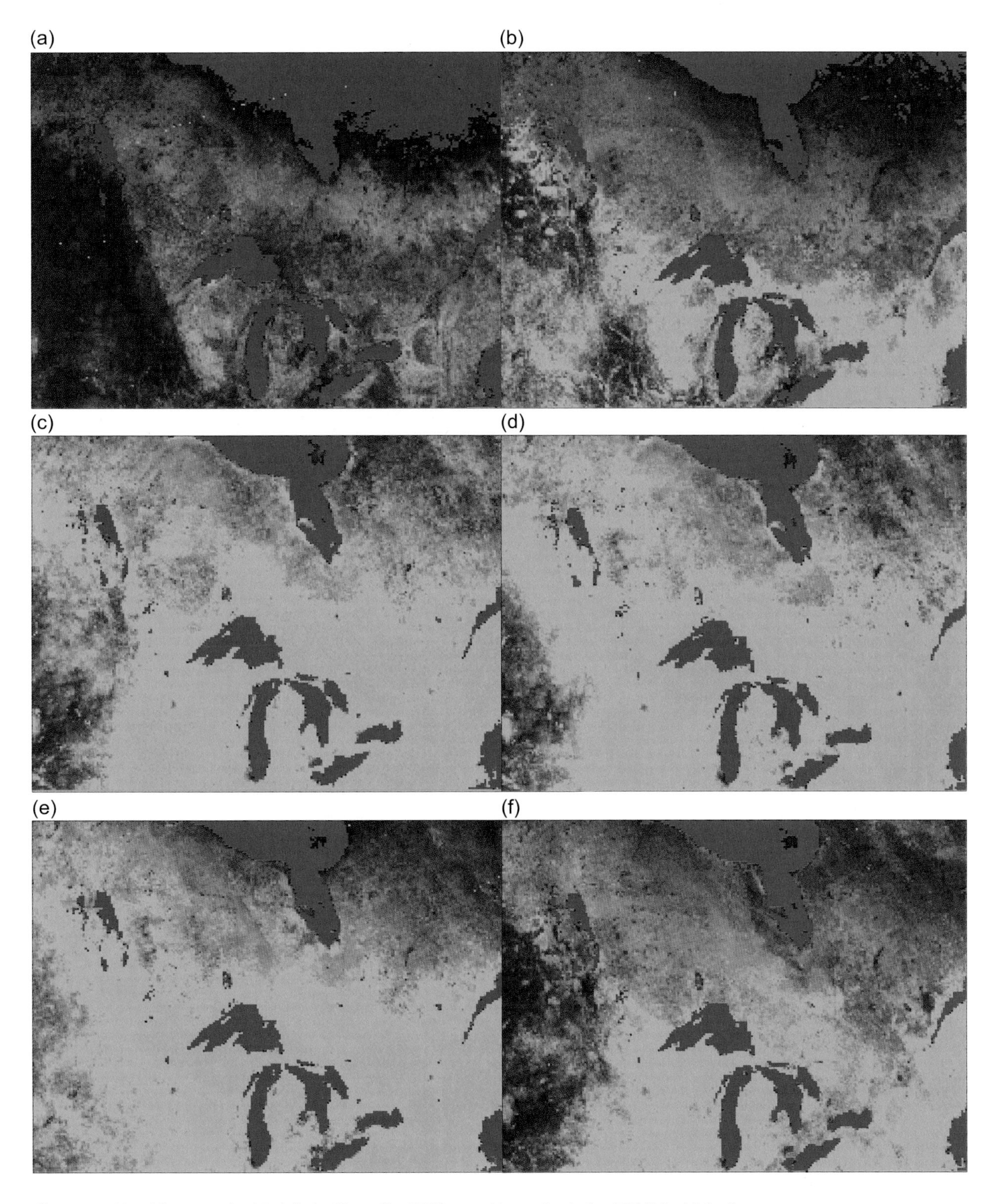

Figure 9.4 Monthly composited Pathfinder Normalized Difference Vegetation Index (NDVI) for (a) April 1987; (b) May 1987; (c) June 1987; (d) July 1987; (e) August 1987; and (f) September 1987.

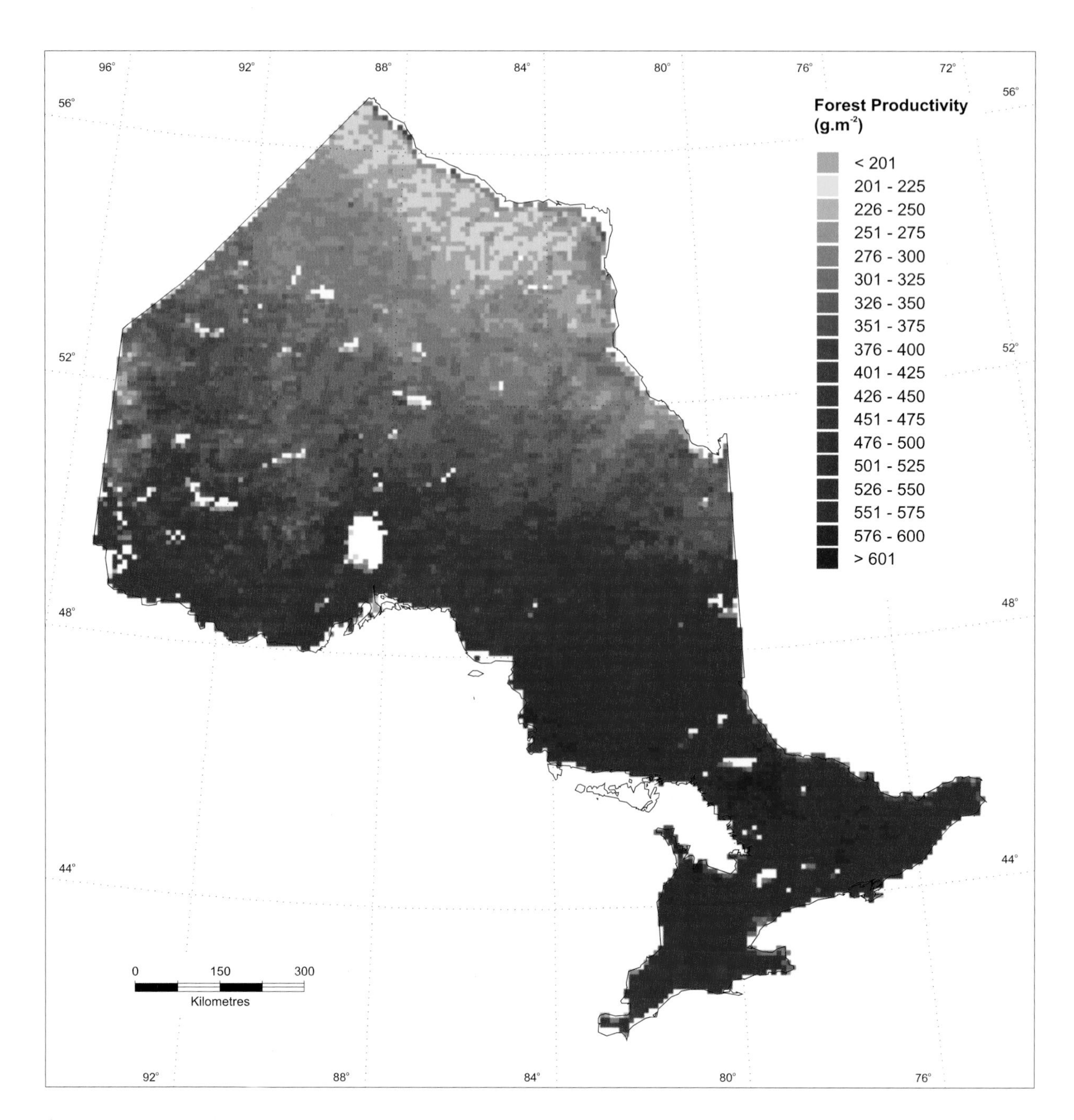

Figure 9.5 Integrated forest productivity (g·m^{-2}) for the 1987 active growing season in Ontario.

growing months of 1987, following equation (9.3). While forest ecosystem NEP are significantly affected by ecosystem respiration losses from soil organic decomposition through the winter months, forest canopy respiration losses are small in winter; therefore, the net carbon exchange for the active growing season is a reasonable index of NPP. The distribution of forest canopy productivity and other forms of biome productivity across the province ranged from near-zero (but with very little area below 50 g to 100 $g{\cdot}m^{-2}$) to approximately 600 $g{\cdot}m^{-2}$ for 1987 (when sampled in 8-km by 8-km blocks). Large water bodies were masked out, so as not to be included in the computations. Comparison of the results with other estimates from the set of global-scale productivity models (Potter et al. 1993; Moore et al. 1995; Hunt et al. 1996) shows consistency in the large-scale patterns of NPP, although small-scale differences in pattern emerge, as expected, from the differences in resolution and approach.

As expected, there is a general increase in productivity from north to south, following temperature, except in the heavily disturbed urban and agricultural areas of southern Ontario. Peak productivity is restricted to areas which have not been as significantly disturbed lying in central and southern Ontario near, and downwind of, the Great Lakes. The limits of disturbance, and of higher productivity, appear to be controlled by geology and topography, as the heavily disturbed urban and agricultural area of southern Ontario extends just to the border of the Canadian Shield. High-productivity areas extend around Georgian Bay, through Algonquin Park and surrounding areas, and into eastern Ontario. They are also found along the north shore of Lake Huron, along the eastern shore of Lake Superior (in the Algoma Highlands), and in more isolated areas of the Temagami district.

DISCUSSION

This chapter summarizes methods used to generate estimates of forest ecosystem productivity and includes a discussion of the major differences in techniques. Plot, stand, and tower-based measurements can provide accurate information at the level of the measurement area, but extrapolation to landscape or regional extents is restricted by the limited representativeness of each site and the unavoidably small number of sites that can be measured or monitored. An important point to be made when extrapolating data generated from plot and stand mensuration to landscapes and regions is the need to incorporate productivity estimates for all types of land cover other than forest.

Process-based ecosystem simulation can be used to simulate landscape-level productivity patterns by linking the model to a GIS database representing the properties of a set of stands or grid cells. The results of such methods can be checked by comparison to available plot and stand-level measurements, but are also ultimately restricted by the limited availability of data on terrain, canopy, soils, and weather needed to drive the models. The latter two variables, soils and weather, are typically the most limiting when data from areas are extended to regions. While the landscape-level simulation methods outlined here, as part of the RHESSys package, can be applied to larger grid cells for direct modelling of the full province, previous work has shown that process to be a potential cause of significant bias in both hydrologic and ecosystem variables (Band 1993; White and Running 1994). In addition, the variation in daily fields of precipitation is very poorly known through areas such as northern Ontario, where meteorological stations are sparse. It is possible that the development of methods to determine spatially distributed meteorological variables, as described by Liu et al. (1997) and Thornton et al. (1997), can provide techniques to improve this situation. Questions are still unanswered, however, regarding the effects of representing large, heterogeneous landscapes by average or modal conditions, instead of accounting for landscape-level heterogeneity, as was achieved by the work with the RHESSys model reported in this chapter.

Top-down estimates of forest ecosystem productivity at the level of the province can be made by combining landscape-level simulations of forest water and carbon budgets; these are used to calibrate the observations of ecosystem activity derived from satellite data, observations which are becoming more readily available. While the productivity patterns for only the first available year of Pathfinder data (1987) are reported here, further analysis is underway to investigate variability among years, in order to see whether there are any significant shifts in the spatial patterns of productivity, either from year to year, or over the full range of the Pathfinder data from 1981 to 1994.

An advantage of the landscape-level methods applied in Ontario is the direct connection through the hierarchy of estimation methods down to scales at which observed ecosystem variables can be used for diagnosis and validation of the larger-scale estimates. This connection across scales has previously not been possible using estimation methods that begin with grids more appropriate to global-scale models.

Extending that type of analysis to include the temporal variations evidenced by inter-annual fluctuations in dominant meteorological patterns requires the use of both multi-year, satellite-derived canopy conditions, which are now available through the Pathfinder program, and also annual surfaces of the key meteorological variables. The current satellite-derived information on land surface conditions will be augmented by the launch of the Moderate Resolution Imaging Spectrometer (MODIS) scheduled for the summer of 1999. This system will have significantly better spatial and spectral resolution, and will provide a richer set of forest ecosystem variables (Running et al. 1994).

This chapter has discussed both the difficulty of extending plot-based measurements of productivity up through regional extents, and the difficulty of validating the large-scale, satellite-based methods. A successful program of monitoring and analyzing the exchange of carbon between forest ecosystem and atmosphere, through estimates of potential timber productivity, may be best approached with a combination of methods drawn from different portions of the bottom-up-to-top-down spectrum. It must be noted, however, that as many jurisdictions shift their monitoring and assessment methods towards the combination of remote sensing, GIS, and modelling approaches, they are reducing their field programs. This reduction may result in another imbalance; that is, a lack of adequate field data to validate and calibrate the estimates derived from large-scale models and remote sensing data.

REFERENCES

Aber, J.D. and C.A. Federer. 1992. A generalized, lumped-parameter model of photosynthesis, evapotranspiration and net primary production in temperate and boreal forest ecosystem. *Oecologia* 92: 463-474.

Agbu, P.A. and M.E. James. 1994. The NOAA/NASA Pathfinder AVHRR Land Data Set User's Manual. Greenbelt, Maryland: NASA, Goddard Distributed Active Archive Center.

Band, L.E. 1993. A Pilot Landscape Ecological Model for Forests in Central Ontario. Forest Fragmentation and Biodiversity Project Report No. 7. Sault Ste. Marie, Ontario: Ontario Forest Research Institute, Ontario Ministry of Natural Resources, 40 p.

Band, L.E. 1994. Development of a Landscape Ecological Model for Management on Ontario Forests: Phase 2 - Extension over an East/West Gradient over the Province. Forest Fragmentation and Biodiversity Project Report No. 17. Sault Ste. Marie, Ontario: Ontario Forest Research Institute, Ontario Ministry of Natural Resources, 41 p.

Band, L.E., F. Csillag, A.H. Perera, and J.A. Baker. 1999. Deriving an Eco-regional Framework for Ontario through Large-Scale Estimates of Net Primary Productivity. Forest Research Report No. 149. Sault Ste. Marie, Ontario: Ontario Forest Research Institute, Ontario Ministry of Natural Resources. In press.

Battaglia, M. and P.J. Sands. 1998. Process-based forest productivity models and their application in forest management. Forest Ecology and Management 102: 13-32.

Bonner, G.M. and S. Magnussen. 1986. Inventory and Growth Predictions of the Petawawa Forest. Information Report PI-X-66. Chalk River, Ontario: Petawawa National Forest Institute, Canadian Forest Service.

Burger, D. 1993. Revised Site Regions of Ontario: Concepts, Methodology and Utility. Forest Research Report No. 129. Sault Ste. Marie, Ontario: Ontario Forest Research Institute, Ontario Ministry of Natural Resources, 21 p.

Cannell, M.G.R. 1982. World Forest Biomass and Primary Production Data. New York, New York: Academic Press, 391 p.

Cramer, W., D.W. Kicklighter, A. Bondeau, B. Moore, G. Churkina, A. Ruimy. 1997. Comparing global models of terrestrial net primary productivity (NPP): Overview and key results. Summary Report No. 30. Potsdam, Germany: Potsdam Institute fur Klimatologie.

Creed, I.F., L.E. Band, N.W. Foster, I.K. Morrison, J.A. Nicolson, R.S. Semkin, and D.S. Jeffries. 1996. Regulation of nitrate-N release from temperate forests: a test of the N flushing hypothesis. Water Resources Research 32: 3337-3354.

Fassnacht, K.S. and S.T. Gower. 1997. Interrelationships among the edaphic and stand characteristics, leaf area index and aboveground net primary production of upland forest ecosystems in north central Wisconsin. Canadian Journal of Forest Research 27: 1058-1067.

Goulden, M.L., B.C. Daube, S. Fan, D.J. Sutton, A Bazzaz, J.W. Munge, and S.C. Wofsy. 1997. Physiological responses of a black spruce forest to weather. Journal of Geophysical Research 102(24): 28,987-28,996.

Goulden, M.L., J.W. Munger, S. Fan, B.C. Daube, and S.C. Wofsyl. 1996. CO_2 exchange by a deciduous forest: response to interannual climate variability. Science 271: 1576-1578.

Gower, S.T., J.G. Vogel, J.M Norman, C.J. Kucharik, S.J. Steele, and T.K. Stow. 1997. Carbon distribution and aboveground net primary production in aspen, jack pine and black spruce stands in Saskatchewan and Manitoba, Canada. Journal of Geophysical Research 102(D24): 29,029-29,043.

Gray, S.L. 1995. A Descriptive Forest Inventory of Canada's Forest Regions. Information Report PI-X-122. Chalk River, Ontario: Petawawa National Forestry Institute, Canadian Forest Service. 192 p.

Harden, J.W., K.P. O'Neill, S.E. Trumbore, H. Veldhuis, and B.J. Stocks. 1997. Moss and soil contributions to the annual net carbon flux of a maturing boreal forest. Journal of Geophysical Research 102(24): 28,805-28,816.

Hills, G. A. 1959. A Ready Reference to the Description of the Land of Ontario and its Productivity. Preliminary Report. Toronto, Ontario: Ontario Department of Lands and Forests. 142 p.

Hills, G.A. 1961. The Ecological Basis for Land-Use Planning. Research Report No. 46. Maple, Ontario: Ontario Department of Lands and Forests. 204 p.

Hunt, E.R. Jr., S.C. Piper, R. Nemani, C.D. Keeling, R.D. Otto, and S.W. Running. 1996. Global net carbon exchange and intra-annual atmospheric CO_2 concentrations predicted by an ecosystem process model and three-dimensional atmospheric transport model. Global Biogeochemical Cycles 10: 431-456.

Jeffries, D.S., J.R.M. Kelso, and I.K. Morrison. 1988. Physical, chemical and biological characteristics of the Turkey Lakes Watershed, Central Ontario, Canada. Canadian Journal of Fisheries and Aquatic Sciences 45 (Supplement 1): 3-13.

Kimball, J.S., P.E. Thornton, M.A. White, and S.W. Running. 1997. Simulating forest productivity and surface-atmosphere carbon exchange in the BOREAS study region. Tree Physiology 17: 589-599.

Landsberg, J.J., S.D. Prince, P.G. Jarvis, R.E. McMurtrie, R. Luxmore, and B.E. Medlyn. 1997. Energy conversion and use in forests: the analysis of forest production in terms of utilisation efficiency. In: H.L. Gholz, K. Nakane and J. Shimoda (editors). The Use of Remote Sensing in the Modeling of Forest Productivity at Scales from the Stand to the Globe. New York, New York: Kluwer Academic Press.

Landsberg, J.J. and S.T. Gower. 1997. Applications of Physiological Ecology to Forest Management. San Diego, California: Academic Press. 354 p.

Liu, J., J.M. Chen, J. Cihlar, and W.M. Park. 1997. A process-based Boreal ecosystem productivity simulator using remote sensing inputs. Remote Sensing and Environment 62: 158-175.

Lowe, J.J., K. Power and S.L. Gray. 1991. Canada's Forest Inventory: 1991. Information Report PI-X-115. Chalk River, Ontario: Petawawa National Forestry Institute, Canadian Forest Service.

Mackay, D.S. and L.E. Band. 1997. Ecosystem processes at the watershed scale: dynamic coupling of distributed hydrology and canopy growth. Hydrological Processes 11(9): 1197-2019.

McKenney, D.W., B.G. Mackey, J.E. McKee, and B.L. Zavitz. 1994. A Representativeness Assessment of Forest Ecosystem and Growth and Yield Plots in Ontario. Northern Ontario Development Agreement Draft Report. Sault Ste. Marie, Ontario: Canadian Forest Service. 29 p. + appendix.

McMurtrie, R.E., D.A. Rook, and F.M. Kelliher. 1990. Modelling the yield of *Pinus radiata* on a site limited by water and nutrition. Forest Ecology and Management 30: 381-413.

Monteith, J.L. 1977. Climate and the efficiency of crop production in Britain. Philosophical Transactions of the Royal Society of London, Series B 281: 277-294.

Morrison, I.K. 1990. Organic matter and mineral distribution in an old-growth *Acer saccharum* forest near the northern limit of its range. Canadian Journal Forest Research 20: 1332-1342.

Morrison, I.K. 1991. Addition of organic matter and elements to the forest floor of an old-growth *Acer saccharum* forest in the annual litter fall. Canadian Journal of Forest Research 21: 443-452.

Nemani, R., S.W. Running, and L.E. Band. 1993. Regional hydro-ecological simulation system: an illustration of the integration of ecosystem models in a GIS. In: M.F. Goodchild, B.O. Parks, and L.T. Steyaert (editors). Environmental Modeling with GIS. New York, New York: Oxford University Press. 297-304.

Ollinger, S.V., J.D. Aber, and C.A. Federer. 1998. Estimating regional forest productivity and water yield using an ecosystem model linked to a GIS. Landscape Ecology 13: 323-334.

Parton, W. J., J.W.B. Steward, and C.V. Cole. 1987. Dynamics of C, N, P and S in grassland soils: a model. Biogeochemistry 5: 109-131.

Plonski, W.L. 1981. Normal Yield Tables (Metric) for Major Forest Species of Ontario. Toronto, Ontario: Ontario Ministry of Natural Resources. 40 p.

Potter, C.S. J.T. Randerson, C.B. Field, P.A. Matson, and P.M. Vitousek. 1993. Terrestrial ecosystem production: a process model based on global satellite and surface data. Global Biogeochemical Cycles 7: 811-841.

Prince, S.D. 1991. A model of regional primary production for use with coarse-resolution satellite data. International Journal of Remote Sensing 12: 1133-1421.

Running, S.W. and J.C. Coughlan. 1988. A general model of forest ecosystem process for regional applications: 1. Hydrological balance, canopy gas exchange and primary production. Ecological Modelling 42: 125-154.

Running, S.W. and R. Hunt, Jr. 1993. Generalization of a forest ecosystem process model for other biomes, BIOME-BGC, and an application for global scale modules. In J.R. Ehleringer and C. Field (editors). Scaling Processes between Leaf and Landscape Levels. San Diego, California: Academic Press. 141-148.

Running, S.W, C.O. Justice, V. Salomonson, D. Hall, J. Barker, Y.J. Kaufmann, A.H. Strahler, A.R. Huete, J.-P. Muller, V. Vanderbilt, Z.M. Wan, P. Teillet, and D. Carneggie. 1994. Terrestrial remote sensing science and algorithms planned for EOS/MODIS. International Journal of Remote Sensing 15: 3587-3620.

Sellers, P.J., F.G. Hall, R.D. Kelly, A. Black, D. Baldocci, J. Berry, M. Ryan, K.J. Ranson, P.M. Crill, D.P. Lettenmaier, H. Margolis, J. Cihlar, J. Newcomer, D. Fitzjarrald, P.G. Jarvis, S.T. Gower, D. Halliwell, D. Williams, B. Goodison, D.E. Wickland, and F.E. Guertin. 1998. BOREAS in 1997: experiment overview, scientific results, and future directions. Journal of Geophysical Research 102(24): 28,731-28,771.

Spanner, M.A., L.L. Pierce, D.L. Peterson, and S.W. Running. 1990. Remote sensing of temperate coniferous forest leaf area index: the influence of canopy closure, understorey vegetation and background reflectance. International Journal of Remote Sensing 11: 95-111.

Steele, S.J., S.T. Gower, J. Vogel, and J.M. Norman. 1997. Root mass, net primary production and turnover in aspen, jack pine and black spruce forests in Saskatchewan and Manitoba, Canada. Tree Physiology 17: 577-588.

Thornton, P.E., S.W. Running, and M.A. White. 1997. Generating surfaces of daily meteorological variables over large regions of complex terrain. Journal of Hydrology 190: 214-251.

Van Cleve, K., C.T. Dyrness, L.A. Viereck, J. Fox, F.S. Chapin III, and W.C. Oechel. 1983. Taiga ecosystem in interior Alaska. BioScience 33: 39-44.

Wang, Y.P. and P.G. Jarvis. 1990. Description and validation of an array model-MAESTRO. Agriculture and Forest Meteorology 51: 257-280.

Waring, R.H. and S.W. Running. 1998. Forest Ecosystems: Analysis at Multiple Scales. New York, New York: Academic Press, 370 p.

White, J.D. and S.W. Running. 1994. Testing scale dependent assumptions in regional ecosystem simulations. Journal of Vegetation Science 5: 687-702.

Whynot, T.W. and M. Penner. 1990. Growth and Yield of Black Spruce Ecosystems in the Ontario Clay Belt: Implications for Forest Management. Information Report PI-X-99. Chalk River, Ontario: Petawawa National Forestry Institute, Canadian Forest Service, 23 p.

Wickware, G. and C.D.A. Rubec. 1989. Ecoregions of Ontario. Ecological Land Classification Series No. 26. Ottawa, Ontario: Environment Canada, Sustainable Development Branch. 37 p.

Yoder, B., M.G. Ryan, R.H. Waring, A.W. Shoetle, and M.R. Kaufman. 1994. Evidence of reduced photosynthetic rates in old trees. Forest Science 40: 513-527.

10 Vegetation Responses to the Managed Forest Landscape of Central and Northern Ontario

TERENCE J. CARLETON*

INTRODUCTION

Worldwide anthropogenic deforestation is causing such a rapid disappearance of forest cover that much international attention is now focused on the remaining extensive forests of the world (Noble and Dirzo 1997). Most of these forests are located in Brazil, Russia, and Canada (Bryant et al. 1997). The less extreme of human activities influence the structure and composition of forests in ways that are not always apparent at first sight. Such activities as the forest management operations of harvesting, thinning, planting, and applying herbicides may have a direct impact. This group includes such subsistence-related activities as gathering firewood, harvesting forest plants for food, collecting ornamental plants, pasturing animals, and obtaining pannage for pigs (Watt 1923; Peterken and Tubbs 1965). Indirect influences on the forest from abiotic sources include insecticide application, atmospheric pollution, and changes to the inherent disturbance regime. Indirect influences from biotic sources include changes to the normal food chain structure, such as the removal of top predators, which allow forest herbivores to flourish, and the unwitting introduction of exotic forest pests and diseases, without the restraint of their subsequent food chain. With the possible exception of a few of the most inhospitable locations on earth, human-related activity has in some way affected the structure and composition of all forests.

As part of one of the remaining transcontinental, continuous frontier forests of the world, the forests of Ontario, Canada, straddle a number of latitudinal bioclimatic zones (Thompson 2000, this volume). The two main management practices that have been applied to the forests of Ontario are fire suppression, which changes the prevailing disturbance regime, and logging. In northern Ontario, these practices were initiated largely in concert, following several extensive, catastrophic fires during the early decades of the 20th century. Over the past two decades, however, logging has rapidly replaced forest fire as the main agent of instantaneous forest biomass loss from any one site in northern Ontario (Hearnden et al. 1992). Fire suppression and logging were less synchronized further south, in the Great Lakes Forest region. Here, loggers moved in to harvest white pine (*Pinus strobus*) and red pine (*Pinus resinosa*) selectively in the 18th and 19th centuries, when neither the logistic support nor the technology existed for aerial fire suppression. To assess the influence of both direct and indirect management practices on Ontario forests requires comparing the past with the present, but the inconsistency in forest management criteria and measures over time makes this difficult. Today our focus is on ecosystem properties and measures, but in the past forest resources were the primary object of scrutiny.

FIRE SUPPRESSION

Implications for Age Structure and Forest Succession

A traditional view of major ecosystem disturbance from

* Faculty of Forestry and Department of Botany, University of Toronto, 33 Willcocks Street, Toronto, Ontario M5S 3B3

fire, massive windthrow, flooding, or logging is that the ecosystem is pushed back to an early successional state and that the system will recover to progress through the same phases of growth and compositional change as had occurred in the previous cycle (Clements 1916). This holistic and highly deterministic view of ecosystem development implies that ecological communities are closely linked to physical environmental conditions and that autogenic succession follows a single, clearly defined pathway. It also implies that the qualitative nature of the disturbance is largely irrelevant; that, for example, the consequence for forest ecosystems of massive windthrow or logging is largely the same as the consequence of wildfire.

It follows, in this view, that the suppression of disturbance will lead to a longer successional sequence, possibly culminating in a self-regenerating community or ecosystem type that is in equilibrium with local or regional environmental conditions. In a forest ecosystem, this means that primary forests, initially conforming to an even-aged structure of shade-intolerant, pioneer species, will lose individual trees as they get larger and older (Oliver and Larson 1990). These trees will be replaced by smaller trees of different species having later successional characteristics, including tolerance of shade and changeable growth rate. The transition to an uneven-aged structure proceeds (Oliver 1981).

It is not my purpose here to develop a discussion on the respective theories of forest succession; indeed, it would be inappropriate. However, it should be borne in mind that the Clementsian model, very briefly outlined above, has formed the framework for rangeland and forest management over much of North America for most of the 20th century. Because this chapter is an examination of the consequences of past forest management practices on current forest vegetation, it is possible to infer, with the benefit of scientific hindsight, the strengths and weaknesses of the Clementsian rubric as it applies to forests of the Great Lakes and boreal regions in Ontario.

Age Structure Prior to Fire Suppression

Few studies explicitly address the forest age structure of the Ontario landscape immediately prior to the onset of fire suppression. Two main approaches have been taken. One is the mapping of past fires from historical records (Donnelly and Harrington 1978; Harrington and Donnelly 1978). The other is a demographic approach based on age structure analysis (Suffling et al. 1982). The mapping approach involves consequent assumptions concerning the re-establishment of forest following fire and tends to underestimate the number of fires, particularly in the Great Lakes, where many are small, sub-lethal, surface events. These fires can provide suitable conditions for regeneration (Carleton 1982; Day and Carter 1991b) that, in turn, influences the regional forest age structure. The demographic approach examines contemporary forest age structure in the landscape, extracts model parameters, and projects, using the model, a predicted past or future age structure.

The assumption inherent in this approach is that the model accurately reflects reality. Suffling et al. (1982) verified the consistency of their model, derived from the negative exponential curve of Van Wagner (1978), by projecting current age structure from reasonable, independently derived estimates of past age structure for polygonal components of the landscape, and comparing the results with current age structure in the Ontario Forest Inventory. The comparison

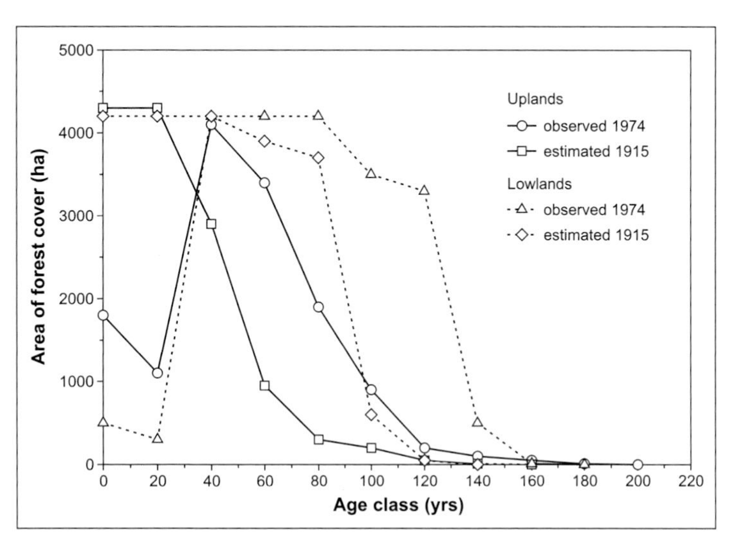

Figure 10.1 Actual contemporary (1974) age structure and estimated age structure of the 1915 forest landscape for a section of northwestern Ontario extending from Pikangikum in the west to Pickle Lake in the east (Suffling et al. 1982).

showed a high degree of consistency between the predicted and observed age structures. In an area of northwestern Ontario extending eastwards from Pikangikum to Pickle Lake, comparing the estimated age structure for 1915 with the projections for 1974 (Figure 10.1) shows that younger age classes represented a much higher proportion of the landscape prior to fire suppression. There is also a lower incidence of extreme old-growth forest in the pre-suppression structure. Consistency in the shape of the modelled age structure distribution between 1915 and 1974 may be a constraint of the model used. It would seem, nevertheless, that before fire suppression the boreal forest complex of northwestern Ontario was approximately thirty years younger than it was during the 1970s.

Contemporary Age Structure

With the age distribution of one pre-fire-suppression forested landscape in mind, it is instructive to examine the contemporary age class structure of the forests in the remainder of Ontario. In Figure 10.2, I have compiled forest resource inventory data of 1986 reported in Gray (1995). The forest age class structure is shown for each forest section within the forest regions described by Rowe (1972). The age structures of most forest sections appear similar to that of the pattern consistent with fire suppression, seen in Figure 10.1. Indeed sections B11 and B14 represent largely the same forest 12 years later, and the modal age class frequency appears, as would be expected, in the next 20-year category. A notable lack of recruitment into the youngest age classes is evident throughout much of the province. This can be attributed, at very least, to a lack of new habitat and suitable regeneration conditions due to the suppression of the natural fire regime.

An exception to the general pattern appears in section B10, the Lake Nipigon area (Figure 10.2): here, young age classes are well represented. For

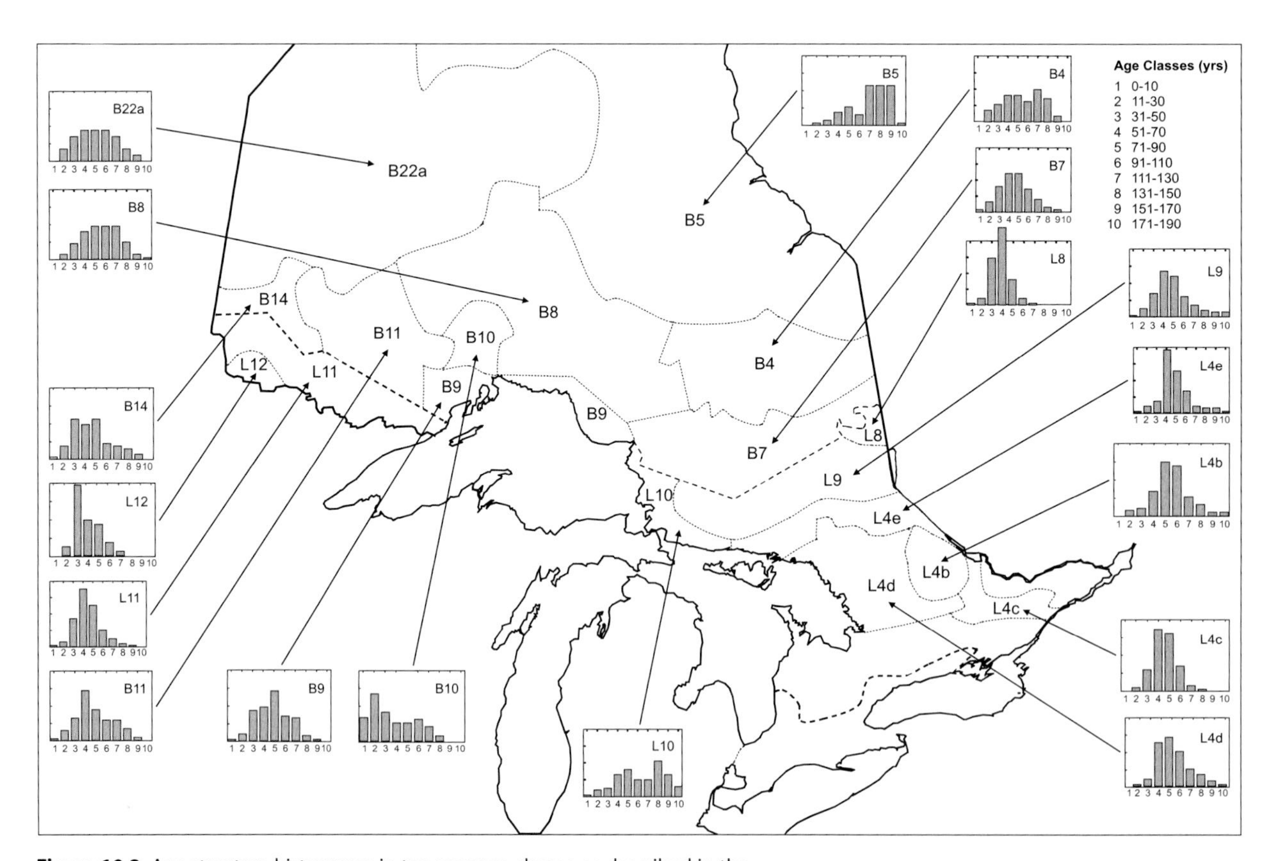

Figure 10.2 Age structure histograms, in ten-year age classes, as described in the 1986 Ontario Forest Resource Inventory, for forest sections of northern and central Ontario. (Data from Gray 1995)

approximately 40 years this comparatively small section was under the control of one forest manager whose harvesting and silvicultural practices, of relatively small-scale strip cutting, were atypical and exceptionally successful (Jeglum 1987). Another section with an age structure anomaly is L8, the Little Claybelt of northeastern Ontario. The forest here is comparatively even-aged: most stands lie in the 30- to 70-year range. Large, destructive fires around the 1920s, the clearing of land for agriculture, and continuing agricultural land use have combined to yield a comparatively young, uniform forest landscape. This pattern is repeated in the southern section, L4C, which is also heavily influenced by agricultural activity.

An extreme representation of old growth is seen in section B5, which includes the James Bay lowlands. In this unmanaged muskeg environment, trees are stunted in growth, yet old. Because of the wetland environment, forest fire rotations of 300 to 400 years are estimated in the region (Suffling et al. 1982). Another section exhibiting abnormally old growth is L10, the area of southeastern Lake Superior and the Algoma highlands. Two old cohorts are apparent in this section. One dates between the 1890s and 1930s, and the older mainly between 1830 and 1850. These cohorts of mixed deciduous and evergreen forest almost certainly represent past disturbance events, but the nature of the disturbances may be anthropogenic, natural, or a combination of both. The earlier dates correspond to the first large influx of European settlement in the region, but it is also known that heavy windstorm activity occurred around the upper Great Lakes during this period. The younger cohort peak corresponds to a period of known windstorm activity. Figure 10.2, therefore, paints a picture of aging forests across the province and an associated lack of young forest.

Does the Evidence of Older Forests Indicate Extended Forest Succession?

Most people travelling in Ontario see the forests from the road. Forest reserve strips approximately 30 m deep are left alongside major highway routes throughout the northern part of the province. Although these reserves retain something of the natural landscape, most observant travellers are aware that much of the tall forest cover disappears beyond the 30-m strip. Ontario's premier transportation corridor starts as Yonge Street, the backbone of Toronto, extends first north then west across the province as Highway 11, and eventually links with the Trans-Canada highway at Nipigon on Lake Superior. Much human settlement and activity can be seen along the length of this route.

As one travels north from the town of North Bay, however, and on through the region of Temagami, the road seems to take on a genuinely wild aspect and tall forest extends well beyond the 30-m roadside strip, until one reaches the Little Claybelt at New Liskeard, a farming centre. The stretch of highway through Marten River, Temagami, and Latchford not only provides a scenic highlight, but crosses the boundary between the Great Lakes and boreal forest bioclimatic zones. Between Temagami and Latchford, within a 5-km stretch of road, white and red pine, black ash (*Fraxinus nigra*), sugar maple (*Acer saccharum*), yellow birch (*Betula lutea*), large-tooth aspen (*Populus grandidentata*), hemlock (*Tsuga canadensis*), and most cedar (*Thuja occidentalis*) disappear from the forested landscape, to be replaced by white spruce (*Picea glauca*), black spruce (*Picea mariana*), jack pine (*Pinus banksiana*), paper birch (*Betula papyrifera*), trembling aspen (*Populus tremuloides*), and balsam poplar (*Populus balsamifera*). Similarly rapid changes in dominant tree composition are evident over the landscape of the Lake Superior shoreline immediately south of Wawa and in the region of Thunder Bay. The Temagami area, which incorporates elements of both the Great Lakes and boreal forest regions (Rowe 1972), is especially interesting because strong attempts have been made to protect it from unsustainable exploitation since the mid-19th century, and fire protection measures have been in place for much the same period.

Under the theory of succession, old, pioneer trees die, then young, more shade-tolerant species take their place in the upper canopy. If the forests of the province are aging, we might expect that the composition is also changing. Most old-growth forests of the Temagami region came into being as a result of fire, and many remain intact because of *The Forest Reserves Act* (1898) (Day and Carter 1991a, 1991b). Examining a breakdown of the forest resources inventory for the region by species and by age class yields an instructive picture (Figure 10.3). White pine and red pine dominate the oldest areas. The highest proportions of poplar, jack pine, and black spruce were recruited at the turn of the 19th century, but after the onset of fire suppression, very little evidence of recruitment appears. What limited recruitment there is

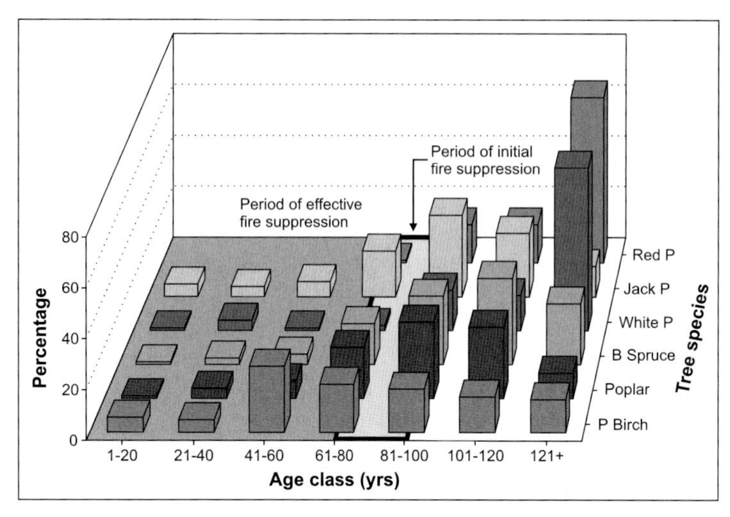

Figure 10.3 Forest Resources Inventory data summarized for the Temagami region of Ontario. Twenty-year age classes of forest are shown by dominant canopy tree species. (RedP = red pine, JackP = jack pine, WhiteP = white pine, Bspruce = black spruce, Poplar = large-tooth aspen and/or trembling aspen, Pbirch = paper birch).

can be attributed to paper birch and also to balsam fir (*Abies balsamea*), a species not shown in the figure because it does not dominate the upper canopy. Successional change, therefore, does not appear in the age structure of the forest inventory (Figure 10.3). Instead a picture emerges of older trees of the same species, and in largely the same proportions, that dominated the landscape prior to fire suppression.

Compositional Change in the Subcanopy

Although the upper canopy trees of the pre-suppression period are still largely in place, the subordinate canopy strata may have changed during the post-suppression period. Resource inventory data, although extremely useful at the landscape scale, cannot readily address this issue. It is necessary to examine stand structure in detail. Conducting a structural analysis of an old-growth forest plot yields a detailed history of forest development and change with time. It follows that the older the stand we investigate, the more of the local site history we can reconstruct. Although we may lose spatial context and generality over the landscape, we gain insight into the processes of succession. Where the trees can be aged by counting tree rings, an accurate time scale of events influencing succession can be obtained.

Figure 10.4 illustrates the vertical profile diagram of a 20-m by 20-m forest plot dominated by red pine on the northeast shore of Lake Herridge, near Highway 11, 20 km south of the town of Temagami. The stand is growing on a steep slope with much rock outcropping and pockets of sandy till to depths of 1 m. A frequency diagram of the age structure by species

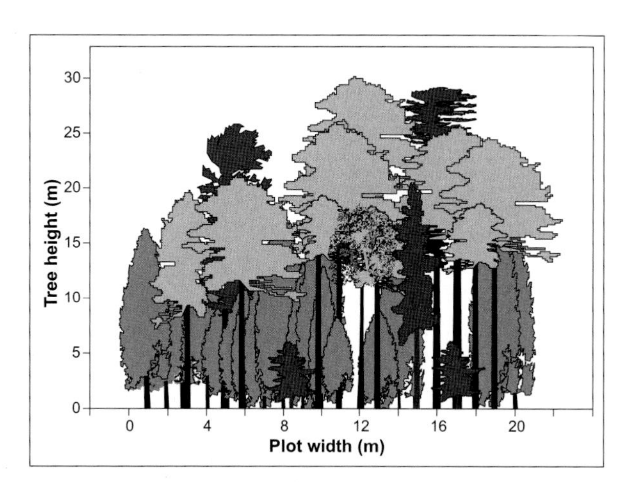

Figure 10.4 Profile diagram, to scale, of all trees in a 20-m by 20-m plot on the northeastern shore of Lake Herridge, 20 km south of Temagami. The upper canopy consists of red pine with two white pine emergents, three white spruce, and a single paper birch.

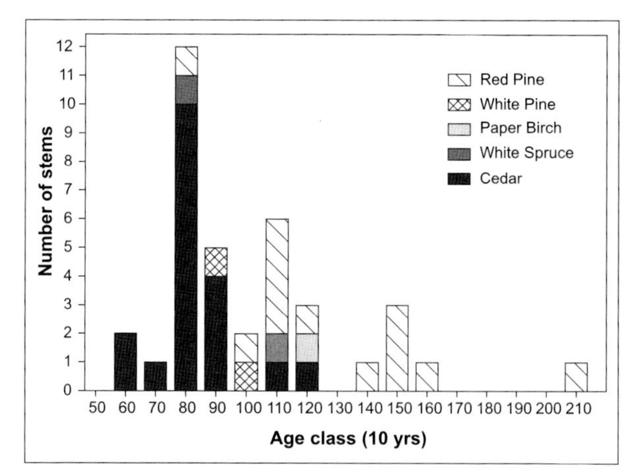

Figure 10.5 Frequency histogram showing the age structure of all trees, by species, in the Lake Herridge forest plot of Figure 10.4.

of all trees in this plot is provided in Figure 10.5. The biggest red pine in the centre is also the oldest tree, at 220 years of age. Dating from about 1770, this tree is the only survivor of its cohort in the plot (Figure 10.4). The main canopy is formed by red pine from a cohort that originated around 1820. Significantly, there is no trace of other tree species from these two establishment events, and the lack of any new trees during the intervening years is strong evidence that the dominant red pines were established immediately following forest fire. However, the stand does support comparatively dense and diverse intermediate and lower strata that include white pine, white spruce, paper birch, and cedar (Figure 10.4). The age structure frequencies clearly show that these strata have their origin after 1870. The preponderance of red pine in the oldest phase indicates that surface fire, a nonlethal underburn, probably stimulated tree seedling establishment by creating appropriate seedbed conditions (Ahlgren 1976; Bergeron and Brisson 1990; Burns and Honkala 1990). Subsequently, in the late 19th century, when fire suppression measures became effective around lakeshores, cedar invaded the understory, along with a little white spruce and white pine. Although the youngest trees examined were 60 years old, younger and smaller trees were present in the plot. These were mostly cedars.

We can conclude from analysis of this one case history that the suppression of fire has had several consequences for the forest. The most obvious is that the dominant red pine is getting larger and older than might otherwise be the case if a natural fire regime prevailed. Secondly, red pine is no longer regenerating. Thirdly, slower-growing, shade-tolerant, fire-susceptible species, such as cedar, invade, become established and grow quite well on suitable microsites. The preponderance of cedar at this particular location is linked to the lakeshore habitat; it is less certain that the species would invade at such a high density elsewhere.

Other old-growth stands around the Temagami region have been similarly investigated, and each yields a comparable, tightly linked pattern of disturbance and successional history (Day and Carter 1991b). However, differences emerge from site conditions. Relatively deep till overlying bedrock fosters the growth of nutrient-demanding broadleaved species. Figure 10.6 illustrates a 234-year-old old-growth pine stand in which sugar maple has become established

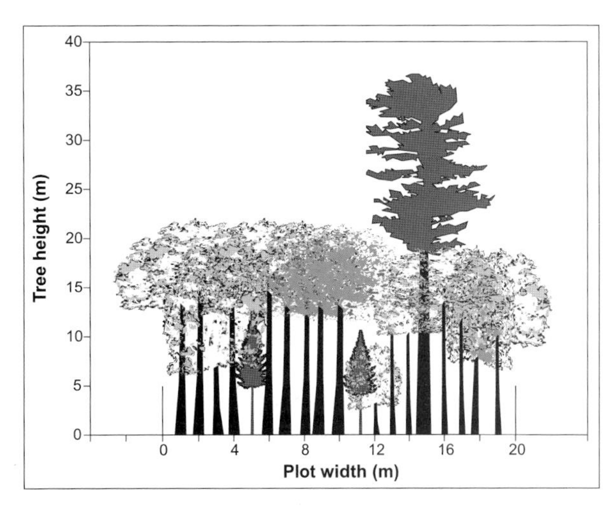

Figure 10.6 Profile diagram of all species in a 20-m by 20-m plot near the Ko Ko peninsula of Lake Temagami. The main canopy is dominated by sugar maple but includes three red maple stems. A single, large white pine is emergent, and there is a sparse sapling stratum of balsam fir.

following fire suppression. The maple now forms the majority of the canopy area, as large, old pine trees have toppled. On such productive sites, old white pine continues to grow upward and eventually outgrows the limits of its own shallow root system to provide anchorage against windstorms (Gilbert 1978; Burns and Honkala 1990).

This stand profile is also a reminder of the dilemma of trying to manage old-growth pine forest. The white pine is essentially "off site." That is, in the absence of disturbance, it would not normally grow to tree size in such a habitat, because broadleaved species would successfully out-compete the comparatively slow-growing conifers. However, white pine appears to be able to disperse into the hardwood complex and establish itself as seedlings, even though the seedlings cannot survive to the sapling stage (Carleton et al. 1996). If past fire events have eliminated susceptible broadleaved species from such sites, a window of opportunity is provided for the establishment of pine through seed dispersal from adjacent areas. Furthermore, the comparative fertility of the site promotes the growth of larger pine trees than elsewhere. The tall, old-growth pines that we see in the Temagami region today, it must be concluded, are largely an artifact of fire suppression measures that have been in place for a comparatively long time. This does not imply any diminution of their worth as forest ecosystems. Not only are these trees aesthetically deeply pleasing, but they contribute to the biodiversity of the region. Furthermore, they present a special opportunity to study old-growth pine forests as ecosystems, ecological communities, and landscape elements, an opportunity that would not otherwise be available under either a natural fire regime or a strict production forest management regime. Both of these regimes impose a relatively short limit on the life span of most pioneer forests.

The evidence from the Temagami region indicates that the onset of fire suppression marks the start of invasion by fire-susceptible species. This invasion produces increased local plant species diversity (that is, alpha diversity) and a shift in composition to include shade-tolerant trees. In the examples cited above, sufficient time has elapsed to confirm that, at least on fertile upland sites, a succeeding canopy does replace the pioneer species canopy. However, it does not necessarily follow that a pioneer canopy is succeeded by a woody understory. The possibility that shade-tolerant species could proceed through a complete generation as a lower stratum is credible, given the potential longevity of some of the canopy pioneers such as white pine, which can live 300 to 400 years.

Figure 10.7 illustrates the profile of a forest dominated by white pine at Galloway Lake in the District

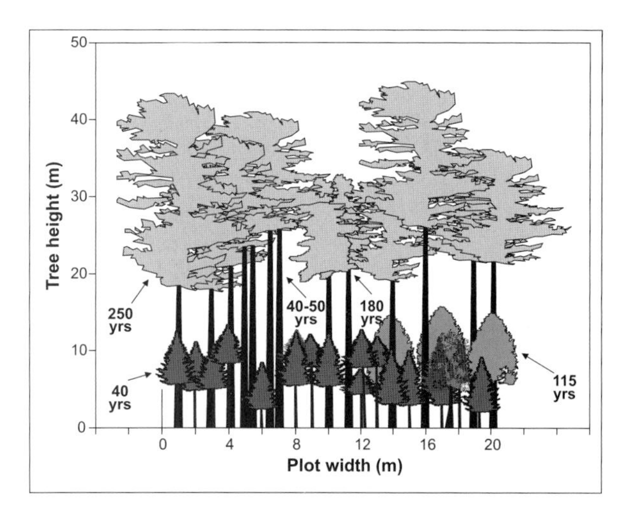

Figure 10.7 Profile diagram of an old-growth white pine stand at Galloway Lake, District of Algoma. Cored-tree age counts are shown for each species in each stratum. The upper canopy is dominated by white pine, and a dense lower stratum of balsam fir and three cedar trees is present.

of Algoma. The uppermost canopy trees are 40 to 45 m tall and approximately 250 years old. A subcanopy of co-dominant white pine is also evident at 180 years of age. The woody understory includes some cedar at 115 years of age and balsam fir trees in the 40- to 50-year age range. Balsam fir rarely lives for more than 90 years (Thompson 2000, this volume). Consequently, the outlook for the balsam fir trees in the understory at Galloway Lake is uncertain. In 40 years or so they will die. If the upper canopy does not open up before that time, they will not be able to initiate cone production, for which full incident light is required. Even if some of the uppermost pines do die during this period, a moderate density of co-dominants is available to take their place.

Understory Vegetation Response to Fire Suppression

Plant species of the forest floor may also exhibit a trend over time toward greater diversity with fire suppression. Reconstruction of the recent history of individual stands, illustrated above, is not possible with herbaceous plants. In fact, there is no real substitute for direct monitoring on a site over time. An indirect substitute is the chronosequence technique, in which the vegetation data from sampled stands are strung together in a sequence of post-disturbance age (Crocker and Major 1955). Unfortunately, many factors influence the variation in understory vegetation among disparate stands, including soil type, forest floor structure, microclimate, woody plant structure/composition, undetectable under-burns, and the historical contingencies of previous dispersal and establishment events. By accounting for these other influences as much as possible, we are likely to obtain a clearer picture, but it will not be easy to achieve.

A structural and compositional survey of 170 stands dominated by white and red pine throughout the Shield portion of Ontario's Great Lakes forest region classified the stands into four site class groups (Carleton et al. 1996). The poorest were typified by very shallow, sandy soils; whereas the best had comparatively deep loamy till soils. On the premise that these categories could account for the site-related influences on understory vegetation, we examined chronosequences within each class. Simple tabular comparisons yielded little insight into compositional change over time. Some species were widespread throughout the sequences, while others seemed to appear at random. Examining data from the 44 stands included in the richest site class, I classified the species into six groups using a multivariate clustering method, which constrained the stand groups to a post-fire age trend from 85 years to 300 years (Carleton et al. 1996).

The result is a compromise between inherent plant associations and stand groups of increasing mean age from 115 years to 274 years (Table 10.1). Although many of the species listed are widespread among the corresponding stand groups, the group to which each species is assigned represents its peak frequency and abundance. The youngest group includes wild strawberry (*Fragaria virginiana*), red maple seedlings, and white pine seedlings. These probably *originated* at a much earlier stage and represent hold-overs from truly

Table 10.1

A classification of understory plant species surveyed in 170 white and red pine-dominated forests throughout the Canadian Shield in Ontario into six forest stand classes.

Group 1 *115 years*	*Group 2* *118 years*	*Group 3* *152 years*	*Group 4* *191 years*	*Group 5* *204 years*	*Group 6* *274 years*
Acer rubrum	*Aster macrophyllus*	*Abies balsamea*	*Acer spicatum*	*Acer saccharum*	*Acer pensylvanicum*
Amelanchier sanguinea	*Oryzopsis asperifolia*	*Coptis trifolia*	*Alnus crispa*	*Dryopteris spinulosa*	*Carex communis*
Carex lucorum	*Pteridium aquilinum*	*Cornus canadensis*	*Aralia nudicaulis*	*Maianthemum canadensis*	*Carex trisperma*
Chimaphila umballata	*Vaccinium myrtilloides*		*Betula papyrifera*		*Clintonia borealis*
Dryopteris intermedia		**Cryptogams**	*Corylus cornuta*	**Cryptogams**	*Oxalis montana*
Fagus grandifolia		*Dicranum flagellare*	*Danthonia spicata*	*Cladonia chlorophaea*	*Picea mariana*
Fragaria virginiana			*Diervilla lonicera*		*Polypodium virginianum*
Galeopsis tetrahit			*Linnaea borealis*		*Salix humilis*
Gaultheria procumbens			*Lycopodium annotinum*		*Sorbus decora*
Hepatica americana			*Lycopodium clavatum*		*Streptopus roseus*
Lonicera canadensis			*Lycopodium obscurum*		*Thuja occidentalis*
Lycopodium complanatum			*Populus tremuloides*		*Trientalis borealis*
Ostrya virginiana			*Rubus pubescens*		*Vaccinium angustifolium*
Picea glauca					
Pinus resinosa			**Cryptogams**		**Cryptogams**
Pinus strobus			*Brachythecium reflexum*		*Bazzania trilobata*
Pyrola asarifolia			*Dicranum montanum*		*Brachythecium curtum*
Pyrola secunda			*Drepanocladus uncinatus*		*Brachythecium curtum*
Quercus rubra			*Hylocomium splendens*		*Brachythecium salebrosum*
Rubus idaeus			*Hypnum pallescens*		*Cladonia coniocrae*
Trillium grandiflorum			*Paraleucobryum longifolium*		*Cladonia squamosa*
Viburnum acerifolium			*Peltigera canina*		*Dicranum ontariense*
Waldsteinia fragaroides			*Plagiothecium laetum*		*Dicranum polysetum*
			Pleurozium schreberi		*Dicranum scopariu*
Cryptogams			*Ptilidium ciliare*		*Pohlia nutans*
Eurhynchium			*Ptilium crista-castrensis*		*Ptilidium pulcherrimum*
Pulchellum			*Rhytidiadelphus triquetris*		
Mnium medium			*Thuidium delicatulum*		

The classification procedure constrains the clustering to an ascending post-fire age trend. Mean stand age for each class is shown.

unshaded conditions. Group 2 includes big-leaved aster (*Aster macrophyllus*), wild mountain rice (*Oryzopsis asperifolia*), bracken fern (*Pteridium aquilinum*), and velvet leaf blueberry (*Vaccinium myrtilloides*). These species are well known to re-sprout following fire; when they occur together in abundance, they are likely to represent the legacy of a past fire regime.

The appearance of several species of Lycopodium in Group 4 is an indication of progressing forest age. Up to five species may appear in some of the oldest mixed forests of Ontario. Although widespread, Canada May lily (*Maianthemum canadense*), the blue bead lily (*Clintonia borealis*), and rosy twisted stalk (*Streptopus roseus*) are conservative, monocotyledonous species that cannot withstand unshaded conditions. Other obligate deep forest species that appear in the old growth groups include wood sorrel (*Oxalis montana*), common polypody fern (*Polypodium virginianum*), and starflower (*Trientalis borealis*). A noticeable trend also emerges among the cryptogams. Few mosses and liverworts appear in the younger stand groups, but with increasing mean age, the number of taxa, including lichens, increases. This analysis yields a trend from a comparatively generalist flora, of plants capable of growing both in shade and open conditions, to a flora of conservative growth patterns and dependence on moist, shaded conditions.

Fire suppression does, therefore, appear to have influenced change in understory composition, but not in any dramatic or deterministic way. Much of this apparent effect is related to ecosystem legacies from past stand growth cycles on a site. As coarse, woody debris persists after a fire or fall-down, the forest floor of a young stand can be very heterogeneous. The diversity of plant species of the forest floor appears to respond primarily to this heterogeneity, so little change in composition is expected with time (Carleton and Arnup 1993). Furthermore, old-growth stands are subject to windthrow, which creates large gaps in the canopy. These microhabitats favour the growth of shade-intolerant and shade-indifferent plant species.

Consequences of Fire Suppression in Boreal Forests

In southern deciduous forests, continuing succession, mediated by gap phase processes, relies on sufficient tree species richness to provide a range of regeneration characteristics. A spectrum of selection pressures and associated adaptations can be linked to regeneration on various substrata within canopy gaps of various size (Drury and Nisbet 1974; Grubb 1977; Runkle 1981; Shugart 1984; Pickett and White 1985). The southern deciduous forest of Ontario contains 63 tree species (Maycock 1963). The Great Lakes forest in Ontario includes 53 of those species (Carleton et al. 1996), but the boreal zone contains only 8 species when specialized riverine habitats are excluded. These few species might be regarded as those which can tolerate the thermal stress of the climate, but here exceptions may prove a rule.

In the river valleys of the James and Hudson Bay drainage basins, debris is exported from seasonally flooded fluvial terraces, and a fresh load of flood silt is deposited on the riverine forest floor. The lowest terraces often support flood-plain forests of elm (*Ulmus americana*), black ash, and cedar. Associated with the trees is an understory flora of very southern wetland affinity, and yet many of these forests are located at or beyond the northern edge of the closed-canopy boreal forest region. If these forest communities were limited by thermal stress, then they could not grow so far north. It also follows that if these species could withstand the temperature regime, we would expect to find them well represented on the moist, fertile, upland clay sites that are abundant throughout northeastern Ontario, but this is not the case. Black ash does appear sporadically on such uplands, but is not abundant.

On the other hand, if we ascribe the richness and southern affinity of these riverine forests to their inaccessible lowland habitat and protection from forest fire, then a consistent view emerges. The flood-related export of forest floor debris, which is not then available for fuel, adds to the protection from fire. This reasoning leads to the conclusion that the boreal flora and vegetation are largely governed by a historical adaptation to fire. The boreal flora is the subset of a much larger pool of forest species that have some mechanism for survival or regeneration in the form of persistent, below-ground plant parts, persistent seed banks above or below ground, or the ability to rapidly re-invade from surviving above-ground patches (Carleton 1979; Rowe 1983). The sharp boundary in tree species richness between the Great Lakes forest landscape and the boreal is likely to be as strongly linked with the respective fire regime of each zone as with the northern tree line (Payette 1992). A further consequence of this hypothesis is that the tree species of the boreal zone possess pioneer, post-fire silvical characteristics (Zasada et al. 1992). Only one species in the Ontario boreal forest, balsam fir, relies solely on establishing beneath an existing tree canopy (Thompson 2000, this volume). To be such a secondary species in a fire-dominated landscape is a risky business. In terms of life history features, it is necessary to establish itself, grow and reproduce before the next fire strikes. Although very shade-tolerant, balsam fir is anomalously short-lived and allocates a large proportion of its assimilate to producing an abundance of small, well-dispersed seed (Burns and Honkala 1990). Most shade-tolerant species are long-lived, have a long juvenile period, and produce a small quantity of large seed that is not well dispersed.

What happens to an upland boreal forest under a regime of fire suppression? Is there sufficient tree species diversity to provide for continuing succession from one canopy dominant to another in all forest types? The classic view is that a spruce-fir association eventually forms a self-replacing, "climax" community over the whole landscape (Shelford and Olson 1935; Shelford 1963). The opposing view is that, in the absence of fire, boreal forests become decadent, fall down and remain as ragged shrub and small tree communities until rejuvenated by fire or by the ploughing of the soil through windfall (Rowe 1961).

Unfortunately, little evidence exists to support or refute this second view. For the most part, it seems that balsam fir forest only fully succeeds a primary canopy of paper birch, or a mixture of birch and white spruce, to produce a continuous secondary canopy (Carleton and Maycock 1978). In most other circumstances, species mixtures of somewhat degraded forest are evident. Unfortunately, sites supporting a post-falldown forest are rarely recorded. They have no immediate commercial value and are, consequently, ignored by forest inventory. They are of little interest to the forest researcher at the stand level because they are no longer closed canopy forest. Such decadent stands, however, are likely to receive increased attention in the future, as the focus of interest shifts to landscape-level processes.

EFFECTS OF LOGGING

Effects of Selective Logging

Selective logging has had two main consequences for the forested landscape of Ontario. The first is the genetic consequence of high-grading what are perceived to be the best trees out of local populations. These are usually the tallest, straightest individuals. Such a tree form may arise from genetic predisposition, from competition with neighbouring trees during growth, or from a combination of the two factors. Site conditions also contribute greatly to the absolute size achieved. Although loss of the tallest and straightest individuals in a population is of concern to the tree breeder and important to both the future economic capacity of the forest and its aesthetic qualities, it is of less concern, from the viewpoint of landscape ecology, than the removal of all members of the local species population that occurred in many mixed forests with large white pine emergents. An acceleration toward dominance by broadleaved, hardwood species of the type seen in Figure 10.6 appears to have been the outcome of that practice (Abrams and Scott 1989).

The second consequence of removing one or a few trees from the upper canopy is to free space and resources for the growth of others. The candidates to fill the space are immediate neighbours, saplings and seedlings beneath the harvested tree(s), and seed that either disperses into the gap or is present as a dormant seed bank near the soil surface. This is exactly the natural process of gap phase replacement that occurs in temperate broadleaved and mixed forest when one or a few neighbouring trees blow over, snap off, or die *in situ*. Consequently, gap filling is a matter of competition among the individuals that either expand laterally or grow up from below. The richness of species in the southern deciduous and Great Lakes forest regions may reflect the characteristic traits of growth and life history of the various species, which are termed their individual "strategy" for coping with canopy gaps of different size in different local environments. However, the spatial contagion that this view entails can also contribute to regional variation in forest composition. The limitations and uncertainties associated with dispersal and establishment, for different strategies in space and time, have been proposed as a prime agent of diversity in forest communities (Grubb 1977). Where the forest is sufficiently rich in tree species, and where the gaps created range in size from the equivalent of a single tree crown to one hectare, selective logging can emulate a natural process in which the integrity of the woody species populations is maintained.

In southern Ontario, however, a focus on removing undesired species has favoured certain species at the expense of others and has created very small gaps that favour shade-tolerant taxa such as sugar maple. Further north, the removal of white pine from mixed Great Lakes forest has been the single most important consequence of selective logging. From the interpretation of Figure 10.6, it follows that broadleaved species have probably replaced pine as the canopy dominant in many areas. Selective removal appears, in many cases, to have accelerated an otherwise natural process of succession under a regime of fire suppression (Abrams and Scott 1989). Unfortunately no explicit, ground-based studies over a broad region have appeared on this topic.

Clear-Cut Logging and Silvicultural Methods

In the early days of logging in northern Ontario, trees were felled by hand and removed to the nearest roadside by horses. Clear-cutting started in the boreal region in the 1920s. Operations were relatively labour-intensive, and horse skidding remained the main method of log transfer to the roadside until the early 1960s. The chainsaw became commonplace in the late 1940s. Then mechanical tree-harvesting machines and, more recently, large feller-bunchers replaced men with chainsaws, for improvements in productivity and safety. Not surprisingly, so-called clear-cuts prior to 1960 involved less than the complete removal of trees from a site, as has been the case since, with mechanical, wheeled skidders. Many uneconomic residual trees were left in the earlier years and these, in many instances, influenced the process of vegetation recovery.

Prior to the 1970s, little effort was made to actively regenerate clear-cut areas. The "natural regeneration" method was assumed to be the best. For comparatively small clear-cuts and patch cuts in southern deciduous and mixed Great Lakes forest, such a strategy was effective, either through re-sprouting or through the natural dispersal of seed from adjacent woodlands. In the boreal forests, however, this was not the case in large, mechanically harvested areas of the late 1960s and early 1970s. Planting programs were expanded greatly in the 1970s. The main seedling stocks were black spruce and jack pine. Jack pine was also directly seeded from the air on cutovers, because abundant seed was available from harvested trees. Spruce seed was far less abundant and more difficult to extract. Consequently, the best use of the valuable seed was to maximize germination, establishment, and about one and a half years of growth in nurseries prior to outplanting.

Extent of Clear-Cut Logging

Figure 5.9 in Perera and Baldwin (2000, this volume) illustrates the extent of clear-cut and selective logging throughout the boreal and northern Great Lakes forests of Ontario. Clearly this activity has affected sizeable areas of the southern half of the boreal zone across the province. The areas affected are especially noticeable in the Greater Claybelt of northeastern Ontario and in regions north and west of Thunder Bay. The most extensive impacts occurred in the two decades following the switch to wholly mechanical logging. Although extensive, the cutting pattern of the 1980s is more fragmented. In the Claybelt, large contiguous patches appear, each as a composite of clear-cuts from different decades. Indeed, the largest composite cutover is half the size of Prince Edward Island. What has been the legacy of this extensive harvesting activity? Do we see a pattern of vegetation recovery that is consistent with recovery after forest fire, or is it qualitatively different? Do we see vigorous young forests, representing an unchanged landscape, or does the landscape differ from that frequently disturbed by forest fire? Few studies have been explicitly conducted to address these questions, although much research is now under way.

In the early 1980s, my co-workers and I began vegetation and soil surveys of post-logging sites in the Claybelt around the town of Kapuskasing. The objective was to identify successional stages and to integrate post-logging forest vegetation into the Forest Ecosystem Classification program (FEC) of the Ontario Ministry of Natural Resources (Jones et al. 1983). The approach entailed following exactly the same survey procedures as had been conducted for FEC in post-fire, commercial forests of the region. The design involved sampling forest plots that represented a range of post-logging stand ages on a similar range of soil types, following the chronosequence method (Crocker and Major 1955). We thus attempted to replicate forest plots of the same age class on each soil type. The survey produced data on 262 post-logging forest plots, as compared to the 250 post-fire plots of the FEC survey. Initial attempts to plot out successional sequences of change in plant species composition over time failed.

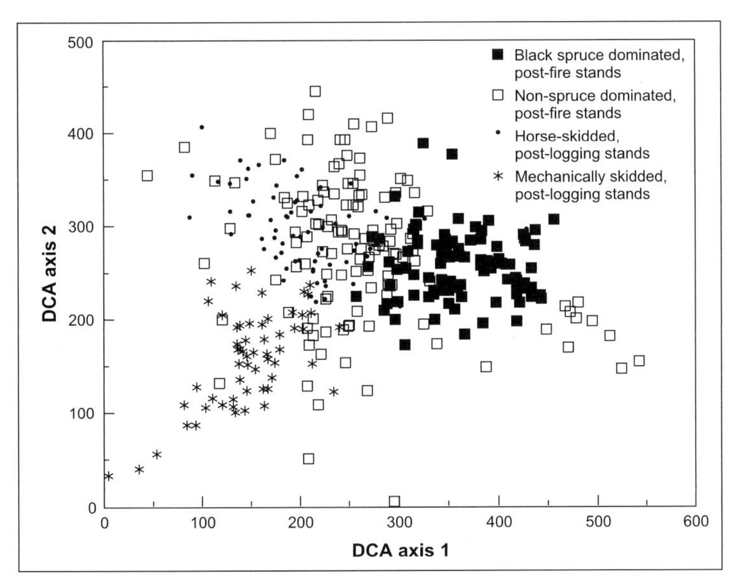

Figure 10.8 Stand ordination diagram, by detrended correspondence analysis (DCA), showing coordinates for 250 upland post-fire (FEC) stands (square symbols) and 132 upland post-logging boreal forest stands (circles and stars) from the Greater Claybelt.

No clear trends emerged with time as the primary axis. However, when less structured analyses were performed in which both data sets were combined, a much clearer picture emerged, as is detailed below.

Recovery of Upland Boreal Forest from Clear-Cut Logging

Figure 10.8 is an ordination diagram of the 250 FEC stands and 132 upland post-logging stands in the Claybelt, showing the two most important axes, based on all tree and tall shrub species abundances. Such diagrams illustrate similarities and dissimilarities in community composition. Stands are represented as points. Those that are close together in the diagram are similar, while those far apart have a very different species composition from one another.

Nutrient-poor, dry, post-fire, jack pine woodland stands appear on the right, and moist, nutrient-rich, broadleaved and mixed forest stands on the left (Figure 10.8). The circumstantial conclusion is that soil properties govern the primary direction of compositional change through the forest complex. More formal methods have confirmed this finding (Carleton and MacLellan 1994). The oldest stands, all post-fire in origin, appear at the top of Figure 10.8, while the youngest, mostly post-logging stands appear at the bottom. The second axis, therefore, is interpreted as a post-disturbance age gradient that strongly influences the woody plant species composition. The type of disturbance, whether fire, logging with horse skidding, or logging with mechanical skidding, is indicated in the use of different symbols to represent the stands. The post-logging stands dominated by black spruce occupy the left-of-centre region of the scattergram. Bearing in mind that all logged stands were dominated by black spruce prior to harvest, one sees that the recovering community composition has been changed considerably by the impact of logging in all but a few cases. The implication from the diagram is that, had these logged stands been burned instead, most would have regenerated to their original composition.

Clearly, the mechanically logged forests appear qualitatively distinct from the others, although there is a small degree of intermingling among the oldest stands with stands of post-fire origin. These stands are dominated by young, pure trembling aspen canopies. In contrast, the horse-skidded stands, in the 25- to 55-year age range, intermingle with the much older, post-fire stands dominated by boreal mixedwood species and balsam fir at the left hand side of the ordination diagram (Figure 10.8). Abundant balsam fir appears to be the common element among stands with either type of forest history. Under natural circumstances, many boreal forest stands support an understory of balsam fir seedlings and saplings. With the comparatively benign impact of horse skidding, the majority of these seedlings and saplings survive after logging as advanced growth. They then grow rapidly to form a new upper canopy. A holistic view of this process is that natural autogenic succession is accelerated by the removal of the primary post-fire canopy (Abrams and Scott 1989). Consequently mid-age, horse-skidded forests come to resemble much older post-fire forests in which natural falldown has occurred. In contrast, balsam fir advanced growth did not survive the logging operations on mechanically skidded sites during the first three decades of such activity. Not only did machinery flatten the balsam fir, but the narrow wheels and metal tracks disrupted the organic surface layers of the forest floor, to expose moist, nutrient-rich mineral soil layers below (Groot 1987). Because all boreal tree seedlings require access to mineral soil for successful early rooting and the avoidance of drought stress, this was not originally perceived as a silvicultural problem. Furthermore, the exposure of nutrient-rich mineral soil was expected to enhance subsequent growth and yield.

As there was little active regeneration of boreal clear-cuts for the first fifteen years of fully mechanized harvesting, few spruce seedlings established naturally, especially on cutovers with fertile soils. The fecund and very effectively dispersed aspens were able to rapidly colonize such sites, however, because the moist, nutrient-rich soil conditions they require had been created by the mechanized logging activity. Their clonal suckering ability added to the ability of aspens to capture a site and severely restrict subsequent tree seedling establishment. Under post-fire conditions, consumption of the organic forest floor is deepest around the bases of the dead black spruce trees. Seeds from serotinous cones fall into these pockets, germinate, grow, and recapture the site for spruce, in a narrow window of opportunity (Black and Bliss 1978; Viereck 1983). On a clear-cut, no instantaneous conifer seed source is readily available just a few days following logging activity. Many of the understory species survive forest fire as underground parts in the residual organic forest floor layers, from which the species resprout and rapidly absorb the nutrients available from ash at the surface. This black spruce ecosystem utilizes nutrients efficiently, because so many are locked up in recalcitrant, organic combination on the forest floor. Indeed, black spruce has been hailed as the most nutrient-stress tolerant tree species on Earth. In contrast, the aspen ecosystem is relatively open and has broadleaved, deciduous litter that is rapidly decomposed, mineralized, and recycled (Viereck et al. 1983). Earthworms are common in the soils beneath aspen stands in the Claybelt, and the carbon/nitrogen (C/N) ratio in the surface layers of the soil is low. The primary constraint on production seems not to be nutrients so much as energy supply; this is a "carbon-limited" ecosystem.

These two very different forest ecosystems, of black spruce and aspen, may represent an interesting case of what ecologists term "multiple stable states"; that is, the situation where more than one ecosystem type can persist under exactly the same set of physical environmental conditions. What remains unclear is the extent to which relatively pure aspen ecosystems will succeed toward a conifer-dominated system over time, in the absence of fire, and the extent to which any such change is affected by nature of the mineral soil and the proximity to conifer seed sources. Some forests may experience continuing forest succession to boreal mixedwood or even to a pure conifer canopy. Evidence in support of this course of events appears in Figure 10.8. It is the older clear-cuts logged by the mechanized method that merge with the boreal mixedwood stands. However, the legacy following the first generation of invading trees may be a landscape of "scrubby" forest if the aspens, themselves, are not harvested as a resource. A complex of small conifer groves amid shrublands and stunted trees may result (Rowe 1961). Already poplars that grew up following the early clear-cuts of the 1920s and 1930s are being harvested for valuable veneer logs. Compounding this process of aspen succession is the prospect of climatic warming during the next half-century. Longer summers will favour the persistence of broadleaved tree species over spruce and may limit the invasion of poplar stands by conifers in general.

Recovery of Lowland Boreal Forest from Clear-Cutting

Wetlands dominated by black spruce occupy a large part of imperfectly drained landscapes such as the Claybelt. Many of these areas are clear-cut during the winter, although some operations continue all year. A higher proportion of trees survive harvest as advanced growth in the lowlands than the uplands, especially during winter logging. The pattern of tree regrowth following pre-mechanical harvesting, therefore, involves takeover of the canopy by balsam fir and spruce advanced growth (Groot 1984; Brumelis and Carleton 1988). For stands subjected to mechanized harvesting, much the same regeneration pattern emerges from lowlands as from uplands, although the proportions of spruce are higher and there is a tendency for a *Sphagnum* moss carpet to regenerate (Brumelis and Carleton 1988). Certain tightly clumped species make appropriate seedbed sites for black spruce (Jeglum 1979). The extent to which the forest floor is re-established is linked to the site nutrient regime and the carbon/nitrogen ratio in the surface soil layers (Brumelis and Carleton 1988). If the site is heavily disturbed during logging activity, then speckled alder (*Alnus rugosa*) frequently sprouts and dominates the area (Vincent 1964). Although a *Sphagnum* carpet can eventually form beneath such a canopy, the process is hindered by the deposition of broadleaf litter. Indications are strong that most logged wetlands eventually return to dominance by black spruce, even though the ecological consequences of peatland disturbance may retard this process by decades (Brumelis and Carleton 1988). In the meantime, alder swales and swamps characterize the sites.

Consequences of Clear-Cutting at the Landscape Scale

From the results of our ground-based survey of point samples described above, conducted within a limited radius of the town of Kapuskasing, we made the prediction that much of the unregenerated, mechanically logged, regional boreal landscape should show conversion from spruce to aspen dominance. The Ontario Independent Forest Audit (Hearnden et al. 1992) examined forest sites that were clear-cut between 1970 and 1990. Over that period, approximately 2.5 million hectares of forest were cut. The audit reports data for the period 1970 to 1985 and accounts for 1,672,265 ha of the 1,909,661 ha clear-cut over this period. The audit confirms that major conversions in forest cover type have occurred. Figures 10.9a and 10.9b, derived from the audit data, illustrate the area of forest in each of five cover types that regenerated from each of those same cover types prior to logging (Hearnden et al. 1992). Where no regeneration efforts have been made, a major shift away from pure conifer-dominated forests toward mixed and broadleaved forest is evident (Figure 10.9a). Most of the black spruce forest has regenerated to mixed softwood, mixed hardwood and softwood, and pure hardwood cover types.

These findings are consistent with our more localized study on the Claybelt, although the report also indicates that, on the less nutrient-rich sites of the Shield country in northwestern Ontario, mixed conifer regeneration frequently occurs. The trends in the data make it clear, however, that a progressive elimination of conifers is likely to occur over more than one rotation of harvesting, regeneration, and subsequent growth. Black spruce has regenerated mostly to mixed softwoods. Mixed softwood forest has regenerated mostly to the mixed hardwood-softwood type. Mixed hardwood and softwood forest, when logged, has regenerated mostly to the same type, but a large proportion also yielded pure hardwoods. Small reversals in this trend appear. A few mixed softwoods regenerated to pure black spruce or jack pine, and some black spruce recurred after clear-cutting. The details of these exceptions to the main trend, toward hardwood dominance, would warrant silvicultural research.

In 1970, 27 percent of the 108,114 ha that were clear-cut were planted or seeded with conifers. This proportion rose, fairly steadily, to 38 percent of the

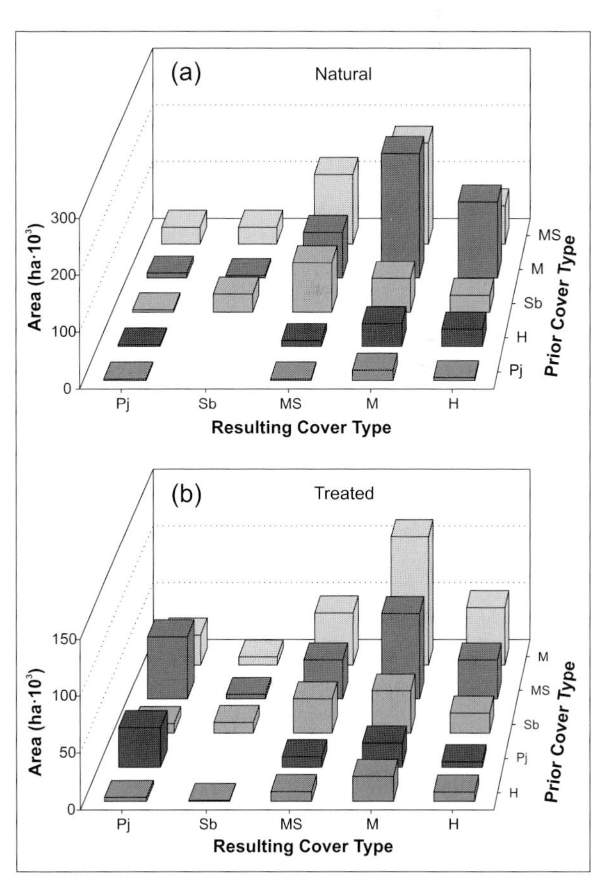

Figure 10.9 Area of forest cover types prior to clear-cut logging in the Ontario boreal forest, in relation to area of forest cover types after logging. Part (a) shows sites that remained unplanted or unseeded after logging, and part (b) shows only sites that were either planted or seeded with conifers after logging. (Data from Hearnden et al. 1992 for the 1970 to 1985 period of mechanical harvesting activity.) Key to forest cover types: M = mixed hardwood-softwood; MS = mixed softwood; Sb = black spruce; Pj = jack pine; H = hardwood.

144,279 ha clear-cut in 1984 (Hearnden et al. 1992). The regeneration picture on the clear-cuts planted or seeded with conifers is very similar to that of untreated cutover sites; the only difference is the increase in jack pine following logging (Figure 10.9b). Because jack pine occupies dry, nutrient-poor sites where it experiences little competitive suppression from aspen, artificial regeneration of this species is generally regarded as a success. That success is increasingly misunderstood: jack pine is currently being planted on many moist, fertile sites that were previously dominated by either spruce or mixed forest. Consequently, the proportion of pine plantations that are successful has declined even though the total area of planting has increased. Evident from the absence of pure, post-harvest spruce forest in Figure 10.9b is the almost complete failure of the spruce plantations established between 1970 and 1985. The proportion that have regenerated to mixed forest and to hardwoods is very high. Indeed, the area and proportion of forest cover that naturally regenerated to black spruce (Figure 10.9a) is greater than where efforts were made to assist in its regeneration (Figure 10.9b). The Ontario Forest Audit, therefore, not only reveals a pattern consistent with our more detailed ecological work on the Claybelt, but also extends a vision of ecosystem conversion to the entire area of the exploited boreal forest across the province. In addition, it graphically illustrates the inertia that seems to exist in the qualitative shift in ecosystem type, from dominance by gymnosperms to dominance by angiosperms, that is precipitated by clear-cut logging activity. The planting of tree seedlings at an ecologically low density, even though the economic and logistic costs have been enormous, has not been sufficient to resist this inertial shift in ecosystem type.

Effects of Clear-Cutting on Understory Vegetation

The foregoing review of forest cover change in the boreal zone, with an emphasis on tree species, illustrates the importance of the nature of the forest floor and the crucial role it plays in governing ecosystem processes and the nutrient regime. Much of the forest floor material derives from understory forest plants, including tree seedlings, shrubs, herbs, bryophytes, and lichens. Indeed, 50 percent of the above-ground plant biomass in a mature black spruce ecosystem may derive from forest floor mosses (Viereck et al. 1983). In concert with the surveys of woody plant regeneration and compositional change in the Greater Claybelt, we have surveyed the forest understory vegetation in the same post-logging stands (Brumelis and Carleton 1989; MacLellan 1991). I have conducted a comparison of the understory composition of these stands with the post-fire stands of the FEC survey. The comparison incorporates the 250 FEC stands from the Greater Claybelt, 45 additional post-fire spruce and poplar stands from our study area around Kapuskasing, 140 upland post-logging stands, and 122 lowland post-logging stands, for a total of 557 stands in the Claybelt. The presence and abundance of 372 plant species are represented in the data.

A number of questions arise in relation to such a comparison: Is the understory species diversity of each stand affected by logging? Is the among-stand understory diversity affected by logging? Is a dramatic change in understory species composition evident with logging, as is the case with the woody vegetation cover?

For the most part, the number of species encountered within the average stand in the post-logging group was not less than in the post-fire group. Thus, local diversity, sometimes termed alpha diversity, does not appear to be affected by the logging process (Brumelis and Carleton 1989; MacLellan 1991).

Figure 10.10 illustrates the stand scatter diagram of the first two axes from a detrended correspondence analysis (DCA) ordination. As with the overstory analysis, if two stand points are close together, they have a similar species composition, but if they are far apart, their composition is very different. Clearly, among-stand diversity (i.e., beta diversity) is greatly affected by logging activity. The post-logging stands are tightly clustered to the left side of the ordination scatter plot, while the post-fire stands are scattered widely (Figure 10.10). The post-fire black spruce stands are indicated in the scatter of points and they also exhibit a very wide spread over the ordination. Consequently, we can conclude that the diversity among stands (beta diversity) is greatly reduced following logging and that little compositional variation appears among the post-logging stands compared with those of post-fire origin. Because the post-logging stands cluster to one end of the first ordination axis, they might be viewed as "abnormal" types, compared with the post-fire stands. However, some young, post-fire, poplar stands mingle with the upland logged

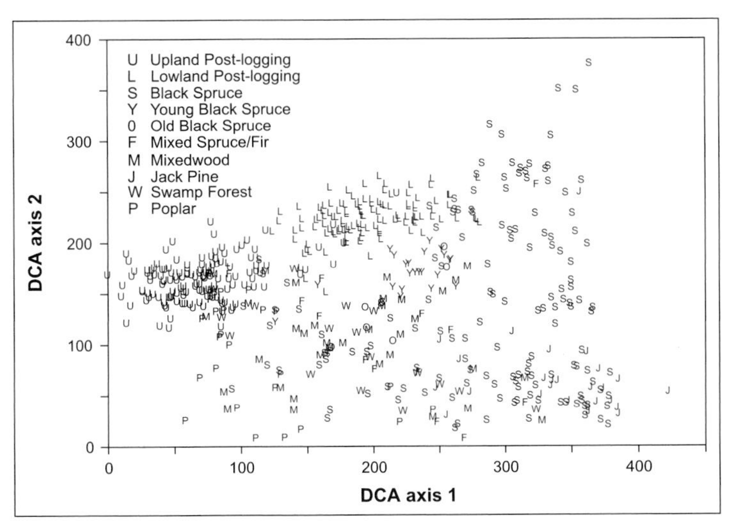

Figure 10.10 The first two DCA ordination axes showing 557 forest stands from the Claybelt for which data on the cover of 372 understory plant species were recorded. The stands surveyed include 295 post-fire stands, 140 upland post-logging stands, and 122 lowland post-logging stands.

stands, and some young post-fire black spruce stands are close to the lowland, logged stands. This effect indicates that the understory vegetation of juvenile forests persists much longer in logged stands than in burned stands. The same features were noted in experiments comparing the effects of cutover and fire treatments on boreal understory vegetation in Manitoba (Ehnes and Shay 1995) and in Quebec (Harvey and Bergeron 1989).

An environmental pattern underlies the ordination diagram in Figure 10.10. Dry, nutrient-poor forest typifies the bottom right. Wet, nutrient-poor bog forest characterizes the top of the stand scatter, and moist, nutrient-rich forests appear at the bottom left. The nutrient-enriching effect of logging activity, therefore, appears in the understory vegetation communities that result. This comes as little surprise, of course, because the FEC system is founded on the presence or absence of understory plant species as indicators of site conditions for forest growth (Jones et al. 1983; Carleton et al. 1985). However, the result also indicates a dramatic change in the composition of post-logging understory plant communities from the community that was present prior to logging.

Table 10.2 lists the most abundant understory plant species in each of the three major data sets in the ordination analysis. The post-fire stands support vegetation that is rich and abundant in mosses. These comprise the feathermosses, including Schreber's feathermoss (*Pleurozium schreberi*), step moss (*Hylocomium splendens*), plume moss (*Ptilium cristacastrensis*), wavy-leaved moss (*Dicranum polysetum*); and the *Sphagna*, including small red peat moss (*Sphagnum nemoreum*), midway peat moss (*Sphagnum magellanicum*), poor fen peat moss (*Sphagnum angustifolium*), common brown peat moss (*Sphagnum fuscum*), Russow's peat moss (*Sphagnum russowii*), and Wulf's peat moss (*Sphagnum wulfianum*). These species are comparatively oligotrophic and receive much of their nutrition from aerial input. Conservative, slow-growing vascular plants are also evident in the post-fire data. These include Labrador tea (*Ledum groenlandicum*) and creeping wintergreen (*Gaultheria hispidula*).

Most of the foregoing species are also well represented in the post-logging wetlands; however, some species, including nodding horsetail (*Equisetum sylvaticum*), Warnstorf's peat moss (*Sphagnum warnstorffi*), Canada blue-joint (*Calamagrostis canadensis*), sheathed sedge (*Carex vaginata*), shaggy moss (*Rhytidiadelphus triquetris*), and ribbed bog moss (*Aulacomnium palustre*), seem not only to recover rapidly from disturbance but also to indicate eutrophied wetland conditions (Brumelis and Carleton 1989). The understory of upland logged stands appears herb-rich and moss-poor, compared with the post-fire stands. Raspberry (*Rubus ideaus*), big-leaved aster (*Aster macrophyllus*), northern blue bells (*Mertensia paniculata*), and three-flowered bedstraw (*Galium triflorum*), considered to be indicators of nutrient-rich soil conditions, typify aspen and mixed forest understory. The feathermosses are especially limited at these sites because of the deposition of broadleaf litter.

The legacy of logging activity on forest floor vegetation appears to be a homogenization of the flora at the landscape scale. Even though individual stands retain moderate diversity, more-of-the-same from one cut area to the next is the overall consequence. The heterogeneity of disturbance associated with the unpredictable variations in fire behaviour, severity, and spatial pattern is largely lost with uniform cutover treatments, and along with it goes diversity in understory plant community types over the landscape. While the woody vegetation shows some very limited signs of potential for recovery of the natural post-fire cover types, our understory results show little evidence of much change in composition after logging. The time scale of 55 years since the onset of clear-cutting in northeastern Ontario may be too short to judge the results, but a parallel study on post-fire succession in the same region, using the same techniques, showed considerable compositional change over the same period (Taylor et al. 1987). It is sobering to bear in mind that many of the understory plants, as individuals, can outlast the trees. Nodding horsetail (*Equisetum sylvaticum*), for example, is an extreme case in point. It can sit, as a large rhizome, up to one metre below the surface, and individuals may be several thousand years old (Beasleigh and Yarranton 1974).

Logging Activity and Environmental Change

On one hand, we are presented with evidence of rapid

Table 10.2

The 20 most abundant species in each of three respective groups of boreal forest stands.

	Post Fire	Post Logged			Post Fire	Post Logged	
Species	*FEC*	*Low*	*Up*	*Species*	*FEC*	*Low*	*Up*
Pleurozium schreberi	38.74	8.93	5.90	*Aralia nudicaulis*	0.69	0.75	3.20
Sphagnum nemoreum	9.40	3.03	0.10	*Mitella nuda*	0.56	1.45	1.81
Sphagnum girgensohnii	7.74	0.28	0.06	*Sphagnum warnstorfii*	0.45	6.48	0.01
Hylocomium splendens	5.88	2.05	0.99	*Clintonia borealis*	0.45	1.03	3.00
Ledum groenlandicum	5.17	8.75	0.29	*Rosa acicularis*	0.44	0.03	4.56
Sphagnum magellanicum	4.96	4.48	0.01	*Anemone quinquefolia*	0.41	0.73	1.22
Ptilium crista-castrensis	4.62	1.13	0.64	*Calamagrostis canadensis*	0.36	6.15	1.13
Sphagnum angustifolium	2.96	2.43	–	*Viola renifolia*	0.35	0.25	1.09
Cornus canadensis	1.98	2.83	3.10	*Carex vaginata*	0.34	2.78	0.06
Dicranum polysetum	1.97	0.90	0.64	*Lycopodium annotinum*	0.32	0.23	1.14
Sphagnum fuscum	1.87	2.25	0.05	*Carex leptalea*	0.29	1.80	0.08
Gaultheria hispidula	1.72	2.05	0.59	*Chamaedaphne calyculata*	0.27	1.75	–
Aster macrophyllus	1.65	0.20	1.75	*Dryopteris disjuncta*	0.27	0.90	1.04
Sphagnum russowii	1.52	0.46	0.01	*Brachythecium triflorum*	0.27	0.18	1.26
Rubus pubescens	1.43	5.30	7.40	*Ribes triste*	0.26	0.50	1.60
Sphagnum wulfianum	1.22	0.73	0.01	*Galium triflorum*	0.27	0.65	1.61
Maianthemum canadense	1.10	0.68	1.55	*Vaccinium myrtilloides*	0.24	1.95	0.88
Alnus rugosa	1.04	23.50	1.04	*Aulacomnium palustre*	0.23	1.53	0.08
Picea mariana	1.00	18.20	1.10	*Epilobium angustifolium*	0.23	0.65	1.25
Coptis trifolia	0.99	2.35	0.93	*Ribes lacustre*	0.19	0.43	1.63
Cladina rangiferina	0.97	0.45	–	*Pyrus decora*	0.18	0.43	1.14
Carex disperma	0.95	2.98	0.24	*Rubus idaeus*	0.17	0.05	3.10
Abies balsamea	0.91	15.20	3.50	*Lonicera canadensis*	0.14	0.46	1.50
Rhytidiadelphus triquetris	0.89	1.98	0.85	*Mertensia paniculata*	0.14	–	1.58
Equisetum sylvaticum	0.88	2.48	0.88	*Viburnum edule*	0.09	0.08	1.06
Smilacina trifolia	0.84	0.88	0.02	*Populus tremuloides*	0.07	0.02	1.14
Carex trisperma	0.76	4.13	0.13	*Salix discolor*	0.03	1.23	0.12
Linnaea borealis	0.76	1.85	1.74	*Viola incognita*	0.02	1.23	1.14

The data includes 250 post-fire stands from the FEC survey of the Ontario Claybelt, 122 post-logging stands from lowland habitats on the Claybelt in the Kapuskasing region, and 140 post-logging upland stands in the same region. For post-logging sites, species abundances are indicated for lowland (Low) and upland (Up) sites separately.

plant community change in Ontario forests, as a result of major disturbance to the ecosystems, but on the other hand, evidence of great inertia emerges. The notion that vegetation and contemporary environmental conditions are in delicate equilibrium is very appealing. The whole basis for ecological site indicator species and for the inference of continuous change in vegetation composition with site is predicated on this deterministic notion.

Increasingly, however, this model is being rejected by ecologists. Indicators of site conditions are often minor species in understory forest communities that show site-related specialization. The dominants in boreal forests might best be characterized as generalists that operate as guilds to capture the site and determine the nature of the ecosystem through a combination of soil conditions and their own growth and decomposition characteristics (Hobbie 1992). Environmental conditions provide a backdrop for the survival and growth of forest organisms, but catastrophic events, whose signatures may remain with the resulting ecosystem for a very long time, are the agents of change (Sprugel 1991). These events dislodge the incumbents and offer other species opportunities to become established. Increasingly, plant ecologists recognize the notion of community inertia and resistance to gradual environmental change. Ecosystem change is usually linked to an identifiable agent such as fire, massive windthrow, ice storms, outbreaks of defoliating insects, or clear-cut logging. These impacts give the ecosystem a "kick uphill," so to speak, out of its locally stable state; then it rapidly changes to some new, relatively stable state.

Climatologists have some catching up to do in this respect. Simplistic models of a balance between climate and vegetation based on heat and moisture alone are not likely to yield useful predictions of future vegetation cover and of the movement in bioclimatic zone boundaries during periods of climate change on time scales of the order of decades (Bonan and Shugart 1989). Such models may be useful, however, over longer periods of time. Many of the catastrophic disturbances to forest ecosystems are induced by climate, but their qualitative impacts and associated vegetation responses may vary considerably. It is in this area of catastrophe that models of climatic impact and anthropogenic impact on forest vegetation converge (Baker 1995). To be useful predictors of global change, continuous climatic variables must be regarded as part of an environmental backdrop of change, either within or beyond the normal range, upon which disturbances operate on forest ecosystems to leave their highly persistent mark (Suffling 1995).

SUMMARY

The ecological legacies of fire-suppression measures across the commercial forests of the province of Ontario have been the following: older forests; an enhanced diversity of woody species that may not persist; succession toward fire-sensitive, shade-tolerant species; and the gradual depletion of fire-tolerant, shade-intolerant conifer species. The legacies of selective logging appear to have been an increase in shade-tolerant tree species and a decline in pioneer conifers. The legacies of clear-cut logging have been a conversion from forests dominated by pioneer, fire-tolerant conifers to pioneer, fire-tolerant broadleaved tree species. The legacy of clear-cutting on the understory vegetation is a greatly reduced beta diversity and the regional extirpation of rare species typical of needle-leaved evergreen forests.

In all cases of anthropogenic impact, pioneer, fire-tolerant, shade-intolerant conifers decline in dominance over the landscape. Deterministic notions of succession toward shade-tolerant plant species, in the absence of disturbance, are largely supported in the Great Lakes forest; however, a large question mark remains on this issue in the boreal region. Clearly, the old, deterministic notions of forest succession do not apply to disturbance. The qualitative nature of the disturbance is very important in effecting ecosystem persistence or change. Uncertainty under a regime of risk is, therefore, the only certain prediction that can be made concerning the forest vegetation of the future. Forest management strategies should be geared to this reality in such a way that management planning not only adapts to change, but is capable of foreseeing the landscape-level consequences of local forest management operations.

A FINAL PERSPECTIVE

Lest we be tempted to assign some value judgement to the inescapable conclusions concerning disturbance, it is worth bearing in mind that many of the pineries around the Great Lakes region resulted from disturbances associated with early European settlement (Heinselman 1981; Whitney 1986). That was an unprecedented event in the history of the region since recession of the last glaciation. Ironically, it is, in some cases, these same old-growth pineries which are now the object of heated debate by conservationists. Mischievously, one might speculate on whether similar debates will become fashionable around the conservation of old-growth poplar forests in northern Ontario 50 to 100 years from now. That is unlikely, because, in contrast to old-growth white pine, poplars are fast-growing, short-lived species, with a high reproductive output and well-dispersed seed, and they are also of comparatively short stature (Burns and Honkala 1990). They represent the weeds among the northern temperate tree species. As with weeds in other ecosystems, their dominance signals the overriding influence of the human species over the landscape of the apparent wilderness.

REFERENCES

Abrams, M.D. and M.L. Scott. 1989. Disturbance mediated accelerated succession in two Michigan forest types. Forest Science 35: 42-49.

Ahlgren, C.E. 1976. Regeneration of red and white pine following wildfire and logging in northeastern Minnesota. Journal of Forestry 74: 135-140.

Baker, W.L. 1995. Longterm response of disturbance landscapes to human intervention and global change. Landscape Ecology 10: 143-159.

Beasleigh, W.J. and G.A. Yarranton. 1974. Ecological strategy and tactics of *Equisetum sylvaticum* during a postfire succession. Canadian Journal of Botany 52: 2299-2319.

Bergeron, Y. and J. Brisson. 1990. Fire regime in red pine stands at the northern limit of the species range. Ecology 71: 1352-1364.

Black, R.A. and L.C. Bliss. 1978. Recovery sequence of *Picea mariana - Vaccinium uliginosum* forest after burning near Inuvik, Northwest Territories. Canadian Journal of Botany 56: 2020-2030.

Bonan, G.B. and H.H. Shugart. 1989. Environmental factors and ecological processes in boreal forests. Annual Reviews of Ecology and Systematics 20: 1-28.

Brumelis, G. and T.J. Carleton. 1988. The vegetation of postlogged black spruce lowlands in central Canada. I. Trees and tall shrubs. Canadian Journal of Forest Research 18: 1470-1478.

Brumelis, G. and T.J. Carleton. 1989. The vegetation of postlogged black spruce lowlands in central Canada. II . Understory plant cover. Journal of Applied Ecology 26: 321-339.

Bryant, D., D. Nielsen, and L. Tangley. 1997. The Last Frontier Forests: Ecosystems and Economies on the Edge. Washington, DC: World Resources Institute. 42 p.

Burns, R.M. and B.H. Honkala. 1990. Silvics of North America. Vol. 1. Agricultural Handbook No. 654. Washington, DC: USDA Forest Service. 675 p.

Carleton, T.J. 1979. Floristic variation and zonation in the boreal forest south of James Bay: A cluster seeking approach. *Vegetatio* 39: 147-160.

Carleton, T.J. 1982. The pattern of invasion and establishment of black spruce (*Picea mariana* [Mill.] BSP.) into the subcanopy layers of jack pine (*Pinus banksiana* Lamb.) dominated stands. Canadian Journal of Forest Research 12: 973-984.

Carleton, T.J., R.K. Jones, and G. Pierpoint. 1985. The prediction of understory vegetation by environmental factors for the purpose of site classification in forestry, an example from northern Ontario using residual ordination analysis. Canadian Journal of Forest Research 15: 1099-1108.

Carleton, T.J. and R. Arnup. 1993. Vegetation Ecology of Eastern White Pine and Red Pine Forests in Ontario. Forest Fragmentation and Biodiversity Project, Report No. 11. Sault Ste. Marie, Ontario: Ontario Ministry of Natural Resources, Ontario Forest Research Institute. 92 p. + appendices.

Carleton, T.J. and P. MacLellan. 1994. Woody vegetation responses to fire versus clear-cut logging. A comparative survey in the central Canadian boreal forest. Ecoscience 1: 141-152.

Carleton, T.J. and P.F. Maycock. 1978. Dynamics of the boreal forest south of James Bay. Canadian Journal of Botany 56: 1157-1173.

Carleton, T.J., P.F. Maycock, R. Arnup, and A.M. Gordon. 1996. In situ regeneration of white and red pine in the Great Lakes region of Canada. Journal of Vegetation Science 7: 431-444.

Clements, F.E. 1916. Plant Succession: An Analysis of the Development of Vegetation. Publication No. 242. Washington, DC: Carnegie Institute. 512 p.

Crocker, R.L. and J. Major. 1955. Soil development in relation to vegetation and surface age at Glacier Bay, Alaska. Journal of Ecology 43: 427-448.

Day, R.J. and J.V. Carter. 1991a. The Ecology of the Temagami Forest Based on a Photointerpretative Survey of the Forest Resource Inventory of the Temagami District. Toronto: Ontario Ministry of Natural Resources. 88 p.

Day, R.J. and J.V. Carter. 1991b. Stand Structure and Successional Development of the White and Red Pine Communities in the Temagami Forest. Toronto: Ontario Ministry of Natural Resources. 203 p.

Donnelly, R.E. and J.B. Harrington. 1978. Forest Fire History Maps of Ontario. Miscellaneous Report FF-Y-6. Ottawa, Ontario: Canadian Forestry Service, Forest Fire Research Institute. 18 p.

Drury, W.H. and I.C.T. Nisbet. 1974. Succession. Journal of the Arnold Arboretum 54: 331-368.

Ehnes, J.W. and J.M. Shay. 1995. Natural Recovery of Logged and Burned Plant Communities in the Lake Winnipeg East Forest Section. Canada-Manitoba Partnerships Agreement Number 1992/93-5009. Winnipeg, Manitoba: Canadian Forest Service. 138 p.

Gilbert, B. 1978. Growth and development of white pine (*Pinus strobus* L.) at Lake Temagami [M.Sc.F. thesis]. Toronto: Faculty of Forestry, University of Toronto.

Gray, S.L. 1995. A Descriptive Inventory of Canada's Forest Regions. Information Report PI-X-122. Chalk River, Ontario: Canadian Forest Service, Petawawa National Forestry Institute. 192 p.

Groot, A. 1984. Stand and Site Conditions Associated with Abundance of Black Spruce Advance Growth in the Northern Clay Section of Ontario. Information Report O-X-358. Sault Ste. Marie, Ontario: Canadian Forest Service, Great Lakes Forest Centre. 15 p.

Groot, A. 1987. Silvicultural Consequences of Forest Harvesting on Peatlands: Site Damage and Slash Conditions. Information Report 0-X-384. Sault Ste. Marie, Ontario: Canadian Forest Service, Great Lakes Forest Centre. 20 p.

Grubb, P.J. 1977. The maintenance of species richness in plant communities: the importance of the regeneration niche. Biological Reviews 52: 107-145.

Harrington, J.B. and R.E. Donnelly. 1978. Fire probabilities in Ontario's boreal forests. In: Proceedings of the Fifth Joint Conference on Fire and Forest Meteorology, March 14-19, 1978, Atlantic City, New Jersey. Boston, Maine: American Meteorological Society. 1-5.

Harvey, B.D. and Y. Bergeron. 1989. Site patterns of natural regeneration following clear-cutting in northwestern Quebec. Canadian Journal of Forest Research 19: 1458-1469.

Hearnden, K.W., S.V. Millson, and W.C. Wilson. 1992. A Report on the Status of Forest Regeneration: Ontario Independent Forest Audit. Sault Ste. Marie, Ontario: Ontario Ministry of Natural Resources. 117 p.

Heinselman, M.L. 1981. Fire and succession in the conifer forests of northern North America. In: D.C. West, H.H. Shugart and D.H. Botkin (editors). Forest Succession - Concepts and Applications. New York, New York: Springer-Verlag. 374-405.

Hobbie, S.E. 1992. Affects of plant species on nutrient cycling. Trends in Ecology and Evolution 7: 336-339.

Jeglum, J.K. 1979. Effects of Some Seedbed Types and Watering Frequencies on Germination and Growth of Black Spruce: A Greenhouse Study. Information Report O-X-292. Sault Ste. Marie, Ontario: Canadian Forest Service. 37 p.

Jeglum, J.K. 1987. Alternate strip clear cutting in upland black spruce. II. Factors affecting regeneration in first cut strips. Forestry Chronicle 63: 439-445.

Jones, R.K., G. Pierpoint, G.M. Wickware, J.K. Jeglum, R.W. Arnup, and J.M. Bowles. 1983. Field Guide to Forest Ecosystem Classification for the Clay Belt, Site Region 3e. Toronto: Ontario Ministry of Natural Resources. 161 p.

MacLellan, P. 1991. Dynamics of post-logged plant communities in the boreal forest of northern Ontario, Canada [Ph.D thesis]. Toronto: University of Toronto, Department of Botany.

Maycock, P.F. 1963. The phytosociology of the deciduous forests of extreme southern Ontario. Canadian Journal of Botany 41: 379-438.

Noble, I.R. and R. Dirzo. 1997. Forests as human-dominated ecosystems. Science 277: 522-525.

Oliver, C.D. 1981. Forest development in North America following major disturbances. Forest Ecology and Management 3: 153-168.

Oliver, C.D. and B.C. Larson. 1990. Forest Stand Dynamics. New York, New York: McGraw-Hill. 467 p.

Payette, S. 1992. Fire as a controlling process in North American boreal forest. In: H.H. Shugart, R. Leemans, and G.B. Bonan (editors). A Systems Analysis of the Global Boreal Forest. Cambridge, UK: Cambridge University Press. 144-169.

Perera, A.H. and D.J.B. Baldwin. 2000. Spatial patterns in the managed forest landscapes of Ontario. In: A.H. Perera, D.L. Euler, and I.D. Thompson (editors). Ecology of a Managed Terrestrial Landscape: Patterns and Processes of Forest Landscapes in Ontario. Vancouver: University of British Columbia Press. 74-99.

Peterken, G. and C.R. Tubbs. 1965. Woodland regeneration in the New Forest, Hampshire, since 1650. Journal of Ecology 2: 159-170.

Pickett, S.A. and P.S. White (editors). 1985. The Ecology of Natural Disturbance and Patch Dynamics. New York, New York: Academic Press. 208 p.

Rowe, J.S. 1961. Critique of some vegetational concepts as applied to the forests of north western Alberta. Canadian Journal of Botany 39: 1007-1015.

Rowe, J.S. 1972. Forest Regions of Canada. Publication No. 1300. Ottawa, Ontario: Canadian Forest Service, 172 p.

Rowe, J.S. 1983. Concepts of fire effects on plant individuals and species. In: R.W. Wein and D.A. MacLean (editors). The Role of Fire in Northern Circumpolar Ecosystems. SCOPE 18. New York, New York: John Wiley. 65-80.

Runkle, J.R. 1981. Gap regeneration in some old-growth forests of the eastern United States. Ecology 62: 1041-1051.

Shelford, V.E. 1963. The Ecology of North America. Urbana, Illinois: University of Illinois Press. 610 p.

Shelford, V.E. and S. Olson. 1935. Sere, climax, and influent animals with special reference to the transcontinental coniferous forest of North America. Ecology 16: 375-402.

Shugart, H.H. 1984. A Theory of Forest Dynamics: The Ecological Implications of Forest Succession Models. New York, New York: Springer-Verlag. 278 p.

Sprugel, D.G. 1991. Disturbance, equilibrium, and environmental variability: What is 'natural' vegetation in a changing environment? Biological Conservation 58: 1-18.

Statutes of Ontario. 1898. *The Forest Reserves Act.* 61 Vict., c. 10.

Suffling, R. 1995. Can disturbance determine vegetation distribution during climate warming? A boreal test. Journal of Biogeography 22: 501-508.

Suffling, R., B. Smith, and J.D. Molin. 1982. Estimating past forest age distributions and disturbance rates in north-western Ontario: a demographic approach. Journal of Environmental Management 14: 45-56.

Taylor, S., T.J. Carleton, and P. Adams. 1987. Understory vegetation change in a *Picea mariana* chronosequence. *Vegetatio* 73: 63-72.

Thompson, I.D. 2000. Forest vegetation of Ontario: controlling factors at the landscape scale. In: A.H. Perera, D.L. Euler, and I.D. Thompson (editors). Ecology of a Managed Terrestrial Landscape: Patterns and Processes of Forest Landscapes in Ontario. Vancouver: University of British Columbia Press. 30-53.

Van Wagner, C.E. 1978. Age-class distribution and the forest fire cycle. Canadian Journal of Forest Research 8: 220-227.

Viereck, L.A., C.T. Dyrness, K. van Cleve, and M.J. Foote. 1983. Vegetation, soils, and forest productivity in selected forest types in interior Alaska. Canadian Journal of Forest Research 13: 695-702.

Viereck, L.A. 1983. The effects of fire in black spruce ecosystems of Alaska and Northern Canada. In: R.W. Wein and D.A. MacLean (editors). The Role of Fire in Northern Circumpolar Ecosystems. SCOPE 18. New York, New York: John Wiley. 201-220.

Vincent, A.B. 1964. Growth and numbers of speckled alder following logging of black spruce peatlands. Forestry Chronicle 40: 515-518.

Watt, A.S. 1923. On the ecology of British beechwoods with special reference to their regeneration. I. The causes of failure of natural regeneration of the beech (*Fagus sylvatica* L.). Journal of Ecology 11: 1-48.

Whitney, G.G. 1986. Relation of Michigan's presettlement pine forests to substrate and disturbance history. Ecology 67: 1548-1559.

Zasada, J, T.L. Sharik, and M. Nygren. 1992. The reproductive processes in boreal forest trees. In: H.H. Shugart, R. Leemans, and G.B. Bonan (editors). A Systems Analysis of the Global Boreal Forest. Cambridge, UK: Cambridge University Press. 85-125.

11 Forest Vertebrate Responses to Landscape-Level Changes in Ontario

DENNIS R. VOIGT,* JAMES A. BAKER,** ROBERT S. REMPEL,***
and IAN D. THOMPSON****

INTRODUCTION

The forested area of Ontario extends over 800,000 km^2 and contains a wide diversity of forest ecosystems and landscape patterns resulting from variations in climate, physiography, and disturbances. These broad-scale patterns and the processes that shape them determine the nature of wildlife habitat. Interactions between forest habitat and wildlife are evident at various spatial and temporal hierarchical scales (Allen and Starr 1982; O'Neill et al. 1986). Understanding the hierarchy of scale is important to understanding wildlife responses. For example, at the broad, provincial scale, climate imposes upper-level constraints on the suitability of habitat for different wildlife species by determining growing season or snow depth. Similarly, geomorphological factors influence forest productivity across the province which, in turn, affects the number of species and their distribution (Thompson 2000b, this volume). Superimposed on these broad-scale patterns is a complex of finer-scale patterns of soils, topography, and weather that modify forest habitat and its suitability for various wildlife species.

The broad-scale processes that limit wildlife populations are part of a system that is always subject to change from disturbances. Evolutionary theory suggests that wildlife populations are adapted to the historical disturbance regimes occurring in the biogeographical region in which they live. Natural and anthropogenic disturbances are the fundamental determinants of the nature of wildlife habitat in forested landscapes. The disturbance regimes in Ontario's boreal and Great Lakes-St. Lawrence forests are inherently different in frequency and intensity. Disturbances such as fire and clear-cut logging, which are common in the boreal forests of northern Ontario, can affect large patches and modify habitat at the landscape level. In the tolerant hardwood and pine forests of central Ontario, however, fire, wind, and selective logging occur in smaller patches and may be less intense in nature. These disturbances tend to occur at the stand level, but they modify the structure of habitat patches and also affect wildlife over large areas.

The purpose of this chapter is to describe responses of wildlife to changes in Ontario's forest landscapes. We restrict our examples to terrestrial vertebrate wildlife, although we recognize that the term "wildlife" encompasses invertebrates and other biota. Insects are likewise excluded; Fleming et al. (2000, this volume) discusses forest insect pests as agents of landscape change. We begin by reviewing how habitat is changed by the major forest disturbances in Ontario and how wildlife responds to habitat. This introduction is

* 1457 Heights Road, Lindsay, Ontario K9V 4R3. Formerly Ontario Ministry of Natural Resources, 300 Water Street, Peterborough, Ontario K9J 8M5

** Ontario Ministry of Natural Resources, 300 Water Street, Peterborough, Ontario K9J 8M5

*** Centre for Northern Forest Ecosystem Research, Ontario Ministry of Natural Resources, Lakehead University, 955 Oliver Road, Thunder Bay, Ontario P7B 5E1

**** Canadian Forest Service, Great Lakes Forest Research Centre, 1219 Queen Street East, Sault Ste. Marie, Ontario P6A 5M7 followed by a short review of how forest landscapes and the disturbances that change them are organized in a spatial and temporal hierarchy, and by a review of landscape-level theories on metapopulations and biogeography. We present hypotheses for broad-scale changes which have occurred in the regional distribution of selected Ontario wildlife species since 1950, taking a top-down approach and considering large-scale constraints. Two hypotheses for changes that have been reported in the range of six wildlife species in Ontario are tested: the hypotheses of habitat change and climate change. Although few studies have been conducted on the response of wildlife population levels to forested landscapes, province-wide density data for moose (*Alces alces*) have allowed us to analyze how broad-scale constraints affect the density of these populations, as well as their response to change in landscape pattern. Studies on songbird response to landscape-level habitat change are also summarized. These studies include empirical evidence for the effects of habitat loss and fragmentation at large scales. Our final discussion focuses on the management approaches and research required at the landscape level to provide for the sustainability of wildlife.

CAUSES OF CHANGE TO FOREST LANDSCAPES AND WILDLIFE HABITAT

Disturbances are now widely recognized to be processes that change forest ecosystems, and thus wildlife habitat, at a variety of scales (Allen and Starr 1982; Pickett and White 1985). A relatively simple definition of disturbance is any change to a particular structure in a system caused by a factor external to that system. (More comprehensive definitions can be found in Pickett and White [1985], Pickett et al. [1989], and White and Harrod [1997].) The general pattern after a disturbance is for ecosystem structure and composition to undergo reorganization and development through a variable succession of seral stages (Bormann and Likens 1979; Holling 1992).

Disturbances that change the structure, composition, and function of forest ecosystems (Pickett and White 1985) change the suitability of habitat for wildlife over time (see DeGraaf and Miller [1996] for a thorough case history from New England). Some of the disturbances in Ontario that result from weather or long-term climate patterns, for example, directly affect populations by effectively lowering the carrying capacity of the habitat. Disturbances such as fire can also kill wildlife directly, but most types of disturbance exert their greatest influence by modifying habitat.

Disturbance types vary widely in their frequency, intensity, and extent. Many natural disturbances, such as fire, windstorms, and ice storms, are caused primarily by climate and involve broad-scale landscape effects (Walker and Walker 1991). The effects of these disturbances on wildlife and habitat depends greatly on their scale, frequency, and intensity. Glaciation that occurred in Ontario 10,000 to 12,000 years ago is a dramatic example of a disturbance with widespread effects. Some disturbances, such as erosion and the freeze-thaw cycle, occur at fine scales but are common and widespread in extent. Others may be rare events, but, because of their intensity, they have long-term effects at the landscape level. For example, a windstorm on July 15, 1996, affected tens of thousands of trees in landscapes across the Great Lakes region from Wisconsin to Ontario. A massive ice-storm in January 1998 caused major changes to the forest canopy structure over a large area of eastern Ontario and southeastern Quebec.

Disturbances such as windstorms and fire can be catastrophic and stand-replacing events at large scales; moreover, even fine-scale disturbances can change habitat dramatically. The scale of herbivory is essentially the bite size of the herbivore in question, from millimetres for insects to centimetres for moose and deer (*Odocoileus virginianus*). However, the intensity of herbivory and its extent, or area of impact, can be so great that landscape-level effects on forest structure and composition result. The annual cycle of foraging by herbivores on particular forest plants occurs within variable patterns of longer-term fluctuation in vegetation cover. For example, the species composition and stem density of conifer forests on Isle Royale in northern Lake Superior have been modified by browsing (Krefting 1974; Risenhoover and Maass 1987;), and both hemlock (*Tsuga canadensis*) and hardwood forests in Pennsylvania show poor regeneration because of deer browsing (Anderson and Loucks 1979). In Ontario, foraging by deer in winter concentration areas can seriously decrease forest regeneration (Voigt and Broadfoot 1995). In some provincial parks of Ontario where ungulates are protected, such as Rondeau Park, Pinery Park, and Algonquin

Park, selective browsing by deer or moose has changed the composition and structure of herbaceous and woody plants over large areas (D. Voigt, unpub.). Such changes may affect the functionality of the habitat for a wide variety of species.

Disturbances occur at different rates in different forest systems. For example, fire return intervals are 60 to 150 years in the boreal conifers and 50 to 150 years in white pine (*Pinus strobus*) systems, but from 300 to more than 2000 years in the Great Lakes-St. Lawrence hardwoods (see Whitney 1986; Frelich and Lorimer 1991; Ward and Tithecott 1993; Baker 1995; Thompson 2000a, this volume). Disturbances such as wind, fire, and clear-cutting may combine to reduce the stand replacement interval. The cumulative effect of these disturbances is to cause different distributions of age classes and patch-size classes in different forest systems. Undisturbed patches are rare in landscapes in which catastrophic disturbances have a short return interval, and such systems contain fewer examples of old-growth forests. Typically, fewer wildlife species dependent on old-growth forest structures and few species requiring large, undisturbed forest patches are found in landscapes that are frequently disturbed.

The anthropogenic disturbances that occur in Ontario's forests may differ in both direct and indirect effects from natural disturbances (for a general review of natural forest disturbances effects, see Lertzman and Fall [1998]). For example, both fire and clear-cutting can remove a large volume of standing live timber, but the subsequent habitat for wildlife may be dramatically different, particularly during the first 30 years after the event. Subsequent successional pathways may also differ, so that the forest systems may change to different states. In central Ontario, areas where white pine was logged during the past century now have less super-canopy pine in the landscape, and many stands have been converted to hardwoods (Aird 1985). In Rondeau Provincial Park, long-term deer grazing has altered plant species composition in some areas (Pearl and Voigt 1995).

Natural and anthropogenic disturbances which seem similar may, in fact, differ in both size and rate of occurrence. For example, selection cuts in the Great Lakes-St. Lawrence forest hardwoods are designed to emulate natural forest-gap dynamics, so as to retain an uneven-aged forest. However, removal of 30 percent of individual trees may occur in a few months over a 20-year cycle; whereas natural gaps develop continuously and remove less than 1 percent of the trees annually (Dahir and Lorimer 1996). Similarly, in the commercially-harvested boreal forests of Ontario, forest fires burn at fluctuating levels of intensity and produce irregular patterns over a period of days; this process is complicated by the provincial fire suppression strategy. In contrast, the clear-cut logging of spruce (*Picea* spp.) and other species occurs over a period of months or years in a particular area, often according to a prescription used in many other areas as well. The frequency or return interval of fires is often different than the return interval of forest harvesting operations (Boychuk and Perera 1997). Boychuk et al. (1995) have shown how variable fire return patterns can be in forest landscapes (see also Li [2000, this volume] and Perera and Baldwin [2000, this volume]). The consequence of this variation for wildlife are mosaics of different forest patch sizes and ages created by disturbance regimes. White and Harrod (1997) provide a further discussion of disturbances in a landscape context and their effects on wildlife.

THEORIES OF WILDLIFE RESPONSE TO FOREST LANDSCAPES AND HABITAT

Wildlife species respond to habitat at different scales. The finest scale at which a species responds, or "makes decisions," is called "grain," and is typically in the order of centimetres to metres. Examples include the decision to eat a particular twig or insect, or to forage in a particular patch. The broadest scale at which a species responds is called the "extent" of that species (Kotliar and Wiens 1990). Extent may correspond to the home range of non-migratory wildlife. In Ontario, home range can be a few square metres or as much as hundreds of square kilometres, and there is a strong relationship, based on energy requirements, between body mass and the size of home range (Harestad and Bunnell 1979; Lindstedt et al. 1986). Neotropical migratory birds in Ontario have an extent measuring thousands of kilometres. Other forest bird species, such as certain winter finches, among them crossbills (*Loxia* spp.), pine siskins (*Carduelis pinus*), and pine grosbeaks (*Pinicola enucleator*), are nomadic and may cover huge ranges over the course of years. Other wildlife species in Ontario also cover a large extent through migration; these include certain bats, deer, caribou (*Rangifer tarandus*), and moose. The home range of members of a species determines their landscape perspective, which may differ greatly from that of humans.

Individuals seek habitat patches in their home range to meet seasonal needs for food and cover, and may need the resources of different types of patches at different points in their life cycle. The patches favoured differ with the species in size, shape, interspersion, stage of development, species composition, and special components such as snags or fallen wood. Deficiencies in a habitat may cause a source-and-sink effect on populations at the landscape level. The success of an individual or population in a habitat determines whether that population is a source of emigrants or a sink requiring immigrants. Since adjacent patches are more accessible than distant patches, the characteristics of contiguous patches are likely to affect a species more strongly. For some species, whether or not certain resource patches are used depends on both the pattern of patches in the landscape and the boundaries between those patches. Some boundaries between patches repel and some are easily crossed.

The success of grain and extent decisions made by individuals in a particular habitat determines their reproductive fitness. The collective performance of an assemblage of individuals (a population), in relation to the carrying capacity of the range, determines rates of birth, death, and recruitment. Carrying capacity (K) is the maximum number of a species that can be sustained within a given area. The rate of population growth for those species that are regulated by factors dependent on population density is determined by the density (N) relative to the carrying capacity; that is, relative density or percentage of K (N/K·100). There is a strong link between population dynamics and habitat, as the quality and quantity of habitat in an area determines carrying capacity.

The estimation of K for ungulates often involves biomass estimation. For other species, it is often estimated from population dynamics. In either case, few studies have considered the fact that food supplies often fluctuate or are even periodic from such causes as variation in seed crops, insect epidemics, or drought. Even fewer studies have employed the concept of relative density, which incorporates the notion that both N and K are constantly subject to change. With respect to the landscape perspective, insufficient attention has been paid to the patchy and dynamic nature of forest habitats. Therefore, it is no surprise to see many authors conclude that models of population dynamics based on "maximum sustained yield," carrying capacity, and the logistic curve, are invalid. In fact, the error lies not in the models, but in a failure to recognize the implicit assumption that K is variable and that the population size relative to K (percentage of K) is a critical consideration. The impact of forest landscapes on wildlife populations is constantly modified, both by interactions among individuals and locally changing habitat conditions, and by interactions between population dynamics and changing carrying capacity.

Hierarchy Theory of Wildlife Response

The time and space relationships of vegetation illustrated in Figure 11.1a show the hierarchical structure of vegetation in forested landscapes. Hierarchy theory contains the following propositions: that, when phenomena are organized in a hierarchy, upper levels in a hierarchy of phenomena constrain lower levels; that upper-level processes are generally slower and lower-level processes faster; and that new properties often emerge at various levels (that is, the whole is greater than the sum of the parts). These propositions have important implications for wildlife. The premise that upper levels constrain lower levels explains why bottom-up approaches, which aggregate fine-scale structures and disturbances, may fail to account for patterns driven by upper levels. The analysis of changes in the distribution of wildlife that we will present in this chapter suggests that there are broad-scale effects of climate and habitat change which imply the need for a top-down approach to predicting wildlife responses. Furthermore, the space and time relationships in atmospheric and disturbance processes (Figure 11.1b) and in the use of space and time by wildlife (Figure 11.1c) allow us to determine interactions between the three sets of processes. Together, these hierarchies of multi-scale components can provide a starting point for the development of models to predict how wildlife respond to forest change, natural disturbances, forestry practices, and atmospheric processes (Holling 1992; Peterson 1996).

Notwithstanding that upper levels constrain lower levels, it is notable in Figure 11.1 how extensively certain processes such as forest harvest and fire interact with other processes. Forest harvest and fire are mesoscale processes, so this observation suggests that such mesoscale effects are critical for planning

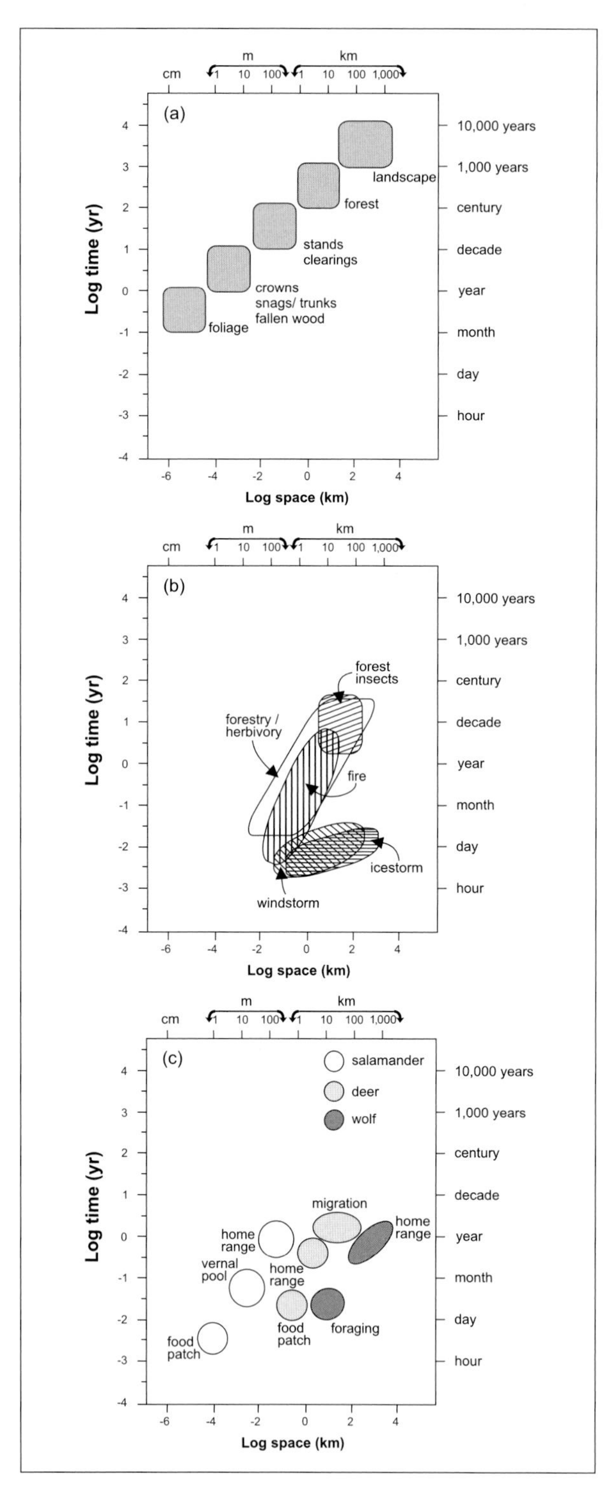

Figure 11.1 The spatiotemporal hierarchy inherent in (a) forest landscapes; (b) natural and anthropogenic disturbances; and (c) habitat use by selected wildlife species. (Adapted from Peterson 1996)

at the landscape level. Mesoscale disturbances such as forest harvest affect the origin and successional stages of stands, and whether the stands are primarily even-aged or not. Macroscale disturbances affect broad landscape patterns, but also patch size. Fine-scale disturbances can change the type and frequency of habitat structures, such as snags and woody debris; some management prescriptions, such as selection logging, make provision for the explicit management of these structural features. In conclusion, the hierarchical organization of forests implies that interactions occur at a variety of scales and among a variety of processes at each level. This is further evidence of the importance of the top-down perspective. For a more detailed summary of hierarchy theory from a wildlife perspective, see King (1997).

Landscape Ecology and Biogeography Theories of Wildlife Response

Biogeography is the study of how, why, and when species came to be arranged as they are (Simberloff 1997). Early theories emphasized colonization and extinction as mechanisms of species distribution. Both areal extent and isolation influence the probability of local extinction, primarily due to stochastic variation at low levels. Landscape ecology and biogeography theories are closely related. Landscape ecology considers factors that change ratios of colonization to extinction, factors such as the size and shape of patches, patch connectivity, and isolation, as well as the initial rarity of the species themselves. Biogeography theory predicts that large patches or continuous habitat should have greater species richness.

Metapopulations are groups of interacting populations, in which the local populations are connected by the dispersion of individuals. Metapopulation theory has developed to explain spatial variation in population dynamics. Reviews by Hanski and Gilpin (1991) and Krohne (1997) show strong links among the theories of metapopulation and island biogeography, and the principles of landscape ecology and conservation biology. Levins (1970) first developed the idea of a "population of populations" and described relationships between extinction and colonization in a set of habitat patches. Source populations are those in which the growth rate is balanced by emigration, and sink populations are those that become extinct without immigration. Source and sink populations undoubtedly occur in habitat patches and landscape types scattered throughout the continuous forest of Ontario. The difference between a source and sink population is primarily due to habitat effects on population dynamics (Pulliam 1988). The short-term size of a population does not reveal whether that population is a source or a sink.

Levins's original model of metapopulations assumed homogeneity among habitat fragments with respect to colonization rates and extinction rates. Since then, a variety of metapopulation models have been developed that accommodate a diversity of patch types, in an effort to simulate the heterogeneity that natural landscapes display, including those in Ontario. The most current ideas of metapopulation involve both gradients and heterogeneity in habitat. These, in turn, underpin many of the current ideas put forward to explain biodiversity patterns and species distributions. Although no populations of forest wildlife in Ontario have been explicitly studied as a metapopulation, species such as white-tailed deer, caribou, marten (*Martes americana*), and beaver (*Castor canadensis*) have characteristics befitting metapopulations at the forest landscape scale. It is also likely that many of the species that thrive in a particular forest stage will occur as a metapopulation if the patches of that stage are scattered across the landscape. As our review of songbirds later in the chapter will show, there are few empirical data describing the effects of patch size or the distance between patches in continuously forested areas.

It is not clear whether landscape patch patterns are more important than basic productivity effects (see Thompson 2000b, this volume) in accounting for species richness, but it does appear that forest composition and structure exert effects on wildlife in forests. Krohne (1997) illustrated five basic landscape patterns for wildlife populations that are distinguished by variation in dispersal, spatial variation in resources (heterogeneity), and hierarchical population dynamics. The recognition that there is such spatial variation in habitat quality and population variables has evolved from a new focus on broad-scale landscape approaches. Models of metapopulations that consider heterogeneity among patches hold promise for predicting how wildlife species respond to habitat changes at intermediate scales. However, these models must be evaluated in relation to higher-level constraints which occur at provincial and regional levels. In the next section, we will examine the effects of some key macroscale factors on regional changes in wildlife distribution.

REGIONAL CHANGES IN THE DISTRIBUTION OF SELECTED VERTEBRATES FROM 1950 TO 1995

The numbers of species occupying any given area of the province may have changed during recent decades, as many vertebrate species have undergone change in their ranges (e.g., De Vos 1964; Outram 1967). In some cases, the changes in range are small and clearly associated with local conditions; for example, the expanded ranges of mourning doves (*Zenaida macroura*) and cardinals (*Cardinalis cardinalis*) may be associated with an increase in the number of backyard bird feeders that enables increased local survival over winter (Cadman et al. 1987). However, many species in the province are known to have contracted or expanded their ranges over considerable areas. Such large changes in the distribution of a particular species would suggest that key factors operating at a regional scale have changed temporally and/or spatially. Thompson (2000b, this volume) shows range changes over the past 50 years for a number of species. In this chapter, we examine factors acting at intermediate scales that may be affecting the regional distributions of several of the species which Thompson considers.

Possible Hypotheses for Changes in Species Range

The factors that cause a change in species range sufficiently distinct to be noted by casual observation act at large spatial scales and over periods of at least a decade or more (e.g., Allen and Hoekstra 1992). There are five hypotheses that might explain the range changes observed in various forested areas of Ontario. The first hypothesis is that climate change, specifically a general warming trend in Ontario since 1960, and particularly since 1980 (see Baldwin et al. 2000, this volume), has resulted in conditions favourable to the expansion of the populations and ranges, particularly for southern species. The second hypothesis is that anthropogenic causes, including logging, fire suppression, and some limited northern agriculture, have sufficiently changed habitats over large areas of the province from conditions which existed before European settlement so that species have responded with range shifts during the period considered. The third hypothesis involves human-caused mortality from hunting and trapping. High mortality rates of certain species during the 1950s and 1960s resulted in the reduction in species distributions; whereas reduced mortality in the past 15 years, under more strict regulations on hunting and trapping, has produced population increases and range expansions. The fourth hypothesis consists of "normal" fluctuations in populations, which may be expected to occur on a regular basis as a result of various extrinsic and intrinsic factors. Such fluctuations can occur simultaneously across the entire province (Finerty 1980; Smith 1983). The last hypothesis has multifactorial causes, and includes the possibility that two or more of the foregoing factors combined to cause species to alter their ranges in Ontario.

Discussion of Hypotheses

Climate Change

Weather may influence animal populations locally from year to year, but longer-term climatic influences may either allow or limit regional expansion in populations. Mortality of ungulates has been attributed to severe winter conditions, which reduce available food sources and cost the animals excessive energy expenditures relative to intake (Passmore and Hepburn 1955; King 1976; Potvin et al. 1981; Wilton 1984; Voigt et al. 1992). The result is increased rates of predation on animals rendered more vulnerable from insufficient food or restricted mobility (Mech 1970; Peterson 1977; Ballard and Van Ballenberghe 1997), particularly at the northern edges of their distributions (Nelson and Mech 1986; Potvin et al. 1988). Population declines and range reductions in moose and white-tailed deer have been attributed to severe winter conditions for three or more years in succession (Wilton 1984; Mech et al. 1987). Messier (1991) suggested that, at low deer numbers, snow is not a limiting factor in deer populations, particularly in the presence of a high density of wolves (*Canis lupus*). All boreal species are adapted in various ways to deep snow and cold temperatures, so it is unlikely that harsh winters alone have a limiting effect on their populations. However, effects might be expected on animals that are not well adapted to deep snow, such as bobcat (*Felix rufus*) (Peterson 1957; McCord 1974), porcupine (*Erethizon dorsatum*) (Banfield 1974), and white-tailed deer.

The severity of winters in Ontario has varied over time (e.g., Thomas 1957) (Figures 11.2 to 11.4); consequently, the influence of deep snow and cold temperatures on wildlife has also changed. Temperatures in the west were similar from the 1950s to the 1970s, but rose in the 1980s, and that warming trend has continued into the 1990s. In eastern Ontario, the 1960s and 1970s were significantly colder than the 1950s and 1980s, and temperatures in the 1980s and 1990s were the highest in 50 years (Figure 11.2). Maximum snow depths in the west were greatest from 1950 to 1980, then declined from the 1980s to the present (Figure 11.3). In the east, snow depths varied from north to south, but generally depths were lowest in the 1950s and 1960s and highest in the 1970s, and have decreased over the past 15 years (Figure 11.3b). The 1970s had the greatest number of winters with snow deeper than the 45-year mean maximum; at least seven out of the ten winters had higher than average snowfall throughout eastern Ontario (Figure 11.3b). The largest increase in decadal maximum snow on the ground was in the Bracebridge-Huntsville-Algonquin Park area, where snow depth had averaged 58 centimetres throughout the period from 1950 to 1969, but suddenly increased to more than 80 centimetres during each of the next decades, with nine of those winters above the 45-year mean.

Habitat Change

Much has already been discussed in earlier chapters in this volume about the effects of anthropogenic changes to Ontario forests and landscapes. The cumulative impacts of logging, fire suppression, agriculture, and settlement have combined to alter the forest landscapes that existed prior to European settlement (Baker 1995; Gluck and Rempel 1996). The rate of change in habitat and the cumulative impact of that change on species is difficult to determine, because of the lack of historical data and the slow rate at which change occurs. Results of comparing distribution changes with habitat change must, therefore, be interpreted cautiously.

There have also been qualitative changes in forest disturbances. Until well into the 1960s, wood was hauled from forests using horses and floated to mills along rivers, or taken by rail (river drives had ended by 1980). Many stands that were too far from water or rail were bypassed. Following the horse logging era,

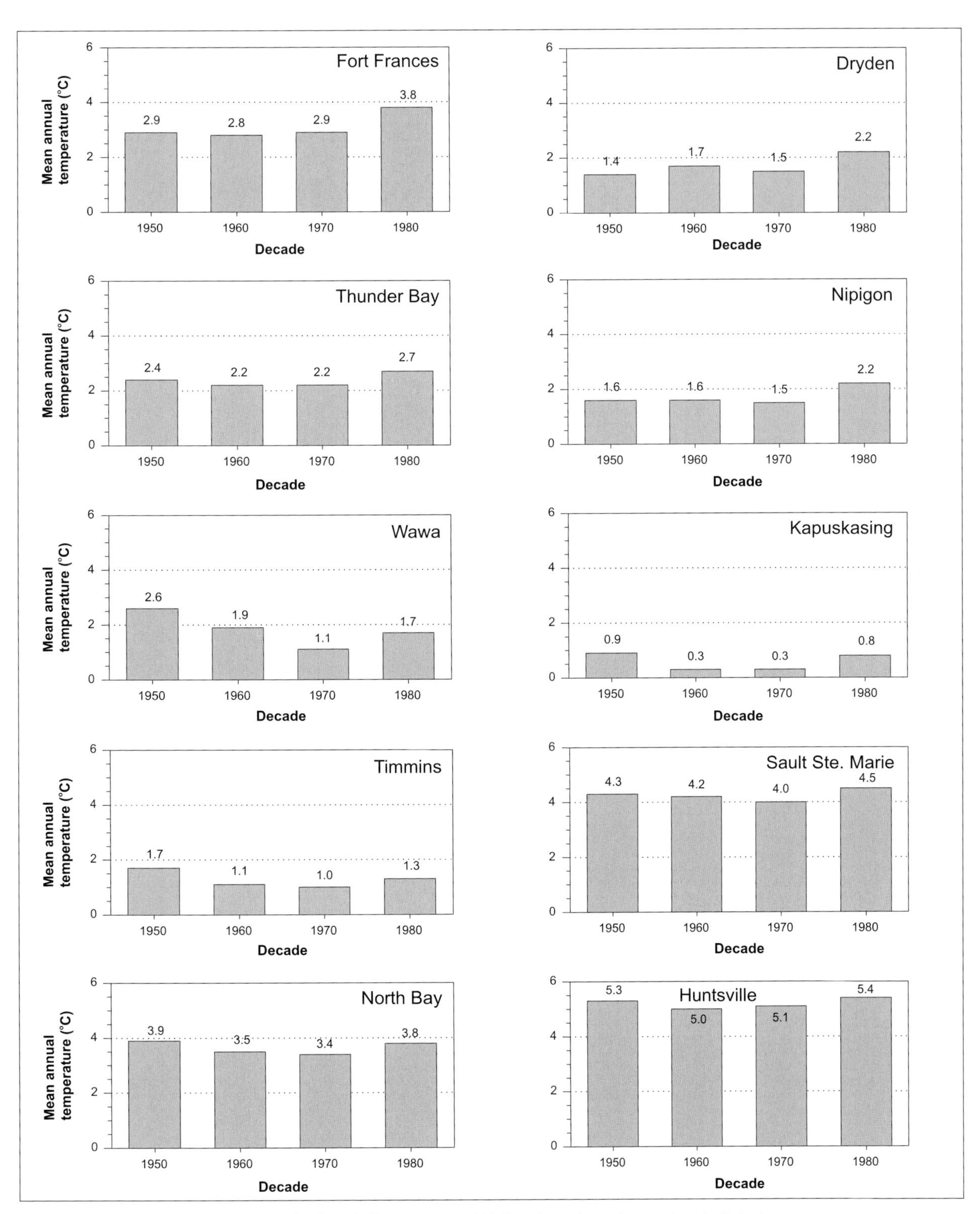

Figure 11.2 Mean annual temperature by decade from 1950 to 1990 for selected weather stations in Ontario. (Data from Environment Canada)

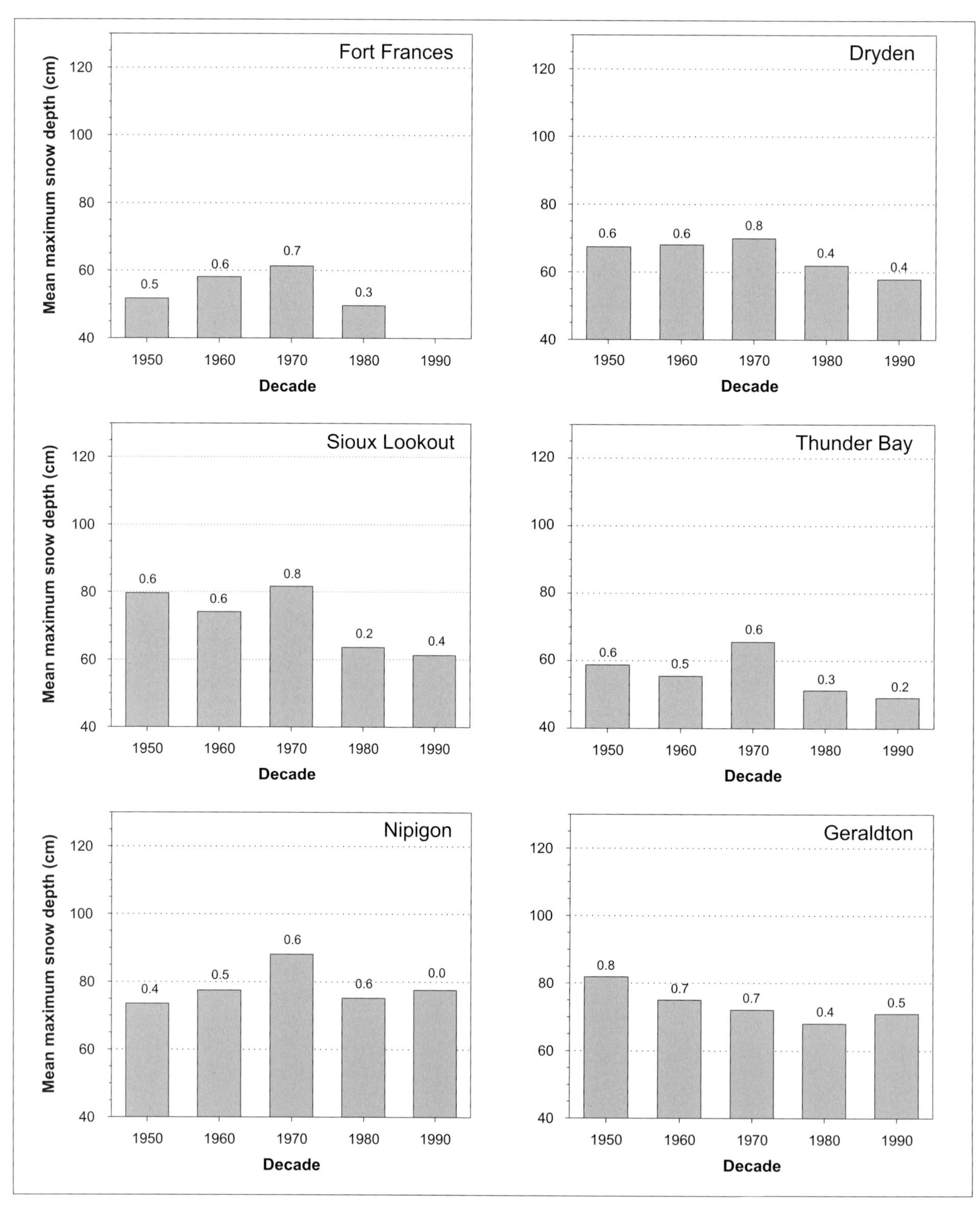

Figure 11.3a Mean annual maximum snow depth by decade from 1950 to 1996 for selected weather stations in western Ontario. The numbers above each bar represent the proportion of the decade when annual snow depth exceeded the 46-year mean value.
(Data from Ontario Ministry of Natural Resources)

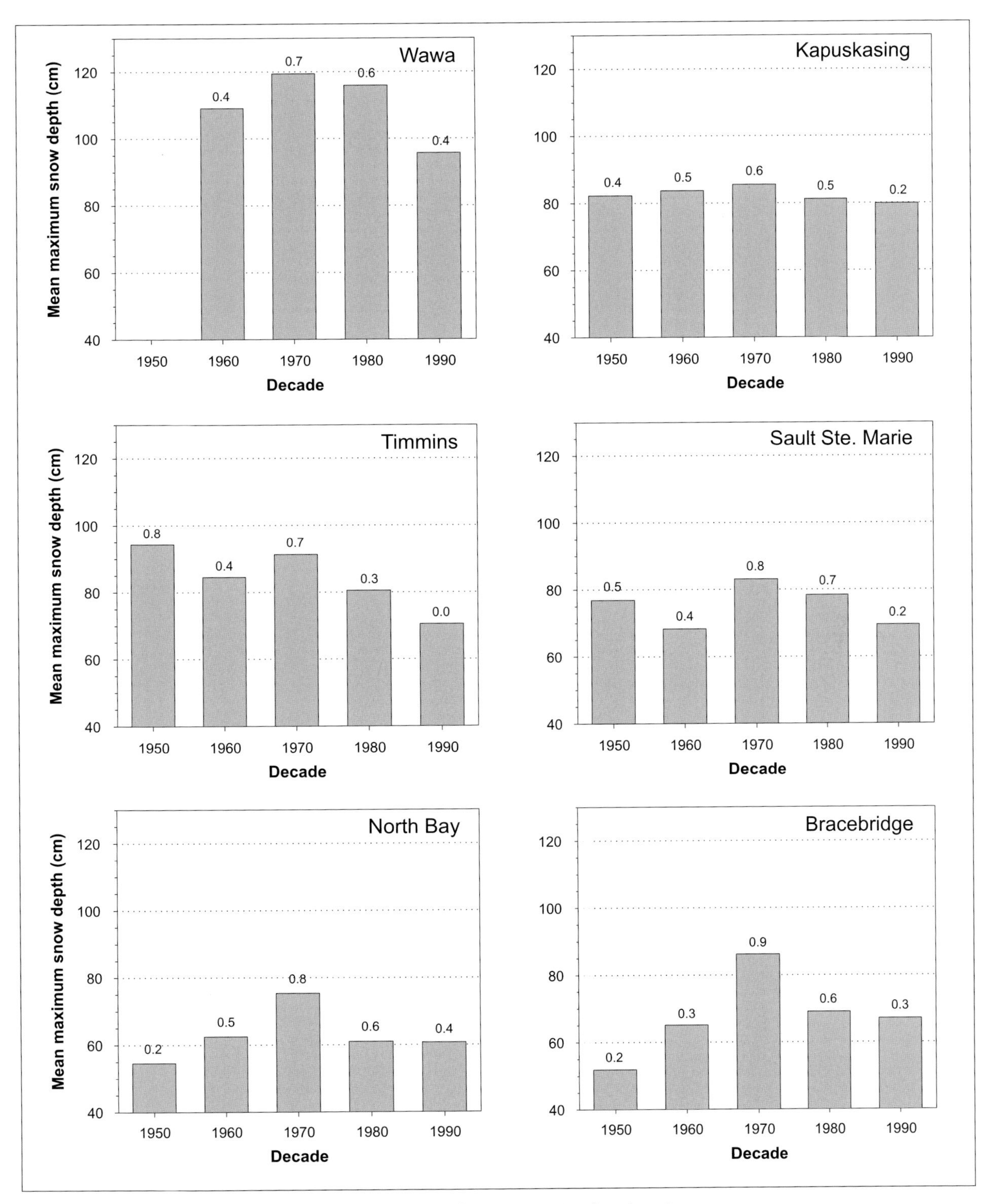

Figure 11.3b Mean annual maximum snow depth by decade from 1950 to 1996 for selected weather stations in eastern Ontario. The numbers above each bar represent the proportion of the decade when annual snow depth exceeded the 46-year mean value.
(Data from Ontario Ministry of Natural Resources)

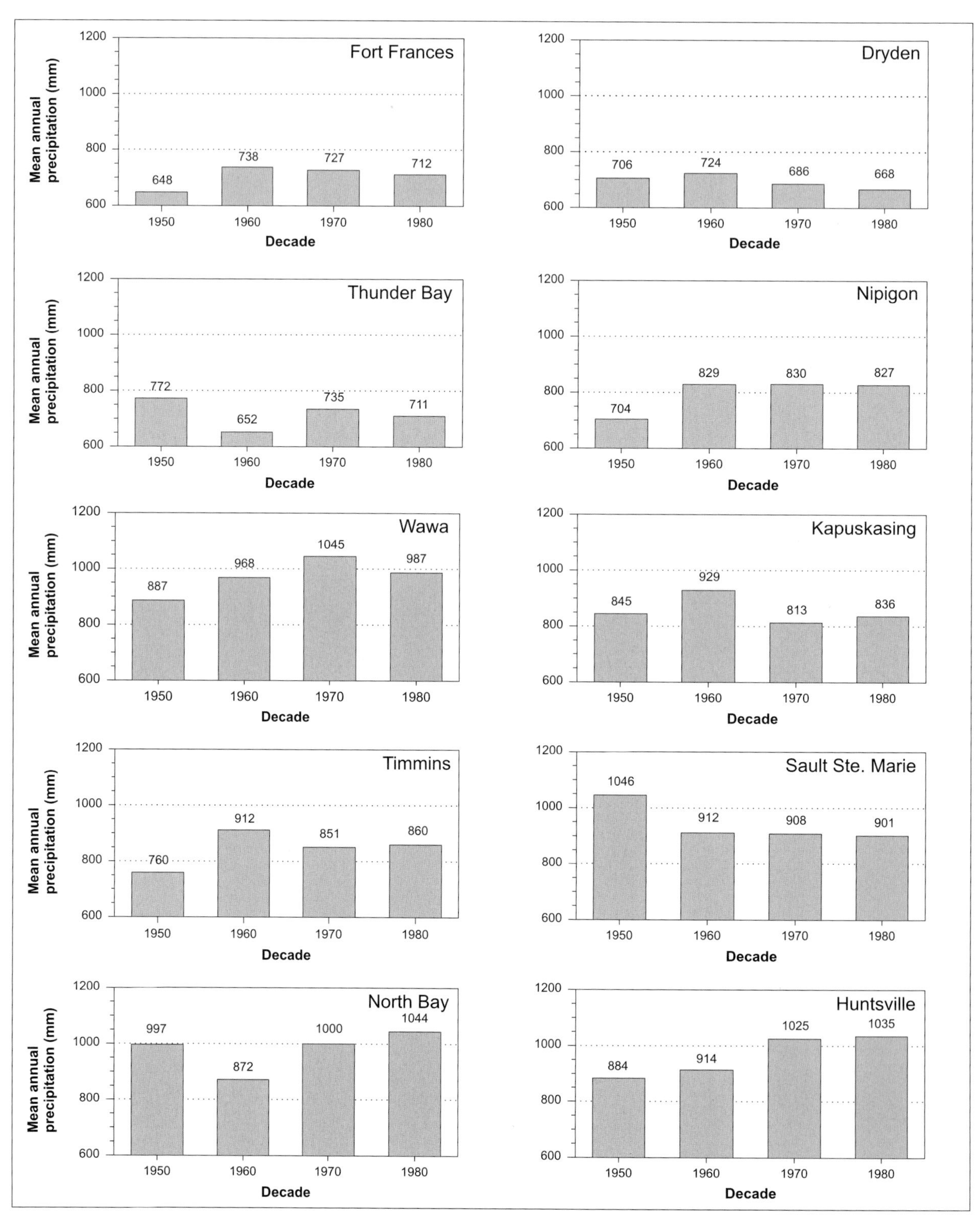

Figure 11.4 Mean annual precipitation by decade from 1950 to 1990 for selected weather stations in Ontario.
(Data from Environment Canada)

the onset of mechanical logging brought an increase in both the size of cutovers and the availability of wood per unit area, because of easier access and more complete wood removal. Perera and Baldwin (2000, this volume) report that the total area logged since mechanical harvesting became widely employed is about double the total area logged in all previous years.

The use of machinery and the development of new forest products have had a qualitative effect on wildlife habitat, relative to the effect of early logging, in part through the removal of trees that would have been bypassed only 50 years ago. To analyze the causes of range changes in Ontario vertebrate species, we have summed the effects over time by pooling the amount of logging disturbance by decade. It should be borne in mind, however, that the period from 1940 to 1960 is not strictly comparable to the decades after that period.

Human-Caused Mortality

The role of hunting and trapping as an additive source of mortality has long been debated; one often wonders why. Unabated killing has been responsible for the decline and even extinction of species through history; examples include the Eskimo curlew (*Numenius borealis*), passenger pigeon (*Ectopistes migratorius*), and Labrador duck (*Camptorhynchus labradorius*). Douglas and Strickland (1987) reported that fisher (*Martes pennanti*) are vulnerable to over-trapping, and Berg and Keuhn (1994) documented trapping as a major factor in the decline of fisher in Wisconsin and Michigan in the early 1900s. Thompson and Colgan (1987) showed that trapping mortality was an additive effect that limited the occurrence of marten occurrence at some sites. In the earliest years of logging, moose, caribou, and deer were hunted to feed workers throughout Ontario, and until the early 1980s there were no restrictions on the numbers of licences issued to hunt moose or deer. In 1981, after carefully analyzing other factors, Ontario biologists reached the province-wide conclusion that over-hunting was limiting moose populations (Thompson and Euler 1987, based on work by A. Bisset, K. Morrison, T. Timmerman, and I. Thompson, Ontario Ministry of Natural Resources [OMNR]). The Ontario Deer Technical Committee reached a similar conclusion for deer populations, although the committee also recognized that hunting had combined with severe winters to reduce herds. Hunting has been suggested as a major factor in the decline of caribou over much of eastern North America (Bergerud 1974).

Certain trapping methods that are now illegal because of their effectiveness or non-selectivity were regularly used in the past. For example, the use of poison baits resulted in multiple catches at a single site. During some periods, the use of poison combined with high unemployment and exceptional fur prices ($345 for a fisher pelt, for example, in the early 1920s [De Vos 1952; Berg and Keuhn 1994, p. 265]) to produce high catches and possibly local extinctions (De Vos 1952). To put that pelt value into perspective, a logger living in a bush camp in 1920 was paid about $50 a month, and an office clerk for the Department of Lands and Forests in Toronto made $125 a month. During that period when fisher fur was so highly valuable, trappers would pursue individual fisher for days (De Vos 1952; Berg and Keuhn 1994).

The number of people living in bush camps and cutting the forests has also changed dramatically over time. During the earliest years of logging, manpower was extremely important in cutting and extracting wood. In 1946, for example, about 40,000 men were employed and living in logging camps in Ontario. Mechanization has reduced the current logging workforce to 5500 (Canadian Forest Service 1995), and few workers remain in bush camps. With thousands of poorly paid workers in the forest, it is reasonable to assume that many trapped indiscriminately and used poison to reduce their effort, prior to the establishment of regulated traplines in 1950. Although data are sometimes poor and are often confounded by other habitat changes that occurred simultaneously, we are persuaded that there is sufficient evidence to implicate hunting and trapping as a major limiting factor in certain wildlife populations, especially historically.

Natural Population Fluctuations

It is well known that periodic fluctuations occur in the animals of Ontario forests (de Vos 1952; Obbard et al. 1987; Thompson and Colgan 1987). Fisher populations, for example, have fluctuated considerably over years and decades (De Vos 1952; Douglas and Strickland 1987; Obbard et al. 1987). The relationship between predators and their prey in forested systems have also been well documented. Recently, Boutin (1995) and Krebs et al. (1995) have shown that, during certain phases of the cycles of snowshoe hare (*Lepus americanus*) and other small mammals, there was interaction between predation and levels of the food supply of the prey species. At a certain population level, predation was responsible for a decline in prey populations that were forced to seek refuge from the predators in suboptimal habitats. It is possible that the range fluctuations that have been observed are the result of such population changes in predator and prey populations. Predator populations normally lag prey populations by one to three years (Krebs 1996). In Ontario, snowshoe hare populations peaked in 1960, 1970, and 1980 (De Vos 1962; Thompson and Colgan 1987; Thompson, unpub.). A peak expected about 1990 did not materialize, or at least was much lower than normal, but the fact that snowshoe hare again became abundant in 1997-98 suggests that, at least in recent decades, the normal cycles may have been altered. In Wisconsin and southern Sweden, as habitat became fragmented, historical cycles in hare and small mammals disappeared, because of altered predation regimes (Keith et al. 1993; Krebs 1996). The two mammalian predators that we will discuss, bobcat and fisher, both prey on snowshoe hare, among other species, and might be expected to respond to hare cycles, at least prior to about 1985.

Multiple Causes

While one factor may be ultimately responsible for changes in animal distributions, the factors discussed above may also produce interactive effects. Many examples have been documented of changes in species numbers caused by human activity, and of the effects of these changes throughout the ecosystem. In British Columbia, for example, Seip (1992) showed that after logging there was an unexpectedly large reduction in caribou numbers in the adjacent, unlogged forests. Logging had created habitat for moose, and wolves (*Canis lupus*) responded numerically to the increased moose numbers; the consequence was higher rates of predation on caribou. In Ontario, there are no long-term data to implicate either predation or early hunting in a decline of caribou. However, it is clear from moose aerial inventories and early population information (Peterson 1955, 1957) that moose numbers increased with logging in Ontario forests from 1900 to about 1960, and then declined with increased hunting over the following 25 years (Thompson and Euler 1987). It is a reasonable hypothesis that wolf populations increased concurrently with moose and that, at least in areas where caribou and moose populations occurred together, wolves limited the range of the caribou (Bergerud 1974). (We will discuss other hypotheses for the decline in caribou following logging.)

Data Available for the Assessment of Range Changes

The data presented by Thompson (2000b, this volume) on range changes in several species in Ontario from 1950 to the present were gathered from several past publications that documented species distributions (Peterson and Crichton 1949; de Vos and Peterson 1951; Peterson 1957; Snyder 1957; de Vos et al. 1959; de Vos 1962, 1964; Outram 1967). Current distributions were obtained from Cadman et al. (1987) and Dobbyn (1994), and through discussion with present and former OMNR staff. Sufficient range change had occurred for white-tailed deer, woodland caribou, fisher, porcupine, and raccoon (*Procyon lotor*) that causes could be analyzed as a binary problem. Other species are treated hypothetically, on the basis of whatever information is known.

Two estimated distributions for the selected species were compiled, based on UTM grid blocks measuring 100 km by 100 km, or 10,000 km^2 in area, for 1960 (or in some cases 1950) and for the present. The area between the two resulting distribution limits represented the changes in range of the species over the 35 to 45 years. Weather and disturbance data were then compiled for each UTM grid cell in the area of the range change. The weather data were obtained from the Atmospheric Environment Service of Environment Canada (AES) and supplemented, particularly in the case of snowfall, with data from the district offices of OMNR. If more than four values in a year were unavailable from either AES or OMNR, the year was omitted. If a value for snow on the ground was missing between December and March, that year was omitted from the calculation of decadal mean. The proportion of each decade when annual snow depth was greater than the 47-year mean (1950 to 1997) was calculated. Weather variables included decadal mean values for the following: annual temperature, July temperature, January temperature, annual precipitation, and maximum snow depth based on monthly measures. The proportion of each decade with an annual maximum snow depth less than the 40-year average maximum and the proportion with more than the 40-year average maximum were also entered into the analyses.

Disturbance data were obtained from the Ontario Forest Research Institute (Perera and Baldwin, unpub.) and included the percentage of each UTM block logged since 1950, and the percentage burned since 1940. Of course, large areas were logged prior to 1950, but those data were not available. A conservative estimate of this area of logging was calculated in the following manner: one-half of the total area of UTM blocks with road access prior to 1950 was assumed to have been previously logged (unless it had already been burned). In other words, half of the area of each accessible block which was neither logged after 1950 nor burned after 1940 was assumed to have been logged before 1950. Within each grid cell, the variables of old and new logged and burned areas provide a gross estimate of forest seral stage that should reflect the general availability of habitat within a grid cell for species that prefer either old forest or younger seral stages. For caribou, we assumed that winter limitations on range restricted habitat to old forests alone. Raccoon, porcupine, grey squirrel (*Sciurus carolinensis*), and fisher were assumed to live where the habitat was at least 30 to 35 years old; therefore, all areas logged or burned prior to 1960 became habitat in 1990, while any areas logged after 1960 were deleted from habitat available to these species. As fisher are known to prefer closed canopy habitats with a conifer component greater than 33 percent and a tree DBH greater than 27 cm (Thomasma et al. 1994), our habitat designations may be rather generous in terms of suitability for fisher.

The species response variable was the presence or absence of the species in a UTM block. For species that had sufficiently altered their ranges between the 1950s or 1960s and the mid-1990s to permit comparison, discriminant function analysis was used to examine the relative importance of two possible causes for the change. These were habitat change, which involved comparing old and new logged and burned areas, and climate change, which compared mean decadal January minimum temperature, mean decadal maximum snow depth, proportion of a decade when the 40- or 50-year mean was exceeded, and decadal mean annual temperature. In the case of fisher, which vacated a wide area of their range during the 1970s and then re-populated much of that area in the mid-1990s, conditions from three periods were compared: the 1950s and 1960s, when fisher were present; the 1970s, when they were absent; and the 1990s, when they had returned to some areas.

Testing of Hypotheses for Individual Species

Acceptance of any of the above hypotheses of changes in the range of wildlife populations in Ontario is hampered by the inconsistent quality of the information. Data on range changes and population levels varies from opinions for some species, such as porcupines and raccoons, to opinions supported by trapping statistics for furbearers, to reasonably good count data for such species as moose and deer. The confidence that can be placed in any analysis obviously reflects the uneven quality of those data. At large spatial scales, we know that changes in forest landscapes and weather patterns have occurred (Figures 11.2 to 11.4). As we noted above, for periods prior to the era of the 1950s and 1960s, mapped data on forest disturbance, particularly for logging, are either poor or unavailable. This is unfortunate, because large areas of Ontario have been logged since the early 1800s, and many areas of central Ontario were logged in the early 1900s, at a time when rail lines were built across the province.

The smaller the area over which we attempt to apply our habitat data, the less certain we can be of exactly how much change has occurred and over what period. Use of the large (10,000-km^2) UTM blocks eliminates some concern over the applicability of both wildlife distributional data and the relevance of the habitat and weather data, because this large scale does not require precise knowledge of when ranges changed or when an area was logged. The large scale is particularly relevant to an understanding of the effects of climate change. On the other hand, many habitat changes are too small to be reflected, or not available, at large spatial and temporal scales; the conversion of stands from conifer dominance to a mixed forest type is one example. Animals often respond to these kinds of local changes, so our broad landscape perspective may be an inappropriate treatment of the problem of habitat selection. Litvaitis (1993) was able to show distribution range and population responses to declines in open and seral forest habitats in the northeastern United States, using a level of data more

Table 11.1

Results of discriminant analyses of changes in range distributions of Ontario forest vertebrate species, between 1960 and 1995.

Species	*Number of UTM blocks*	*Significant Variables*[1]	*Individual F*	*Discriminant F*	*P*
White-tailed deer	5	No. severe winters	16.22		
		Habitat	7.52	8.22	0.020
Woodland caribou	11	Habitat	40.72		
		Annual temperature[2]	3.07	14.40	<0.001
Fisher	13/5[3]	No. severe winters	22.08		
		Habitat	6.05	9.71	<0.001
Porcupine	16	No. severe winters	39.43	16.19	<0.001
Raccoon	7	No. severe winters	17.15	9.05	0.003

Variables were calculated for decades, applied to UTM grid cells, and entered into the analysis for the period during which the decline (or increase) occurred. If the decline occurred across more than one decade, the two decadal values were averaged.

1 Number of severe winters = proportion of winters/decade with >45 year mean snow depth.

2 Annual temperature = mean annual temperature for the decades considered

3 Fisher vacated 13 formerly occupied blocks between 1960 and 1980, and re-occupied 5 of these by 1995.

precise than those in our analyses. At the level of precision we have attempted, it is sufficient if habitat change enters the model as a significant variable. From that, we know that habitat change may well be highly important, and that it is worthy of further consideration at smaller scales.

White-Tailed Deer

Our analysis suggested that multiple severe winters were primarily responsible, directly or indirectly, for population and range declines in deer between 1960 and 1980 (Table 11.1). The number of severe winters per decade in the 1950s was significantly lower than in the 1970s across Ontario: the mean values were 0.56 for the 1950s and 0.82 for the 1970s ($P < 0.001$). Significantly deeper snow in the 1970s than in the 1950s was recorded in the eastern part of the province, at Huntsville and North Bay, but not in the west, at Sioux Lookout or Dryden (Figure 11.3a, b).

The conclusion that severe winters drove down deer numbers is supported in the east by data from King (1976) and Wilton (1984), and from unpublished OMNR records that show massive deer mortality from starvation during winter in the early and mid-1970s. King (1976) suggested that there was a reduction in the Ontario deer herd of at least 13 percent in 1970-71 alone, and Wilton (1984) showed that the number of deer in Algonquin Park declined by about 80 percent between 1969 and 1974. A peak in the wolf population in Algonquin Park occurred just prior to the decline in deer, suggesting that predation, at least in Algonquin, may also have played a role in the decline (Wilton 1984). During severe winters, however, predation may be only compensatory, because of the large numbers of deer dying. The current northward expansion of deer in both the eastern and western sectors of the province (Thompson 2000b, this volume) is correlated with the recent period (from 1990 to 1997) of lower-than-average depth of snow on the ground, during which only one year has exceeded the 45-year mean (Figure 11.3a, b).

In the west during the 1970s, snow was not as deep as in the east, certainly not deep enough to precipitate the type of massive mortality from starvation observed in the areas of North Bay and Huntsville. Nonetheless, eight out of ten winters in the 1970s, and six out of ten in the 1960s, had more than the average maximum snow depth, compared to three out of ten during the 1950s (Figure 11.3a, b). Western winter temperatures are lower than those in the east (Voigt et al. 1992; Baldwin et al. 2000, this volume), and these harsher-than-normal winters appear ultimately responsible for the western decline. It is probable that there were higher wolf densities in the west than in the east, supported by higher moose densities within the range of deer during the 1960s. OMNR aerial survey data for 1957, for example, show that the district around the town of Sioux Lookout in northwestern Ontario had a moose density of 0.22 per km^2, but the district around North Bay only 0.06 per km^2.

It is possible, therefore, that deer became highly vulnerable to predation in the moderately deep snow in the west.

The second factor that was important in the decline in deer range across the province was a decline in early seral habitats, from a mean of 46 percent in 1960, to 34 in 1970. Although a change in estimated deer habitat of 12 percent seems minimal, it explains a small portion of the variability in the model and suggests support for the habitat hypothesis at the landscape level. Large areas logged or burned during the period from 1920 to 1940 had aged sufficiently by 1970 to no longer constitute favourable deer habitat. This is particularly true in the central areas of Ontario which were logged earliest, and from which hemlock was preferentially removed. Both the logging and removal of hemlock eliminated preferred winter cover on 51 percent of the deer yards in Algonquin Park, between 1920 and 1970 (Wilton 1984).

Our analyses support two hypotheses: that deer range changed in response to a reduction in available habitat or to severe weather which continued over long periods. Weather caused direct mortality, but excessive predation in the west and the reduction of shelter in the east through the long-term loss of hemlock in traditional deer yards may have exacerbated the effect of weather.

Woodland Caribou

Analysis of the blocks that were previously occupied by woodland caribou, but are no longer, suggested that habitat decline explained most of the change (Table 11.1). As noted above for deer, habitat change can act in two ways: areas can become unsuitable for the species because of a lack of preferred habitat components or the system dynamics may change with respect to competing species or predators. Logging roads may also have given predators (and humans) greater mobility and greater access to the forest interior, thereby putting caribou at further risk.

Bergerud (1974) rejected the habitat-change hypothesis on the basis of observed increases in caribou herds in northern Ontario and in Quebec, even in the presence of logging and wildfires. He noted that some species of terrestrial lichens, which are eaten by caribou in winter, grow best in open habitats. However, the lichens that are preferred by caribou are predominantly *Cladina* spp. (Schaefer and Pruitt 1991; Thomas et al. 1994). *Cladina* spp. lichens are most abundant in old stands dominated by conifers, particularly black spruce (*Picea mariana*) and jack pine (*Pinus banksiana*) (Baldwin and Sims 1989); caribou, likewise, are primarily associated with these older forest types in Ontario (Darby and Duquette 1986; Racey et al. 1989). *Cladina* is most available to animals in winter, under a canopy cover that reduces snow depth (Darby and Duquette 1986; Cumming and Beange 1987); Schaefer and Pruitt (1991) supported that finding and rejected dismissal of the importance of lichens. It is these old forest habitats that are removed by logging for a substantial period of time. Our analysis supports the habitat decline hypothesis for caribou (Table 11.1).

The other argument made by Bergerud (1974), that predation can cause declines in caribou populations, is supported in certain areas (e.g., Seip 1992), but not in others; for example, the George River and Porcupine caribou herds have increased their numbers in the presence of predation during the past two decades. The question remains whether the loss of habitat affects caribou directly, or whether the effect comes indirectly through an increase in predation by wolves, as moose (and perhaps deer) begin to occupy logged areas, and wolves respond numerically. The timing of the disappearance of the caribou is important to the question. If the predation hypothesis were true, then declines in caribou should have occurred when moose populations peaked and supported the highest wolf numbers; that is, during the 1960s and 1970s. While this may be the case in some areas (such as south of Longlac), it appears from Armstrong et al. (1998) that caribou abandoned most areas prior to 1960; that is, possibly before moose and wolves had become abundant. There is insufficient data to refute the predation hypothesis definitively, but most of the evidence supports the habitat change hypothesis. Although there is room for debate over which were the proximate causes of the caribou retreat and which the ultimate causes, the most convincing arguments relate to changes in landscape structure.

The second term in the discriminant function is the decline in mean annual temperature from 1.24 in the 1950s to 0.97 in the 1970s. Although a decline in temperature could affect plant productivity, caribou range occurs throughout Canada, even in the coldest areas on the continent. We believe that a decline in temperature is probably a statistical anomaly with respect to caribou and that it probably has no role in the biology of caribou, either directly or indirectly. Given the upper-level effect of climate on species distributions in general, however, climate change should not be omitted in the consideration of factors that have affected caribou.

Fisher

Most of the decline in fisher range during the 1980s, and all of the recent expansion that we analyzed, was found in UTM blocks in eastern Ontario, except for one block northeast of Thunder Bay. The observed changes in fisher distribution appear to be primarily related to climate change (Table 11.1). The number of winters with deep snow was the most important discriminating variable during both the contraction and the expansion of fisher range (Figure 11.3a, b). Deep snow may affect the use of habitat by fisher and their foraging success (Raine 1983), but not if the snow is heavy or compacted (Powell 1994).

Winter foraging by fisher takes place in areas and habitats where porcupines are common (Powell 1994), although in western Ontario porcupines do not occur over most of the fisher's range. Snowshoe hare is also a common prey species of fisher (de Vos 1952; Clem 1975; Powell 1979), and it is the main prey in areas outside porcupine range (e.g., Raine 1987). Fisher seem to cycle with snowshoe hare in some areas, with a two- to three-year lag (Bulmer 1974; Powell 1979). A third major component of the fisher diet is carrion from ungulates, particularly deer that occur together with fisher over much of their range (Keuhn 1989). Clem (1975) reported little deer carrion in the fisher diet from Algonquin Park during the winter of 1973-74, but that was the year after the major deer die-offs had occurred, so deer carrion may not have been available.

The highest catches of fisher in Ontario, per unit area, come consistently from the areas of Kenora and Fort Frances in the west of the province, and the areas of Parry Sound and Pembroke in the southeast (Thompson 2000b, this volume). This fact suggests that fisher populations are highest where all three of their main foods, snowshoe hare, porcupine, and deer carrion, occur together. Porcupine populations declined over much of northern Ontario in the 1970s (Thompson 2000b, this volume), and the hare cycle in 1974 and 1975 was at its low. The decline in these two species coincided with the broad decline in fisher populations in the east; this suggests that the fisher were responding to the simultaneous collapse in populations of two of their main prey species, in an area outside of deer range. An increase in snowshoe hare alone in the mid-1980s was apparently insufficient to result in a large range expansion, but did apparently support the fisher population in the eastern Claybelt for a few years longer, until the hare populations fell there again.

The expansion of fisher range in the mid-1990s appears to be a response to recent expansions in the populations of porcupine and snowshoe hare, both in the east and the west. The observed statistical relationship between fisher range and climate (Table 11.1) may actually be an indirect response to an ameliorating climate which favours the fisher's preferred prey species, the porcupine, currently on the increase across the province. In the west, fisher likely cycle with snowshoe hare in the areas of Red Lake and Sioux Lookout, in the absence of porcupines, as they do in Manitoba (Raine 1983). There was thus no dramatic decline in fisher range in the west during the 1970s; in fact, peaks in catches occurred there in 1963, 1973, and 1980 (Thompson 2000b, this volume), which correspond approximately to the hare cycle in that area.

The second variable that improved the discriminant model was habitat. Three mean habitat values track fisher population trends over three decades: a habitat value of 63 percent occurs at the period of lowest population, from 1950 to 1960; a habitat value of 45 percent occurs at the period of lowest population, the 1970s; and a habitat value of 53 percent occurs in the 1990s, when fisher populations have reached a high level once again. Either fisher or porcupine, or both, may be responding to habitat change over broad areas of eastern Ontario. The available data lend support to the hypothesis that fisher respond both to climate change and broad habitat change, but probably because these variables influence the rise and fall of porcupine populations.

Moose

Moose populations have not shown large changes in distribution since the 1950s; however, the number of sightings has increased dramatically along the southern limit of their range in the 1990s. There were insufficient data from this southern area to test hypotheses, but good density data were available for the remainder of the range. We will discuss moose population responses to both regional and landscape-level changes in Ontario.

The annual home range of a moose averages about 36 km^2; this large range explains, in part, why moose respond to large-scale, landscape-level effects of timber management. A number of factors, across multiple spatial scales, influence reproductive success and survival in moose, and hence the distribution of the species. Studies of moose have successfully demonstrated a close link between vegetation structure and moose population response. Moose depend on young, woody browse created by new vegetation disturbances as a major source of forage (Peterson 1955; Krefting 1974), but at the same time they must adapt to the ever-present danger of predators. To satisfy both needs, moose forage in young growth that is often adjacent to mature forest, where they can obtain a certain degree of protective cover from predators. Consequently, the distribution of moose is related to the amount of forest disturbance, to the amount of "edge" between forage areas and forest cover, and to the interspersion and juxtaposition of vegetation age classes. These linkages between landscape patch pattern and vegetation composition in moose habitat make the species an excellent subject for the study of comparative effects of alternative landscape patterns. Factors that influence moose survival range from the spatial patterns of broad climatic zones, to the location of good-quality browse within a single plant. Together, the factors limiting the survival of moose form a hierarchy of constraints, which include limiting factors that are independent of population density, and regulating factors that depend on population density (Figure 11.5). The principles of landscape ecology require wildlife managers to consider this hierarchy of constraining factors, and the interactions of factors among spatial scales.

Regional Scale Factors in Moose Response

At the regional level, broad patterns of climate have a direct influence on the survival of moose through bioenergetic constraints. Moose must expend energy for movement in deep snow, for example, and must regulate body temperature in both hot and cold weather. When calories used exceed calories gained, an energy deficit occurs that results in lower body mass, reduced reproductive success, and possibly failure to survive extreme weather events (Schwartz et al. 1987). Cederlund and Bergström (1991) found that loss of body mass in moose calves over the winter was correlated with the severity of winter climate, and especially with snow depth. Predator avoidance strategies are also affected by climate, as both moose and its predators have evolved abilities to cope with snow, including morphological adaptations (Telfer and Kelsall 1984). Climate also indirectly influences the distribution of moose through bioenergetic constraints on food availability and quality. The length of the growing season, summer temperatures, and the amount and timing of precipitation, all influence the productivity of primary browse species. A moose consumes up to 20 kg of browse per day, less in winter than in summer (Verme 1970); hence, lower densities and production rates of available browse constrain the success of moose. These effects are illustrated by a recent multivariate study of climate, landscape, and hunting influences on moose distribution in Ontario (Rempel et. al., in preparation). The highest rates of moose population growth were most strongly associated with areas that had higher mean temperatures in both the hottest months and the coldest months, higher precipitation in the warmest quarter, and lower cumulative snow-depth and precipitation in the coldest quarter. These particular variables were selected for the study because of an a priori expectation that these climate parameters would represent bottlenecks for both moose and its primary browse species.

Band et al. (1999) recently developed a hierarchical ecoregional classification based on estimates of net primary productivity (NPP) (see Band 2000, this volume). The classification is constrained by geoclimatic variables, and provides a spatial delineation of regions with different levels of primary productivity. The boundaries of the ecoregions align with moose density relatively well on a north-south basis, but less well on an east-west gradient (Figure 11.5). This finding suggests that primary productivity is partially correlated with moose density, on the basis of latitudinal factors, but that other variables discriminate within that gradient. The question of whether moose are regulated by food, hunting, or natural predation is of considerable relevance to resource managers.

The influence of climate is not contained within a single spatiotemporal scale. Meteorologists are still discovering and trying to understand the dynamics of various climatic cycles, including the relatively recently

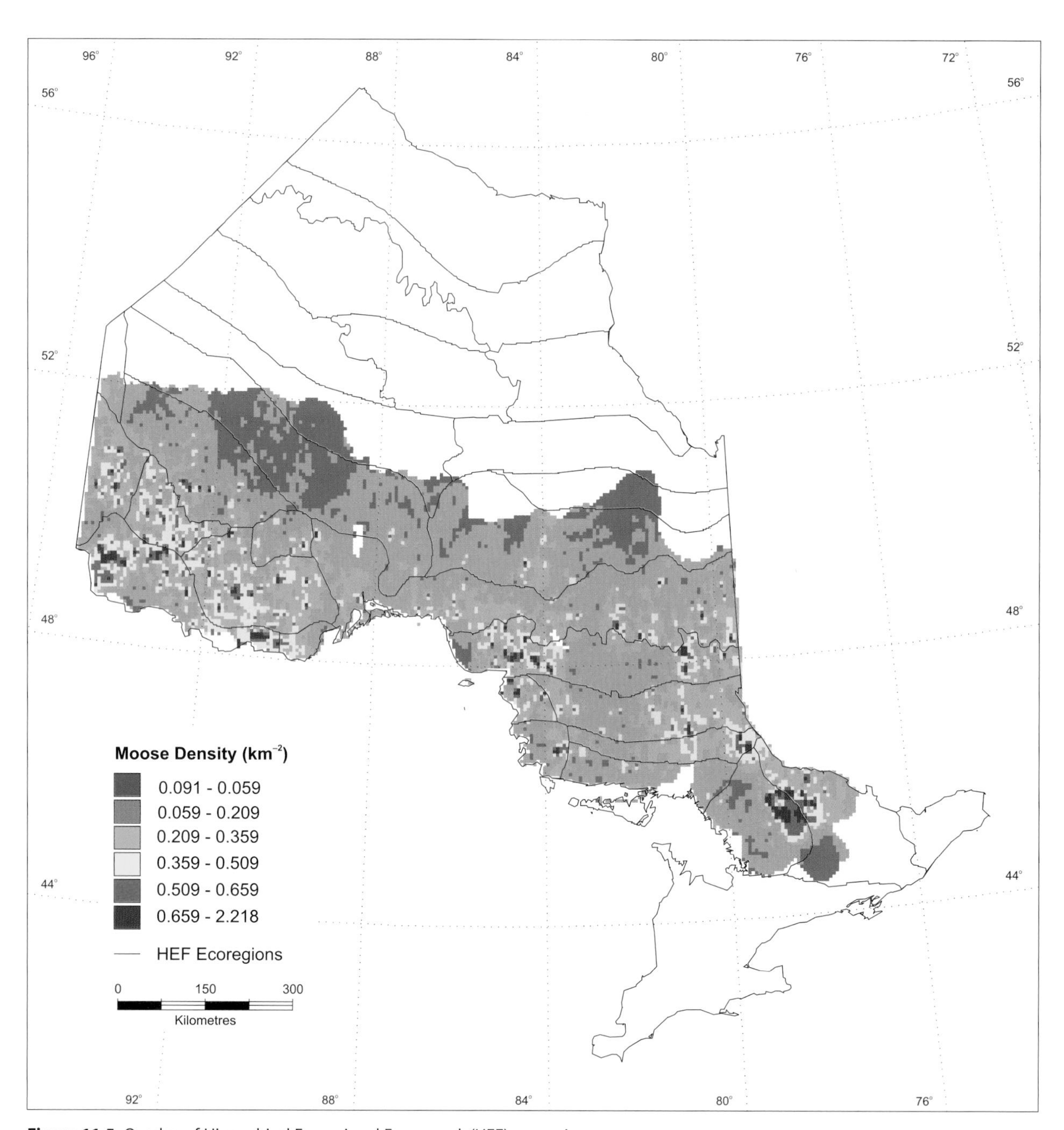

Figure 11.5 Overlay of Hierarchical Ecoregional Framework (HEF) ecoregions, representing isoclines of net primary productivity, on moose density surface (1990-95). (Ecoregional lines adapted from Forest Landscape Ecology Program 1998; moose density adapted from McKenney et al. 1998)

discovered cycles of El Niño and La Niña. Successive annual periods of drought, heavy snow accumulation, high summer temperatures, and cold winter temperatures may influence the reproductive success and survival over winter of moose in Ontario (Thompson 1980). Mech et al. (1987) and Thompson (1980) argued that cumulative winters with deep snow resulted in decreased reproductive success of moose on Isle Royale in Lake Superior and in northeastern Ontario. In terms of the hierarchy of constraining factors, these events are outliers residing in a temporal range of two to ten years, but they influence spatial extents from 1000 km^2 to 10,000 km^2 (Figure 11.6). Although there is evidence that these events have a significant effect on moose populations in the short term, their importance on the long-term viability of the population is debatable.

Landscape-Scale Habitat Factors in Moose Response
Landscape-scale factors, which are nested within regional-scale factors, can include both slow-speed variables, such as the distribution of geological landforms, and variables with intermediate speed and extent, such as natural disturbance patterns (Figure 11.1) (see Perera and Euler 2000, this volume). For example, the Claybelt in the east and pockets of clay south of Lac Seul in the west have higher levels of soil productivity than do the surrounding areas. Rates of moose population growth were higher in these areas than in surrounding, less productive areas. Landscape patterns created by disturbance regimes, however, are perhaps the most important habitat factors at the landscape scale. For example, Rempel et al. (in preparation) found that moose density increased more strongly in wildlife management units which had higher disturbance patch densities and higher levels of interspersion of young and old patches, but also a smaller mean patch size and less interior forest. The analysis was confounded, however, because this general increase in forest edge (resulting from the

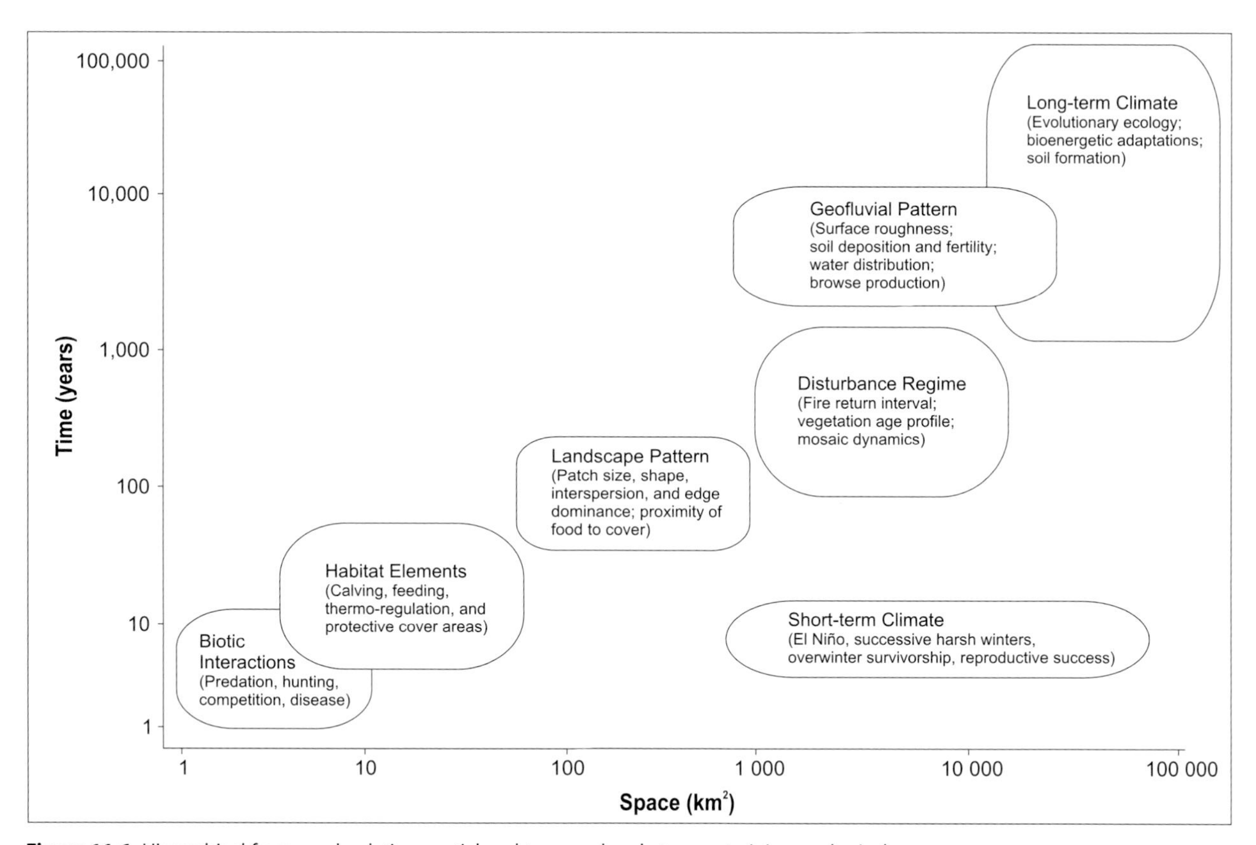

Figure 11.6 Hierarchical framework relating spatial and temporal scale to constraining ecological factors in moose ecology.

application of Ontario's moose-habitat guidelines) occurred simultaneously in the late 1980s with a general reduction in hunting pressure, which resulted from the adoption of a selective harvest strategy. It is likely, however, that several factors are at play. The overall increase in Ontario's moose herd probably results from generally lower hunting pressure, while the differential increase in density among wildlife management units results from finer-scale differences in moose habitat, from reduced hunting pressure, and from improved climate conditions over the past several decades.

Hierarchy theory suggests that forest patterns may be structured hierarchically, with patches nested within patches. For example, when fire frequency is low, then finer-scale gap phase dynamics caused by the fall of individual trees may structure landscapes within a coarser-scale fire disturbance pattern. Such "textural discontinuities," as Holling (1992) describes them, may be important habitat features to which animals have adapted. Elkie (1998) recently demonstrated that northern Ontario forests may, indeed, be structured hierarchically, and that the hierarchical patterns of northeastern and northwestern Ontario are different. Ecosystem-based management must strive to characterize these complex patterns and to ensure that habitat management does not simplify the overall structure. For example, it may be appropriate in northeastern Ontario forests to use uneven-aged silvicultural practices to emulate gap-phase disturbance, but only within a broader pattern that emulates large, but infrequent, burns.

Within the hierarchy of landscape patterns, density-dependent regulating factors may interact with density-independent limiting factors. In Ontario, hunting is density-dependent, because the number of permits issued is based on density estimates from the moose population surveys of previous years. To study the interaction of the two classes of factors, Rempel et al. (1997) compared landscapes with and without recent vegetation disturbance to landscapes with and without hunter access (Figure 11.7). The landscapes included areas logged in small, dispersed clear-cuts and in large, continuous clear-cuts, areas burned by wildfire, and areas without any recent stand-level disturbance. The analysis showed that moose density only increased in landscapes with recent vegetation disturbance, but that, even with disturbance, density did not increase where road-access resulted in high rates of hunting. Earlier studies found similar results. Thompson and Euler (1987), Euler (1983), and Eason (1985) found that a general increase in forest edge (resulting from application of the moose habitat guidelines) and a general reduction in hunting pressure have resulted in an increase in Ontario's moose herd. By supporting all three hypotheses for changes in the distribution of moose populations (habitat change, climate effects, and human-related mortality), these analyses suggest that moose distribution and abundance is determined by a combination of those factors rather than by any one factor alone.

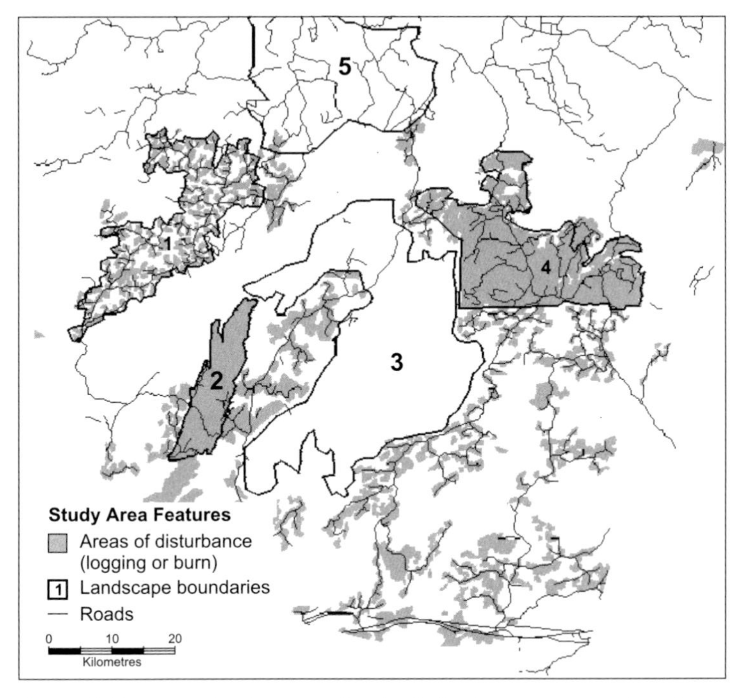

Figure 11.7 Map (derived from satellite data) of a study area where the effects of disturbance and hunting on moose population density were assessed (Rempel et al. 1998). Shading indicates areas of disturbance (logging or burn), and the lighter lines indicate roads. Numbers and darker lines indicate landscape boundaries of (1) modified clear-cut; (2) wildfire burn; (3) uncut wilderness with no roads; (4) progressive clear-cut with restricted hunter access; and (5) uncut landscape with roads.

It thus appears that there is a strong interaction between density-independent landscape disturbance patterns and density-dependent hunting pressure, although in unmanaged systems, natural predation may be the dominant regulating factor. The detailed patch patterns created by vegetation disturbance may be relatively unimportant, and moose numbers may increase equally well in naturally disturbed landscapes as in logged landscapes. This is an important consequence for ecosystem-based management, because it demonstrates that emulating patterns of natural disturbance will not harm, and may, in fact, benefit moose populations. The wildfire burn included in this study, however, also affected density-dependent factors, in that both the blowdown that occurred on the site prior to the wildfire and the dense regrowth of jack pine after the fire impeded the ability of hunters to approach their prey and to take a clear shot, and thus decreased kill rates. Ecosystem-based forest management must also emulate this interaction with regulating factors, perhaps by relating the control of hunting to vegetation structure, thus affecting hunter success rates.

Bobcat

The small shift in the range of bobcats could not be analyzed statistically. The factors that limit the northern distribution of this species appear to be winter snow conditions and the presence of its favoured prey, snowshoe hare (McCord 1974). A recent moderate expansion observed in the areas of Sault Ste. Marie and Sturgeon Falls (Thompson 2000b, this volume) is probably a response to low snow depths in the 1990s (Figure 11.3b). Bobcat range in the west appears to have declined over the past 30 years, probably as a response to the abandonment of numerous early-settlement farms in the Lake of the Woods region, where former snowshoe hare habitat has now mostly disappeared with the aging of successional forests.

Porcupine

The reduction of porcupine range in the 1970s appears related to the high maximum snow depths of that decade (Table 11.1; Figure 11.3a, b). Further support for the climate change hypothesis comes from the general absence of porcupines in the Kapuskasing area, where snow depths have not diminished, but have averaged 80 cm to 83 cm through the 1980s and 1990s; however, at Geraldton, where snow depths have declined to less than 72 cm in the 1990s, porcupines have become reasonably common. In the west, the relatively small expansion in range north and east of Fort Frances also appears to be related to reduced snow depths. Snow depths in the west, however, are generally much lower than in the east (Figure 11.3a, b), so other factors must also be affecting porcupines in the west, but these were not reflected in our analyses. For example, possible effects of habitat change in the west may be masked by the high loadings in the discriminant function for decreased snow depth from the eastern Ontario UTM blocks.

Grey Squirrel

The range expansion of grey squirrels could not be tested statistically. On the basis of food and habitat requirements, the range expansion in this species is likely now at its maximum limit, and the rate has apparently been on the decline since 1970. The grey squirrel has been in Rainy River, Fort Frances, and Sault Ste. Marie since the early 1960s. Although it has moved into areas along the north shore of Lake Huron, it has been unable to expand much farther northwards. The Great Lakes forest biome ends less than 60 km north of the three locations named above, and with it the mast from white pine, walnut (*Juglans* spp.), oak (*Quercus* spp.), and beech (*Fagus grandifolia*) that forms the bulk of the diet of grey squirrels (Banfield 1974). The fact that populations of this species are apparently higher within the towns at the limits of their range than in the surrounding forests suggests that an important supplement to their diets comes from food sources provided by humans.

Raccoon

Our analyses suggest that much of the expansion in raccoon range since the 1970s has been related to an improvement in winter climate that brought less snow and warmer temperatures (Table 11.1; Figures 11.2 and 11.3). As raccoons are dormant during the winter, this effect of weather may be related to the amount of time that raccoons must go without food during dormancy, and thus to their physical condition prior to winter. Raccoons eat a variety of seeds, insects, crayfish, and small rodents, none of which are available in winter. The moderating climate in the 1980s and 1990s may have permitted a northward movement in response to shorter periods with snow cover, later freezing of waterbodies, and earlier access to food resources in the spring than in the 1970s. Another important factor in the northward movement of raccoons is the relatively recent availability of hybrid corn, which is grown as far north as Earlton and New Liskeard. Corn is a staple of the diet of raccoons in many agricultural areas (Banfield 1974; Kaufmann 1982; Sanderson 1987), and the arrival of stable populations of raccoons in the eastern Claybelt and in the Sault Ste. Marie area corresponded to the growing of hybrid corn as a cash crop in those areas. A reduction of raccoon populations in the United States has been related to the loss of stream habitat and to hunting (Preble 1941), but neither of these factors would have limited an earlier range expansion in Ontario.

Sandhill Crane

We did not statistically test the range expansion of sandhill cranes (*Grus canadensis*). These birds nest in bogs within the forest matrix and require isolation from disturbance (Tebbel 1981). After nesting, the birds rear their young in upland meadows or agricultural fields. Range expansion has likely resulted from a combination of generally increasing populations (Tebbel and Ankney 1982) and the end of an extended period of logging in the Algoma region. The early logging operations were labour-intensive and noisy and may have disturbed breeding pairs. The population increase follows lows caused by humans through shooting and habitat destruction (Cadman et al. 1987). Up until the mid-1960s, wetland habitats may have been lost through the regular damming of streams and lake outlets to increase spring water levels for river-driving logs to the mills.

Distributional Changes in Other Species

Several species that do not use forests have greatly expanded their ranges over the past three to four decades. These include the house finch (*Carpodacus mexicanus*), the northern cardinal, the American crow (*Corvus brachyrhynchos*), and the mourning dove. Range expansion in these species is likely related to the ameliorating climate and to fragmentation in southern forests. Several other species have undergone population declines, but not necessarily declines in their breeding range. These include the wood thrush (*Hylocichla mustelina*), the hooded warbler (*Wilsonia citrina*), the northern oriole (*Icterus galbula*), the chestnut-sided warbler (*Dendroica pensylvanica*), the black-throated blue warbler (*Dendroica caerulescens*) (Sauer et al. 1997), and the eastern cottontail (*Sylvilagus floridanus*) (OMNR unpub.).

Although no causes are known for certain, decreases in habitat on breeding grounds or wintering grounds, or both, are suspected in all cases of songbird decline (Terbourgh 1989). Declines in the populations of eastern cottontails and woodchucks (*Marmota marmota*) are not well documented, but are widespread and dramatic. These declines coincide with changes in land use, but perhaps more importantly with increases in predator populations. Coyote (*Canis latrans*) and red fox (*Vulpes vulpes*) populations increased when low fur prices and mild winters coincided in the 1980s. The distribution of these species is very dynamic over large areas. Recently, mange has been reported in coyotes throughout southern Ontario. Rabies, on the other hand, a major cause of fox mortality, has been virtually eliminated by OMNR control programs.

Although no documented evidence exists for a major decline in the populations and distribution of red crossbills (*Loxia curvirostra*) in Ontario, we believe that a substantial historical decline in their range and abundance has occurred. Dickerman (1967) suggested that the subspecies *Loxis curvirostra neogaea* had nearly become extinct by 1900 because of the decline in white pine and eastern hemlock from logging. He also suggested that increases in numbers of *Loxis curvirostra neogaea*, decades later, resulted from the return of some mature forests of white pine and eastern hemlock in the northeastern United States. Old-growth conifer forests are important habitat for crossbills, because of relative crop stability and high seed production (Benkman 1993). Stands dominated by Ontario's old-growth white pine, red pine (*Pinus resinosa*), eastern hemlock, black spruce, red spruce (*Picea rubens*), and white spruce (*Picea glauca*) have all declined in total area (Aird 1985; Eagar and Adams 1992; Baker et al. 1996; OMNR 1996, n.d.), and that decline has probably affected populations of crossbills breeding in Ontario.

RESPONSES BY BIRDS TO LANDSCAPE-LEVEL HABITAT CHANGE

Landscape Pattern Effects

Changes in landscape pattern from the loss and fragmentation of forest habitat in agro-urban landscapes have been implicated as a potential cause of declines of neotropical forest breeding birds. Metapopulations in fragmented habitats are affected by three key factors: loss of original habitat, reduction in patch size, and increased isolation of patches (Krohne 1997). Evidence from studies of the effects of habitat loss and fragmentation in agro-urban landscapes has led to speculation that similar effects might occur in forested landscapes (Harris 1984; Hunter 1990; McGarigal and McComb 1995), where forest management creates an increase in edge through the juxtaposition of undisturbed and disturbed patches, and young and older forest age classes. As many of the studies in agro-urban landscapes pertain to landscapes which were forested earlier in history, we will briefly review their findings as a context for considering the implications of habitat loss and fragmentation in continuous forested landscapes.

There is considerable empirical evidence of local impacts on neotropical migrant birds from habitat loss and fragmentation effects in agro-urban landscapes, including southern Ontario (Galli et al. 1976; Lynch and Whitcomb 1978; Whitcomb et al. 1981; Ambuel and Temple 1983; Brittingham and Temple 1983; Wilcove 1985; Freemark and Merriam 1986; Urban et al. 1987; Harris 1988; Robbins et al. 1989a, b; Schmiegelow 1990; Van Horn 1990; Freemark and Collins 1992; Friesen et al. 1995; Austen et al. 1996; Henschel 1997). The fact that many neotropical migrants nest on or near the ground makes them susceptible to predation and also to brood parasitism from the brown-headed cowbird (*Molothrus ater*) (Gates and Gysel 1978; Chasko and Gates 1982; Brittingham and Temple 1983; Temple and Cary 1988; Johnson and Temple 1990; Hoover et al. 1995; Robinson et al. 1995). These effects have been linked to conflicting evidence for regional, long-term population declines in neotropical migrants (Askins et al. 1990; James et al. 1996). Regional declines in these species might be expected to occur in agro-urban landscapes as a consequence of declines beyond a certain threshold in amount of habitat. Fragmentation may, however, have an impact over and above the reduction of forest habitat (Andrén 1994; Fahrig 1997).

Large-scale habitat loss and fragmentation has had a notable effect on the spotted owl (*Strix occidentalis*) (Thomas et al. 1990; Verner et al. 1992). Recent studies comparing the effects of habitat loss and fragmentation on the spatial patterns of bird populations (Andrén 1994; McGarigal and McComb 1995; Fahrig 1997) indicate that habitat loss is the primary factor. Andrén (1994) suggests that, where habitat loss is between 10 and 30 percent, fragmentation can have a complementary impact to habitat loss, but above 30 percent loss, fragmentation is of minor importance. Fahrig (1997) identifies 20 percent of the breeding habitat as the threshold for ensuring species survival, regardless of the degree of fragmentation, if each patch is large enough to support breeding individuals. Her simulation results are consistent with many earlier studies (e.g., Lande 1987; Lamberson et al. 1992; Lawton et al. 1994; Schneider and Yodzis 1994; McGarigal and McComb 1995; Hanski et al. 1996). Thompson and Welsh (1993), however, have warned that habitat quality may ultimately affect animal populations in second-growth forest habitats.

Numerous studies have measured the effects of clear-cutting on wildlife: Wedeles and Van Damme (1995) review the subject as a whole; Lidicker and Koenig (1996) report on the effects of edge; and Schmiegelow et al. (1997) discuss residual patches. Predictions from these small-scale studies, which concern patches up to 100 ha in size, are difficult to extrapolate to landscapes extending over thousands of hectares, the normal spatial scale of forest management in the boreal forest. Two recent landscape-scale studies of breeding birds in forested landscapes of the eastern United States (Flather and Sauer 1996; James et al. 1996) demonstrated the weaknesses of extrapolating knowledge of small-scale processes to landscape scales. The occurrence of non-linear results at large scales from changes at small scales is predicted by hierarchy theory (Allen and Hoekstra 1992; Holling 1992).

Although other, unknown effects may be important, neither short-term experimental studies conducted in Alberta (Schmiegelow et al. 1997) nor comparative studies conducted in Oregon (McGarigal and McComb 1995) provide strong evidence to support the hypothesis that fragmentation has an impact on birds in continuous forests. Not all bird species exhibiting preferences for particular forest types and seral stages in Oregon (that is, habitat specialists) were affected by habitat loss and change in landscape structure (McGarigal and McComb 1995). Most habitat specialists studied by these authors were associated with the most fragmented habitats; nevertheless, the tree species composition of those habitats explained more of the variation than did the degree of fragmentation. Similarly, no relationship was found to exist between patch size and the richness or abundance of bird species favouring old-growth forest at the centre of remnant patches of coastal temperate rain forests in British Columbia (Schieck et al. 1995). Otto (1996) found no evidence of spatial pattern in forest composition or bird community structure in small study areas of about 3000 ha in New Brunswick. Basing our concept of fragmentation in continuous forests on agro-urban systems may, however, be an oversimplification (Haila et al. 1994; Schmiegelow et al. 1997). The natural disturbance dynamics of the boreal forest provide a mosaic of vegetation types and age classes, and it is this complexity (Frelich and Reich 1995; Bergeron and Harvey 1997) that needs to be maintained by forest management.

Evidence from studies of forest bird responses to variation in landscape patterns suggest that there is much yet to be learned about the long-term, landscape-level effects of forest management activities on habitat suitability. In particular, to identify the fragmentation of Ontario's continuous forests described by Perera and Baldwin (2000, this volume) as having a potentially negative impact on birds requires an understanding of how fragmentation affects important processes. Predation on ground-nesting birds by raccoons and feral cats is not an important factor, as these species are not abundant in coniferous forests. In most continuous forest areas of Ontario, the lack of parasitic nesting species means that at least one of the mediating processes for fragmentation effects that occurs in agro-urban landscapes is absent. Perhaps more importantly, much fragmentation in forested areas is only of short duration, since disturbed forests regenerate rapidly. Other processes associated with fragmentation in continuous forests may await discovery. But the explanation of impacts on neotropical forest nesting birds in the fragmented forests of agro-urban landscapes cannot be adopted as a prediction of potential long-term impacts of the type of fragmentation that occurs in continuous forests.

Gurd (1996) showed that, at a regional scale in Ontario, there are differences in the presence or absence of birds between landscapes disturbed by fire and landscapes subjected to clear-cutting. He found that the proportion of area-insensitive species (i.e., species unaffected by forest fragmentation) negatively associated with clear-cut landscapes was significantly greater than the proportion of area-sensitive species (i.e., species sensitive to forest fragmentation). Furthermore, a significantly greater proportion of short-distance migrants and resident species than of neotropical migrants were negatively associated with clear-cutting. Such results suggest that the response of forest breeding birds to the landscape configurations associated with clear-cutting are complex, and that the effects of fragmentation of the continuous forest by clear-cutting cannot be identified as either harmful or helpful to birds in general, or to any particular species of birds. These results, along with those of Flather and Sauer (1996) and James et al. (1996) from the northeastern USA, point to the need for caution in drawing general conclusions about landscape-level changes and the responses they may evoke from forest breeding birds.

Forest Composition Effects

In Ontario, only a few studies of birds and their habitats in continuous forests have been conducted, primarily by Daniel Welsh and his colleagues at the Canadian Wildlife Service (Welsh 1987; Spytz 1993; Kirk et al. 1996; Welsh and Lougheed 1996; Kirk and Welsh, unpub.). Welsh (1987) demonstrated that in the boreal mixedwood forest of Ontario, certain breeding bird species are associated with seral forest stages. He also demonstrated that neo-tropical migrant species that were affected by forest fragmentation in the eastern United States (Terbourgh 1989) were breeding successfully in isolated forest fragments of less than 10 ha in Ontario's boreal mixedwood forest. In the conifer-dominated Claybelt of eastern Ontario, Welsh and Lougheed (1996) reported a positive association between breeding bird species and gradients in nutrient regimes which corresponded to conifer, mixedwood, and deciduous vegetation types. Both of these studies suggest plasticity in habitat associations by age and forest type for many of these passerine bird species.

Welsh and Kirk (unpub.) measured the relative abundance of forest breeding birds in forest stands dominated by white pine across the province. They found that species diversity was negatively associated with the amount of white pine in a stand; that is, that stands containing predominantly pine contained fewer bird species than mixed stands. They also found a clear distinction between bird communities associated with pine stands in the fragmented landscapes of eastern and southern Ontario and those of the continuously forested regions. Species such as the brown-headed cowbird were a part of the bird communities in the fragmented forests of the agricultural landscapes of southern and eastern Ontario, but not in the continuous-forest landscapes sampled in central and northwestern Ontario. Conversely, some bird species associated with the continuous forest were not associated with the fragmented forests of the southern and eastern pine stands; these include the black-throated blue warbler, the black-throated green warbler (*Dendroica virens*), and the Blackburnian warbler (*Dendroica fusca*).

In conclusion, the results from studies of songbirds indicate that landscape pattern has an effect on populations where fragmentation is high (for example, in agricultural ecosystems). There is less evidence for any effect of fragmentation in continuously forested areas, although the type of disturbance (that is, forest harvest or wildfire) may have an effect at the regional scale. Many of the predictions from landscape ecology remain untested for continuously forested areas. The results of our review do indicate, however, that differences in landscape context and tree species composition within and among geographic regions influence bird community composition.

LANDSCAPE-LEVEL MANAGEMENT OF FOREST WILDLIFE

To this point, we have outlined what is known of the relationship among the range of certain wildlife species in Ontario and the actual or potential factors that cause species ranges to change. Now we address the approaches taken to wildlife management in Ontario, in the face of that knowledge, and examine the assumptions behind a recent major shift in the foundation of provincial wildlife management.

Current Wildlife Management Approaches in Ontario

The hierarchical nature of forest wildlife systems and the objective of emulating natural disturbance patterns are two starting points for the multi-scale management of forest landscapes to benefit wildlife. Although there is no lack of evidence to support a top-down approach or of theories for taking that approach, the task of addressing complexity and cross-scale concerns across landscapes remains problematic. A variety of approaches have evolved in Ontario over the past two decades to deal with wildlife concerns in forested areas.

Featured Species

Ontario's featured species approach attempts to ensure that forest management provides habitat for selected species (OMNR 1990; Voigt et al. 1997). This approach does not explicitly prescribe management at the landscape level, except insofar as the two species featured first, white-tailed deer and moose, have large home ranges. Deer traditionally migrate up to 100 km between their summer and winter ranges (Broadfoot et al. 1996), and the guidelines specify provision of habitat at the landscape scale. Although deer and moose are generalists, habitat provision for them might also provide habitats for an estimated 70 percent of other species of wildlife that are not specifically considered in the guidelines for moose and deer (Baker and Euler 1989). This speculation may be invalid for some areas and patterns; for example, mature conifer may provide adequate wintering areas for deer, but not all suitable wintering areas are necessarily mature conifer. In addition, wintering areas for deer are traditional-use areas that constitute only 10 to 15 percent of the landscape, and provision for mature forest may not be made beyond these limited areas.

Certain forest types, structures, and patterns are not required under deer and moose habitat guidelines. Partly in recognition of this, guidelines have been developed for pileated woodpeckers (*Dryocopus pileatus*) and marten as featured species (Naylor et al. 1997; Watt et al. 1997), in order to ensure the provision of mature and old-growth forests at the landscape scale. Here also, it is implicit that the guidelines will provide habitat for additional species. Threatened or endangered species are also featured, and other species may be locally designated as featured.

Coarse and Fine Filters

An evolving approach that is not exclusive of the featured species approach involves the use of coarse and fine filters (Hunter 1990; Naylor 1994). The coarse-filter approach attempts to provide a representative range of forest ecosystem types, which should supply a diversity of habitats for wildlife at a variety of spatial scales through time. Although the use of this filter clearly involves planning at the forest landscape scale, just what arrangement of diverse habitats is appropriate remains a question. If the coarse-filter approach uses the habitat management guidelines for species such as deer or moose, both of which have broad habitat needs, then it does not differ from the featured species approach. What might be best, moreover, for some species or groups of related species (termed "guilds") might not be best for others. For this reason, a fine-filter approach may also be invoked. The fine-filter approach is designed to provide the habitat management requirements for particular species that are judged to be inadequately managed by the coarse-filter approach. In Ontario, the fine-filter approach is used primarily for threatened and endangered species, and for other species of concern at either the provincial or local level (OMNR 1995). Other approaches that have gained some acceptance outside Ontario, such as the guild approach or the approach of indicator species (Wedeles et al. 1991), do not necessarily address needs at the landscape level.

Habitat Suitability and Landscape Analysis Models

Attempts are being made to mitigate the potential impacts of habitat change from forest management in Ontario. Habitat associations of bird species and of other vertebrates, which are derived partly from empirical associations with forest ecosystem types (as described in Welsh and Lougheed 1996) and partly from the expert opinion of wildlife biologists, have been incorporated into habitat suitability indices. These associations are integrated into a Strategic Forest Management Model (SFMM) (Watkins and Davis 1997) that is used to forecast changes in both wood supply and accompanying habitat supply over two future rotations, or 200 years. Figure 11.8 provides an example of changes in habitat supply forecast for seven wildlife species. These forecasts are used to compare alternative forest harvesting scenarios within a forest management unit. The scenario can then be selected which is forecast to minimize habitat loss, while supplying sufficient wood for harvesting.

Forecasts of future habitat supply, and by implication, of impacts on populations of species using these habitats, provide only working hypotheses, and remain to be tested experimentally. Another weakness of the SFMM tool is that it does not, at this time, forecast the spatial configuration of habitat. At least for birds, the landscape context of habitat quantity and quality influences local abundance, as Venier (1996) showed from studies of birds and forest landscapes in western Ontario. There is insufficient knowledge, however, about the relationship between the amount and quality of that habitat and its relative abundance at landscape scales to make confident predictions about long-term consequences for particular spatial configurations on habitat quantity and quality in managed forests.

To deal with the spatial configuration question and potential adverse impacts of forest fragmentation, a spatial analysis package known as LEAP II (Perera et al. 1998) is used by forest managers. By incorporating the FRAGSTATS program (McGarigal and Marks 1995), LEAP II allows planners to select among a variety of landscape structural indices for patchiness, shape, and isolation for use in comparing the current landscape to a planned future landscape and choosing among alternative planning scenarios. As the relationship between changes in landscape structure and effects on wildlife is only partly understood, the general objective is to keep future planned landscape patterns as close as possible to the spatial heterogeneous patterns of natural landscape patterns (OMNR 1996). Other spatial models that use habitat suitability indices are being developed (Naylor et al. 1998) to help planners of forest harvesting minimize the impact on specific wildlife species.

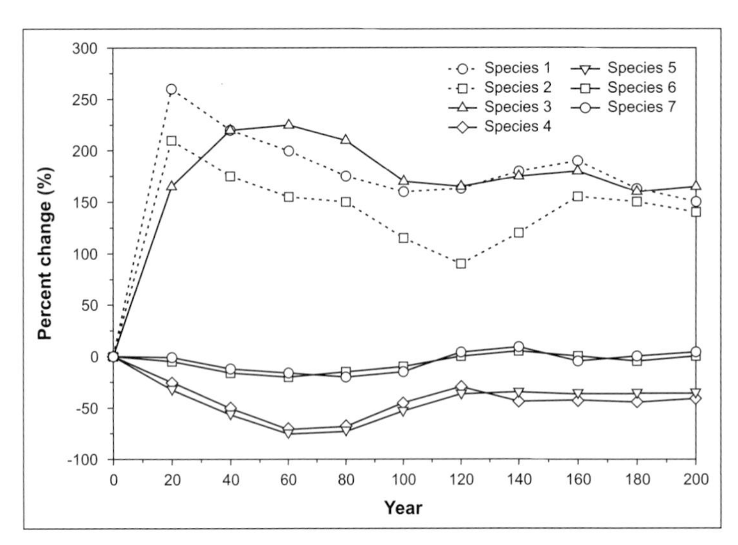

Figure 11.8 Habitat supply forecasts for seven wildlife species of Ontario generated using the Strategic Forest Management Model.

Emulating Natural Disturbance: The Grand Assumption

The *Crown Forest Sustainability Act* (1994) requires industrial forest management on crown land to emulate the landscape patterns created by natural disturbances such as fire, insects, and wind. To emulate does not mean to produce exactly the same patterns, but rather to provide the same qualities those patterns confer or the same processes they promote (Thatcher 1971). The grand assumption behind the adoption of this goal is that ecosystem processes, such as nutrient and energy transfer and the presence of animal and plant communities, will be maintained by emulating the patterns produced by natural disturbances (Thompson and Welsh 1993; Bunnell 1995). Industrial forest management is a mechanical process, however, and fire is a chemical process. Current silviculture in the boreal forest may favour the succession of similar stand types rather than the diversity of types that results from natural disturbances (Bergeron and Harvey 1997). Carleton and MacLellan (1994) found that logged stands were disproportionately succeeded by deciduous species, as compared to fire-disturbed stands in Ontario's boreal forest. If wildlife species respond to both the configuration and composition of their habitat, then it is necessary to consider both in landscape management for wildlife.

But what spatial patterns and amounts of habitat are required in a forested landscape to maintain species, and how should we expect species to respond to patch dynamics? The boreal forest landscape dynamics are determined by wildfire disturbance and by insect infestation. The amount of old growth and young stands in the forest depends on the fire frequency (Baker 1995; Johnson 1996). In the boreal forest, where the fire-return interval is 60 to 150 years, the normal percentage of old growth has been estimated at 5 to 10 percent (Johnson 1996). Patches of older forest occur within a landscapes dominated by younger forests, because the infrequent fires burn over large areas and spread to other, small areas that were burned previously. The consequences of spatial patterns of fires in which large, infrequent fires burn over smaller patches of landscapes are consistent with the spatial dynamics simulated by Boychuck and Perera (1997) and Li et al. (1997). Both older forests and patches of particular forest types occur within a larger matrix of varied age classes and types exposed to frequent burning. This dynamic creates considerable heterogeneity in the landscape (Haila et al. 1994; Bergeron and Harvey 1997).

Results for forest birds, such as those of Otto (1996) and Schieck et al. (1995), are consistent with a hypothesis that many forest habitat specialists are adapted to these spatial and temporal dynamics. Habitat specialists have evolved in landscapes containing a shifting mosaic of patches of various types and age classes, where edges abut non-preferred forest types and age classes (McGarigal and McComb 1995; Schieck et al. 1995). We should not be surprised if some wildlife species, such as neotropical forest breeding birds, that may be considered potential candidates for population decline under large-scale industrial forestry (Mönkkönen and Welsh 1994; Kirk et al. 1996; Welsh and Lougheed 1996), may be able to cope with landscape spatial patterns configured by managed disturbance, provided that a diversity of forest types is maintained on the landscape (Bergeron and Harvey 1997). There may be many other species, however, that are unable to cope with the landscape patterns produced by industrial forestry. It will be necessary to identify these species and the appropriate temporal and spatial scales of management necessary to maintain them on the landscape (Holling 1992; Haila et al. 1994).

FUTURE DIRECTIONS IN WILDLIFE RESEARCH AND MANAGEMENT

Our analyses and review of the available literature suggest that factors operating both independently and synergistically at different spatial and temporal scales affect wildlife distributions in Ontario. Climate limits species ranges both directly and through NPP. The presence or absence of a species may be locally affected by habitat availability and suitability, as well as by cumulative, slow habitat change over time resulting from anthropogenically altered processes. What we have described above invites uncertainty as to the appropriate approach for managing habitat and the long-term effects on wildlife that may result from habitat change caused by both natural and anthropogenic disturbances. We do know that an increasing amount of edge and fragmentation is being produced by forest management, as compared to natural disturbance (see Perera and Baldwin 2000, this volume). From the standpoint of landscape pattern alone, there would seem to be little uncertainty about which direction to take to reduce whatever risk of long-term change to wildlife there may be from the current fragmented patterns. From the standpoint of wildlife response to these current patterns, however, our ability to predict future responses of many wildlife species at regional and landscape scales is limited.

Clearly, we cannot stop the train of management and wait to reduce uncertainty. Consequently, the prudent response is an active adaptive management process, coupled with studies based on the predictions from current landscape theories. Such an approach must involve an adaptive cycle of forest planning and monitoring for learning about the interaction of wildlife response and landscape pattern. OMNR has begun to implement two programs to deal with this uncertainty, a wildlife population monitoring program, conducted at regional and landscape scales, and an active adaptive management program to test alternative hypotheses for managing landscape configuration.

One purpose of the Ontario population monitoring program is to monitor trends in populations of selected species (McLaren et al. 1998) in order to determine to what degree these trends may be the consequence of forest management activities. It will be necessary to measure changes in populations in each major ecodistrict in the province (Hills 1961), because the potential cumulative effects of forest management may act differently in each zone. Metapopulation theory and landscape structure have significant implications for monitoring populations. Those monitoring studies that operate in a vacuum with respect to habitat or which continue only for short periods will fail to provide an understanding of why some populations persist and others do not. Ideas of metapopulation demonstrate that we can have suitable habitat with a missing species, but we may also observe a species in unsuitable habitat (Hanski and Gilpin 1991). Short-term appraisals of species presence and absence erroneously equate the quality of very different habitats. Increasing both the area and the time-period sampled and stratifying the area monitored according to disturbance or structure will reduce the probability of poor predictions. If the monitoring program is designed to evaluate a priori hypotheses about the effects of forest management, relative to the effects of "natural" change, the potential impacts of forest management should be more easily detected.

OMNR's adaptive management program for forest lands and wildlife addresses not only whether forest management has any effect, but why any observed effects may have occurred. The impacts of forest management at small scales are reasonably well known, but the cumulative impacts of changing landscape structure, including the impact of roads, are only speculation at this point. Natural disturbance, whether from fire, insects, wind, or other causes, creates considerable variation in landscape patterns. Logging activities also create variation across harvested landscapes (Rempel et al. 1997). A major hypothesis to be tested is that the wildlife response to managed disturbance does not differ from the response to natural disturbance during the same period of time following disturbance.

It will also be necessary to test the result of continuing with the status quo in producing fragmented patches, and the effects of changing forest composition in relation to landscape configuration. Mönkkönen and Welsh (1994) have suggested that resident birds of the boreal forest may be affected by long-term, large-scale changes to landscape structure. Attempts to ensure that landscape patterns created by forest management activities emulate the range of sizes and configurations created by natural disturbances such as forest fires may not be enough, or conversely even necessary, to maintain forest ecosystem processes. Perhaps the long-term maintenance of ecosystem processes on managed landscapes will require the creation of sizes and configurations that are different from natural patterns, or may permit different configurations to be used as long as forest composition is maintained. To succeed in sustaining ecosystem processes under forest management will require a much better understanding of the relationship between landscape patterns and processes as these affect species distributions and interactions among species. It is clear that hypotheses need to be evaluated at multiple scales, from the landscape down to the individual stand.

Multi-scale measures of biological diversity may be a necessary component of the evaluation of management alternatives. Biodiversity has emerged in recent years as an indicator useful in planning for ecosystem sustainability. The definition of biodiversity as "variety in life" has an appealing simplicity, but fails to convey the multi-scale nature of biodiversity that Noss (1990) has demonstrated. The list of indicators which Noss formulated to assess diversity at the regional scale corresponds to the variables of composition, structure, and function that are the subject of this chapter. Many studies of biodiversity have used relatively fine-scale measures in a particular ecosystem and at a specific scale. Cornell and Lawton (1992) suggested, however, that regional species richness controls local species richness, and that searching for correlates of biodiversity at one given scale may be misguided. The implication is that broad-scale biogeographical processes control community structure to a greater degree than do fine-scale local processes. But there are lower-level factors at work that are, nonetheless, important. One of these factors is the keystone species, or a species which appears to influence biodiversity to a degree out of proportion to its numbers. An example of a keystone species in Ontario is beaver. This species creates the wetlands, openings, and meadows that a variety of species require as habitats. The coarse-filter approach described above aims for diverse representation, and the assumption is that if the appropriate habitat is provided, then the diversity of wildlife species will follow, at least at the landscape scale. The uncertainty of this assumption, coupled with the uncertainty of the "grand assumption," has led some to argue for the indicator species approach. McLaren et al. (1998) argue for the use of indicator species in a monitoring program, not so much to assess other species as to assess changes in habitat availability. Given our uncertainty about many species and interactions, the indicator-species approach seems appropriate for landscape-level monitoring.

The programs of population monitoring and adaptive forest management will combine to tease apart the effects of management from the "normal" effects of natural disturbance and of other environmental factors, such as climate, on wildlife in particular and biodiversity in general. The major value of an adaptive management approach to policy evaluation is the anticipation of unexpected results (see Baker 2000, this volume). It is quite possible that there will be many effects on wildlife that are unexpected; however, by having such a program in place, we will have a much better chance of detecting effects on species and processes that might otherwise have been overlooked. This approach allows management agencies to anticipate rather than simply react. Through understanding these relationships we will be better able to know how and why to change our habitat management approach as the future unfolds, even if we are not any better at predicting the future.

A FINAL PERSPECTIVE

The foregoing discussion of what will be needed in the future to manage forested landscape for wildlife illustrates the difficulties that arise from a lack of knowledge of complex systems (Holling 1992). Complexity is recognized as a feature of many systems, and perhaps of their sustainability (Pickett et al. 1985; Lewin 1992; Casti 1994). Our poor understanding of the complexity and uncertainty involved in dealing with many individual species or populations emphasizes the need for a landscape-level approach to managing forest ecosystems (Allen and Hoekstra 1992). Bissonette (1997) also argues that ecological complexity is often not understood by reductionist approaches and that studies at larger scales have an important role in providing context (see also Pickett et al. 1997).

The results of our landscape-level analysis of selected Ontario wildlife species clearly shows broad-scale constraints on the distribution of wildlife and upper-level factors that are operating to change wildlife distribution. The discussions of factors shaping the ecology of forest birds and of moose reveal that a variety of regional and landscape factors affect populations, but also that many conservation biology theories remain untested for continuously forested landscapes. Effects on forest ecosystems from anthropogenic disturbances are likely to be strongest at intermediate scales, where disturbances such as forestry are prevalent. Planning at both the provincial level and at the level of local management units often misses this important middle scale. The complex and dynamic systems which we endeavour to manage appear to be hierarchically organized; therefore, we recommend a top-down approach to both the planning and the evaluation of alternative management approaches for addressing uncertainties about the sustainability of forest landscapes.

ACKNOWLEDGEMENTS

The following former and current OMNR personnel contributed their knowledge of changes in species distributions: Evan Armstrong, Al Bisset, Neil Dawson, Gord Eason, Mike Eliuke, Linda Ferguson, Mick Gauthier, Greg Gillespie, Fred Johnston Sr., Scott Jones, Ken Koski, Dave Lyons, Vic Miller, Ben Miron, Brian Naylor, Milan Novak, Don Perry, Norm Quinn, Gerry Racey, Bruce Ranta, Will Samis, Barry Snider, Milan Vukelich, Bernie Wall, Randy Wepruk, Mike Wilton, and Orville Wohlgemuth. Darcy Ortiz of the Canadian Forest Service compiled the weather data, and Rob Warren provided the OMNR snow depth data.

REFERENCES

Aird, P.L. 1985. In Praise of Pine: The Eastern White Pine and Red Pine Harvest from Ontario's Crown Forests. Report PI-X-52. Chalk River, Ontario: Petawawa National Forestry Institute, Canadian Forest Service. 11 p.

Allen, T.F.H. and T.W. Hoekstra. 1992. Toward a Unified Ecology. New York, New York: Columbia University Press. 384 p.

Allen, T.F.H. and T.B. Starr. 1982. Hierarchy: Perspectives for Ecological Complexity. Chicago, Illinois: University of Chicago Press. 320 p.

Ambuel, B. and S.A. Temple. 1983. Area-dependent changes in the bird communities and vegetation of southern Wisconsin forests. Ecology 63: 1057-1068.

Anderson, R.C. and O.L. Loucks. 1979. White-tailed deer influence on the structure and composition of Tsuga canadensis forests. Journal of Applied Ecology 16: 855-861.

Andrén, H. 1994. Effects of habitat fragmentation on birds and mammals in landscapes with different proportions of suitable habitat: a review. *Oikos* 71: 355-366.

Armstrong, E., G. Racey, and N. Bookey. 1998. Landscape-level considerations in the management of forest-dwelling woodland caribou in northwestern Ontario. Paper to the North American Caribou Conference, March, 1998, Whitehorse, Yukon.

Askins, R.A., J.F. Lynch, and R. Greenberg. 1990. Population declines in migratory birds in eastern North America. In: D.M. Power (editor). Current Ornithology. New York, New York: Plenum Press. 1-57.

Askins, R.A. and M.J. Philbrick. 1987. Effect of changes in regional forest abundance on the decline and recovery of a forest bird community. Wilson Bulletin 99: 7-21.

Austen, M.J.W., C.M. Francis, D.B. Burke, and M.S.W. Bradstreet. 1996. Effects of forest fragmentation on woodland birds in southern Ontario. In: Report on the Effects of Forest Fragmentation on Woodland Biodiversity in Southern Ontario and Recommendations for Woodland Conservation. Port Rowan, Ontario: Long Point Bird Observatory. 85 p. + appendices.

Baldwin, K.A. and R.A. Sims. 1989. Common Forest Plants in Northwestern Ontario. Sault Ste. Marie, Ontario: Forest Resource Development Agreement, Forestry Canada. 344 p.

Baker, J.A. and D. Euler. 1989. Featured Species Management in Ontario. Ontario Ministry of Natural Resources, Toronto, Ontario. Unpublished report. 19 p.

Baker, J.A., T. Clark, and I.D. Thompson. 1996. Boreal mixedwoods as wildlife habitat: observations, questions and concerns. In: C.R. Smith and G.W. Crook (editors). Advancing Boreal Mixedwood Management in Ontario. Sault Ste. Marie, Ontario: Canadian Forest Service and Ontario Ministry of Natural Resources. 41-52.

Baker, W.L. 1995. Longterm response of disturbance landscapes to human intervention and global change. Landscape Ecology 10: 143-159.

Baldwin, D.J.B., J.R. Desloges, and L.E. Band. 2000. Physical geography of Ontario. In: A.H. Perera, D.L. Euler, and I.D. Thompson (editors). Ecology of a Managed Terrestrial Landscape: Patterns and Processes of Forest Landscapes in Ontario. Vancouver, British Columbia: University of British Columbia Press. 12-29.

Ballard, W.B. and V. Van Ballenberghe. 1997. Predator/prey relationships. In: A.W. Franzmann and C.C. Schwartz (editors). Ecology and Management of the North American Moose. Washington, DC: Smithsonian Institute Press. 247-273.

Band, L.E. 2000. Forest ecosystem productivity in Ontario. In A.H. Perera, D.L. Euler, and I.D. Thompson (editors). Ecology of a Managed Terrestrial Landscape: Patterns and Processes of Forest Landscapes in Ontario. Vancouver, British Columbia: University of British Columbia Press, Vancouver, British Columbia, Canada. 163-177.

Band, L.E., F. Csillag, A.H. Perera, and J.A. Baker. 1999. Deriving an Eco-regional Framework for Ontario through Large-Scale Estimates of Net Primary Productivity. Sault Ste. Marie, Ontario: Ontario Ministry of Natural Resources, Ontario Forest Research Institute. In press.

Banfield, A.W.F. 1974. The Mammals of Canada. Ottawa, Ontario: National Museum of Natural Sciences. 438 p.

Benkman, C.W. 1993. Logging, conifers, and the conservation of crossbills. Conservation Biology 7: 473-479.

Berg, W.E. and D.W. Keuhn. 1994. Demography and range of fishers and American martens in a changing Minnesota landscape. In: S.W. Buskirk, A.S. Harestad, M.G. Raphael, and R.A. Powell (editors). Martens, Sables and Fishers: Biology and Conservation. Ithaca, New York: Cornell University Press. 262-271.

Bergeron, Y. and B. Harvey. 1997. Basing silviculture on natural ecosystem dynamics: an approach applied to the southern boreal mixedwood forest in Quebec. Forest Ecology and Management 92: 235-242.

Bergerud, A.T. 1974. Decline of caribou in North America following settlement. Journal of Wildlife Management 38: 757-770.

Bissonette, J. 1997. Scale-sensitive ecological properties: historical context, current meaning. In: J. Bissonette (editor). Wildlife and Landscape Ecology. New York, New York: Springer-Verlag. 3-31.

Bormann, F.H. and G.E. Likens. 1979. Pattern and Process in a Forested Ecosystem. New York, New York: Springer-Verlag. 565 p.

Boutin, S. 1995. Population changes of the vertebrate community during a snowshoe hare cycle in Canada's boreal forest. *Oikos* 74: 69-80.

Boychuck, D. and A.H. Perera. 1997. Modeling temporal variability of boreal landscape age-classes under different fire disturbance regimes and spatial scales. Canadian Journal of Forest Research 27: 1083-1094.

Boychuk D., A.H. Perera, M.T. Ter-Mikaelian, D.L. Martell, and C. Li. 1995. Modeling the effect of scale and fire disturbance patterns on forest age distribution. Forest Fragmentation and Biodiversity Project Report Number No. 25. Sault Ste. Marie, Ontario: Ontario Ministry of Natural Resources, Ontario Forest Research Institute. 59 p.

Brittingham, M.C. and S.A. Temple. 1983. Have cowbirds caused forest songbirds to decline? Bioscience 33: 31-35.

Broadfoot, J. D., D.R. Voigt, and T.J. Bellhouse. 1996. White-tailed deer, *Odocoileus virginianus*, summer dispersion areas in Ontario. Canadian Field-Naturalist 110: 298-302.

Bulmer, M.G. 1974. A statistical analysis of the 10-year cycle in Canada. Journal of Animal Ecology 43: 701-718.

Bunnell, F.L. 1995. Forest-dwelling vertebrate faunas and natural fire regimes in British Columbia: patterns and implications for conservation. Conservation Biology 9: 636-644.

Cadman, M.D., P.F.J. Eagles, and F.M. Helleiner. 1987. Atlas of the Breeding Birds of Ontario. Waterloo, Ontario: University of Waterloo Press. 617 p.

Canadian Forest Service. 1995. The state of Canada's forests. Ottawa, Ontario: Natural Resources Canada and Queen's Printer. 112 p.

Carleton, T.J. and P. MacLellan. 1994. Woody vegetation responses to fire versus clear-cut logging: a comparative survey in the central Canadian boreal forest. Ecoscience 1: 141-152.

Casti, J.L. 1994. Complexification. New York, New York: Harper-Collins. 320 p.

Cederlund G. and R. Bergström. 1991. Trends in the moose-forest system in Fennoscandia, with special reference to Sweden. In: R.M. DeGraaf and R.I. Miller (editors). Conservation of Faunal Diversity in Forested Landscapes. New York, New York: Chapman and Hall. 265-281.

Cederlund, G.N., H.K.G. Sand, and A. Pehrson 1991. Body mass dynamics of moose calves in relation to winter severity. Journal of Wildlife Management 55(4): 675.

Chasko, G.G. and J.E. Gates. 1982. Avian habitat suitability along a transmission-line corridor in an oak-hickory forest region. Wildlife Monographs 82: 1-41.

Clem, M.K. 1975. Interspecific relationship of fishers and martens in Ontario during winter. In: R.L. Phillips and C.J. Jonkel (editors). Proceedings, 1975 Predator Symposium, University of Montana, Missoula, Montana. 165-182.

Cornell, H.V. and J.H. Lawton. 1992. Species interactions, local and regional processes, and limits to the richness of ecological communities: a theoretical perspective. Journal of Animal Ecology 61: 1-12.

Cumming, H.G. and D.B. Beange. 1987. Dispersion and movements of woodland caribou near lake Nipigon, Ontario. Journal of Wildlife Management 51: 69-79.

Dahir, S.E. and C.G. Lorimer. 1996. Variation in canopy gap formation among developmental stages of northern hardwood stands. Canadian Journal of Forest Research 26: 1875-1892.

Darby, W.R. and L.S. Duquette. 1986. Woodland caribou and forestry in Northern Ontario, Canada. Rangifer Special Issue No. 1: 87-93.

DeGraaf, R.M. and R. I. Miller. 1996. The importance of disturbance and land-use history in New England: implications for forested landscapes and wildlife conservation. In: R.M. DeGraaf and R.I. Miller (editors). Conservation of Faunal Diversity in Forested Landscapes. New York, New York: Chapman and Hall. 3-35.

De Vos, A. 1952. The Ecology and Management of Fisher and Marten in Ontario. Wildlife Series Technical Bulletin No. 1. Toronto, Ontario: Ontario Department of Lands and Forests. 90 p.

De Vos, A. 1962. Changes in the distribution of mammals and birds in the Great Lakes area. Forestry Chronicle 38: 108-113.

De Vos, A. 1964. Range changes of mammals in the Great Lakes region. American Midland Naturalist 17: 210-231.

De Vos, A., A.T. Cringan, J.K. Reynolds, and H.G. Lumsden. 1959. Biological investigations of traplines in northern Ontario. Wildlife Series Technical Bulletin No. 8. Toronto, Ontario: Ontario Department of Lands and Forests. 48 p.

De Vos, A. and R.L. Peterson. 1951. A review of the status of woodland caribou in Ontario. Journal of Mammalogy 32: 329-337.

Dickerman, R.W. 1967. The old northeastern subspecies of red crossbill. American Birds 41: 189-194.

Dobbyn, J. 1994. Atlas of the Mammals of Ontario. Toronto, Ontario: Federation of Ontario Naturalists. 120 p.

Douglas, C.W. and M.A. Strickland. 1987. Fisher. In: M. Novak, J.A. Baker, M.E. Obbard, and B. Malloch (editors). Wild Furbearer Management and Conservation in North America. Toronto, Ontario: Ontario Trappers Association and Ontario Ministry of Natural Resources. 511-529.

Eagar, C. and M.B. Adams. 1992. Ecology and Decline of Red Spruce in the Eastern United States. London, UK: Springer-Verlag. 417.

Eason, G. 1985. Overharvest and recovery of moose in a recently logged area. *Alces* 21: 55-75.

Elkie, P.C. 1998. Analysis of hierarchical characteristics of landscapes in Ontario: Detecting emergent levels of organization [M.Sc.F. thesis]. Thunder Bay, Ontario: Faculty of Forestry, Lakehead University. 74 p.

Euler, D. 1983. Selective harvest, compensatory mortality and moose in Ontario. *Alces* 19: 148-161.

Fahrig, L. 1997. Relative effects of habitat loss and fragmentation on population extinction. Journal of Wildlife Management 61: 603-610.

Finerty, J.P. 1980. The Population Ecology of Cycles in Small Mammals: Mathematical Theory and Biological Fact. Cambridge, Massachusetts: Yale University Press. 234 p.

Flather, C.H. and J.R. Sauer. 1996. Using landscape ecology to test hypotheses about large-scale abundance patterns in migratory birds. Ecology 77: 28-35.

Fleming, R.A., A.A. Hopkin, and J.-N. Candau. 2000. Insect and disease disturbance regimes in Ontario's forests. In: A.H. Perera, D.L. Euler, and I.D. Thompson (editors). Ecology of a Managed Terrestrial Landscape: Patterns and Processes of Forest Landscapes in Ontario. Vancouver, British Columbia: University of British Columbia Press. 141-162.

Freemark, K. and B. Collins. 1992. Landscape ecology of birds breeding in temperate forest fragments. In: J.M. Hagan III, and D.W. Johnston (editors). Ecology and Management of Neotropical Migrant Landbirds. Washington, DC: Manomet Bird Observatory, Smithsonian Institute Press. 443-454.

Freemark, K.E. and H.G. Merriam. 1986. Importance of area and habitat heterogeneity to bird assemblages in temperate forest fragments. Biological Conservation 36: 115-141.

Frelich L.E. and C.G. Lorimer. 1991. A simulation of landscape-level stand dynamics in the northern hardwood region. Journal of Ecology 79: 223-233.

Frelich, L.E. and P.B. Reich. 1995. Spatial patterns and succession in a Minnesota southern-boreal forest. Ecological Monographs 65: 325-346.

Friesen, L.E., P.F.J. Eagles, and R.J. MacKay. 1995. Effects of residential development on forest-dwelling neotropical migrant songbirds. Conservation Biology 9: 1408-1414.

Forest Landscape Ecology Program. 1998. Forest Landscape Ecology Program, Ontario Ministry of Natural Resources. 1998. Ontario's Forest Fire History: An Interactive Digital Atlas [CD-ROM]. Sault Ste. Marie, Ontario: Ontario Forest Research Institute, Ontario Ministry of Natural Resources.

Galli, A. E., C.F. Leck, and R.T.T. Forman. 1976. Avian distribution patterns in forest islands of different sizes in central New Jersey. Auk 93: 356-364.

Gates, J.E. and L.W. Gysel. 1978. Avian nest dispersion and fledging success in field-forest ecotones. Ecology 59: 871-883.

Gluck, M.J. and R. S. Rempel. 1996. Structural characteristics of post-wildfire and clearcut landscapes. Environmental Monitoring and Assessment 39: 435-450.

Gurd, D.B. 1996. Avian occurrence in Ontario boreal forests subjected to disturbance from timber harvest and fire [M.Sc. thesis]. Guelph, Ontario: University of Guelph. 105 p.

Haila, Y., I.K. Hanski, J. Niemelä, P. Punttila, S. Raivio, and H. Tukia. 1994. Forestry and the boreal fauna: matching management with natural forest dynamics. *Annales Zoologici Fennici* 31: 187-202.

Hanski, I. and M.E. Gilpin. 1991. Metapopulation dynamics: brief history and conceptual domain. In: M.E. Gilpin and I. Hanski (editors). Metapopulation Dynamics. London, UK: Academic Press. 3-16.

Hanski, I., A. Moilanen, and M. Gyllenberg. 1996. Minimum viable metapopulation size. American Naturalist 147: 527-541.

Harestad, A.S. and F.L. Bunnell. 1979. Home range and body weight - a reevaluation. Ecology 60: 389-402.

Harris, L.D. 1984. The Fragmented Forest: Island Biogeography and the Preservation of Biotic Diversity. Chicago, Illinois: University of Chicago Press. 211 p.

Harris, L.D. 1988. Edge effects and conservation of biotic diversity. Conservation Biology 2: 330-332.

Henschel, C.P. 1997. Using null models to test for effects of habitat fragmentation on species richness of forest song birds (M.Sc. thesis). Guelph, Ontario: University of Guelph. 44 p.

Hills, G.A. 1961. The Ecological Basis for Land-Use Planning. Research Report No. 46. Toronto, Ontario: Ontario Department of Lands and Forests. 204 p.

Holling, C.S. 1992. Cross-scale morphology, geometry, and dynamics of ecosystems. Ecological Monographs 62: 447-502.

Hoover, J.P., M.C. Brittingham, and L.J. Goodrich. 1995. Effects of forest patch size on nesting success of Wood Thrushes. Auk 112: 146-155.

Hunter, M.L. 1990. Wildlife, Forests, and Forestry: Principles of Managing Forests for Biological Diversity. Englewood Cliffs, New Jersey: Prentice-Hall. 370 p.

James, F.C., C.E. McCulloch, and D.A. Wiedenfield. 1996. New approaches to the analysis of population trends in land birds. Ecology 77: 13-27.

Johnson, E.A. 1996. Scales and processes in forest fire ecology. In: Sustainable Forest Management Network of Centres of Excellence. Conference Summary's: Forging a Network of Excellence, Sustainable Forestry Partnerships, March 8-10, 1996, Sustainable Forest Management Network, Edmonton, Alberta. 12-15.

Johnson, R.G. and S.A. Temple. 1990. Nest predation and brood parasitism of tallgrass prairie birds. Journal of Wildlife Management 54: 106-111.

Kaufmann, J.H. 1982. Raccoon and allies. In: J.A. Chapman and G.A. Feldhamer (editors). Wild Mammals of North America. Baltimore, Maryland: John Hopkins University Press. 567-585.

Keith, L.B., S.E.M. Bloomer, and T. Willebrand. 1993. Dynamics of a snowshoe hare population in fragmented habitat. Canadian Journal of Zoology 71: 1385-1392.

Keuhn, D.W. 1989. Winter foods of fishers during a snowshoe hare decline. Journal of Wildlife Management 53: 688-692.

King, A. 1997. Hierarchy theory: a guide to system structure for wildlife biologists. in J. Bissonette (editor). Wildlife and Landscape Ecology. New York, New York: Springer-Verlag. 185-214.

King, D.R. 1976. Estimates of white-tailed deer population and mortality in central Ontario, 1970-72. Canadian Field-Naturalist 90: 29-36.

Kirk, D.A., A. W. Diamond, K. A. Hobson, and A. R. Smith. 1996. Breeding bird communities of the western and northern Canadian boreal forest: relationship to forest type. Canadian Journal of Zoology 74: 1749-1770.

Kotliar, N. B. and J. A. Wiens. 1990. Multiple scales of patchiness and patch structure: a hierarchical framework for the study of heterogeneity. *Oikos* 59: 253-260.

Krebs, C.J. 1996. Population cycles revisited. Journal of Mammalogy 77: 8-24.

Krebs, C.J., S. Boutin, R. Boonstra, A.R.B. Sinclair, J.N.M. Smith, M.R.T. Dale, K. Martin, and R. Turkington. 1995. Impact of food and predation on the snowshoe hare cycle. Science 269: 1112-1115.

Krefting, L.W. 1974. Moose distribution and habitat selection in north central North America. Nature Canada 101: 81-100.

Krohne, D. T. 1997. Dynamics of metapopulations of small mammals. Journal of Mammalogy 78: 1014-1026.

Lamberson, R.H., K. McKelvey, B.R. Noon, and C. Voss. 1992. Reserve design for territorial species: the effects of patch size and spacing on the viability of the northern spotted owl. Conservation Biology 6: 505-512.

Lande, R. 1987. Extinction thresholds in demographic models of territorial populations. American Naturalist 130: 624-635.

Lawton, J.H., S. Nef, A.J. Letcher, and P.H. Harvey. 1994. Animal distributions: patterns and processes. In: P. J. Edwards, R. M. May, and N. R. Webb (editors). Large-scale Ecology and Conservation Biology. Boston, Maine: Blackwell. 41-58.

Lertzman K and J. Fall. 1998. From forest stands to landscapes: spatial scales and the roles of disturbance. In: D. Peterson and T. Parkers (editors). Ecological Scale: Theory and Applications. New York, New York: Columbia University Press. 339-368.

Levins, R. 1970. Extinction. In: M. Gerstenhaber (editor). Some Mathematical Questions in Biology. Second Symposium on Mathematical Biology. Lectures on Mathematics in the Life Sciences, Volume 2. Rhode Island: American Mathematical Society. 77-107.

Lewin, R. 1992. Complexity: Life at the Edge of Chaos. New York, New York: MacMillan. 208 p.

Li, C. 2000. Fire regimes and their simulation with reference to Ontario. In A.H. Perera, D.L. Euler, and I.D. Thompson (editors). Ecology of a Managed Terrestrial Landscape: Patterns and Processes of Forest Landscapes in Ontario. Vancouver, British Columbia: University of British Columbia Press. 115-140.

Li, C., M. Ter-Mikaelian, and A. Perera. 1997. Temporal fire disturbance patterns on a forest landscape. Ecological Modelling 99: 137-150.

Lidicker, W.Z. and W.D. Koenig. 1996. Responses of terrestrial vertebrates to habitat edges and corridors. In: D.R. McCullough (editor). Metapopulations and Wildlife Conservation. Island Press, Washington, DC, USA. 85-109.

Lindstedt. S.L., B.J. Miller, and S. W. Buskirk. 1986. Home range, time, and body size in mammals. Ecology 67: 413-418.

Litvaitis, J.A. 1993. Response of early successional vertebrates to historic changes in land use. Conservation Biology 7: 866-873.

Lynch, J.F. and R.F. Whitcomb. 1978. Effects of insularization of the eastern deciduous forest on avifaunal diversity and turnover. In: A. Marmelstein (editor). Classification, Inventory, and Analysis of Fish and Wildlife Habitat. Publication OBS-78716. Washington, DC: US Fish and Wildlife Service. 461-489.

McCord, C.M. 1974. Selection of winter habitat by bobcats on the Quabbin Reservation. Journal of Mammalogy 55: 428-437.

McGarigal, K. and W.C. McComb. 1995. Relationships between landscape structure and breeding birds in the Oregon coast range. Ecological Monographs 65: 235-260.

McGarigal, K. and B. Marks. 1995. FRAGSTATS: Spatial Pattern Analysis Program for Quantifying Landscape Structure. General Technical Report PNW-GTR-351. Portland, Oregon: USDA Forest Service. 122 p

McKenney, D.W., R.S. Rempel, L.A. Venier, Y. Wang, and A.R. Bisset. 1998. Development and application of a spatially explicit moose population model. Canadian Journal of Zoology 76: 1922-1931.

McLaren, M.A., I.D. Thompson, and J.A. Baker. 1998. Selection of vertebrate wildlife indicators for monitoring sustainable forest management in Ontario. Forestry Chronicle 74: 241-248.

Mech, L.D. 1970. The wolf: The Ecology and Behavior of an Endangered Species. Garden City, New York: The Natural History Press. 384 p.

Mech, L.D., R.E. McRoberts, R.O. Peterson, and R.E. Page. 1987. Relationship of deer and moose populations to previous winters' snow. Journal of Animal Ecology 56: 615-627.

Messier, F. 1991. The significance of limiting and regulating factors on the demography of moose and white-tailed deer. Journal of Animal Ecology 60: 377-393.

Mönkkönen, M. and D.A. Welsh. 1994. A biogeographical hypothesis on the effects of human caused landscape changes on the forest bird communities of Europe and North America. *Annales Zoologica Fennici* 31: 61-70.

Naylor, B.J. 1994. Managing wildlife habitat in red and white pine forests of central Ontario. Forestry Chronicle 70: 411-419.

Naylor, B.J., J.A. Baker, D.M. Hogg, J.G. McNicol, and W.R. Watt. 1997. Forest Management Guidelines for the Provision of Pileated Woodpecker Habitat. Toronto, Ontario: Ontario Ministry of Natural Resources. 26 p.

Naylor, B. J., D. Kaminski, and P. Kovitz. 1998. User's Guide for Ontario Wildlife Habitat Analysis Model and OWHAM Tool. Ontario Ministry of Natural Resources, South Central Science and Technology Unit, Unpublished Technical Report. 9 p.

Nelson, M.E., and L.D. Mech. 1986. Mortality of white-tailed deer in northeastern Minnesota. Journal of Wildlife Management 50: 691-698.

Noss, R.F. 1990. Indicators for monitoring biodiversity: A hierarchical approach. Conservation Biology 4: 355-363.

Obbard, M.E., J.G. Jones, R. Newnham, A. Booth, A.J. Satterthwaite, and G. Linscombe. 1987. Furbearer harvest in North America. In: M. Novak, J.A. Baker, M.E. Obbard, and B. Malloch (editors). Wild Furbearer Management and Conservation in North America. Toronto, Ontario: Ontario Trappers Association and Ontario Ministry of Natural Resources. 1007-1043.

O'Neill, R.V., D.L. DeAngelis, J.B. Waide, and T.F.H. Allen. 1986. A Hierarchical Concept of Ecosystems. Princeton, New Jersey: Princeton University Press. 253 p.

Ontario Ministry of Natural Resources 1990. Management of Timber for Featured Wildlife Species. Wildlife Branch, Policy 6.04.01. Toronto, Ontario: Ontario Ministry of Natural Resources, Wildlife Branch. 4 p.

Ontario Ministry of Natural Resources. 1995. Forest Operations and Silviculture Manual. Toronto, Ontario: Ontario Ministry of Natural Resources. 64 p.

Ontario Ministry of Natural Resources. 1996. Forest Resources of Ontario 1996. Toronto, Ontario: Queen's Printer for Ontario. 86 p.

Ontario Ministry of Natural Resources. n.d. A Conservation Strategy for Old Growth Red and White Pine Forest Ecosystems in Ontario. Toronto, Ontario: Ontario Ministry of Natural Resources. 1 p.

Otto, R.D. 1996. An evaluation of forest landscape spatial pattern and wildlife community structure. Forest Ecology and Management 89: 139-147.

Outram, A.A. 1967. Changes in the mammalian fauna of Ontario since Confederation. Ontario Naturalist. September: 19-21.

Passmore, R.C. and R.L. Hepburn. 1955. A Method for Appraisal of Winter Range of Deer. Research Report No. 29. Maple, Ontario: Ontario Department of Lands and Forests. 7 p.

Pearl, D.L. and D.R. Voigt. 1995. Interactions between Deer and Vegetation in Southern Ontario: Monitoring and Restoration of Overgrazed Plant Communities in Pinery and Rondeau Provincial Parks. Technical Report TR-010. Toronto, Ontario: Ontario Ministry of Natural Resources, Southern Region Science and Technology Transfer Unit. 49 p.

Perera, A.H. and D.J.B. Baldwin. 2000. Spatial patterns in the managed forest landscapes of Ontario. In: A.H. Perera, D.L. Euler, and I.D. Thompson (editors). Ecology of a Managed Terrestrial Landscape: Patterns and Processes of Forest Landscapes in Ontario. Vancouver, British Columbia: University of British Columbia Press. 74-99.

Perera, A.H. and D.L. Euler. 2000. Landscape ecology in forest management: an introduction. In: A.H. Perera, D.L. Euler, and I.D. Thompson (editors). Ecology of a Managed Terrestrial Landscape: Patterns and Processes of Forest Landscapes in Ontario. Vancouver, British Columbia: University of British Columbia Press. 3-11.

Perera, A.H., F. Schnekenburger, and D. Baldwin. 1998. LEAP II: A Landscape Ecological Analysis Package for Land Use Planners and Managers. Forest Research Report No. 146. Sault Ste. Marie, Ontario: Ontario Ministry of Natural Resources, Ontario Forest Research Institute. 88 p.

Peterson, G. 1996. Forest History, Fire Spread and Climate Change: Cross-Scale Dynamics in Boreal Forests. Sustainable Forestry Partnerships, Conference Summaries. Edmonton, Alberta. 169 p.

Peterson, R.L. 1955. North American Moose. Toronto, Ontario: University of Toronto Press. 280 p.

Peterson, R.L. 1957. Changes in the mammalian fauna of Ontario. In: F.A. Urquhart (editor). Changes in the Fauna of Ontario. Toronto, Ontario: Royal Ontario Museum and University of Toronto Press. 43-58.

Peterson, R.L. and V. Crichton. 1949. The fur resources of the Chapleau District. Canadian Journal of Forest Research 27: 68-84.

Peterson, R.O. 1977. Wolf ecology and prey relationships on Isle Royale. National Park Service. Scientific Monograph Series No. 11. 210 pages.

Pickett, S.T.A. and P.S. White (editors). 1985. The Ecology of Natural Disturbance and Patch Dynamics. Orlando, Florida: Academic Press. 208 p.

Pickett, S.T.A., J. Kolasa, J.J. Armesto, and S.L. Collins. 1989. The ecological concept of disturbance and its expression at various hierarchical levels. *Oikos* 54: 129-136.

Pickett, S.T.A., M. Shachak, R.S. Ostfeld, and G.E. Likens. 1997. Toward a comprehensive conservation theory. In: S.T.A. Pickett, R.S. Ostfeld, M. Shachak, and G.E. Likens (editors). The Ecological Basis of Conservation: Heterogeneity, Ecosystems and Biodiversity. New York, New York: Chapman and Hall. 384-399.

Potvin, F., J. Huot, and F. Duchesneau. 1981. Deer mortality in the Pohenegamook wintering area, Quebec. Canadian Field-Naturalist 95: 80-84.

Potvin, F., H. Jolicoeur, and J. Huot. 1988. Wolf diet and prey selectivity during two periods for deer in Quebec: decline vs. expansion. Canadian Journal of Zoology 66: 1274-1279.

Powell, R.A. 1994. Effects of scale in habitat selection and foraging behavior of fishers in winter. Journal of Mammalogy 75: 349-356.

Powell, R.A. 1979. Ecological energetics and the foraging of the fisher. Journal of Animal Ecology 48: 195-212.

Preble, N.A. 1941. Raccoon management in central Ohio. Release No. 161. Ohio Wildlife Research Station. 9 p.

Pulliam, H.R. 1988. Sources, sinks, and population regulation. American Naturalist 132: 652-661.

Racey, G.D., T.S. Whitfield, and R.A. Sims. 1989. Northwestern Ontario forest ecosystem interpretation. Technical Report No. 46. Forest Resource Development Agreement. Thunder Bay, Ontario: Ontario Ministry of Natural Resources, Northwestern Ontario Forest Technology Development Unit. 90 p.

Raine, R.M. 1983. Winter habitat use and responses to snow cover of fisher and marten in southeastern Manitoba. Canadian Journal of Zoology 61: 25-34.

Raine, R.M. 1987. Winter food habits and foraging behaviour of fishers and martens in southeastern Manitoba. Canadian Journal of Zoology 65: 745-747.

Rempel, R., D. McKenney, A. Perera, J. Baker, A. Bisset, and D. Voigt. In preparation. Effects of landscape pattern, climate, and hunting on distribution and abundance of moose.

Rempel, R. S., P.C. Elkie, A.R. Rodgers, and M.J. Gluck. 1997. Timber-management and natural-disturbance effects on moose habitat: landscape evaluation. Journal of Wildlife Management 61: 517-524.

Risenhoover, K.L. and S.A. Maass. 1987. The influence of moose on the composition and structure of Isle Royale forests. Canadian Journal of Forest Research 17: 357-364.

Robbins, C.S., D.K. Dawson, and B.A. Dowell. 1989a. Habitat area requirements of breeding forest birds of the middle Atlantic states. Wildlife Monographs 103: 1-34.

Robbins, C.S., J.R. Sauer, R.S. Greenberg, and S. Droege. 1989b. Population declines in North American birds that migrate to the Neotropics. Proceedings of the US National Academy of Sciences 86: 7658-7662.

Robinson, S.K., F.R. Thompson III, T.M. Donovan, D. Whitehead, and J. Faaborg. 1995. Regional forest fragmentation and the nesting success of migratory birds. Science 267: 1987-1990.

Sanderson, G.C. 1987. Raccoon. In: M. Novak, J.A. Baker, M.E. Obbard, and B. Malloch (editors). Wild Furbearer Management and Conservation in North America. Toronto, Ontario: Ontario Trappers Association and Ontario Ministry of Natural Resources. 487-499.

Sauer, J.R., J.E. Hines, G. Gough, I. Thomas, and B.G. Peterjohn. 1997. The North American Breeding Bird Survey Results and Analysis. Version 96.3. Laurel, Maryland: Patuxent Wildlife Research Center.

Schaefer, J.A. and W.O Pruitt. 1991. Fire and woodland caribou in southeastern Manitoba. Wildlife Monograph No. 116. 39 p.

Schieck, J., K. Lertzman, B. Nyberg, and R. Page. 1995. Effects of patch size on birds in old-growth montane forests. Conservation Biology 9: 1072-1084.

Schmiegelow, F.K.A. 1990. Insular Biogeography of Breeding Passerines in Southern Ontario Woodlots: A Rigorous Test for Faunal Collapse [M.Sc. thesis]. Guelph, Ontario: University of Guelph. 53 p. + appendices.

Schmiegelow, F.K.A., C.S. Machtans, and S.J. Hannon. 1997. Are boreal birds resilient to forest fragmentation? An experimental study of short-term community responses. Ecology 78: 1914-1932.

Schneider, R. R. and P. Yodzis. 1994. Extinction dynamics in the American marten (*Martes americana*). Conservation Biology 8: 1058-1068.

Schwartz, C.C., W.L. Regelin and A.W. Franzmann. 1987. Seasonal weight dynamics of moose. Swedish Wildlife Research Supplement 1: 301-310.

Seip, D.R. 1992. Factors limiting woodland caribou populations and their interrelationships with wolves and moose in British Columbia. Canadian Journal of Zoology 70: 1494-1503.

Simberloff, D. 1997. Biogeographic approaches and the new conservation biology. In: S.T.A. Pickett, R.S. Ostfeld, M. Shachak, and G.E. Likens (editors). The Ecological Basis of Conservation: Heterogeneity, Ecosystems and Biodiversity. New York, New York: Chapman and Hall. 274-284.

Smith, C.H. 1983. Spatial trends in Canadian snowshoe hare, *Lepus americanus*, population cycles. Canadian Field-Naturalist 97:151-160.

Snyder, L.L. 1957. Changes in the avifauna of Ontario. In: F.A. Urquhart (editor). Changes in the Fauna of Ontario. Toronto, Ontario: Royal Ontario Museum and University of Toronto Press, Toronto, Ontario. 26-42.

Spytz, C.P. 1993. Cavity-nesting bird populations in cutover and mature boreal forest, northeastern Ontario [M.Sc. thesis]. Waterloo, Ontario: University of Waterloo. 88 p.

Statutes of Ontario. 1994. *Crown Forest Sustainability Act*. S.O. 1994, c. 25.

Tebbel, P.D. 1981. The status, distribution, and nesting ecology of sandhill cranes in central Ontario [M.Sc. thesis]. London, Ontario: University of Western Ontario. 66 p.

Tebbel, P.D. and C.D. Ankney. 1982. The status of sandhill cranes in central Ontario. Canadian Field-Naturalist 96: 163-166.

Telfer, E.S. and J.P Kelsall. 1984. Adaptations of some large North American mammals for survival in snow. Ecology 65: 1828-1834.

Temple, S.A., and J.R. Cary. 1988. Modeling dynamics of habitat-interior bird populations in fragmented landscapes. Conservation Biology 2: 340-347.

Terbourgh, J. 1989. Where Have All the Birds Gone? Chicago, Illinois: Princeton University Press. 207 p.

Thatcher, V.S. (editor). 1971. The New Webster Encyclopedia Dictionary of the English Language. Chicago, Illinois: Consolidated Book Publishers.

Thomas, D.C., E.J. Edmonds, and W.K. Brown. 1994. Comparative diet of woodland caribou populations in west-central Alberta. Unpublished report. Alberta Fish and Wildlife Department, Edmonton, Alberta. 21 p.

Thomas, M.K. 1957. Changes in the climate of Ontario. In: F.A. Urquhart (editor). Changes in the Fauna of Ontario. Toronto, Ontario: Royal Ontario Museum and University of Toronto Press. 59-75.

Thomas, J. W., E. D. Forsman, J. B. Lint, E. C. Meslow, B. R. No, and J. Verner. 1990. A Conservation Strategy for the Northern Spotted Owl. Report of the Interagency Scientific Committee to Address the Conservation of the Northern Spotted Owl. Portland, Oregon: USDA Forest Service. 427.

Thomasma, L.E., T.D. Drummer, and R.O. Peterson. 1994. Modeling habitat selection by fishers. In: S.W. Buskirk, A.S. Harestad, M.G. Raphael, and R.A. Powell (editors). Martens, Sables, and Fishers: Biology and Conservation. Ithaca, New York: Cornell University Press. 297-315.

Thompson, I.D. 1980. Effects of weather on productivity and survival of moose in northeastern Ontario. Proceedings of the North American Moose Conference 16: 463-481.

Thompson, I.D. 2000a. Forest vegetation of Ontario: factors influencing landscape change. In: A.H. Perera, D.L. Euler, and I.D. Thompson (editors). Ecology of a Managed Terrestrial Landscape: Patterns and Processes of Forest Landscapes in Ontario. Vancouver, British Columbia: University of British Columbia Press. 30-53.

Thompson, I.D. 2000b. Forest vertebrates of Ontario: patterns of distribution. In: A.H. Perera, D.L. Euler, and I.D. Thompson (editors). Ecology of a Managed Terrestrial Landscape: Patterns and Processes of Forests in Ontario. Vancouver, British Columbia: University of British Columbia Press. 54-73.

Thompson, I.D. and P.W. Colgan. 1987. Numerical responses of marten to a food shortage in Ontario. Journal of Wildlife Management 51: 824-835.

Thompson, I.D. and D. Euler. 1987. Moose habitat in Ontario: A decade of change in perception. Swedish Wildlife Research (Viltrevy) Supplement 1 :181-194.

Thompson, I.D. and D.A. Welsh. 1993. Integrated resource management in boreal forest ecosystems: Impediments and solutions. Forestry Chronicle 69: 32-39.

Urban, D.L., R.V. O'Neill, and H.H. Shugart, Jr. 1987. Landscape ecology, a hierarchical perspective can help scientists understand spatial patterns. Bioscience 37: 119-127.

Van Horn, M. A. 1990. The relationship between edge and pairing success of the ovenbird (*Seiurus aurocapillus*) within Missouri forest fragments. Auk 112: 98-106.

Venier, L.A. 1996. The effects of amount of available habitat in the landscape on relations between abundance and distribution of boreal forest songbirds [Ph.D. thesis]. Ottawa, Ontario: Carleton University, Department of Botany. 195 p.

Verme, L.J. 1970. Some characteristics of captive Michigan moose. Journal of Mammalogy 51: 403-405.

Verner, J., K.S. McKelvey, B.R. Noon, R.J. Gutiérrez, G.I. Gould, Jr., T.W. Beck, and J.W. Thomas (technical coordinators). 1992. The California Spotted Owl: A Technical Assessment of its Current Status. General Technical Report PSW-GTR-133. Albany, California: USDA Forest Service.

Voigt, D.R. and J.B. Broadfoot. 1995. Effects of cottage development on white-tailed deer, *Odocoileus virginianus*, winter habitat on Lake Muskoka, Ontario. Canadian Field-Naturalist 109(2): 210-204.

Voigt, D.R., J.D. Broadfoot, and J.A. Baker. 1997. Forest Management Guidelines for the Provision of White-tailed Deer Habitat. Toronto, Ontario: Ontario Ministry of Natural Resources. 33 p.

Voigt, D.R., M. Deyne, M. Malhiot, B. Ranta, B. Snider, R. Stefanski, and M. Strickland. 1992. White-tailed Deer in Ontario: Background to a Policy. Draft document. Ontario Ministry of Natural Resources, Wildlife Policy Branch, Toronto, Ontario. 83 p.

Walker, D.A. and M.D. Walker. 1991. History and pattern of disturbance in Alaskan arctic terrestrial ecosystems: a hierarchical approach to analysing landscape change. Journal of Applied Ecology 28: 244-276.

Ward, P.C. and A.G. Tithecott. 1993. The Impact of Fire Management on the Boreal Landscape of Ontario. Publication No. 305. Sault Ste. Marie, Ontario: Ontario Ministry of Natural Resources, Aviation, Flood, and Fire Management Branch. 12 p.

Watkins, L. and R. Davis. 1997. SFMM Tool Version 1.4: Description and Users Guide. Sault Ste. Marie, Ontario: Ontario Ministry of Natural Resources. 22 p.

Watt, W.R., J.A. Baker, D.M. Hogg, J.G. McNicol, and B.J. Naylor. 1997. Forest Management Guidelines for the Provision of Marten Habitat. Toronto, Ontario: Ontario Ministry of Natural Resources. 26 p.

Wedeles, C.H.R., P.N. Duinker, and M. J. Rose. 1991. Wildlife-Habitat Management Strategies: A Comparison of Approaches for Integrating Habitat Management and Forest Management. Report prepared for the Ontario Ministry of Natural Resources by ESSA Ltd., Richmond Hill, Ontario. 66 p.

Wedeles, C.H.R. and L. Van Damme. 1995. Effects of clear-cutting and alternative silvicultural systems on wildlife in Ontario's boreal mixedwoods. NODA/NFP Technical Report TR-19. Sault Ste. Marie, Ontario: Canadian Forest Service. 56 p.

Welsh, D.A. 1987. The influence of forest harvesting on mixed coniferous-deciduous boreal bird communities of white and red pine communities in Ontario. *Oecologica* 8: 247-252.

Welsh, D.A. and D.A. Kirk. n.d. Breeding Bird Communities of White and Red Pine Communities in Ontario. Unpublished report. Canadian Wildlife Service, Nepean, Ontario. 55 p.

Welsh, D.A. and S.C. Lougheed. 1996. Relationships of bird community structure and species distributions to two environmental gradients in the northern boreal forest. Ecography 19: 194-208.

Whitcomb, R.F., C.S. Robbins, J.F. Lynch, B.L. Whitcomb, M.K. Klimkiewicz, and D. Bystrak. 1981. Effects of forest fragmentation on avifauna of the eastern deciduous forest. In: R.L. Burgess and D. M. Sharpe (editors). Forest Island Dynamics in Man-Dominated Landscapes. New York, New York: Springer-Verlag. 125-205.

White, P. and J. Harrod. 1997. Disturbance and diversity in a landscape context. In: J. Bissonette (editor). Wildlife and Landscape Ecology. New York, New York: Springer-Verlag. 128-159.

Whitney, G.G. 1986. Relation of Michigan's presettlement pine forests to substrate and disturbance history. Ecology 67:1548-1559.

Wilcove, D.S. 1985. Nest predation on forest tracts and the decline of migratory songbirds. Ecology 66: 1211-1214.

Wilton, M.L. 1984. How the moose came to Algonquin. *Alces* 23: 89-106.

SECTION III:

Policy and Planning for Forest Landscape Management

Historical forest management policy contains many changes in approach, and the current landscape is as much a product of this diverse history as of such natural forces as fire. Ontario's current forest management policy contains many of the same principles being developed in forest management across North America.

Section III reviews past and current forest management policy in Ontario, identifies challenges that can be foreseen in implementing current policy, and discusses approaches to forest land management that offer a reasonable process for responding to these challenges.

• Chapter 12 details the development of forest use and management in Ontario from before European settlement to the early 1990s, providing a close examination of the political process and socio-economic climate of each era that gave rise to forest policy development. • Chapter 13 takes up the story from passage of the *Crown Forest Sustainability Act* in 1994, an event which has reset the targets and methods of Ontario forest management planning in a fundamental way. The chapter reviews basic concepts and key provisions of the *CFSA* and discusses the first forest management plans prepared under the Act.

The following chapters provide a broad perspective on the processes of policy formation and strategic land use planning, relative to forest management. • Chapter 14 discusses various modalities of strategic planning and identifies process characteristics required to plan for future forest land use, under economic conditions that cannot be known at present. • Chapter 15 discusses adaptive resource management as a response to the uncertainty inherent in forest management planning and a process for integrating the findings of science into Ontario forest management policy.

The final chapter, • Chapter 16, summarizes the main themes of the book, discusses particular challenges facing forest managers and planners in Ontario, and identifies information requirements which researchers must meet to support the new processes of forest management planning.

12 Ontario Forests and Forest Policy Before the Era of Sustainable Forestry

A. ERNEST EPP*

THE GROWTH OF THE FORESTS

Although human activity has changed the forests dramatically during the past two centuries, this era of agricultural settlement and industrial development represents little more than a moment in geological time. Over the past 20,000 years, however, the forest endowment that Ontarians take for granted has been greatly affected by climatic change (Pielou 1991). A succession of ice ages, and the warmer periods that occurred between them, led to startling changes in the forests of North America. The Wisconsinan glaciation of the late-Pleistocene period, for example, allowed boreal forest to grow far down the Appalachian Mountains. At its maximum extent some 18,000 to 16,000 years ago, this boreal forest extended from the Atlantic coast to the Mississippi valley across what is now the northern United States (Harris 1987). As the glaciers melted and largely disappeared during an abrupt warming about 10,000 years ago, the boreal forest grew northward across the present New England states, southern Ontario, and the American Midwest.

The first boreal forest that succeeded the glaciers consisted largely of white spruce (*Picea glauca*), whereas the later boreal forest consisted of black spruce (*Picea mariana*) (Pielou 1991). This northward "movement" of the boreal forest was accompanied by the development of muskeg, which came to cover half of northern Ontario. As the climatic warming reached a height in the hypsithermal period some 7000 to 3000 years ago, the white pine (*Pinus strobus*) that later characterized the Great Lakes-St. Lawrence Forest grew almost as far north as James Bay, and eastern white cedar (*Thuja occidentalis*) flourished in the Claybelt. (The successive phases of forest growth are clearly depicted in Plate 4 of *the Historical Atlas of Canada: I. From the Beginnings to 1800* [Harris 1987].) Only in the last 3000 years has Ontario's forest endowment achieved its present array of deciduous (Carolinian) forest in southern Ontario, the Great Lakes-St. Lawrence Forest on the southern parts of the Precambrian Shield, and the boreal forest covering much of northern Ontario (Rowe 1972).

The conditions that greeted humans as they returned to these recently glaciated areas are hard to imagine (Pielou 1991). The forest returned very slowly, and there is evidence in Lake Superior of great dust storms during the period from 9500 to 6000 BP (before present). The hypsithermal period reached its peak in northern Ontario about 6000 years ago. During those years, white pine grew as much as 200 km north of its present limits (Pielou 1991). Evidence that the forest grew beyond the present tree line is provided by the remains of a forest fire that have been dated to about 3500 BP. In such conditions of climatic warmth, grass and forest fires were frequent phenomena and imperilled the lives of the Aboriginal people moving into these territories. The Plano culture that developed between 10,500 and 8000 BP was dependent in part on the hunting of large animals, including bison (*Bos americanus*) and caribou (*Rangifer tarandus*) (Harris 1987). These denizens of the forest were killed to

* Department of History, Lakehead University, 955 Oliver Road, Thunder Bay, Ontario P7B 5E1

obtain meat for food, hides for clothing, bones for tools, and sinews as fasteners (Dawson 1983; Dickason 1992; Wright 1995).

The climatic tendency during much of the last 4000 years has been increasing coolness and wetness (Harris 1987; Pielou 1991). The main effect has been the spread of muskeg, which was related to the carving out by earlier glaciation of low areas in which plants grew and died. Cooling conditions forced a retreat of white pine and an advance of jack pine (*Pinus banksiana*), which had reached its northernmost limit by 2400 BP. Black spruce, meanwhile, grew well in the muskeg areas. Cooler summers allowed the black spruce to flourish, spreading as it did by the natural cloning process known as layering (Pielou 1991). Conditions improved somewhat during the Little Climatic Optimum, however, which reached its peak about 1800 years ago. Studies in northern Wisconsin suggest that forest fires occurred frequently during this period (Pielou 1991). It is possible that some of these fires were set by Aboriginal people in the same way that fire was used to manage the forest in other areas (Canadian Forest Service 1997).

This was also the period of the Initial Woodland or Laurel culture. In northern Ontario, the people of the Laurel culture continued to live by the hunting of beaver (*Castor canadensis*), caribou, and moose (*Alces alces*). Ceramics were used for storage and for cooking, although no local plants were cultivated. The medicinal value of various plants was gradually discovered during these centuries of Aboriginal life in the forest (Dickason 1991). Forest resources were also used to make birchbark canoes and containers, snowshoes, toboggans, and skin clothing. The limited food resources of the boreal forest were supplemented by fishing in the lakes and rivers, especially in such rivers as the St. Mary's River and the Rainy River (Holzkamm, Lytwyn, and Waisberg 1991).

THE TERMINAL WOODLAND PERIOD

The period immediately preceding European settlement in what is now southern Ontario has been characterized as the Little Ice Age. During this period, which may have begun soon after AD 1000 and ended about AD 1750, the Algonkian peoples around the Great Lakes lived through the period anthropologists have labelled "Terminal Woodland" and entered the historical period resulting from European activity in the New World (Dawson 1983). New types of pottery appeared during this period, such as those among the Ojibwa that have been labelled "Blackduck." The means of subsistence remained largely unchanged, however, as these forest residents continued to depend on the resources of the forest for their sustenance and survival (Johnston 1995). Where food resources were abundant, a richer economy and culture could be achieved. It is possible that the pictographs that constitute such a treasure of Aboriginal art were painted during this era (Harris 1987). Still another important people, the Iroquoian-speaking Wendats (or "Hurons," as the French called them) of the Georgian Bay area, cut fields out of the Laurentian forest and cultivated maize and other crops to enrich their diets (Trigger 1976; Trigger 1985; Harris 1987; Dickason 1992). Even before European newcomers in the sixteenth century began to offer European manufactured goods in trade for the furs that could be obtained from forest animals, these various Aboriginal groups were trading with each other. The Wendats, for example, exchanged their maize for northern forest products such as meat, furs, and ritual objects.

The European fur trade, which began in the St. Lawrence watershed during the last third of the sixteenth century and on James Bay a century later, increased the value of the forest significantly (Innis 1956). It did so, of course, by substantially changing the relationship between the Aboriginal peoples and the creatures of the forest (Martin 1978; Bishop 1981). This relationship had been a profoundly spiritual one in which humans and animals lived together in mutual dependence. The hunter sought to attune himself to the spirits of the animals and even more to the master-spirit Keeper of the Game. The animal was asked to sacrifice itself to the needs of the hunter and his family. The hunter kept faith with the animal by killing it in a respectful manner and treating the remains properly. This complex of customs and taboos made the forest a spiritual place. It should be said that there is debate about various aspects of Aboriginal use of resources (Rogers 1986). The spiritual relationship between humans and animals can still be appreciated in effigy mounds and pictographs left by the Aboriginal people (Dewdney and Kidd 1962; Dewdney 1970).

When European traders offered manufactured goods that enhanced the quality of life, they also encouraged the killing of animals, not to meet the immediate needs of the hunter and his family but rather to obtain the furs needed to buy these trade goods. Metal cutting edges certainly opened new possibilities, not least of all in the processing of furs. Much of the trade, however, came to be in cloth and clothing items (Anderson 1994). The ironic result was an exchange of pelts, which were used to make clothing in Europe, for cotton and wool cloth that Aboriginal people used to dress themselves in the New World. Given the importance of the fashionable beaver hat to the fur trade, one might well argue that the prime beneficiary of the trade was the Aboriginal hunter and his family. On the other hand, the forest lost something of its spiritual nature for many Aboriginal people. These people began to see the forest in the commercial way that became dominant in the industrial era (Martin 1980).

Until at least the middle of the 17th century, most of the territory that now constitutes Ontario remained the domain of various First Nations (Trigger 1976; Trigger 1985). Although Jesuit missionaries and their servants lived among the Wendats, the fur trade largely remained in the hands of Wendats, who travelled far into what is now northern Ontario (Harris 1987). The lakes and rivers provided the routes along which they travelled to obtain furs and transport them to the European posts. The forest was the home of the animals that were killed for pelts, and the forest provided the material for making canoes. Fuel was also obtained from the forest, as were the materials for every kind of shelter. The dispersion about 1650 of the Wendats and their neighbours by war parties of the Iroquoian-speaking Five Nations did not change the relationship between humans and the forest.

European traders, originally as unlicensed *coureurs de bois* and later as *voyageurs* employed by Montreal merchants, took up the work the Wendats had done earlier (Rich 1966). For a time, they faced the competition of the Algonquian-speaking Odawa and Ojibwa, but the Montrealers took over the trade of the Canadian Shield far more rapidly than the Hudson's Bay Company moved beyond the Bays to the north. For more than a century (1670-1774), the Hudson's Bay Company depended on Cree middlemen to bring furs from their forest suppliers to the posts on Hudson and James Bay (Ray 1974). Agriculture in what is now southern Ontario did decline for a time, but the region did not become the hunting territory of the Five Nations, as historians used to believe. In fact, the "war the Iroquois did not win" at the end of the 17th century saw Algonquian-speaking people winning battles and settling between Lake Huron and Lake Ontario (Eid 1979; Schmalz 1991).

THE SETTLEMENT ERA, 1784-1875

Human occupation across all of what is now Ontario remained a variety of Aboriginal villages and European posts until the fourth quarter of the 18th century. However, anticipations of the agricultural future of southern Ontario could be seen at both ends of that region: in the Montreal area, where the French began to farm in the 1640s, and at Detroit, where European settlement gradually developed through the 18th century. The outbreak of the American Revolutionary War in 1775 and the flight of the Loyalists to safety in British territory launched a new era of agricultural settlement (McCalla 1993). The United Empire Loyalists were almost entirely farmers, although some merchants and professionals settled at Kingston and in the Niagara peninsula. The Loyalist demand for freehold tenure and a government of their own choosing led the British government in 1791 to cut the colony of Upper Canada (later Ontario) out of the French domain obtained by war in 1759-60 and to give the new colony its own government. The coming of the United Empire Loyalists also forced the British government to negotiate with the Aboriginal inhabitants in order to acquire land – including that of the loyal Mohawks – for Loyalist settlement (Craig 1963; Gates 1968).

Once Aboriginal title had been ended, clearing of the forests of southern Ontario began in earnest (Kelly 1975). The forest provided the material for the cabins and homes the Loyalists built. Whether it was a log cabin or a frame building, the structure required the cutting of trees and the preparation of logs or boards. Sawmills were constructed at the waterfalls on various streams as quickly as were the grist mills that converted grain into flour (Gentilcore, Measner, and Walder 1993, Plate 16). The timber and lumber demands created by settlement gave new value to the forest endowment of southern Ontario. It is worth making this point about value obtained from the forest, even in the late 18th century, in order to minimize the brute reality of cutting a farm out of the forest. Clearing the land was a terribly wasteful process when trees were girdled or cut down green, piled up so that they could dry well, and then burned to remove this impediment to farming activities. The ashes were valuable, of course, especially in the 19th century when they were purchased by merchants and processed for export to the British Isles. They also provided the first cash crop for many settlers.

Still, this process of clearing land in order to grow crops was carried on with little regard for the trees. In fact, the forest was generally seen as a threat to civilized life, which was inhabited by wild animals and which inhibited travel to one's neighbours and to the towns where products could be sold and goods purchased. Ironically enough, the Ontario Bicentennial publication on the white pine provides relevant quotations of these pioneer attitudes (Morse 1984). The clearing of land was also a slow and difficult process. Study of census records suggests that farmers forced to cut trees themselves and to clear the land did very well to "improve" one hectare a year (Russell 1982, 1983). The fact that a lifetime's work might produce a field of only 24 ha makes it easy to appreciate how great a challenge the forest was for the settler. The forest should not be seen only in this negative light, however, since it was an essential source of fuel and enabled farmers to spend some of their winter time cutting staves for sale. Pigs could root about in the forest, and cows grazed among the trees, although a grassy clearing surely provided more nourishment than did dense forest (Gentilcore, Measner, and Waldner 1993, Plate 14).

THE SQUARE TIMBER TRADE

If settlers cleared the Great Lakes-St. Lawrence Forest quite slowly, the square timber trade that began in the Ottawa River valley in the first decade of the 19th century removed pine trees much more rapidly (Lower 1973; Morse 1984; Reid 1989; McCalla 1993). The origin of this great trade was ironic. The residents of the densely settled British Isles had long since cut most of the forests on their islands. The timber demands of the Royal Navy remained, however, especially during the drawn-out struggles of the French Revolutionary and Napoleonic Wars. When Napoleon sought to defeat the British by cutting off their European trade by his Continental System in 1805, he imperilled the British import of timber and other naval stores from the Baltic region and Scandinavia. The British government responded to this threat by offering merchants a bounty in the form of a preference on timber from British North America, and British importing houses sent representatives to St. John (New Brunswick) and Quebec. The organization of timber production in New Brunswick, Lower Canada (later Quebec), and Upper Canada followed quickly. The business thus begun survived the end of the Napoleonic Wars in 1815 and continued in the Ottawa River watershed through the rest of the 19th century. This trade flourished by exploitation of the great stands of red and white pine that surrounded the Ottawa River and its tributaries. The initial cutting was of red pine (*Pinus resinosa*); white pine did not come to the fore until about 1845. Organization of this production and financing of the trade required resources that only a few could muster. Rivalry between French Canadians and Irishmen for work as lumbermen, however, led to the "Shiner's War" of the 1830s and posed a real challenge to social order in that part of Upper Canada (Cross 1973; Reid 1989).

The economic impact of the square timber trade was substantial, although much of it was felt in Lower Canada, after the timber rafts had been floated down to Quebec City and the "sticks" loaded into ships for export to British ports (Lower 1973; McCalla 1993). An annual average of 3.8 million board feet of timber had been cut in the second half of the 1830s, but timber production reached 20 million board feet in 1845-46. Exports from Quebec (coming from both of the Canadas) reached peaks of some 708,000 m^3 in 1854, 1857, and 1861; the all-time high was 991,000 m^3 in 1863. More than half of the timber makers appear to have come from Lower Canada, and even those that did not are likely to have spent some of their earnings in Lower Canada before returning to their homes. Employment in the timber trade was probably in the range of 1500 men during the 1830s, but it rose towards a high of 8000 in the 1840s. On the other hand, much of the pork and flour required to feed the lumbermen, like the feed required by their oxen and horses, may have been provided by Upper Canada farms. The animals themselves were likely bred on the farms of both Lower and Upper Canada. The axes and other equipment used to cut down the trees and square the timber are less likely to have come from Upper Canada.

THE PRODUCTION OF LUMBER

The parallel growth of a trade in planks, or "deals," as they were called in England, led to the construction of large sawmills (Gentilcore, Measner, and Walder 1993; McCalla 1993). The Hamiltons built a deal mill at Hawkesbury soon after the Napoleonic Wars ended in 1815 and could produce 300,000 deals annually by the mid-1830s. They had doubled the capacity of their mill by 1847, and there were a further 11 deal mills in the Ottawa Valley by that time. All of this forest production generated revenue for the Upper Canadian government; the charges reached almost £8000 in 1831, exceeded that level throughout the years 1836 to 1840, and must have risen even higher during the 1840s and the following decades. The operation of other sawmills whose product was exported to the United States intensified the pressure on the forests of southern Ontario (Lower 1938; Gentilcore, Measner, and Walder 1993, Plate 16). This pressure became even more intense during the period of the Reciprocity Treaty, 1854 to 1866, the latter part of which coincided with the Civil War and great US demand for Canadian supplies. However, the high point of the period for exports (over 500 million board feet) occurred in 1870-71 (McCalla 1993).

As settlement spread across the fertile areas of southern Ontario, a process that continued through the middle of the 19th century, the forests of this region gradually disappeared (Head 1975; Kelly 1975). One of the political grievances of the time, the Clergy Reserves authorized by the *Constitutional Act* of 1791, aroused anger as much for the fact that these lands were not being taken up by pioneers, who preferred to own the land they cleared, as for the fact that the proceeds of the lease or sale of these Reserves were initially monopolized by the Anglican Church (Wilson 1968). When the Clergy Reserves were abolished in 1854, these areas of forest also began to fall to the farmer's axe. The introduction of farm machinery in the middle decades of the century forced farmers to remove even the stumps that had remained in the fields earlier (Pomfret 1976). Farm improvement thus destroyed much of the southern Ontario forest, although the wise farmer retained a woodlot as one of the assets of his farm (Gentilcore, Measner, and Walder 1993, Plate 14).

USES OF THE FOREST IN THE 19TH CENTURY

The agricultural settlement of Upper Canada represented a challenge to British policy regarding the forests and forced a significant development in colonial policy (Nelles 1974). Imperial law had long provided that the pine trees of North America were reserved to be made into masts for the ships of the Royal Navy. This policy had been applied to the American colonies, and it was declared for the colony of Quebec after the British conquest. Enforcement of this policy in the southern parts of Upper Canada could have prevented the settlement already described. When colonial officials declined to enforce the law, however, the pine trees in the settlement area were cut and used in the diversity of ways already noted. The southern areas of southern Ontario thus continued to produce both pine sawlogs and wheat through the third quarter of the 19th century (Head 1975; McCalla 1993). While the square timber trade was facing reductions of preferences during the 1840s, as Britain moved toward free trade, the export of sawlogs to Oswego, New York (not to mention other ports), was growing from 2 million board feet in 1840 to 60 million board feet in 1850. An industry concentrated along Lakes Erie and Ontario at mid-century had spread throughout southern Ontario by 1871.

The export of Ontario forest products to the United States was a response to urban growth there and the housing demands of a rapidly growing population. This initial American assault on the Canadian forest might also be seen as an extension of the American frontier into British North America. In the southern areas of Ontario, the American pattern of private ownership of forest lands prevailed, and imperial forest policy was ignored. As settlement moved toward the rocky regions of the Canadian Shield, however, settlers came to be regarded as fraudulent applicants who claimed the land only to cut the trees on it before moving on to another district (Nelles 1974). It is possible that some of them followed the southern model in clearing the land of its trees as a necessity of farm improvement, only to discover that the land they had cleared was not suited to farming. For the timber companies of the Ottawa valley, however, any threat to "their" forests was reason enough to press for an Ontario variant on the imperial policy. The ultimate result was Ontario's policy combining Crown ownership of forested land with a specified right to cut timber on that land granted by the government to individuals and corporations in return for specific payments.

As the square timber trade gradually changed after the Napoleonic Wars from supplying the navy to meeting civilian needs in an industrializing Britain, the initial imperial and military system in the colonies also changed (Nelles 1974). In 1826, the Executive Council of Upper Canada opened the military licensing system to commercial timber operators, who could obtain licenses to cut timber on Crown land in return for the deposit of a personal bond and their promise to pay for the timber they cut according to a set schedule of duties. The first statutory provision for forest policy, the *Crown Timber Act* of 1849, came from the first "Responsible Government" cabinet of the province and enshrined the Canadian combination of Crown ownership of land and commercial removal of timber growing on that land. Initially, security of tenure for the licence holder depended on his swift removal of the timber and surrender of the land for agricultural settlement. This same Canadian government ordered a reconnaissance of the upper Great Lakes country in 1849 and the negotiation by W.B. Robinson of treaties with the First Nations of Lake Huron and Lake Superior that opened the door to the mineral and forest exploitation of much of northern Ontario (Morris 1880).

The statutory expectation that land cleared of timber would be turned into farms demonstrated the extent to which forest operations and agricultural settlement went hand in hand through much of the 19th century. The use of wood for fuel, in towns and cities as well as on farms, provided a further demonstration of the general value of the forest. Maintenance of a woodlot on each well-established farm revealed the farmer's recognition of this same reality. A fuelwood survey in 1886 estimated that consumption had peaked in 1881 at 5.4 million cords (19.6 million m^3) of firewood. The fuelwood crisis then beginning to develop in southern Ontario was only averted by the import of coal from the United States and the gradual conversion of railway locomotives and other steam engines to the use of coal (Gentilcore, Measner, and Walder 1993, Plate 46; K.A. Armson, pers. comm., January 1998).

Improvement of roads to the ports and other towns of Upper Canada was an enormous challenge as long as the forest prevailed. Ironically enough, one of the responses to this problem drew on the forest in the form of logs laid side by side to create "corduroy" roads on which wagons and coaches bumped along above the muddy ground in which they would otherwise have been mired. These wagons and carriages, as well as the sleighs used during the winter, were all made of wood (Small 1884). Transportation improvements in the middle decades of the 19th century gave increased value to forest products (McCalla 1993). Shipbuilding had drawn on the lumber production of the colony from early in its history, of course, and forest products such as staves had been among the most important articles of freight carried on the schooners and barges of the lower Great Lakes and the St. Lawrence River. The use of steam-powered vessels from the 1830s created a new demand for wood as fuel for their engines (McCalla 1993). Railway locomotives likewise used wood as fuel for some time before the railway companies turned to imported coal (Gentilcore, Measner, and Walder 1993, Plate 46). Railway construction also required enormous numbers of wooden ties as the base on which rails were laid (and relaid) (Bertrand 1997). Every rail line was matched by a telegraph line, which similarly required large quantities of wooden poles to support the wires carrying the messages. The invention of the telephone later in the 19th century added to this demand, as did the construction of transcontinental railway lines in the 1880s and after the turn of the century. The impact of the railway on the forest extended to the construction of rolling stock made in part of wood. Ironically, one of the other impacts of the railway on the forest involved the fires started by sparks from locomotive stacks and the braking of iron wheels.

Transportation-related uses of wood added much value to forest production. Other industrial uses had a more varied effect (Small 1884). Southern Ontario was a significant producer and exporter of squared oak (McCalla 1993). It also exported smaller quantities of tanbark. Both of these had industrial potential within Ontario, too. The production of cabinets, furniture, and musical instruments, especially pianos, depended on the output of the Southern Ontario hardwood forest (Small 1884; Parr 1990). These industries developed in south-central Ontario, especially in such towns as Hanover, where German craftsmen applied their skills to wood in order to create fine furniture that would enhance the homes of middle and upper class Canadians. Not unrelated to this activity was the production of hardwood flooring, moulding, and other wooden items required to finish the fine home (Small 1884). This output also depended on mills where wood was cut and shaped by specialized machinery. Still another kind of "forest," with its own distinctive products, were the fruit orchards of southern Ontario.

By way of contrast, the discovery that paper could be made from wood fibres by large-scale, mechanized processes led to the construction of large mills where wood was reduced to its constituent fibres by chemical or mechanical means and processed into various grades of paper (Zaslow 1971; Sinclair 1987). The main product was newsprint, which fed the steam-powered presses on which the mass newspapers of the late 19th century were printed. The development of the pulp and paper industry began in the south, but grew largely in the north of the province after the turn of the century, where both the Great Lakes-St. Lawrence forest and boreal forest found new use by an industry more notable for the high capitalization of its plant than for the number of employees required to work these machines (Gentilcore, Measner, and Walder 1993). The employment demands of the industry remained large in the forest, of course, as long as pulpwood was cut by axe and crosscut saw. Such methods of cutting prevailed until after the middle of the 20th century (Radforth 1982, 1987, 1995).

Forest industry statistics illustrate these changes in the uses made of the forest (Drummond 1987). Although square timber production totalled 949,000 m^3 in 1870 and climbed to 1.47 million m^3 in 1880, it had collapsed to only 108,000 m^3 by 1900. Total lumber production in Ontario was reported in 1900 to be 3200 m^3 (1,353,575 thousand feet board measure [MFBM]), and increased to 4800 m^3 (1,642,191 MFBM) in 1910. Only 65.1 cords of pulpwood were reported as cut in 1900, but production reached 90.7 cords in 1910-11 and climbed to 140.3 cords in 1911-12. The value of pulp and paper production grew by 9.2 percent annually over the period from 1890 to 1929, with most of that growth occurring after 1912. All of these uses of wood were already to be found in H.B. Small's *Canadian Forests: Forest Trees, Timber and Forest Products* (Small 1884), although the wood being pulped at that time was aspen rather than spruce.

RECREATIONAL USE OF THE FOREST

This focus on the cutting of trees and the various uses to which wood was put in the 19th century should not obscure the fact that the forest also came to be appreciated aesthetically in the latter part of that century (Jasen 1995). The first area to arouse aesthetic interest was the Thousand Islands of the upper St. Lawrence River and Lake Ontario. An appreciation of forests developed on the upper Great Lakes, as artists travelled to these areas by steamer to paint scenery that the Group of Seven would later make even more famous. Southern Ontarians came to their most avid appreciation of the forest in the near-northland, which had long been regarded primarily as territory where "settlers" obtained grants of forest land in order to cut the timber and then abandoned the "homestead" as land unsuited to farming. As rest cures in the wilderness became a form of therapy for city-dwellers in the late 19th century, the Muskoka region took on a new meaning and value (Smith 1990). James Dickson's *Camping in the Muskoka Region* encouraged canoeing, fishing, and camping there as early as 1886 (Lambert with Pross 1967).

The reservation of land around the Niagara Falls in the mid-1880s provided an international demonstration of the aesthetic importance of forest lands. Lord Dufferin, the Governor-General, used the occasion of a visit to New York State in 1878 to encourage "restoring the locality to its pristine condition of wild and secluded beauty" (Hardy 1880). When New York and Ontario representatives, the latter led by Premier Oliver Mowat, met at Niagara Falls in September 1879, they agreed "to restore and preserve as nearly as possible the natural characteristics of the site ... but not lay out a park or an artificial enclosure" (Hardy 1880). The Ontario government believed that "preservation of the site of Niagara Falls [was] national, international, or even cosmopolitan, in its nature" and passed legislation in the hope that the Dominion government would create the reserve. Only after the Dominion government failed to do so and New York State had acted by 1885 to create "The State Reservation of New York" did Ontario appoint a Board of Commissioners for the Queen Victoria Niagara Falls Park. Their Third Annual Report described the work done "to restore to some extent the scenery around the Falls of Niagara to its natural condition, and to preserve the same from further deterioration; as well as to afford travellers and others facilities for observing the points of interest in the vicinity" (Board of Commissioners for the Queen Victoria Niagara Falls Park 1889).

The ultimate policy response to the perceived need for recreational facilities was the creation of Algonquin Park in a central part of the Great Lakes-St. Lawrence Forest. Establishment of national parks in both the United States and Canada inspired the original suggestion (described in the next section) that the Ontario government reserve an area of forest as "a

National Forest and Park to be called Algonkin Forest and Park" (Kirkwood 1886) Kirkwood expressed concern about protecting the headwaters of the streams down which the Ottawa valley lumbermen were floating logs and stressed the role of forests in maintaining the ecology of the area. When Algonquin Park was created by passage of the *Algonquin Park Act* in 1893, the *Act* described it as a park for the public, a reservation for fish and game, and a health resort and pleasure ground for the benefit, advantage and enjoyment of the people of the province. Not until 1902 did George Bartlett make a canoe trip across the Park from north to south, however, clearing the portages as he went and making the Park more accessible to the public (Lambert with Pross 1967).

Although hunting was prohibited in Algonquin Park, a good deal of effort was required to stop illicit fur traders and poachers from preying on the protected wildlife. The opportunity to hunt or fish drew sportsmen into northern Ontario from the middle of the 19th century. One may wonder whether these hunters saw the creatures of the forest as their Aboriginal predecessors had done. Something of what remained of the age-old symbiosis with the creatures may have been conveyed by Native guides in such areas as Temagami, which people from southern Ontario also discovered in the late 19th century (Hodgins and Benidickson 1989). It should not be forgotten, of course, that the fur trade remained an important commerce for those who lived in the forests of northern Ontario (Zaslow 1971).

THE BEGINNINGS OF FOREST CONSERVATION, 1870-1900

Another phenomenon of the last third of the 19th century was a growing appreciation of the ecological importance of forests. The Vermont naturalist, George Marsh, made a powerful impression with his book, *Man and Nature*, first published in 1865 (Lambert with Pross 1967; Gillis and Roach 1986). Marsh's emphasis on the climatic consequences of deforestation was keenly appreciated by some observers of southern Ontario, who found the denuding of the moraine ridges of southern Ontario as great an environmental loss as was the cutting of trees in the rocky Muskoka region. John Astin Warder's reforestation efforts in Ohio aroused interest in Canada as well as the United States. Some of the concerned Ontarians joined the American Forestry Association organized by Warder and others in 1875. Official support for these efforts was obvious in the *Report on the Necessity of Preserving and Replanting Forests*, compiled by R.W. Phipps and published in 1883 "at the instance of the Government of Ontario" (Phipps 1883).

Reforestation efforts in southern Ontario were also encouraged by the Ontario Agricultural College. William Brown, Professor of Agriculture at the College, read a paper entitled "The Application of Scientific Arboriculture to Canada" at the Montreal meeting of the British Association for the Advancement of Science in 1884 in which he asserted, "A peopled agricultural country is an impossibility without trees" (Brown 1884). It should be noted, however, that he referred to British instances of reforestation, specifically the Seafield estates in Banff and Invernesshire and the Invercauld estates in Aberdeenshire, rather than contemporary American efforts. The reforestation activity that the Ontario Fruit Growers Association encouraged during these years found a useful guide in T.B. White's 1886 pamphlet, *Arboriculture and Agriculture, or Forestry and Farming in Ontario* (White 1886).

A leading figure in the development of silviculture in Canada was the Quebec seigneur-politician, Henri Gustave Joly, who had reported on forestry to the Parliament of Canada in 1878 and addressed the matter of returns on forest tree culture in 1882 (Joly 1878, 1882). Joly played an important part in Quebec's taking up the Midwest American holiday, Arbour Day, in order to educate children and the community about the importance of trees (Joly 1878). Ontario was only a little behind the province of Quebec in these efforts to enhance its endowment of what had until recently been viewed chiefly as a barrier to agricultural settlement. In efforts to deal with the forest fire threat, Quebec passed its *Forest Fire Prevention Act* in 1870 and Ontario, an *Act to Preserve Forests from Destruction by Fire* in 1878 (Lambert with Pross 1967).

Depletion of the forests of the Ottawa Valley, which had for so long been a source of employment and wealth, also began to arouse concern as early as the 1870s. In two pamphlets published in 1872 and 1876, James Little asserted that the United States was rapidly depleting its forest endowment and warned of a coming rush into the Canadian forest that would decimate this forest, too (Gillis and Roach 1986). Little advocated strict government regulation of forest activities, under which each year's cutting should not exceed the annual growth of trees. This pioneering statement of a sustained-yield approach to the forest struck a sympathetic chord with some of the large Ottawa Valley lumbermen, but most people remained complacent about the "endless" Ontario forest. By 1880, nonetheless, a common concern about the forests had developed among the members of the Ontario Fruit Growers Association, many of whom were the scientific farmers of their time, and the timber barons of the Ottawa Valley (Gillis and Roach 1986).

The Canadian members of the American Forestry Association were probably as surprised as their American colleagues to be invited to an American Forestry Congress in Cincinnati, Ohio, in April 1882 (Gillis and Roach 1986). An Ontario delegation was soon organized, however, including four members of the Ontario Fruit Growers Association and an entomologist, Professor William Saunders of the Ontario Agricultural College. Attending from Montreal were James Little and his son, William, and A.T. Drummond, a lawyer-financier and author of articles about forestry (Drummond 1879). James Little had, in fact, been asked to be the keynote speaker and used the occasion to reiterate his concern about the future of both the American and Canadian forests. Little and D.W. Beadle, secretary of the Ontario Fruit Growers Association, were appointed to a committee to consider the future of the Congress (and conceivably of the American Forestry Association). They suggested that the meetings of the American Association for the Advancement of Science and the Society for the Promotion of Agricultural Science, to be held in Montreal that August, might serve as the occasion for another Forestry Congress.

The Forestry Congress that met in Montreal in August 1882 carried the Canadian forest conservation movement vigorously forward (Gillis and Roach 1986). Whereas, Canadian attendance at the Cincinnati gathering involved chiefly the Ontario delegation, William Little's efforts now brought major figures of the Canadian forest industry into the Congress. The Quebec Limitholders' Association, whose members were facing increasing competition in the forests from the Roman Catholic Church's colonization movement, was soon involved. Among these men were such notable Ottawa Valley timber barons as the Honourable George Bryson, Sr., John R. Booth, and Peter White, MP. The last spoke for the industry politically, as the Conservative Member for Renfrew North.

Chosen as chair of the Congress was H.G. Joly, who was keenly aware that settlers' activities in the forests and the forest fires blamed on them were the leading concerns of the timber companies. The Congress struck a forest fire committee that included Bryson, Joly, and White, and this committee recommended that governments deal with the problem of forest fires by establishing new regulations. Among the most significant suggestions were that land not suited to agriculture should be set aside in forest reserves, that the burning of brush not be allowed when there was a danger of forest fires, and that a force be established to enforce these regulations. The fact that the Quebec government had sent its Minister of Crown Lands and several officials to the Montreal Congress helps to explain why that government responded to these recommendations by establishing the first Canadian forest reserve in the Ottawa Valley and providing authority to hire forest rangers (paid by the holders of the forest limits) to patrol the limits. Unfortunately, a change of government led to the cancellation of the forest reserve soon afterward (Gillis and Roach 1986).

THE ESTABLISHMENT OF FOREST RESERVES

The Ontario government acted somewhat more slowly than the Quebec government, but it achieved much more when it did act (Lambert with Pross 1967; Gillis and Roach 1986). In 1883, the Mowat government created the new position of Clerk of Forestry in the Department of Agriculture and the Arts. The appointment to this position of Robert W. Phipps, a writer on forest matters, represented government support for the efforts of the Ontario Fruit Growers Association in encouraging tree planting and good woodlot management in southern Ontario. The lumbermen's concerns required more study, and Aubrey White, formerly a woods ranger and Crown Lands Agent at Bracebridge, spent until March 1885 satisfying himself that the recommendations of the Congress were good ones. The government then agreed to appoint fire rangers to patrol the forests during the fire season, to educate the public about the risks of fire, and to organize fire-fighting efforts when they were required. The 37 men recommended by the forest companies and hired in the summer of 1885 were the first of many such guardians of the forest to be paid jointly by the companies and the government.

The following year, 1886, saw another important recommendation from Alexander Kirkwood, whose advice on northern settlement a few years earlier had included a hair-raising description of how to burn windrows of cut trees and brush in the heat of summer (Kirkwood and Murphy 1878). Kirkwood now turned his pen to advocacy of an "Algonkin National Forest and Park" (Kirkwood 1886). In deliberate Ontario style, however, Algonquin Park was not created until 1893, 11 years after the convening of the Montreal Forestry Congress (Lambert with Pross 1967; Gillis and Roach 1986; Killan 1993). However, a second park, Rondeau, was created only a year later to protect the best Ontario example of the Carolinian forest on the north shore of Lake Erie, where a caretaker had been guarding the forest since 1876 (Lambert with Pross 1967; Killan 1993).

The latter years of the 1890s saw the Ontario government giving still more attention to the forests (Lambert with Pross 1967). A year after Phipps died in 1894, another journalist was appointed as Ontario's Clerk of Forestry. The transfer of Clerk Thomas Southworth to the Crown Lands Department enabled him to advocate better forest practices by the lumbermen, as well as the tree planting and good woodlot management that were the concerns in southern Ontario. His 1896-97 survey of white pine regeneration provided an example of such action, although the report was primarily important for its conclusion that white pine would regenerate naturally if protected from human depredation (Gillis and Roach 1986). The appointment in 1897 of a Royal Commission on Forest Protection in Ontario, chaired by progressive lumberman E.W. Rathbun and including both Kirkwood and Southworth, revealed a growing provincial interest in forest conservation. The Rathbun Commission recommended an extension of the system of fire rangers, but under the control of the Department of Crown Lands, and the setting aside of forest areas as reserves. Passage of the *Forest Reserves Act* in 1898 gave Southworth the authority to establish the Eastern, Mississaga, Temagami, and Sibley Forest Reserves (Lambert with Pross 1967).

The Clerkship of Forestry was converted into a Bureau in 1898, with Southworth as director. Within a year, he and his chief, Aubrey White, had completed and published a *History of the Crown Timber Regulations from the Date of the French Occupation to the Year 1899* (Lambert with Pross 1967). One of Southworth's continuing concerns, the need to establish both private woodlots and public forest reserves, found a new expression with the discovery of the northern Claybelt. He now began urging the provincial government to create forest reserves in that area in order to control the cutting of the forest. As noted earlier, Algonquin Park had been established for a variety of uses. The historians of the Temagami region have observed that, "although popular writers often associated Algonquin Park with the later system of forest reserves and even viewed Algonquin as the first provincial forest reserve, experienced observers generally distinguished between the two concepts: parks such as Algonquin emphasized watershed management, game preservation, and recreation alongside lumbering, while forest reserves were primarily expected to ensure the perpetual use by the industry of successive crops of pine timber" (Hodgins and Benidickson 1989, p. 71).

One of the forest reserves, Temagami, quickly became an area of multiple use of the forest. The reserve was created in 1901 to protect an area of virgin white and red pine forest thought to be one of the best stands of timber in the province. The existence within it of a First Nation which had missed the negotiation of the Robinson Huron Treaty in 1850 and remained without a reserve was largely ignored by the Ontario government. The beauty of the region impressed every visitor, however, and canoeing its waterways became a very popular activity. Recreational canoeing led naturally to the establishment of camps by both Americans and Canadians, who saw canoe tripping as a character-building activity for youth. Construction of the Timiskaming and Northern Ontario Railway, which reached the Northeast Arm of Lake Temagami in 1903, and later of the Ferguson Highway, which was opened to Lake Temagami in 1927, made the region readily accessible. A reserve that included both great pine forests and substantial pulpwood stands thus became a region of value both to those who appreciated the forest and to those who profited from this appreciation (Hodgins and Benidickson 1989).

GROWTH OF THE PULP AND PAPER INDUSTRY

While the Ontario government was acting to reserve areas of forest in various parts of the province, it was also making concessions of forest to a new industry eager to exploit the previously neglected stands of spruce (Lambert with Pross 1967; Gillis and Roach 1986). Cutting of wood for pulp and paper mills was first reported by the Crown Lands Department in 1892.

A pioneering agreement was arrived at that year by the government of Ontario and the Sault Ste. Marie Company. In return for spending $200,000 by December 1895 to build a mill that would employ 300 men for at least 10 months of the year, this company was given permission to cut wood on 50 sq. mi. of forest "running back upon either side of one or more rivers flowing into Lake Superior ... west of Sault Ste. Marie" (Lambert with Pross 1967, p. 251). The latter provision reflected the use of river runs as the mode of transportation of pulpwood; a grant as large as 50 sq. mi. was required to obtain the large capital invested in the mill.

By the turn of the century, similar agreements provided for mills in five other locations: at Renfrew and Sturgeon Falls, and on the Mattawa, Spanish, and Nipigon Rivers. The terms of these large grants – and the frequent need to extend their terms in order to give the holders time to obtain the required financing and build their mills – aroused the attention of the Conservative Opposition, which promised to change the system if they formed the government. When James P. Whitney led the Conservatives to power in 1905, his Lands, Forests and Mines Minister Frank Cochrane cancelled five pulpwood concessions and offered these resources for public sale. Unfortunately, only two of the areas sold by tender in 1906. Another attempt to sell the other three in 1907 also failed. Not until the US market for newsprint was opened more substantially to Canadian production after 1911, and that despite the Canadian rejection of the Reciprocity Agreement, did the pulp and paper industry come into its own in Canada (Epp 1973). It has been argued, however, that the commercial and manufacturing policies of both the United States and Canada were less important to the growth of the newsprint industry in Canada than was the American demand for newsprint and the Canadian capacity to meet this demand (Dick 1982).

Although delayed in the pulp and paper industry, the American assault on the Canadian forest aroused a significant policy response from the Ontario government just before the 19th century ended (Lambert with Pross 1967; Nelles 1974; Gillis and Roach 1986). The problem related to US interests which had the right to cut sawlogs in Ontario and floated these raw materials to their sawmills in Michigan. Ontario's Georgian Bay forests thus generated their most valuable returns in the United States. As early as 1886, US interests controlled as much as 1750 million board feet of standing timber on Georgian Bay. Canadian lumbermen began to press George Ross's Liberal government for action to ensure that sawlogs were processed in Canada. The US Congress reacted in 1897 with a duty of $2 per thousand feet on all sawn lumber entering the United States; a further provision of this Dingley tariff would add the amount of any export duty Canada placed on sawlogs on top of this US lumber duty. When it became clear that the Canadian government had been rendered impotent, the Ontario Lumbermen's Association turned to the Ontario government. The Ross government responded with a "manufacturing condition," under which all timber cut on Crown lands had to be sawn into lumber in Ontario. As exports dwindled to sawlogs cut on private and Indian lands, Michigan operators built mills on Georgian Bay, and Ontario mill owners enjoyed a boom that had recently appeared quite impossible.

PROFESSIONAL FORESTERS AND THE CONSERVATION MOVEMENT, 1900-20

Although scientific forestry developed more slowly in Canada than in the United States, this new profession began to be recognized in Ontario early in the 20th century (Lambert with Pross 1967; Gillis and Roach 1986). The Canadian government played an important role in this development, a result of the fact that the Department of the Interior was responsible for the use of natural resources throughout the vast territory that had once been Rupert's Land (although most of this territory was in the prairie West, it also included the northern parts of northern Ontario). A Forestry Commissioner had been appointed as early as 1884 and the Canadian government soon began to establish forest reserves to limit the tree-cutting activities of settlers, real and claimed. A Forestry Branch was created in 1899 and a Canadian Forestry Association organized in 1900 (Sisam 1961; Foster 1978).

The German-born and -trained B.E. Fernow was a leading force in the development of forestry in Canada. Dr. Fernow had attended both the Cincinnati and Montreal Forestry Congresses, and he addressed a public meeting of the Royal Society of Canada in 1894 on the importance of forestry education (Sisam 1961). These comments aroused interest at Queen's University, where Fernow was asked to lecture again in 1901. Queen's desire to establish a school of forestry led the university in 1903 to invite Fernow to give a series of lectures on forestry in Canada. As the official historian of the Department of Lands and Forests has observed (Lambert with Pross 1967, p. 186), these lectures "caused a sensation in forestry circles" as they "ranged over every aspect of Canadian forestry – including resource management, tree-cropping, silviculture, lumbering, forest economy, wood characteristics, forestry practice, etc. – in such a way as to provide, when published later in book form (80 pages), a succinct manual that has been used for the education of subsequent generations of foresters and others."

In the meantime, faculty at the University of Toronto wished to see a provincial school of forestry established in their own institution (Sisam 1961, p. 13-15). President Dr. James Loudon informed the Canadian Forestry Association at its fifth annual meeting in 1904 that the Senate of the University of Toronto had already authorized a three-year program in forestry. Although the Association declined to favour any university, it did urge the Ontario government "to make an appropriate grant for the operation of a provincial school or schools of forestry." The Royal Commission appointed in 1905 to study the University of Toronto recommended a year later that a forestry school be established at that university. Having been told that "a great work in forestry can be done in this Province by the University, provided it receives the cooperation of the Government," the Board of Governors used its powers under the *University of Toronto Act* (1906) to create a Faculty of Forestry on March 28, 1907. This became the first forestry school in Canada, with Dr. Fernow as its first dean.

This development occurred despite the somewhat mixed experience the Ontario government had with its first professional forester (Lambert with Pross 1967). Dr. Judson F. Clark was appointed to the Bureau of Forestry in 1904 as Ontario's first Provincial Forester. A graduate of the Ontario Agricultural College, Clark had gone on to graduate work with Fernow at Cornell University and completed a doctorate before joining the faculty there. Although Clark initially focused his energies on southern Ontario, he was critical of existing forest practices. In a presentation to a national forest conference, he urged that the holders of forest licenses be required to clean up their limits in order to lessen the risk of fire and to assist in regeneration. He went on to criticize the Doyle Scale which was commonly used to measure timber. The suggestion that Ontario should adopt his own "International Scale" in order to obtain a better return on its forest resources received no support from Aubrey White while (or because?) it enraged the forest operators. When Whitney's new Conservative government returned the Bureau to the Department of Agriculture in 1905 as the Bureau of Colonization and Forestry, Clark resigned to join a lumber company in British Columbia.

In the meantime, the Canadian Forestry Association was developing an interest in forest conservation. One of the best means to this end was the Canadian Forestry Convention it convened in Ottawa in 1906 (Foster 1978; Gillis and Roach 1986). In his opening speech, Prime Minister Sir Wilfrid Laurier expressed concerns that echoed the "wise use" concepts of Gifford Pinchot, briefly Chief Forester in the US Department of Agriculture and then Chief of the Bureau of Forestry in the US Department of the Interior. As the honoured guest at this Convention, Pinchot had every opportunity to emphasize that "forestry with us is a business proposition" and to declare that, "if our forests are to stand unused there, if all we get out of them is the knowledge that we have them, then, so far as I am concerned, they disappear from my field of interest" (Foster 1978, p. 35). The principle that "use is the end of forest preservation, and the highest use" (p. 35) provided the rationale for forest reserves in Canada as well as the United States. It had also guided the establishment of Algonquin Park where, as the historian of the Department of Lands and Forests has observed, "the precaution was taken of disposing of all merchantable timber remaining in the Park area" when the Park was being created (Lambert with Pross 1967, p. 173).

When US President Theodore Roosevelt decided to convene a National Conservation Conference in February 1909, he invited the Canadian government to participate (Foster 1978; Hall 1985). The Laurier government responded by sending a delegation that included Clifford Sifton, Member of Parliament (MP) for Brandon and Prime Minister Laurier's Minister of the Interior from 1896 to 1905. A Declaration of Principles adopted by the Conference stated that the "following measures should be encouraged: afforestation, reforestation, protection of forested land at the headwaters of streams, public ownership of forested land, protection of land better suited for forest growth than for other purposes, reforestation of private land, increase in fire protection services, strict regulation of cutting" (Foster 1978, p. 36). This summary of attitudes towards the forest, Canadian as well as American, suggested the direction that the Commission of Conservation created by Parliament that spring would take (Hall 1985). Commission Chair Sifton told the Canadian Club in 1910 that "conservation means the utilization of our resources in a proper and economical way" for the benefit of Canadians (Foster 1978, p. 41).

Ontario participation in the work of the Commission of Conservation began at the first meeting, when Lands, Forests, and Mines Minister Frank Cochrane addressed the "Conservation of Natural Resources" in Ontario and spoke particularly of the forest reserves. K. Evans addressed the situation of "Fish and Game" in Ontario (Canada Commission of Conservation 1910). The participation of members of the Faculty of Forestry could be seen in such reports as Fernow's 1916 discussion of "Silvicultural Problems in Forest Preserves," which included the observation that, "while the apparent economy, in relying on Nature's ability to establish a new crop, is in favor of natural regeneration, [thus] avoiding the cash outlay necessary to start the crop by artificial means ... in the end the latter often proves the cheaper" (Canada Commission of Conservation 1916, p. 72). Fernow had already, under the auspices of the Commission of Conservation, organized the Trent Watershed Survey in 1912 and a survey of conditions in the new agricultural settlement in the Claybelt in 1913. The Trent Watershed Survey, carried out by Drs. C.D. Howe and J.H. White of the Faculty of Forestry, anticipated the watershed surveys made by the Department of Planning and Development after the Second World War. Out of this activity would grow the Conservation Authorities and protection for more of the Ontario forest (K.A. Armson, pers. comm., January 1998).

CONCERNS ABOUT FOREST REGENERATION

Fernow's observation on regeneration anticipated articles by his colleague, Dr. Howe. A report on "Forest Regeneration on Certain Cut-Over Pulpwood Lands in Quebec" focused on the neighbouring province but arrived at a conclusion of great interest to Ontarians: "white pine is not reproducing itself" (Canada Commission of Conservation 1918, p. 59). Only a year later Howe asked, "How Shall we Make our Forests Safe for Trees?" (Canada Commission of Conservation 1919). After "repeating Senator Edwards's statement that over twenty times as many [trees] have been killed by fire as have fallen before the lumberman's axe" (Canada Commission of Conservation 1919, p. 177), Howe asserted that "fires on cut-over lands or on recuperating old burns are the most destructive to future values." His survey of 32,000 hectares of "cut-over and burned-over pine lands in the central portion of 'old' Ontario" (p. 177) had revealed 272 young pines per ha in areas burned once but only 35 pine trees per ha on areas burned twice. A third fire reduced the average to 17 trees and a fourth fire, to 7 trees. In a further discussion of spruce budworm attacks and the "following-up diseases, a beetle and a weevil, a heart rot, and a sap rot," Howe asked whether "our treatment of the forests ... the accumulation of slash and the increased number of injured and weakened trees as the result of logging operations" (p. 178) might not be causing these diseases. From his study of the limited regeneration of spruce in cut-over areas – Howe concluded that "only 7 trees will remain to take the place of the 26 per acre removed by the logging operations" – he asserted that "Forests Cannot Regenerate Themselves" (pp. 180-181). The last parts of this seminal paper urged the Ontario government to protect the forest against fires and advocated the reinvestment of forest revenues in an industry that Howe saw as creating over $400 million of wealth annually.

Howe's comments on regeneration, which were clearly oriented towards the growth of spruce forests for the pulp and paper industry, found an ironic complement in pioneering regeneration efforts by the Abitibi Paper Company (Armson 1993). The forest limits on which this company was based had been granted in 1912, and its Iroquois Falls mill and Twin Falls hydroelectric station were in operation by 1916. Only a year later, while the Great War continued in Europe, the company established a nursery near its power plant and began to experiment with trees purchased from the United States. The regeneration program envisaged the planting of over 4 million seedlings on areas burned in 1916 and 1921. Unfortunately, the soils and conditions of the Claybelt led to a 90 percent mortality rate among the trees, and the nurseries were abandoned in 1928. A survey 20 years later of 38 plantations, totalling 156 hectares of land on which 467,000 trees had been planted, found only 60,766 trees growing satisfactorily but noted that the white spruce had done best.

The contributions of professional foresters were only gradually accepted by the Ontario government (Lambert with Pross 1967). Their experience with Dr. Clark made them cautious. The next professional forester to find a permanent place in the public service was E.J. Zavitz, a graduate of both McMaster University and the Ontario Agricultural College who earned his Master of Science in forestry at the University of Michigan before beginning to lecture in forestry at the Ontario Agricultural College in 1908. Especially interested in farm forestry, Zavitz hoped to reforest parts of southern Ontario such as the Normandale plains of South Norfolk and areas of Simcoe County where deforestation had produced waste lands. In 1908, his *Report on the Reforestation of Waste Lands in Southern Ontario* was published (Zavitz 1908). That same year the Ontario Agricultural College acquired one hundred acres of abandoned farm land in Norfolk County and Zavitz began to experiment with large-scale reforestation there. After Zavitz was appointed to the Department of Lands, Forests and Mines in 1912, he continued his reforestation efforts in southern Ontario while becoming involved in fire protection work.

The death of Aubrey White in 1915 did not lead to an immediate promotion for Zavitz, but he continued his labours with the support from 1917 to 1919 of Dr. J.H. White, the first graduate of the University of Toronto Faculty of Forestry (Lambert with Pross 1967). They clarified the postwar program of the Forestry Branch as involving three policies: reforestation in southern Ontario, fire protection in northern Ontario, and the gathering of information about the timber supply of the province. When the *Forest Fires Protection Act* of 1917 extended the fire protection service, it became the responsibility of the new Provincial Forester, E.J. Zavitz. Not until 1926, however, did Zavitz become Deputy Minister of Forestry on an even footing with the Deputy Minister of Lands, Forests and Mines.

The American influence that led to Canada's creation of a Commission of Conservation had a more permanent impact in the establishment of Quetico Park (Lambert with Pross 1967; Killan 1993). Minnesotans had long been resisting timber baron E.W. Backus's forest and waterways ambitions on their side of the international boundary. The Member of the Provincial Parliament (MPP) for Rainy River, W.A. Preston, agreed with Minnesota's Forestry Commissioner, General C.C. Andrews, that the moose living in Quetico needed protection. In 1908, Arthur Hawkes, publicity agent for the Canadian Northern Railway (which traversed the Rainy River valley), carried an Ontario promise to St. Paul, Minnesota: if Minnesota would create a game and forest reserve on its side of the border, Ontario would do the same. Deputy Minister White's concern for pine stands in the area led to both the promise and the Order-in-Council of April 1, 1909, creating a Quetico Forest Reserve to match Minnesota's Superior National Forest. Four and a half years later, Quetico was made a provincial park in order to protect the animals living in the reserve. Although Lambert later described the Quetico area as "a heavily forested region, peopled in 1913 largely by Ojibway Indians who still practiced their primitive arts, living mainly by hunting, fishing and trapping, and adhered to their traditional tribal religion," these Aboriginal people soon found themselves being driven from their homeland (Lambert with Pross 1967, p. 284). This dispossession of the Aboriginal residents required ameliorative action by the Ontario government eight decades later, as it also did in Algonquin Park (McNab 1991; Killan 1993).

THE FORESTS BECOME CONTROVERSIAL, 1915-30

The ten years (and more) preceding the elevation of Zavitz to Deputy Minister of Forestry in 1926 was a period of intense controversy about Ontario forest management (Lambert with Pross 1967). Whitney's Conservatives had criticized the agreements that the Liberal government began making in the 1890s to develop the pulp and paper industry, but they found their own alternative of selling pulpwood concessions by public tender relatively unsuccessful. Consequently, as the industry began to develop during the second decade of the century, the Conservative government of William Hearst began to make its own agreements with those wishing to exploit the forests. The minister responsible was G. Howard Ferguson; his Deputy Minister, the former Conservative MPP for Algoma, Albert Grigg, was appointed after White's death and held the position until 1921.

As the author of *G. Howard Ferguson: Ontario Tory* (Oliver 1977, pp. 97-98) has stated in regard to the timber commission that later examined this period: "The testimony of such lumber barons as E.W. Backus of Backus-Brooks and Jim Mathieu of Shevlin, Clarke, a friend of Ferguson and MPP for Rainy River, showed how flagrantly the lumbermen of the far northwest had ignored Crown land regulations when they conflicted with their selfish interests." He added that "early in the hearings it became evident that it had been totally unrealistic to expect the White-Cochrane reforms of an earlier day, predicated as they were on close administrative control over an enormous and far-flung field force, to function effectively, particularly in the war period, which had so depleted the ranks of the provincial service." The fact that at least 3100 km^2 of forest lands were granted without any public competition or tender during the last year of Hearst's government fed charges that a "timber ring" had been operating in northwestern Ontario in the interest of the ruling party, if not of the minister himself (Lambert with Pross 1967, p. 263).

A new era appeared to have dawned when the United Farmers of Ontario (UFO) defeated Hearst's Conservatives in 1919 and formed a Farmer-Labour government led by E.C. Drury (Johnston 1986). Premier Drury was a keen supporter of southern Ontario reforestation, and his government established two new nurseries at Midhurst and Orono (in addition to the one at St. Williams established in 1908). They also established two new provincial parks on Lake Erie, at Long Point in 1921 and at Presqu'ile in 1922. (Deputy Minister Grigg, not incidentally, was replaced in 1921.) Drury's government also had the opportunity to begin using aircraft in forest surveys and protection efforts. The first operations in protection, air sketching, and aerial photography were undertaken in 1921, in conjunction with the Canadian Air Board. The work of this first season was promising and in 1922 a larger program was undertaken which included the survey of a large part of the province adjacent to James Bay, covering an area of [35,119 km^2] (Richardson 1927). This project represented the first major forest inventory done in Ontario (K.A. Armson, pers. comm., January 1998).

The UFO premier was also determined to learn the truth about forest administration under Hearst and Ferguson. As Drury appointed a timber commission to look into the awarding of pulpwood and timber contracts, Ferguson prepared to defend himself and save his political career (Lambert with Pross 1967; Oliver 1977). Justices W.R. Riddell and F.R. Latchford of the Supreme Court of Ontario soon discovered that large areas of forest had been granted without tender to various individuals and companies. As already noted, among the leading beneficiaries were the Backus-Brooks and Shevlin, Clarke companies, which were large operators in northwestern Ontario. Ferguson had been particularly busy granting timber rights without the required process of public tender during the months preceding the 1919 election. The suspicion that these grants had been made in return for financial support for Ferguson and/or the Conservative Party was inevitable (Oliver 1977). Some of the guilty parties recognized the advisability of confessing their violation of the regulations and making restitution to the government. Controversy swirled around the Commission, however, and strange facts came to light (including an earlier association by Commission counsel, R.T. Harding, with the Backus-Brooks interests). As the historian of the department has conceded, "The enquiry uncovered many shady and irregular deals but did not succeed in bringing home the responsibility for them" (Lambert with Pross 1967, p. 195). Ferguson insisted that he had done no wrong and was able to continue as leader of the Conservative Party.

Ironically, the Conservatives went on to win the 1923 election and Ferguson became Premier of Ontario. For a brief period in 1926, Ferguson even served as Minister of Lands, Forests and Mines. It was the Ferguson government, then, that made Zavitz the Deputy Minister of Forestry and expanded such activities as the county forests program, first authorized by the *Counties Reforestation Act* of 1911. Ferguson also strengthened the administration of the Department of Lands and Forests by creating the position of Inspector of Crown Timber Agents and Supervisor of Operations and appointing Major J.I. Hartt, "a hard-headed and competent public servant," to that position (Oliver 1977, p. 209). The Ferguson government recognized the value of aircraft for forest protection and surveying by establishing the Ontario Government Air Service in 1924. This Provincial Air Service was by 1929 operating 22 float-planes from its headquarters in Sault Ste. Marie and maintaining district bases at Sudbury, Orient Bay (on Lake Nipigon), and Sioux Lookout, and sub-bases at another nine locations across northern Ontario (Richardson 1927). The Ferguson government also passed a *Forestry Act* in 1927 which authorized a Forestry Board "to study all questions in connection with the planting, growth, development, marketing and reproduction of pulpwood" (section 17).

Appointed to this Forestry Board, in addition to Deputy Minister Zavitz and Dr. Howe of the University of Toronto, were B.F Avery, forester for the Spanish River Company at Sault Ste. Marie, J.A. Gillies, a lumberman based at Braeside, and H.D. Schanche, Abitibi's Chief Forester at Iroquois Falls (Departmental Report 1929; Armson 1993). Given the studies that Howe had carried out in the preceding decade and the costly experience that Abitibi was gaining, it was hardly surprising that the Board made "sweeping recommendations for tree-planting in the north, administration of the Province's lands by the Forestry Branch, and so on" (Lambert with Pross 1967, p. 201). Oddly enough, the departmental historian forty years later regarded the Board "as going too far too fast"; he also described "surviving members of the Department [as] inclined to write off the [*Forestry, Provincial Forests, and Pulpwood Conservation] Acts* as 'window-dressing', intended more for political showmanship than for actual administration." Lambert's colleague, Pross, clearly disagreed with this assessment, made by one of the other researchers involved in the project (Pross 1967). Efforts under the *Pulpwood Conservation Act* to obtain extensive data from the pulp and paper companies – this Act gave the Minister "the right and authority to direct and control the cutting from year to year, by any company, for the purpose of conserving the source of supply and placing Ontario on a sustained-yield basis" – were forward-looking and far more than window-dressing (Oliver 1977, p. 343).

THE PROFITABILITY OF THE FOREST INDUSTRY

Whatever one may think of Ferguson's forest administration, there can be no doubt of his determination in the late 1920s to restore the profitability of the Canadian pulp and paper companies (Oliver 1977). In later years he boasted to President F.J. Sensenbrenner of the Spruce Falls Pulp and Paper Company: "I have had much to do from the very beginning with the initiation of almost all the newsprint undertakings in Ontario and I have sought to lend what assistance I might, consistent with the protection of the province, to the placing upon a permanent basis [of] these great undertakings" (Oliver 1977, p. 208). In negotiations with the *New York Times* to establish the Spruce Falls company, Ferguson declared that the provincial government was "in a way the largest shareholders, because we contributed the power and the timber at a very reasonable price, that will undoubtedly enable your organization to flourish" (Oliver 1977, p. 211). The newsprint industry had expanded greatly after US publishers obtained almost free entry for their vital raw material. Investment in mills continued through the 1920s, while the price of newsprint declined from year to year (Bladen 1956; Dick 1982). Attempts to maintain the price failed time and again as one company or another signed a contract for less than that price and encouraged another price slide. These efforts at price maintenance were limited, of course, by the illegality of such collusion in restraint of trade under the laws of both Canada and the United States.

Those financially interested in the pulp and paper industry responded to these difficulties by combining mills under the control of a few companies. The hope was undoubtedly that an oligopoly could achieve price stability. Among the most important results of this array of mergers was the Abitibi Power and Paper Company, which controlled hydroelectric power facilities and pulp and paper mills from the lower St. Lawrence River to eastern Manitoba (Bladen 1956). These large companies also organized a Canadian Newsprint Company which "pooled the orders and allocated tonnage to the member mills on a rated daily basis" (Bladen 1956, p. 194). The continuing problem of companies breaking rank eventually led Ferguson to join Quebec Premier Alexandre Taschereau in providing government support for a newsprint cartel, the Newspaper Institute. The companies involved – which did not include the International Paper, Ontario Paper, or Spruce Falls Pulp and Paper companies – agreed to let the two premiers "act as arbitrators in the event of any dispute arising" over the prorating of production and sales (Bladen 1956; Oliver 1977, p. 348). This support by two provincial governments of a combination in restraint of trade certainly demonstrated the importance of this forest industry to the province of Ontario. The US Federal Trade Commission stated on July 2, 1930, that the cartel had forced the Hearst newspapers to pay an extra $5 a ton for their newsprint (Oliver 1977).

While Ferguson was doing his best to make the newsprint industry profitable, his government was also encouraging entrepreneurship in the provincial parks. Concerned by years of budgetary deficit, the Ferguson government sought to obtain new revenues by leasing lots within the parks. "From 1923 to 1928, two new lodges, four youth camps, and dozens of cottages appeared in Algonquin [while] the number of cottages in Rondeau rose from sixty in 1928 to 268 in 1931" (Killan 1992, p. 29). The commissions still responsible for Long Point and Presqu'ile were even more susceptible to this conversion of parks into cottage country. (Long Point had so many cottages by 1949 that the original site had to be abandoned and a new Long Point Park created!)

The "utility and profit" principles governing park policy did not go unchallenged. In 1928, Minnesotans organized the Quetico-Superior Council to protect the recreational and wilderness properties of the Rainy Lake watershed. They were determined to prevent E.W. Backus from building a series of dams that would raise the water level on the border lakes by as much as 24 m. Although it conceded that logging should continue, the Quetico-Superior Council "demanded ... a reworking of the wise-use equation to give primacy to the recreational and scenic values along the waterways of Quetico-Superior [specifically by having] all shorelines reserved and protected from logging, flooding, and other forms of exploitation" (Killan 1992, p. 30). Although the Council failed to obtain any treaty between Canada and the United States containing such a guarantee, "by 1941 shoreline reservations had been placed on all canoe routes in Quetico Park and adjacent sections of the Superior National Forest" (p. 30).

As the 1920s drew to a close, the work of the Department of Lands and Forests was presented to the public in two useful books. A.H. Richardson wrote the first, *Forestry in Ontario*, in which he discussed the following subjects: the forest itself, forest protection, air operations, reforestation, and forest surveys. The statistics provided in five appendices were preceded by Richardson's conclusions that the forests of Ontario were not inexhaustible but that about 49 percent of the forest was made up of mature stands of timber ready to be cut (Richardson 1927). This "comforting fact," as he called it, was underscored by the Department's 1931 publication, *The Forest Resources of Ontario 1930*, prepared by J.F. Sharpe and J.A. Brodie and reporting all of the surveys done since 1912. As the departmental historians have commented, "For the first time the Province's timber resources as a whole could be seen in an authoritative perspective" (Lambert with Pross 1967, p. 198).

THE GREAT DEPRESSION AND SECOND WORLD WAR

If there had been any doubt about the importance of the forest industries to more than just the woodsmen and millhands of northern Ontario, that doubt was dispelled as the Wall Street Crash of 1929 ushered in the Great Depression. Demand for newsprint and other forest products fell rapidly, and the great combinations of the late 1920s could not earn enough to pay interest on their bonds, much less dividends on the millions of dollars in stock that had been issued (Bladen 1956). The stock became worthless as the bond holders asserted their right to run the companies and take whatever profits could be earned as returns on their own financial instruments. Abitibi Power and Paper was only one of the companies run by receivers on behalf of the bond holders, and it did not come out of receivership until after the Second World War (Bladen 1956).

After the new Conservative Prime Minister, R.B. Bennett, appointed Ferguson as Canada's High Commissioner in London, George S. Henry was left to struggle with the economic and social problems of the Great Depression. One of his responses to high unemployment at the Lakehead was to suspend the *Crown Timber Act*'s prohibition on exports of wood to the United States, provided it was not used to make newsprint there (Saywell 1991). In as dramatic a style as Bennett had used in the 1930 federal election, Ontario Liberal leader Mitchell F. Hepburn promised in the 1934 provincial election to deal with the depression by "mak[ing] our natural resources available to enterprise" in order to "revive our forest industries and restore the Provincial revenues" (Lambert with Pross 1967, p. 335).

The forest operators appeared, as soon as Hepburn had taken office, to seek concessions "in the way of reductions, not only in Crown dues, but also on bonuses ... on a mutually co-operative basis ... and only where the operator undertakes to place a certain quota of workers in the bush, on the drive and in the sawmill" (as the Hepburn government explained them) (Lambert with Pross 1967, pp. 335-36). The government also sought to end bitter strikes in the woods by requiring the operators to increase wages, to reduce the charges in their company stores, and to comply with health regulations in their camps. The *Woodsmen's Employment Act* also empowered the Minister to investigate a dispute at any stage of difficulties in industrial relations (Lambert with Pross 1967). Similarly, in order to assist settlers who were selling pulpwood from their own or other patented lands, the

Hepburn government passed the *Settlers' Pulpwood Protection Act*. Minister of Lands and Forests Peter Heenan also persuaded the pulp and paper companies to increase their prices for pulpwood (Saywell 1991).

Perhaps the most remarkable demonstration of the importance of the forests to the province of Ontario was provided by passage in 1936 of the *Forest Resources Regulation Act* (Lambert with Pross 1967; Saywell 1991). The problem was that large areas of northern Ontario forest were held by companies that would not (or could not) cut wood on them. The Great Lakes Company, for example, held rights to no less than 23,085 km^2 of northwestern Ontario forest. The government's slogan was "get the men to work," in order to reduce the enormous burden of unemployment relief. Having obtained the power to negotiate with the companies, Heenan reduced the holdings of the bankrupt Great Lakes Company to 3668 km^2. The lands freed by these negotiations were conceded to new companies, some of which were given permission to cut for export in further violation of the "manufacturing condition" in the *Crown Timber Act* (Wightman and Wightman 1997, pp. 236-237). As a result, between 1934 and 1936, the number of forest operators increased from 149 to 227, the number of bush camps from 282 to 557, and the number of forest industry employees from 11,184 to 23,140 (Lambert with Pross 1967).

Hepburn's efforts in support of the forest industries also led him into a renewal of the cartelization efforts of the preceding decade. In cooperation successively with Quebec Premiers Taschereau and Maurice Duplessis, Hepburn supported the newsprint manufacturers in their "proration" of newsprint orders among the manufacturers (Bladen 1956). However, such companies as Ontario Paper and Spruce Falls Power and Paper, which produced exclusively for the *Chicago Tribune* and the *New York Times*, sought to maintain their freedom of action (and avoid prosecution). A more serious challenge came from the Great Lakes Company which, newly out of receivership in 1938, undercut the price the cartel had established (Lambert with Pross 1967). After a good deal of hesitation, the Hepburn government used its authority under the *Forest Resources Regulation Act* to penalize the company to the extent of $500,000, in order to force compliance with the agreement (Saywell 1991). A similar dispute, arousing the same response, occurred with Backus's Minnesota and Ontario Company, which operated mills at Fort Frances and Kenora.

FROM CONSERVATION TO PRESERVATION

While the Ontario government saw the forest primarily as a source of business activity, the Ontarians who organized the Federation of Ontario Naturalists in 1931 finally gave organizational expression to the preservationist wing of the conservation movement. The first publication of the Federation in 1934, *Sanctuaries and the Preservation of Wild Life in Ontario*, stated that the need for nature preservation did not depend solely on the economic value of the natural life believed worthy of conservation (Federation of Ontario Naturalists 1958). By calling for the preservation of representative samples of natural conditions, both plants and animals, the Federation of Ontario Naturalists was far ahead of both the public and the government in applying the insights of the new science of ecology. However, the acceptance of this objective by Algonquin Park Superintendent Frank A. MacDougall provided a striking demonstration of how far in advance of prevailing attitudes this public servant was (Killan 1992; Killan 1993).

The conflict between the utilitarian and preservationist wings of the conservation movement came to a head in Algonquin Park in 1938. The Hepburn government's emphasis on forest operations to create employment had expanded logging in the Park; there were as many as 14 licence holders with 1500 men at work already in the winter of 1934-35. At the same time, construction of Highway 60 enabled large numbers of people to visit the Park in search of recreation. The Federation of Ontario Naturalists expressed public horror at destructive logging practices by calling for an end to logging in Algonquin Park altogether. Although "wise use" foresters would not adopt such a policy, Park Superintendent MacDougall had been developing a policy "in favour of scientific research as a basis for resource management, scenic and wildlife protection, recreation, and visitor education" (Killan 1992, p. 31). Since his multiple use policy "meant that wherever a clash occurred between recreationist and logger, scenic protection and recreational values should become the primary variables in management policy," MacDougall was able in 1939 to institute a "standard shoreline reserve policy on all canoe routes and portages" in Algonquin Park (p. 31). He took a further preservationist initiative the following year by "setting aside two small areas of pine forest as nature preserves" (Killan 1992, p. 31).

It was hardly surprising in the circumstances of the Great Depression that public concern developed over Ontario's forest policy and the activities of the "timber wolves" (Lambert with Pross 1967; Bertrand 1997). The strong actions of the Hepburn government angered many in the industry. One of the most vocal critics of the government, C.W. Cox, was both the Liberal MPP for Port Arthur and a forest operator whom the president of the Fort William Liberal Association described as having "too much to answer for" (Lambert with Pross 1967, p. 343). (These two constituencies were based, cheek by jowl, in what is now the city of Thunder Bay.) When the northwestern Ontario section of the Canadian Society of Forest Engineers suggested that "the present policy (or rather lack of true forest policy) be thoroughly overhauled by competent foresters and economists," the need for investigation and debate could no longer be denied (Lambert with Pross 1967, p. 343). George Drew, leader of the Conservative Opposition and a former Securities Commissioner, also attacked the Hepburn government for abusing its power over the forest industry and called for a legislative investigation of the Department of Lands and Forests.

A Select Committee of the Provincial Parliament was struck on April 27, 1939, but did not begin hearings until January 1940. Led by the Conservative members, the Select Committee then focused on the forest policies the government should follow rather than on the organization of the department, which one of the witnesses (J.C.W. Irwin) then and the departmental historians (Lambert with Pross 1967) later regarded as the source of the problems. The Conservative minority report recommended that a commission be established to take forest matters out of politics. Hepburn himself recognized that there were administrative problems in the Department of Lands and Forests and obtained the resignation of Peter Heenan, Minister since the formation of his government. Hepburn appointed N.O. Hipel to replace Heenan and chose Frank A. MacDougall as the new deputy minister. Hipel and MacDougall carried out a thorough reorganization of the department during 1941 (Lambert with Pross 1967).

PROGRESS DURING THE WAR

The most important reform in Ontario forest administration during these war years was the creation of a Timber Management Division. As the departmental historians have observed, "the achievement of a sustained yield from a forest managed with increasing care given to the individual tree by a professional service became the declared goal of the Department" (Lambert with Pross 1967, p. 390). However, this determination to manage the forests soon came up against weaknesses in personnel within the Department of Lands and Forests. The Hepburn government also clarified the policy of consultation and negotiation required to arrive at agreements with forest operators, especially the pulp and paper companies, "to make it clear that the Department was not being run by the timber interests" (Lambert with Pross 1967, p. 391). The 1943 provincial election, which gave a minority mandate to Drew's Conservatives, did not lead to the creation of any timber resources commission as their minority report (and election platform) had advocated. Drew's government, which went on to achieve a majority in the 1945 election, seemed content with the reforms that Hipel and MacDougall had made in the Department of Lands and Forests. The Drew government did cancel some pulpwood concessions, and it negotiated the construction (or completion) of mills at Marathon (then known as Peninsula), Red Rock, and Terrace Bay on Lake Superior (Lambert with Pross 1967).

One of the real gains of these war years and the existence of the Drew government was the establishment in 1944 of two more parks in northern Ontario, Sibley Provincial Park on the peninsula that holds the Sleeping Giant rock formation (and whose name was given to the whole park 40 years later) and Lake Superior Provincial Park northwest of Sault Ste. Marie (Lambert with Pross 1967). Sibley Forest Reserve had been considered for a national park during the preceding decade, but hunters' fears that the area would be closed to them if it were a national park led to discussion of a provincial park. The hostility that had developed between Hepburn and Liberal Prime Minister W.L.M. King may have been a reason that the Dominion assistance hinted at was not obtained. In any case, Drew's government had the pleasure of creating Sibley Provincial Park soon after it came to power (Killan 1993). MacDougall also extended his shoreline protection policy to Quetico Park soon after he became deputy minister in 1941. Coincidentally with the creation of these new parks, MacDougall also designated a 7770-ha wilderness area in his beloved Algonquin Park to be used for research in wildlife and silviculture (Killan 1992).

POSTWAR CONCERN ABOUT THE FORESTS

Reorganization of the Department of Lands and Forests was hardly enough to solve such continuing problems as the use of sawlogs as pulpwood. The unwillingness of the pulp and paper companies to share the fruits of their limits with sawmill operators continued to force forest issues into the political arena. These issues helped to explain Drew's decision in 1946 to appoint Major-General Howard Kennedy as a one-person Royal Commission of Enquiry into forest policy. Having asserted his support for sound forestry principles, Kennedy presented a large number of recommendations (Ontario Royal Commission on Forestry 1947). He recommended that provincial forest policy favour the lumber industry and called for punishment for the waste of wood in the woods. A substantial chapter devoted to "Private Lands" indicated the importance of reforestation in southern Ontario. An even longer chapter on "Forest Protection" underscored this continuing concern and included recommendations for strengthening the ability of the Department of Lands and Forests to protect the forest. Kennedy also had a good deal to say about strengthening the capacity of the department to deal with the large forest companies. Forest research and public education should also be encouraged. Visits to northern Europe would provide opportunities to learn from the Swedes.

In regard to issues that the Conservatives had raised while they were still the Opposition, Kennedy recommended against the establishment of any forest resources commission to replace the Department of Lands and Forests (Ontario Royal Commission on Forestry 1947). He did not believe that the government could escape its responsibility for forest policies and practices. In order to assert public control over the forests, however, Kennedy proposed the suspension for ten years of all existing rights to timber supplies. He also recommended the establishment in each watershed of an independent timber cutting company which would fairly apportion the cut among the various users. These new organizations would be expected to avoid the waste that he had observed the existing companies to perpetrate. He also recommended against the export of pulpwood cut on Crown lands that had become so common during the Great Depression. Kennedy's recommendation to divest the pulp and paper companies of their rights to forest lands, however, proved too much for Drew's Conservative government (Lambert with Pross 1967).

The government's need to know the quantity of timber to be found in the limits had been recognized even before the war ended. Trained personnel were required to manage the timber lands of the province, and a Forest Ranger School was established at Dorset in anticipation of the large numbers of veterans looking for peacetime employment (Lambert with Pross 1967). Once the government knew what was in the forests, it could encourage a fuller utilization of the annual growth. Kennedy believed in the principle of sustained yield and had advocated a system of regulations under which the amount of timber cut each year would not exceed the new growth in the forests. Application of this principle would clearly require all the information that departmental foresters were gathering about the forest resources of the province (Lambert with Pross 1967). The Kennedy Commission thus gave the Department of Lands and Forests an immense boost by legitimizing scientific forest management. The professional staff expanded rapidly and the "timber wolves" had to start acting in a more responsible manner (W.D. Addison, pers. comm., February 4, 1998). The need for more forest technologists also led to the opening in 1948 of a training program at the newly established Lakehead Technical Institute (Braun and Tamblyn 1987).

The Select Committee had recommended in 1941 that a comprehensive forest inventory be developed, and J.A. Brodie of the Timber Management Division was now authorized to achieve it (Lambert with Pross 1967). Two approaches were used: aerial photography by the Photographic Survey Company of 320,000 km^2 of Ontario forest and timber cruises by parties of five cruisers and a cook sent into 38,400 km^2 of more accessible forest. A fifth of the total area had been photographed by April 1947, and 11,360 km^2 of planimetric maps had been produced. By 1951, the Photographic Survey Company had photographed 326,000 km^2 of forest, and the Department itself another 69,600 km^2. That year, the government extended the project to cover southern Ontario, as the federal government agreed to pay half the cost of this enormous mapping project. Thus, the Forest Resources

Inventory of Ontario was gradually and painstakingly built up. How this inventory could be used to advance forest management soon became clear.

The Drew government had realized that professional help was needed in developing the forest management capacity of the Department of Lands and Forests (Lambert with Pross 1967). It turned for assistance to Professor D.M. Matthews of the University of Michigan, who had developed an international reputation in administration of forest resources. Over two and a half years before his death in 1948, Matthews advised the government in the establishment of management units, beginning at such points as Kirkwood and Severn River north of Orillia, and using these unlicensed areas to learn how to supervise forest activities. He also encouraged a departmental assertion of control over the pulp and paper companies, beginning with the Abitibi Power and Paper Company. With the company still in receivership at the end of the war, those interested in its financial re-establishment were told that they would only receive the rights to timber limits vital to the financial future of the company if they agreed to the proper management of these limits. In March 1945, the Abitibi Power and Paper Company agreed that it would, within five years, submit "an estimated inventory ... of the timber on the concession area by species and size classes and [provide] information with respect to the forest types and general age-classes of such timber" (Lambert with Pross 1967, p. 408). The company was also to indicate its long-term cutting plans on the map and to obtain the approval of the minister before carrying these plans out. A *Forest Management Act* was passed in 1947 to impose these timber management requirements on all of the companies.

The Timber Management Division was given additional means, following passage of the *Forest Management Act*, to ensure that the new regime in the forests succeeded (Lambert with Pross 1967). M.B. Morison, a forester with the Dominion Forest Service, was hired to supervise industry submissions of forest management plans. The companies needed advice on how to meet the new expectations, and Morison prepared a manual to guide company employees in preparing these plans. The companies thus discovered their own need for foresters but were forced, even before they had them, to develop proper operational plans. Within a year, the Department of Lands and Forest could report that most holders of rights to Ontario timber were developing their own forest inventories and management plans. The large mapping projects that the department was simultaneously pursuing helped to provide these companies with some of the information they needed. These maps enabled the Timber Management Division to deal with companies from a new position of strength. This strength was gradually extended to the field, as trained and experienced foresters exercised the government's power in the forests. By March 31, 1950, the Department of Lands and Forests could report that 20 of the 69 companies required to provide forest management plans had done so, and that these plans covered 38,000 km^2 of Ontario forest lands.

The efficacy of these new management plans was most clearly demonstrated in the Petawawa Management Unit within Algonquin Park. The initial units at Kirkwood and Severn involved areas that had been largely cut over, where the department could only mount "rescue operations ... to capture whatever could be saved from the depradations of the past before a weedy second or third growth completely destroyed an area" (Lambert with Pross 1967, p. 410). The Petawawa Management Unit involved an area where red and white pine had been cut in the 19th century, but 240,000 ha of healthy mixed stands were now coming to maturity. Work on a management plan for this area began in 1945 but was not completed until 1951. This plan envisaged the cutting of mature trees to encourage the growth of smaller red and white pine thus freed from competition by the large trees. A road network would facilitate this harvesting process, since the Department intended the result to be long-term cutting in this management unit. Within the same year, the Department could report that "the cooperating companies ... demonstrated their ability to work under regulations which provide for the protection and development of future crops on the same area on a comparatively short cutting cycle" (Lambert with Pross 1967, p. 410).

Despite these advances in forest management, the pulp and paper companies often continued to operate in the forests as they wished. Ironically enough, the cutting practices of these large forest operators were challenged in 1951 by the director of woodlands research of their own Canadian Pulp and Paper Association (Swift 1983). Alexander Koroleff was keenly aware of the fact that cutting methods affected the regeneration of the forest. From his European experience, he knew that tree cutters were also tree breeders. The practices of the large Canadian companies, which he characterized as "nomadic logging," were all too likely to inhibit regeneration of the forests. Koroleff called for sensitivity to the varied characteristics of forest areas in attempting to avoid such consequences (Swift 1983, p. 170). If Koroleff's critical comments were in order in the earlier era of horse logging, they became even more relevant in the face of the increasing mechanization of forest operations (Radforth 1987, pp. 179-200).

The replacement of the axe and crosscut saw by the motor-driven chainsaw occurred rather slowly after the Second World War. Once the woodsmen discovered that the chainsaw increased their output and consequently their earnings, however, they became eager proponents of technological change. The development of skidders and other machines did even more to change the nature of cutting operations in Ontario's forests. The consequences for forest employment were astounding: where 3000 workers were required in 1950 to cut and deliver 100,000 cunits of wood, only 500 workers were required to supply the same volume in 1975 (Hearnden 1977). Of great significance for the waterways of northern Ontario was the replacement of river drives and lake booming of logs by truck transport of these raw materials of industry. The result was a marked decline in water pollution but an increase (at least for a time) in degradation of the land (Swift 1983). As the construction of roads was brought under control by the Department of Lands and Forests, however, this construction added greatly to the transportation infrastructure of northern Ontario. The forest was opened to sportsmen and the general public in a way it never had been before. No longer was it necessary to charter an aircraft to reach many beautiful and productive lakes, although fly-in fishing operations continued in more isolated areas. Of course, some of the operators on the middle range of lakes found themselves forced to seek more isolated lakes in order to continue their business.

A governmental unwillingness to challenge some forest companies became distressingly clear in one case of water pollution (Swift 1983). After years of suffering from pollution of the Spanish River by the Espanola mill of the Kalamazoo Vegetable Parchment Company, landowners along the river obtained an injunction in 1949 under the *Lakes and Rivers Improvement Act* to restrain the company from such pollution. The company appealed this injunction all the way to the Supreme Court of Canada. While the matter was still before the courts, Leslie Frost's Conservative government amended the law to require judges to consider the economic consequences of their rulings before issuing injunctions. The Supreme Court justices, noting that this amendment had been passed after the issue of the injunction they were considering, upheld the injunction against the Kalamazoo Vegetable Parchment Company. The Frost government then passed an act invalidating the injunction and giving the company permission to continue polluting the Spanish River. As the author of *Cut and Run: the Assault on Canada's Forests* concludes, this American company offered its ultimate thanks to a responsive Ontario premier by appointing Frost to its Board of Directors following his retirement from active politics in 1961 (Swift 1983).

FOREST POLICY IN THE 1950S

The forests of northern Ontario gradually came under the new management regime through the 1950s (Lambert with Pross 1967). As stronger field staff worked with the foresters employed by the limit holders or planned the areas which were the immediate responsibility of the Department, the Forest Resources Inventory became more accurate and complete and the management plans more refined. By 1954, there were 123 management units, 36 the immediate responsibility of large holders of limits and the other 87 Crown units for which the Department of Lands and Forests was responsible. Within the latter, however, independent operators were responsible for 130 km^2 or more of forest, and each of them was required to supply a management plan like that of the large holders of limits. Since the original work on the Forest Resources Inventory had envisaged a revision of the work at ten-year intervals, this effort began in 1957. By 1966, the timber lands of the province had been further organized into 148 management units, not counting the large areas held by agreement with the Crown.

The ultimate test of governmental stewardship of Ontario's forests occurred in regard to the large limit holders (Lambert with Pross 1967, p. 414). The Conservative government of Leslie Frost, who became premier in 1949, passed an amended *Crown Timber Act* in 1953 under which the old agreements with limit holders were replaced by licences issued under Order-in-Council for a maximum of 21 years. These licences were made subject to the forest management provisions of the *Act* and were thus much simpler than the old agreements, which had resulted from individual negotiations between company and government. The *Act* specified not only the practices to be followed but also the penalties for companies that wasted timber in their cutting operations (Lambert with Pross 1967).

Only a year later, Frost's Lands and Forests Minister W.S. Gemmell issued a White Paper entitled *Suggestions for a Program of Renewable Resources Development* (Lambert with Pross 1967). This statement of government policy suggested that the Forest Resources Inventory could not only serve the interests of the forest operators but also be used to integrate "the many and varying uses of land for forests and recreation with their use for wild-life; the use of streams and lakes for hydro developments with their use for log driving and fishery management" (Department of Lands and Forests 1954, p. 2). This optimism about the use of land and water may have been influenced by the pessimism that the White Paper expressed regarding the limitations of Ontario's forest resources. The existing stands of red and white pine, for example, would be gone by 1971 at the rate of cutting then in effect. The White Paper even alluded to "many indications that the cut of spruce pulpwood in Ontario may soon reach its peak or a position similar to that reached by the white pine saw log industry in 1908" (p. 4). These gloomy prognostications underscored the need for sustained-yield forest management in Ontario.

Gemmell's White Paper estimated that Ontario had a potentially productive forest area of 343,000 km^2, 46 percent of which consisted of mature forest ready to be cut (Lambert with Pross 1967). The total volume of merchantable timber was calculated to be 1.75 billion m^3, 1.1 billion m^3 of which was softwood and 0.65 billion m^3 hardwood. Among the softwoods, spruce and red and white pine were being over-cut in the accessible areas of forest, whereas stands farther north were becoming mature or overmature. Although spruce constituted much less than half of the allowable cut, three-quarters of the pulpwood being cut was spruce. Making the situation even more serious was the fact that the succession trends involved in natural regeneration of the spruce forest produced "inferior" species such as poplar. As regards hardwoods, these already consisted largely of poplar (*Populus tremuloides*) and white birch (*Betula papyrifera*), which were of limited interest to either the pulp and paper companies or the sawmill operators. A sense of urgency was created by the conclusion that Ontario "could only provide for present requirements and normal expansion for a period of twenty years" (Lambert with Pross 1967, p. 415). The White Paper argued for effective forest management, which might save the situation in regard to spruce even if it was already too late for red and white pine. A sustained-yield policy would require the full allowable annual cut to be cut annually (or over five-year periods, on the average, to allow for market fluctuations).

The White Paper also addressed the situation of the sawmill operators (Lambert with Pross 1967). Given the holdings of large forest limits by the pulp and paper companies, most of the sawlogs were also being cut by these companies. Since the pulp and paper companies were able to realize a greater return from these logs as pulpwood, it was hardly surprising that sawmill operators had great difficulty in obtaining any of these raw materials for their own mills. The government consequently declared that, "in view of the rather parlous situation facing the sawmilling industry and its incidental effect on the people of Ontario, who own the forests but who will be dependent upon distant sources for lumber despite high and rising freight rates, measures will be taken over the next few years to provide for the continued existence of the sawmilling industry within the province" (Lambert with Pross 1967, p. 416). A relocation of some mills would be likely, and the mills depending on white pine would have to turn to other kinds of trees. The government also encouraged pulp and paper companies to make long-term agreements with the sawmill operators to provide sawlogs from their pulpwood concessions. Only by such measures, in addition to obtaining sawlogs from unlicensed areas, could the sawmilling industry survive while red and white pine grew to sawlog size. The White Paper expressed concern about the period from 1970 to 1990 unless action was taken immediately. While hoping to avoid strict regulation of the size of pulpwood, the White Paper suggested application of the "manufacturing condition" principle that maple (*Acer saccharum*) and yellow birch (*Betula lutea*) sawlogs should not be exported. Clearly, the plight of the sawmill operators called for the fullest possible utilization of hardwoods and the most strenuous efforts to ensure regeneration of red and white pine.

The concern regarding future wood supply expressed in the White Paper led the Department of Lands and Forests to undertake serious forest regeneration efforts (Lambert with Pross 1967). "Project Regeneration" was launched in 1958 as "a pilot programme of silvicultural work involving application of treatments, such as soil preparation by hand tool or mechanical means to provide conditions for natural seeding; experimentation with herbicide sprays to eliminate weed trees choking out better varieties; and a change of cutting methods to give young trees a better chance of growing in" (Lambert with Pross 1967, p. 417-18). These efforts were applied to some 5700 ha in the first year of the project. By 1965, almost 23,000 ha had been scarified to prepare the ground for planting and natural or artificial seeding, 1956 ha had been seeded, and 29,650 hectares had been planted with trees. A Silviculture Section was created in April 1959 and made responsible for maintaining the Forest Resources Inventory, planning regeneration efforts, planting trees, and improving the forest generally. Creation of this new section followed on the amalgamation in 1957 of the Timber Management Division and the Reforestation Division into a Timber Division that was responsible for most aspects of forest administration. In fact, only forest protection and forest research retained separate organizations within the Department.

The 1960s, a period in which John Robarts headed the Conservative government of Ontario, saw significant investments in both the pulp and paper and the sawmill industries (Lambert with Pross 1967). Pulp and paper mill investments were facilitated both by new estimates of the allowable cut of spruce and balsam (*Abies balsamea*) and by the fact that much of the spruce and balsam forest was seen as nearing the mature stage (or already overmature) and needing to be cut. New investments by such companies as Domtar Newsprint Limited and Great Lakes Paper Company increased the annual cut by almost 3,600,000 m^3. The government sought, however, to limit the construction of new mills because of its recognition that, following the cutting of this glut of overmature spruce and balsam, there would come a period of short supply – unless mills turned to sources of pulpwood that at the time were less desirable. The possibility that this might happen was suggested by the way in which new sawmills turned to jack pine and spruce for their raw material. This shift, at least in part, of the "centre of gravity," as the White Paper put it in 1954 (Lambert with Pross 1967, p. 416) was epitomized by the construction of sawmills integrated with pulp and paper mills. These increasingly automated mills cut lumber from small logs and converted the residue into chips for pulping. Such combined operations increased the efficiency of forest operations. These advances in forest operations still left large stands of "weed trees," particularly poplar and birch, to be used before full utilization of Ontario's forest endowment could be said to have been achieved.

THE DEVELOPMENT OF PARKS, 1953-73

Although only a small number of parks had been created before the 1950s, Ontario found itself discovering the recreational value of the forests in a new way in that decade (Lambert with Pross 1967). A historian of Ontario parks has spoken of "rapid population growth, urbanization, a rising standard of living, increased levels of leisure time, more personal mobility, tourism, and an increasingly younger, more educated population [as combining] to sustain thirty years of explosive growth in outdoor recreation in Ontario and a concomitant demand for parkland" (Killan 1992, p. 31). The Frost government was slow to recognize these postwar realities, however, and actually turned responsibility for the parks in southern Ontario over to the Department of Municipal Affairs between 1949 and 1952.

The Department of Lands and Forests began to respond to the social need in 1953, however, when its seven regional foresters went on a tour of national, state, and municipal parks in the United States. They came back with the realization that Ontario was far behind the Americans in park development (Killan 1993). One of them, E.L. Ward of North Bay, stated the brute reality: "Where population increases locally or through an influx of tourists, and the area of publicly-owned land decreases ... [there] comes the inevitable clash between the land-owner and the potential recreationist" (Lambert with Pross 1967, p. 476). The Frost government responded to the recreational need in 1954 by creating a Division of Parks within the Department of Lands and Forests (Lambert with Pross 1967). The Division was authorized to establish a network of parks around the shores of the lower Great Lakes, close to the growing mass of Ontario's population. The Frost government had been advised by the touring foresters that it might have to spend money acquiring land on which to establish parks and was prepared to do so.

The foresters had also emphasized the importance of long-range planning of parks. When a new *Provincial Parks Act* was passed in 1954, it gave the Department of Lands and Forests responsibility for northern parks and possibly Rondeau and Ipperwash. Long Point and Presqu'ile Parks were still to be administered by commissions (on the pattern of the Niagara Parks Commission). In 1958, however, an amended Act gave responsibility for all the provincial parks (but not the Niagara Parks Commission or the St. Lawrence Development Commission) to the Department of Lands and Forests (Lambert with Pross 1967). More significant to the public than these administrative rearrangements was the creation of more parks. The number of provincial parks had grown to 72 by 1960; during the following decade, it expanded to 108 (Killan 1992). Transfer of the Conservation Authorities branch to the Department of Lands and Forests in 1962 enabled the Department to encourage both forest conservation and recreational activity by the Authorities, especially in southern Ontario (Lambert with Pross 1967).

During these years, the provincial government slowly developed a policy, as the historian of the park system has put it, "to accommodate the resource extractors in the midst of hordes of recreationists, while at the same time advancing protectionist objectives" (Killan 1992, p. 32). A "Back to Nature" strategy announced in 1954 led to the phasing out of leaseholds in the southern Ontario parks. In 1956, the government banned prospecting, staking, and mining in all the parks. In 1961, the Parks and Timber divisions of the Department of Lands and Forests agreed on guidelines, based on the principles that MacDougall had developed in Algonquin Park in the 1930s, under which "waterways and portages, public use areas, scenic vistas, and nature reserves [were protected] from the loggers" (p. 32). Lack of a comprehensive parks policy led to protests in 1958, when the Minister allowed public use of four logging roads to the perimeter of Algonquin Park, and in 1959, when Deputy Minister MacDougall gave the National Research Council of Canada permission to use 2200 hectares of Algonquin Park to build a radio observatory (Killan 1993).

The failure of the ministerial Ontario Parks Integration Board, established in 1956, to enunciate a general parks policy aroused both the Quetico Foundation and the Federation of Ontario Naturalists to action (Killan 1993). In October 1958, the Quetico Foundation asked for both a general policy statement and the reservation of another 1,335,000 ha of parkland in northern Ontario. Two months later, the Federation issued the *Outline of a Basis for a Parks Policy for Ontario*. This document "denounced the dominance of recreation and utilitarianism in existing provincial parks policies and demanded that more consideration be given to the protection of natural areas" (Killan 1992, p. 32). That objective could only be attained by ending logging, trapping, sport hunting, and "any other activities detrimental to ecological values and nature appreciation" within the provincial parks (p.32). A series of demands by Premier Frost finally forced the Parks Integration Board to issue a policy statement in March 1959 that would guide the development of the park system for a decade.

The Federation of Ontario Naturalists had also urged the Ontario government to follow the example of the British Nature Conservancy in setting aside "a representative system of nature reserves" (Killan 1992, p. 32). The Frost government provided a limited response to the last suggestion by passing the *Wilderness Areas Act* in 1959. Unfortunately, section 3 of this Act provided full protection of "wilderness areas" only for those under 1 sq. mi. (260 ha) in size. The 35 wilderness areas established by 1961 included 25 nature reserves, a number of them within existing parks (Killan 1993). However, five of them became the basis of full-fledged parks in the late 1960s, including Cape Henrietta Maria (Polar Bear Provincial Park). The Pukaskwa Wilderness Area established in 1964 was conveyed into federal hands in 1978 to create the national park of the same name. The latter action finally gave Ontario a national park of some size.

Both the public and the provincial government came to realize the need for a variety of parks and hence for an articulated park system. This understanding was significantly influenced by the work of the Outdoor Recreation Resources Review Commission (appointed by the US Congress in 1958), especially its 1962 final report, *Outdoor Recreation for America*. The Canadian Federal-Provincial Parks Conference had developed its own standard classification of parks by 1965. A strengthened Ontario Parks Branch was thus able to propose the five categories of parks which the Robarts government adopted in 1967: primitive, wild river, natural environment, recreation, and nature reserve (Ontario Department of Lands and Forests 1967). This government also launched both a study into Ontario tourism and outdoor recreation planning and a study of the Niagara Escarpment. In 1968, it created Ontario's first Primitive Park – Polar Bear – which at 18,000 km^2 was the second largest park in Canada. The conversion of five wilderness areas into nature reserves and the creation of two new reserves gave the Federation of Ontario Naturalists and Ontarians generally additional reasons for joy (Killan 1992).

As the departmental historians saw it, the "development of parks [came] more and more under the influence of the modern concept of multiple land-use" (Lambert with Pross 1967, p. 495), which the Forestry Study Unit defined as "the deliberate and carefully planned integration of various uses of land so as to interfere with each other as little as possible, with due regard to their order of importance in the public interest." The competition that complicated the integration of these varied uses was clearly expressed when the Ontario Federation of Anglers and Hunters criticized road-building and other consequences of cutting operations in Algonquin Park. This protest had little effect, as the district forester continued to be responsible for both the protection of aesthetic values and the continuance of cutting operations. The departmental history was written in the midst of this struggle over wilderness, in which a multiple use policy initially applied to limit logging was used by the forest companies to check the preservationists. As the parks historian has stated, "Forestry practices had undergone a massive transformation since the Second World War [and] the teamster, winter camp, and spring river drive had been replaced by a highly mechanized and automated year-round operation" in which "the scream of chain saws and the roar of engines ... reverberated through the bush" (Killan 1992, p. 37). He underscored the environmental impact of the new industrial methods by adding, "worse still for wilderness users, the forest industry required permanent road systems through formerly inaccessible areas to accommodate the massive trucks now required to haul the logs to the mills" (p. 37).

Two new organizations took up the struggle for Ontario wilderness, with passage of the US *Wilderness Act* in 1964 to inspire them. Although the National and Provincial Parks Association of Canada had been established in 1963, it was the naming of Gavin Henderson as executive director in 1965 that made it an organization worthy of its later name, the Canadian Parks and Wilderness Society. Even more important was the Algonquin Wildlands League, which "sprang into prominence [in June 1968] under the leadership of such worthies as Douglas Pimlott, Abbott Conway, Walter Gray, and Patrick Hardy" (Killan 1992, p. 37; Warecki 1992). These organizations "generat[ed], for the first time, widespread public support for wilderness preservation" and "succeeded in making the management of the province's large parks a litmus test of the Robarts and Davis governments' commitment to environmentalism" (Killan 1992, pp. 37-38). Their strength was never more clear than in Premier William Davis's decision, just before he launched his first attempt at an electoral mandate in 1971, to sign an agreement with the federal government to establish Ontario's first large national park, Pukaskwa on Lake Superior. This action also responded to the call of the Canadian Audubon Society for each province to create a national park in honour of Canada's Centennial (Killan 1992).

The failure of the Algonquin Wildlands League to obtain an end to logging in Algonquin Park contrasted sharply with the successful struggle over Quetico (Killan 1993). Northwestern Ontarians, "who had been part of the post-war outdoor recreation boom, who treasured their parks and natural areas, and who feared the threat of air and water pollution," joined with protesters across the province in insisting that the policy of multiple use in Quetico be terminated (Killan 1992, p. 38). As one observer has commented, "The war over Quetico, with its demonstrations and subsequent public hearings across the province, was a defining event for both the forest industry in Ontario and the parks movement in Ontario ... Quetico changed the way people looked at the forest industry and its exaggerated claims ... Quetico [also] placed wilderness parks in general, and Quetico in particular, at the symbolic heart of the parks system" (B. Addison, pers. comm. February 4, 1998).

The speed with which Premier Davis responded to the recommendations of the Quetico Advisory Committee – he ordered a moratorium on logging in Quetico Park on May 13, 1971 – demonstrated the impact of the struggle (J. Foulds, pers. comm., November 28, 1997). The Davis government went on to designate Quetico a Primitive Park in 1973 (Killan 1993). Not only was the cutting of timber ended in this park, but the concern (expressed more than a decade earlier by the Federation of Ontario Naturalists) that "recreational pursuits which demand the use of mechanical power should be strongly discouraged" (Lambert with Pross 1967, p. 494) was applied to Quetico in a prohibition of power boats. Quetico thus reverted to the travel by canoe that had characterized Aboriginal water technology for centuries, if not millennia, and had also been used by the EuroCanadian fur traders of the 18th and 19th centuries as they traversed these waters (Nute 1941). It was ironic, however, that Aboriginal guides would later be allowed to use motorboats on certain lakes when they were guiding visitors, and that this loophole has grown since then (W.D. Addison, pers. comm., February 4, 1998).

Although most of northern Ontario remained dedicated to industrial use of the forest, successive governments did seek advice on other uses of the forest. The Frost government had responded to early demands for multiple use of the forest by a 1956 Order-in-Council that established land zoning committees in sixteen of the Lands and Forests districts. Each committee was chaired by the district forester and included the local MPP and representatives of the Northern Ontario Tourist Outfitters Association, the Ontario Federation of Anglers and Hunters, and the forest industries. Early in 1962, Robarts' new Conservative government reconstituted these committees as district advisory committees on recreational land use planning. The recommendations of these committees could be given effect under a 1958 amendment to the *Public Lands Act* which empowered the Minister to zone land for recreational purposes (Lambert with Pross 1967). The Davis government went further in creating a Provincial Parks Advisory Council in 1974 to focus public opinion on the existing parks and on potential additions to the system (Killan 1993).

THE STRUGGLE OVER WOOD SUPPLY, 1970-92

As park concerns took on a new urgency in the 1970s, so also did concern about the future supply of timber and the success of forest regeneration. The latter had gradually become a concern, first of the government and later of the forest industries, during the postwar period. This concern was reflected in the fact that the Timber Management Division, rather than the Reforestation Division, was made responsible for "Project Regeneration" in the late 1950s. By amendment of the *Crown Timber Act* in 1962, the Robarts government accepted responsibility for forest management on Crown lands. The Division of Forests, however, which was charged with both "the production and harvest of timber and the provision of woodland for recreational opportunities," only slowly developed its silvicultural capacity (Armson 1976, p. 7).

The federal government supported these efforts following a federal-provincial conference on forestry and passage of the *Canada Forestry Act* in 1949. Under the authority of this Act, Louis St. Laurent's Liberal government negotiated a number of agreements with Ontario to support development of the Forest Resources Inventory and reforestation activities (from 1951 to 1964) and forest research (1952). Federal forest research was focused under the 1952 agreement in two laboratories, the one at Sault Ste. Marie being concerned with forest insects and the other, at Maple, with forest pathology. Ontario maintained these laboratories, while the federal government supplied the personnel and equipment and ensured that the results were relevant to Ontario. The federal change of government in 1957 brought the even more concerned Conservative government of John G. Diefenbaker to power. Agreements to assist with forest fire protection and construction of forest access roads were negotiated in 1957 and 1958 respectively. In 1960, the Diefenbaker government repealed the *Canada Forestry Act* but established a federal Department of Forestry to administer the federal-provincial agreements. In 1962, this Department was given still another agreement to supervise which dealt with stand improvement activities. By 1967, Ontario was party to no fewer than 32 federal-provincial resource agreements, over half of which concerned its Department of Lands and Forests (Lambert with Pross 1967).

Ontario foresters had meanwhile come to recognize that forest regeneration could not be left either to nature or the forest companies (Lambert with Pross 1967). The search for a white pine that could survive white pine blister rust, for example, or for management methods that would enable yellow birch to regenerate had to be carried out by the Department of Lands and Forests. Similarly, it was Departmental staff who experimented with growing seedlings in tubes so they could be planted in mid-summer of the year of their germination in areas affected by forest fire. By 1966, the Department was able to plant these tubed seedlings by using a planting gun. That year it planted about 26 million trees on over 9700 ha of land in its 16 northern districts. Supporting regeneration by such means was an important achievement, but it did not come cheaply, and the limited survival rate increased the cost per mature tree (C.A. Benson, pers. comm., January 1998). The cost of artificial regeneration concerned the government and led to a series of assessments, beginning with the Brodie Study Unit Report of 1967 (Wightman and Wightman 1997).

Concern about the supply of wood fibre for pulp and paper mills gradually intensified through the 1970s. The Lumber and Sawmill Workers Union, as well as the companies, were concerned when multiple use was rejected for Quetico Park (J. Foulds, pers. comm., November 4, 1997). As long as there were areas of accessible forest not yet granted to any company, however, even the ambitious operators could avoid panic. Great Lakes Forest Products managed to acquire rights to very large forest tracts, including one northeast of Sioux Lookout in 1973 (Wightman and Wightman 1997). The American-owned Kimberly-Clark, which operated the mill in Terrace Bay, managed to obtain 11,000 km^2 of forest north of Nakina in 1974. When the British-owned Reed Paper, which operated a mill in Dryden, similarly sought rights to forest lands north of the 50th parallel, the protests of environmentalists and First Nations people led the Davis government to appoint a Royal Commission on the Northern Environment in 1977 (Ontario Royal Commission on the Northern Environment 1985). Industry concern about the future wood supply reached a peak in 1980, when F.L.C Reed and Associates released a report, *Wood Fibre Supply and Demand in Northern Ontario*, commissioned by ten pulp and paper companies (Wightman and Wightman 1997).

A report of the Ontario Economic Council, *A Forest Policy for Ontario*, had posed the issues in early 1970. On the one hand, "under present conditions, as indicated by allowable annual cuts, some two-thirds of Ontario's renewable forest resources are going unused" (Ontario Economic Council 1970, p. 3). On the other hand, while "wood-using industries in other parts of Canada and of this continent are expanding rapidly[,] no new pulp and paper mill has been built in Ontario for some 21 years [and] lumbering and many of the secondary segments of the wood-using industry reflect a similar lack of growth" (p. 3). If this represented a potential for economic growth, the report also asserted that "many companies have ... failed over the years to carry out adequate reforestation and regeneration programs and, in particular, research into and development of improved tree strains, fertilization, and harvesting productivity" (p. 4). The first of nine recommendations of the Council was that "a significantly greater degree of integration of wood supply for pulp and paper and sawmilling operations is essential if returns from forest resources are to be optimized" (p. 4) and the third recommendation was that "consideration should be given to the establishment of a system of land tenure which would create for the companies involved a greater interest in the economics of silviculture and land management" (p. 4).

The existence of more than 42 million ha of productive forest area, 90 percent of it Crown land, constituted an enormous economic opportunity and a no less enormous forest management challenge. The newly created Ministry of Natural Resources made its own assessment in a 1972 statement, *Forest Production Policy Options for Ontario*, which estimated that about 24 million ha of forest would be required to produce the 12 million cunits of wood that might be cut annually by 2020. Five options were stated, ranging from no investment at all, through regeneration spending at the current level of $8.8 million, at increased levels of $15.3 million or $29.8 million, up to the most expensive option, annual silvicultural expenditures of $44.5 million. As two critics of forest policy have observed, the choices facing the Davis government in 1972 "ranged from 'orderly liquidation' of the remaining commercial timber, through somewhat increased regeneration efforts to permit continued cutting at the then current 17.6 million cubic metres yearly rate, to very intensive timber management to achieve a sustained annual harvest of over 45 million cubic metres" (Dunster and Gibson 1989, p. 18). The Davis government opted for the intermediate objective of increasing the annual cut from the 1970-71 level of 17.6 million m^3 to 25.8 million m^3 by 2020 and increasing spending on regeneration somewhat.

Although the Ministry of Natural Resources referred to the recreational, environmental, and social values of the forests in its *Forest Production Policy Options for Ontario*, the title of the document indicated a preoccupation with timber management for the wood-using industries. Discussion of the various options was based on the assumption that only 52,610 ha of forest land cut each year would regenerate naturally. The current level of silvicultural expenditure would regenerate another 55,038 ha annually. This "forest base of some [107,000 ha regenerated] annually [represented] slightly more than half of the annual acreage [being] cut at [the time] and [did] not take into account other areas lost to production through parks, fires, roads, power lines, failures in regeneration, etc." (OMNR 1972, p. 37-38). Levels of production then current would actually require the regeneration of 152,000 ha annually, whereas the treatment of 100,000 ha was estimated to require almost a doubling of expenditures ($15.3 million). Any plan to double production by 2020 would require the doubling of this last figure.

Under the inflationary pressures facing governments in the mid-1970s, Natural Resources Minister Leo Bernier might well wonder whether the Ministry was spending its forest management dollars well. The study that Professor K.A. Armson carried out during 1975-76 recognized that "we are coming to the end of the natural exploitable forests of Ontario" (Armson 1976, p. 1). In fact, silvicultural efforts had been applied to 769,000 ha by 1973 and "a total of [66,445 ha] was given regeneration treatment [during 1974-75], of which [49,038 ha] were planted or seeded" (p. 4). Armson observed, however, that "if some significant progress has been made in treatment of areas for regeneration there are still obstacles to the implementation of forest management in Ontario" (p. 4). The "most serious impediment" was "the fact that the forest is viewed as a resource to be exploited." If the Ministry were to be organized for proper management of the forests, it needed "a statement of policy together with a direction provided by leadership, adequate support, and a competent professional and technical staff to undertake the necessary program" (p. 139). Among the specific needs were a strengthening of the Division of Forestry and the hiring of professional foresters and technical staff to work in the district offices where they could refine the Forest Resources Inventory and integrate silvicultural projects and management planning.

The Armson study appeared in close conjunction with the report of the Timber Revenue Task Force and enunciation of a proposed policy on clear-cutting. The Ontario Forest Industries Association responded to Armson by saying that it agreed "with most of his recommendations pertinent to the industry" (Ontario Forest Industries Association 1977a, p. 2). Action was needed "if it is true that production will meet allowable cut within the next few years," and the Ministry could "be assured of the continuing cooperation of the industry" (p. 27). The Association rejected the Timber Revenue Task Force's proposal for "a single tenure charge ... [which] should be increased annually at the rate of 10 per cent [*sic*]" as revealing a pre-occupation with revenue and a failure to realize that "tenure is an instrument of policy which can serve a number of purposes, one of which is to derive revenue; another is to 'affect forest management'" (Ontario Forest Industries Association 1977b, p. 5). The Association also objected to the proposal to limit the size of clear-cuts, because it was not convinced that "excessively large clearcuts do not regenerate as well as more protected smaller cuts." If "wildlife and aesthetic values" were to determine this matter, "the forest industry should not be the one group fully burdened with the costs of achieving such values" (p. 7).

The Ontario Forest Industries Association went on to concede that both "Government and Industry and, to an increasing extent the public, now recognize that a satisfactory level of [forest] management has not been achieved" (Ontario Forest Industries Association 1977b, p. 10). In particular, the government (which had accepted responsibility for management in 1962) had failed "in obtaining successful regeneration of too much of our logged over lands" and there was clearly "a need to increase the growth on our productive forest lands to provide for the future development of the forest industry" (p. 10). Premier Davis responded to this challenge in his 1977 "Brampton Charter" by declaring that, if re-elected, his Conservative government would plant two trees for every tree that was cut (Spears 1979; J. Foulds, pers. comm., November 4, 1997). His government also responded to the assertion that "the profitability of the pulp and paper industry for 1967-74 was significantly less than total manufacturing, total industrial or natural resource industries such as metal mines and primary metals" (Ontario Forest Industries Association 1977b, p. 3). Financial assistance seemed necessary since "expenditures just for air and water pollution abatement have cost Ontario's pulp and paper industry some $155 million in the past decade and a half and a further estimated one billion dollars will be needed to meet the requirements over the next five to ten years" (p. 4).

The most significant proposal of the Ontario Forest Industries Association was "an amendment to present legislation, transferring responsibilities for forest management of certain Crown lands to industry at the option of the licensee" (Ontario Forest Industries Association 1977b, p. 13). The Association insisted on three conditions of such a change: "a defined basis of tenure, ... provision of suitable incentives to encourage the licensee to undertake management, [and] an effective process to monitor and review achievement in management" (pp. 13-14). Where Armson had recommended that "the Government of Ontario consider a form of licence, under the authority of the Minister of Natural Resources, for a period of 15 years," with review "at five year intervals" and licences to "remain in force for a further fifteen years from the time of review" (Armson 1976, p. 27), the Ontario Forest Industries Association suggested that the "Forest Management Licence, with 'evergreen' conditions, ... should be for 20 years with the period of review set at five-year intervals" (Ontario Forest Industries Association 1977b, p. 15).

In the midst of these timber supply and forest management debates between government and industry, one northwestern Ontario First Nation was able to join the ranks of the timber suppliers (Blair 1984). Although Aboriginal men had often been employed by the forest companies, the Gull Bay First Nation faced the possibility in the early 1970s that harvesting of trees in the vicinity of their Lake Nipigon reserve could damage their hunting and trapping without bringing them any employment opportunities. Chief Tim Esquega's determination to find employment for his people and forester John Blair's ability to persuade both the Ministry of Natural Resources and the Great Lakes Paper Company Limited to allow the Gull Bay Development Corporation (now Kiashke River Native Development Inc.) to cut timber that had already been granted to Great Lakes Paper created the opportunity. The readiness of Northern Wood Preservers in Thunder Bay to lease three skidders to the Gull Bay Development Corporation and to buy its production at the roadside enabled the Gull Bay community to begin forest operations in October 1974. Blair's proposal that cutting take place on a "modified clear-cut basis (alternate 2.4 [ha] rectangles or 4 [ha] square blocks) to provide a pleasing environment that would support wildlife populations essential for trapping and hunting" (Blair 1984, p. 15) was an imaginative response to the multiple-use needs of the community. It also facilitated continuance of these cutting operations when they were found to be occurring in a park reserve.

The developing crisis in wood supply encouraged the pulp and paper companies to move toward fuller utilization of the forest. This process was assisted by the readiness of American consumers to accept newsprint with hardwood content. The addition of poplar to the mix of trees going into the mills in the early 1980s was a milestone for the forest industries of Ontario. This advance in pulp and paper production was a response to one of several challenges facing the pulp and paper industry in the 1970s and early 1980s. The 1970s saw sharp petroleum price increases as a result of the activities of the Organization of Petroleum Exporting Countries. These oil price shocks increased the cost of transportation and thus of the wood fibre that companies were having to transport over increasing distances to their mills. The requirement to reduce air and water pollution had forced the companies into the large investments already noted. The Ontario government provided financial assistance to some of the companies, although the assistance would later be criticized as having been unnecessary for this profitable industry (Bonsor and Anderson 1985). Whether the companies required public investment at this time may be debatable; what was beyond debate was the fact that these investments invariably reduced the labour forces of the mills as the industry became ever more capital-intensive. The Boise Cascade strike at the end of the decade, in which the Lumber and Sawmill Workers Union fought the proposal that workers should be required to purchase wheeled skidders as a condition of employment, indicated another way by which the pulp and paper companies could reduce both their investment needs and their operating costs (J. Foulds, pers. comm., November 4, 1997).

As these concerns about the declining quantity and increasing cost of wood fibre intensified for the pulp and paper companies, the Davis government moved towards a new contractual arrangement with these companies. After two decades of taking primary responsibility for forest regeneration and spending increasing amounts of money in the process, the government decided to shift the legal responsibility to the forest operators. The negotiation of Forest Management Agreements was influenced by the realization that harvesting activities had a significant impact on the regeneration of the forest. Large clearcuts inhibited the natural regeneration of the forest and required a variety of artificial regeneration efforts. As long as the taxpayer was covering the cost, whether through the Ministry's own budget or through Canada-Ontario Agreements, the companies had only a limited stake in better silviculture. Although the governments initially covered much of the cost of silvicultural activity under the Forest Management Agreements, the future would see the industry paying more of the cost.

The Reed report had called on the Ontario government to increase its efforts in forest regeneration, but there were significant counter-pressures on the Davis government (Wightman and Wightman 1997). The New Democratic Party had been emphasizing the responsibility of the forest companies since the declaration to that effect by leader Stephen Lewis during the 1975 election campaign, a declaration that produced a great deal of debate within party and caucus over how this responsibility could be made effective (J. Foulds, pers. comm., November 4, 1997). Given the Davis government's own desire to make those cutting the trees responsible for the regeneration of the forest, Natural Resources Minister James Auld had the *Crown Timber Act* amended in 1979 to authorize the negotiation of Forest Management Agreements with each company. The New Democrats convinced the Davis government that the 5-year reviews of these 20-year agreements should be made available to the public (in return for not pressing the Premier on the tree-planting promise in his Brampton Charter). This, they hoped, would encourage the companies to make more effective use of the public funding that they received for their regeneration efforts. It might also make woodland managers more sensitive to the consequences of their harvesting practices.

The concern about wood fibre supply expressed so forcefully in the Reed report in 1980 could not be ignored by the Ministry of Natural Resources. The question inevitably reflected on the accuracy of the Forest Resources Inventory developed during the 1950s and regularly updated since then. The appointment of Professor Armson, first as a consultant and then as the Provincial Forester, enabled the advisor of 1976 to ensure that the Ministry took his advice on forest management. The Ministry launched a computerized model in 1981 that was designed to use the Forest Resources Inventory as a basis for alternative scenarios of timber harvest, regeneration silviculture, and forest protection. It also launched a new aerial photography program in which infrared film provided a cheaper means of measuring the success of regeneration efforts. A central concern was whether the planting and seeding activities being carried out with public funds under the Forest Management Agreements were actually regenerating the forest. Public concern was intensified by the publication of two critiques of Canadian forest practices, Jamie Swift's *Cut and Run: The Assault on Canada's Forests* in 1983 and Donald MacKay's *Heritage Lost: The Crisis in Canada's Forests* in 1985.

The provincial political landscape changed in 1985 with the defeat of Frank Miller's Conservatives and the formation of David Peterson's Liberal government. As one item of the political accord between the Liberals and New Democrats that ended 42 years of Conservative rule, Dr. Gordon L. Baskerville, Dean of Forestry at the University of New Brunswick, was asked to carry out an independent audit of the Ontario forests (Wightman and Wightman 1997). Baskerville's audit focused on forest management and led to a number of recommendations for improvement of the procedures that the Ministry of Natural Resources was following. Although Baskerville did not focus on the state of the forests, he did warn of supply shortages that would arise between 2010 and 2030. He also recognized the public concern about the Forest Resources Inventory. The Ministry of Natural Resources responded to the Baskerville audit in 1986 with an *Action Plan on Forest Management* that recognized a need to re-establish confidence in the Ontario Forest Resources Inventory data used in forest management planning (Rosehart et al. 1987).

Peterson's Minister of Natural Resources, Vincent Kerrio, acted on this Ministry recognition by appointing an independent committee "to provide the Minister of Natural Resources with a definitive assessment of the Forest Resources Inventory and to recommend a process for verification of the accuracy of the inventory on an on-going basis" (Rosehart et al. 1987, p. 25). Headed by Dr. Robert C. Rosehart, President of Lakehead University, this Ontario Forest Resources

Inventory Committee consisted of Dr. O. Brian Allen of the University of Guelph, Dr. J.R. Carrow, Dean of Forestry at the University of Toronto, Monte Hummel of the World Wildlife Fund Canada, G.R. Seed of Great Lakes Forest Products, and Gregg Sheehy of the Canadian Nature Federation. The Committee "found that Ontario's Forest Resources Inventory (FRI) is sufficiently accurate when used for the purpose originally intended," and it asserted that "public confidence in FRI data is generally felt to be low because of misapplication of FRI data by some users, particularly the smaller industrial forestry operators" (Rosehart et al. 1987, p. iii). An Inventory developed "for macro-planning forest management purposes" (p. iii) needed to be tested by such detailed means as operational cruising of forest stands. When the results of five operational cruises were compared with the relevant FRI data, the average absolute differences ranged from 6 percent to 123 percent, but the Committee concluded that "for township and larger areas the FRI differs in an absolute sense by about 20 [percent] when compared against the OPC survey results" and saw "such accuracy [as] acceptable for broad macro-planning purposes" (p. 10).

The importance of site-specific forest management received increasing emphasis during these years from George Marek and other foresters. At the same time, the penchant of the companies to clear-cut large areas of forest came under increasing attack from some foresters and many environmentalists. One of the most eloquent critics attacked the "cut and run" mentality of Ontario's (and Canada's) forest companies (Swift 1983). Another quoted the harsh assessment of Ken Hearnden, an industry forester before he joined the Faculty of Forestry at Lakehead University: "From every aspect, the Boreal Forest scene shows all the hall-marks of a long-standing policy and practice of unmodified forest exploitation and the liquidation of our best natural growing stock" (MacKay 1985, p. 4). As Hearnden saw the situation, most foresters had functioned as technicians rather than professionals in supporting industrial use of the forests (Hearnden 1985). Thus, as the Ministry of Natural Resources confronted the challenge of the *Environmental Assessment Act* passed in 1975, questions about its regeneration practices as well as about the consequences of timber harvesting became ever more urgent.

Still another change of government, the election of Bob Rae's New Democrats in 1990, opened the door to a further examination of forest regeneration. The Independent Forest Audit that Minister of Natural Resources Bud Wildman announced on April 18, 1991, was to be made by Kenneth Hearnden, the already-quoted former Director of Lakehead University's School of Forestry, Susan V. Millson, co-owner of Millson Forestry Services in Timmins, and William Wilson, Chief of the Rainy River First Nation and Board Chair of the National Aboriginal Forestry Association. Their procedure was to select test plots across northern Ontario in the largest areas of harvesting between 1970 and 1985, with the aim of comparing the harvested areas to the areas treated and untreated by human activities and by describing the current plant species composition. Their *Report on the Status of Forest Regeneration* confirmed the vitality of the forest when they concluded that virtually all naturally regenerated stands met or exceeded provincial minimum regeneration stocking standards when all boreal tree species were considered. However, they concluded that regeneration treatments for spruce and jack pine must be planned and implemented on productive upland cutover sites formerly occupied by natural stands of those species in order to avoid the conversion of the sites to mixedwood and hardwood stands dominated by poplar and other species. If regeneration of the original forest was the objective, then further management activities could not be avoided. In an interview shortly after the release of the report, Hearnden pointed out that the committee had not been asked whether a supply of wood existed to sustain Ontario's pulp and paper mills. Consequently, "the broader question of future timber supplies also remain[ed] unanswered" (Sanders 1992, p. 1).

The changes in pulp and paper operations noted earlier increased the challenges facing lumber producers and other forest industries. Obtaining enough sawlogs had been a challenge for the sawmill operators since at least the 1940s. The efforts of successive governments to ensure that sawlogs were used (at least in part) to produce lumber rather than being reduced to fibre in the pulp mill had been only partially successful. The sawmill operators had been more successful in using "weed" trees to produce lumber. Equally significant through the 1970s was the development of other uses for these trees. The production of plywood and chipboard from poplar marked an important move towards full utilization of the forest. The use of these new products in house construction also reduced market pressure on the sawmills and thus on the forest. These changes in technology and market demand were accompanied in northwestern Ontario by the combination of many sawmills into the Buchanan group of companies, which was based in Thunder Bay but operated most of the important sawmills in the region (Wightman and Wightman 1997).

The Buchanan group and its employees and contractors thus became the chief northwestern Ontario protagonists when the US government, responding to the demands of the American forest industry, began to attack Canadian lumber exports in the early 1980s (Anderson 1986). Each of these US attempts to reduce the export penetration of the Canadian industry as it advanced towards (and even beyond) one-third of the US market was based on the charge that the Canadian industry was subsidized, especially by the low stumpage fees it was paying for its timber. The Canadian industry fought each of these attacks with considerable success (McCloy 1986). The battles were significant factors, however, in motivating Brian Mulroney's federal Conservative government to launch negotiations in 1985 with the US government in the hope of achieving a conflict-resolution mechanism that the US government would respect. Ironically, however, Mulroney's Finance Minister Michael Wilson conceded the battle to the US lumber industry in 1987 by placing an export tax on lumber to compensate for low stumpage charges and to force provincial governments to increase these charges for the trees cut on Crown lands.

THE BATTLE OVER LAND, 1980-83

The Ministry of Natural Resources had laboured throughout the 1970s both to ensure that the production targets of industry could be met and to provide for a fair sharing of the forests by all of the interests. Each of the Ministry's regional offices was required to plan the regeneration work required to meet the production targets set by the government. The urgency of these efforts and the optimism of this "implementation schedule" was obvious in the fact that the production policy envisaged cutting trees within 50 years of their planting, even though the northern Ontario forest rotation period was generally regarded as 60 to 80 years (J.F. Adderley, pers. comm., January 1998). These and the other concerns of the Ministry – "promoting and regulating aggregate extraction and mining, providing for recreation and environmental preservation through parks and other protected areas, managing fish and wildlife resources and serving hunting, fishing and trapping interests, and enhancing other outdoor recreation opportunities" (Dunster and Gibson 1989, p. 20) – were all considered in the Strategic Land Use Planning (SLUP) exercise by which the Ministry sought to provide targets for each of these activities in all of its 47 administrative districts.

As two critics of the planning process in the Ministry of Natural Resources later observed, the "Strategic Land Use Plans ... along with associated resource management plans, were envisioned as the central elements of a top-down planning process, translating provincial goals and objectives for timber production, fishing and hunting opportunities, mine development, parks and recreational activities, etc., into more specific decisions on land allocation and conflict resolution at the regional and district levels" (Dunster and Gibson 1989, p. 91). As they also observed, "the actual work of land use planning ... was largely an internal exercise within MNR ... [and] opportunity for public comment was provided [only] after the planning options had been developed" (Dunster and Gibson 1989, p. 92). Conflict inevitably arose among various users of the forest when the Ministry's proposals were made public. The Ministry had originally intended to develop methods of reconciling the contenders and formalizing the allocations of land to various uses. Preoccupation with timber production biased the planning exercise, however, and district staff found themselves unable to respond sympathetically to others wanting to enjoy the forest. Growing knowledge of the varied demands being made on the northern Ontario forest only complicated the issue for Ministry planners.

It was in this context of struggle that the early 1980s saw significant park development across Ontario (Killan 1993; Wightman and Wightman 1997). The Davis government was well aware of the public demand for more parks and of the forest industry's opposition to these proposals. The Provincial Parks Advisory Council, created in 1974 by the Davis government, provided a conduit for the public demand. From 1976 to 1978, the Advisory Council worked with the Ministry of Natural Resources on its Strategic Land Use Planning for Crown lands. This process had by 1981 generated a list of 245 potential parks. Meanwhile, the professional park administrators in the Ministry of Natural Resources had been at work developing the existing system of parks. "The highlight of their efforts occurred in 1978 with the approval by the Davis cabinet of an official parks policy and the completion of the so-called 'blue book,' the *Ontario Provincial Parks Planning and Management Policies* manual" (Killan 1992, p. 39).

While a committee chaired by Richard Monzon was re-examining the creation of more parks, Provincial Parks Advisory Council chair G.B. Priddle denounced the varied ways in which Ontario parks policy was "being circumvented, circumscribed and sabotaged to the point that it will soon be meaningless" (Hodgins and Benidickson 1989, p. 279). Priddle asserted that "the Monuments, Cathedrals and Art Galleries of the country are our natural and unspoiled places" (p. 279) and urged the immediate establishment of both the Lady Evelyn-Smoothwater and Ogoki-Albany Wilderness Parks, located respectively in northeastern and northwestern Ontario. Meanwhile, Timber Branch and Lands and Parks personnel in the Temagami district office of the Ministry of Natural Resources continued to pursue their contradictory policies. The Monzon report, which was given to Natural Resources Minister Alan Pope in March 1982, supported the establishment of Lady Evelyn-Smoothwater Wilderness Park, among other proposals. Pope now called on the public to indicate support for the establishment of more parks.

Development of the existing park system and proposals for additional parks aroused strong opposition in northern Ontario. This was especially true "when the MNR attempted to dovetail the regional park systems plans into the broader Strategic Land Use Planning (SLUP) program, a remarkably ambitious attempt to sort out the future uses of Ontario's vast Crown land base" (Killan 1992, p. 42). An intense battle ensued in which the forest industries, sometimes supported by anglers and hunters, tourist operators, and First Nations groups, urged the government not to create these parks (Swift 1983). As the historians of the Temagami region have observed, "immediately after the 1981 election, Premier Davis called a temporary halt to all park expansion and, in a sense, to the whole elaborate SLUP process that had encouraged broadly based public participation" (Hodgins and Benidickson 1989, p. 279).

Thus, through 1982, "in 141 district open houses and seven Minister's forums, ... a coalition of eight environmentalist groups ... faced a formidable and well-financed anti-parks coalition made up of such groups as the Ontario Forest Industries Association, the Ontario Trappers Association, the Prospectors and Developers Association, and the Ontario Federation of Anglers and Hunters" (Killan 1982, p. 42). The last was "a highly regarded conservationist organization which had resolved to oppose any new provincial park in which hunting was prohibited" (p. 42). Caught in the middle was the Davis government, especially "Alan Pope, the new Minister of Natural Resources, who was determined to bring SLUP to a swift and successful conclusion" (p. 42). Pope felt able by June 1983 to add Lady Evelyn-Smoothwater and five other wilderness parks to the Ontario system (Hodgins and Benidickson 1989; Killan 1992). However, another 90 candidate parks, including 76 potential nature reserves, had been abandoned or awaited future protection only as areas of natural and scientific interest (Killan 1992).

The *Temagami District Land Use Guidelines* issued soon after Pope announced his decision included the new park. However, the "skyline reserve [around Lake Temagami] and the narrower shoreline reserves for Temagami and adjacent lakes were reconfirmed [but] unilaterally down-graded in the MNR guidelines and described merely as 'areas of concern'" (Hodgins and Benidickson 1989, p. 280). In addition, the new production guidelines for Temagami included a renewed commitment to sustained-yield forest management but also proposed an increase in the annual cut of conifers from 93,000 m^3 to 255,000 m^3 by the year 2000 and of hardwood by 400 percent. These increases would require more roads, with the Red Squirrel Lake Forest Access Road being the most important extension. This transportation "linkup would enable the Liskeard Lumber mill at Elk Lake to obtain logs from the Diamond Lake-Lake Temagami area [and] Milne Lumber in Temagami [to] gain direct access to mature pine and large spruce in Shelburne and Acadia townships" (Hodgins and Benidickson 1989, p. 281).

The Ministry of Natural Resources, in support of the lumber company, started road work even before there had been an environmental assessment of the undertaking (Hodgins and Benidickson 1989). A right-of-way of 15 km was cleared in 1984, even though part of it ran alongside the main canoe route between Diamond and Wakimika Lakes. While canoeists were protesting the disorder created and the debris left by the clearing, 42 years of Conservative rule of Ontario drew to a close. David Peterson appointed Vincent Kerrio as his Natural Resources Minister in June 1985, and the new Minister decided to subject construction of the Red Squirrel Lake road to a full-scale assessment under the *Environmental Assessment Act*. The Ministry's response was profoundly revealing: it insisted that its "undertaking" consisted only of road building, not the timber production that the road was designed to make possible. As Dunster and Gibson have observed (1989, p. 102), "the Ministry insisted that ... 'other timber management operations such as harvesting and forest renewal are not part of the undertaking,' even though the road was intended to facilitate logging and the acceptability of logging the forest made accessible by the road was the major focus of public controversy." Government reviewers unfortunately accepted this narrow view and, despite the protests of many environmentalists, the Peterson government refused to order a full environmental assessment. The protest reached a political height when Opposition Leader Bob Rae joined the blockade of the road and was arrested with other protesters (Killan 1992). The Peterson "government's acceptance of MNR's environmental assessment and the denial of a hearing in this case suggest[ed] that there [was] not yet much government commitment to integrated assessment" (Dunster and Gibson 1989, p. 103).

Although the Davis government had approved 155 new parks in 1983, many of them were much smaller than the Parks Branch had originally proposed. This concession to the opposing interests was compounded by the provision in the *Provincial Parks Act* that permitted activities such as cutting, hunting, and even mining inside these parks. The Sierra Club, along with the Federation of Ontario Naturalists and the National and Provincial Parks Association, returned to the fray. Not until 1988 was the battle won when Peterson's Liberal government banned these activities in the parks (Killan 1993). The struggle over land entered a new phase when the World Wildlife Fund Canada and the Canadian Parks and Wilderness Society launched an "Endangered Spaces" campaign in September 1989 in support of the World Conservation Strategy to protect one-eighth of the land and water from industrial use (Killan 1993). Although that strategy did not go as far as Monte Hummel, Canadian head of the Fund, wished, it gradually came to be infused with a preservationist concern to protect significant examples of every biome (Mowat 1990). The 1992 statement of Ontario Provincial Parks Policy by the New Democratic government, which provided for wilderness parks, nature reserves, natural environment parks, and waterway parks, in addition to historical and recreation parks, reaffirmed the policy development of the preceding decades (Killan 1993).

SUMMARY

The forests of Ontario have found many uses during the past two centuries. Through much of the 19th century, forests were cut more often to clear land for agriculture than to provide square timber and lumber to British and American markets. Among the varied users of the forest, the pulp and paper industry has played the leading role in the 20th century. Ensuring that forests cut for pulpwood or sawlogs would regenerate after these harvesting operations was gradually recognized as calling for silvicultural efforts. These efforts were encouraged by the professional foresters trained in faculties of forestry, both within and outside Ontario, who gradually enhanced the forest management capacity of both the Ontario government and the forest companies. Although sustained-yield forest policies were under discussion by the middle of the 20th century, ensuring that forest practices were placed on a sustainable basis remained a real challenge. The last third of the century has been marked by concerns about the future supply of timber and the success of regeneration efforts, and these concerns have led several governments to institute audits of forest management and surveys of the forests. The concerns were intensified by the demands of many people, in northern Ontario as well as in the south, that areas of forest be set aside in parks to be enjoyed and in reserves that preserved examples of the various biomes of the province. Although these concerns intensified over the years and governmental responses to them gradually developed, the process of environmental assessment to which the Ministry of Natural Resources was subjected in the 1980s tested the Ministry severely, and found it initially wanting. Out of this process, however, as the next chapter will indicate, came policies and practices of sustainable forestry that opened a new era in the history of the Ontario forests and their management.

ACKNOWLEDGEMENTS

Earlier versions of this paper were read by J.F. Adderley, W.D Addison, K.A. Armson, C.A. Benson, P. Duinker, J. Flowers, J. Foulds, K.W.

Hearnden, P. Jasen, R. Klein, J. Naysmith, M.O. Nelson, M.F. Squires, and three anonymous reviewers. I am grateful for the comments that each provided. These readers have given me a deeper appreciation of the complexities of forest use and management. They have also tempered my tendency to judge those at work in or about the forest. The interpretations remain my own, and I accept full responsibility for them.

REFERENCES

Anderson, D.L. 1994. The flow of European trade goods into the Western Great Lakes Region, 1715-1760. In: The Fur Trade Revisited: Selected Papers of the 6th North American Fur Trade Conference, Mackinac Island, Michigan, 1991. East Lansing and Mackinac Island, Michigan: Michigan State University Press and Mackinac State Historical Parks. 93-115.

Anderson, F.J. 1986. Trade barriers and Canadian forest products: opening comments and review. In: Tariffs and the Canadian Forest Industry. Proceedings of the 19th Annual Lakehead University Forestry Association Forestry Symposium, Lakehead University, Thunder Bay, Ontario, January, 1986. 11-16.

Armson, K.A. 1976. Forest Management in Ontario. Toronto, Ontario: Ministry of Natural Resources. 185 p.

Armson, K.A. 1993. Summary of nursery and early plantings 1917-1927. Abitibi Power and Paper Company Ltd. Iroquois Falls, Ontario. Unpublished paper.

Bertrand, J.P. 1997. Timber Wolves: Greed and Corruption in Northwestern Ontario's Timber Industry, 1875-1960. Thunder Bay, Ontario: The Thunder Bay Historical Museum Society Inc. 164 p.

Bishop, C.A. 1981. Northeastern Indian concepts of conservation and the fur trade: a critique of Calvin Martin's thesis. In: S. Krech III (editor). Indians, Animals and the Fur Trade: A Critique of *Keepers of the Game*. Athens, Georgia: University of Georgia Press. 39-58.

Bladen, V.W. 1956. An Introduction to Political Economy. Toronto, Ontario: University of Toronto Press. 317 p.

Blair, J.H. 1984. Producing and Providing: The Story of Kiashke River Native Development Inc. N.p.: Royal Commission on the Northern Environment. 50 p. + appendices.

Board of Commissioners for the Queen Victoria Niagara Falls Park. 1889. Third Annual Report. Sessional Paper No. 37. Toronto, Ontario: Queen's Printer.

Bonsor, N. and F.J. Anderson. 1985. The Ontario Pulp and Paper Industry: A Regional Profitability Analysis. Toronto, Ontario: Ontario Economic Council. 104 p.

Braun, H.S. and W.G. Tamblyn. 1987. A Northern Vision: The Development of Lakehead University. Thunder Bay, Ontario: Lakehead University. 232 p.

Brown, W. 1884. The application of scientific and practical arboriculture to Canada. Synopsis of paper read before the British Association for the Advancement of Science, Montreal, August 1884.

Canada Commission of Conservation. 1910-18. Reports. Ottawa: King's Printer.

Canadian Forest Service. 1997. The state of Canada's forests 1996-1997. Publication Fo-6/1997E. Ottawa, Ontario: Canadian Forest Service. 128 p

Craig, G.M. 1963. Upper Canada: The Formative Years. Toronto, Ontario: McClelland and Stewart. 331 p.

Cross, M.S. 1973. The Shiner's War: social violence in the Ottawa Valley in the 1830's. Canadian Historical Review 54(1): 1-26.

Dawson, K.C.A. 1983. Prehistory of Northern Ontario. Thunder Bay, Ontario: Thunder Bay Historical Museum Society. 44 p.

Dewdney, S. 1970. Dating Rock Art in the Canadian Shield Region. Toronto, Ontario: Royal Ontario Museum. 71 p.

Dewdney, S. and K.E. Kidd. 1967. Indian Rock Paintings of the Great Lakes. Toronto, Ontario: N.p.: University of Toronto Press for the Quetico Foundation. 201 p.

Dick, T.J.O. 1982. Canadian newsprint, 1913-1930: national policies and the North American economy. Journal of Economic History 42(3): 659-687.

Dickason, O.P. 1991. "For every plant there is a use": the botanical world of Mexicans and Iroquoians. In: K. Abel and J. Friesen (editors). Aboriginal Resource Use in Canada: Historical and Legal Aspects. Manitoba Studies in Native History. VI. Winnipeg, Manitoba: University of Manitoba Press. 11-34.

Dickason, O.P. 1992. Canada's First Nations: A History of Founding Peoples from Earliest Times. Toronto, Ontario: McClelland and Stewart. 559 p.

Dickson, James. 1886. Camping in the Muskoka Region. Toronto, Ontario: C. Blackett Robinson. 164 p.

Drummond, A.T. 1879. Canadian timber trees: their distribution and preservation. Report of the Montreal Horticultural Society and Fruit Growers' Association. Montreal, Quebec: "Witness" Printing House. 18 p.

Drummond, I. 1987. Progress without Planning: The Economic History of Ontario from Confederation to the Second World War. Toronto, Ontario: University of Toronto Press. 502 p.

Dunster, J.A. and R.B. Gibson. 1989. Forestry and Assessment: Development of the *Class Timber Assessment for Timber Management* in Ontario. Toronto, Ontario: Canadian Institute for Environmental Law and Policy. 189 p. + exemption orders.

Eid, L.V. 1979. The Ojibway-Iroquois war: the war the Five Nations did not win. Ethnohistory 17(4), 297-324.

Epp, A.E. 1973. The paper makers. In: Cooperation Among Capitalists: The Canadian Merger Movement 1909-13 [Ph.D. thesis]. Baltimore, Maryland: Johns Hopkins University. 244-96.

Federation of Ontario Naturalists. 1958. Outline of a Basis for a Parks Policy for Ontario. Toronto, Ontario: Federation of Ontario Naturalists.

Foster, J. 1978. Working for Wildlife: The Beginning of Preservation in Canada. Toronto: University of Toronto Press. 283 p.

Gates, L.M. 1968. Land Policies of Upper Canada. Toronto, Ontario: University of Toronto Press. 378 p.

Gentilcore, R.L., D. Measner, and R.H. Walder (editors). 1993. Historical Atlas of Canada. II: The Land Transformed 1800-1891. Toronto, Ontario: University of Toronto Press. 207 p.

Gillis, R.P. and T.R. Roach. 1986. Lost Initiatives: Canada's Forest Industries, Forest Policy and Forest Conservation. Westport, Connecticut: Greenwood Press. 314 p.

Hall, D.J. 1985. Clifford Sifton. Volume II: A Lonely Eminence 1901-1929. Vancouver: University of British Columbia Press. 432 p.

Hardy, A.S. 1880. Report Respecting Recent Proceedings in Reference to the Niagara Falls and Adjacent Territory. Sessional Paper No. 51. Toronto, Ontario: Queen's Printer. 16 p.

Harris, R.C. (editor). 1987. Historical Atlas of Canada. I: From the Beginning to 1800. Toronto, Ontario: University of Toronto Press. 198 p.

Head, C.G. 1975. An introduction to forest exploitation in nineteenth century Ontario. In: Perspectives on Landscape and Settlement in Nineteenth Century Ontario. Toronto: McClelland and Stewart. 78-112.

Hearnden, K.W. 1977. Forest resources and development in Northwestern Ontario. Address to the Northwestern Ontario Municipal Association, Thunder Bay, Ontario. Unpublished. 18 p.

Hearnden, K.W. 1985. Forestry practice in Canada – an ethical dilemma. Presentation to a Symposium on Occupational Ethics, March, 1985, Lakehead University, Thunder Bay, Ontario. Unpublished. 4 p.

Hodgins, B.W. and J. Benidickson. 1989. The Temagami Experience: Recreation, Resources, and Aboriginal Rights in the Northern Ontario Wilderness. Toronto, Ontario: University of Toronto Press. 358 p.

Holskamm, T.E., V.P. Lytwyn, and L.G. Waisberg. 1991. Rainy River sturgeon: an Ojibway resource in the fur trade economy. In: K. Abel and J. Friesen (editors). Aboriginal Resource Use in Canada: Historical and Legal Aspects. Manitoba Studies in Native History VI. Winnipeg, Manitoba: University of Manitoba Press. 119-140.

Innis, H.A. 1956. The Fur Trade in Canada: An Introduction to Canadian Economic History. Toronto, Ontario: University of Toronto Press. 442 p.

Jasen, P. 1995. Wild Things: Nature, Culture, and Tourism in Ontario 1790-1914. Toronto, Ontario: University of Toronto Press. 193 p.

Johnston, B. 1995. The Manitous: The Spiritual World of the Ojibway. Toronto, Ontario: Key Porter Books. 270 p.

Johnston, C.M. 1986. E.C. Drury: Agrarian Idealist. Toronto, Ontario: University of Toronto Press. 311 p.

Joly, H.G. 1878. Report on Forestry and Forests of Canada. Ottawa, Ontario: House of Commons Sessional Paper No. 9. Appendix I. 20 p.

Joly de Lotbiniere, Sir. H.G. 1882. The returns of forest tree culture. Report of the Montreal Horticultural and Fruit Growers' Association. 11 p.

Kelly, K. 1975. The impact of nineteenth century agricultural settlement on the land. In: Perspectives on Landscape and Settlement in Nineteenth Century Ontario. Toronto, Ontario: McClelland and Stewart. 64-77.

Killan, G. 1992. Ontario's provincial parks, 1893-1993: "we make progress in jumps." In: Islands of Hope: Ontario's Parks and Wilderness. Willowdale, Ontario: Firefly Books. 20-45.

Killan, G. 1993. Protected Places: A History of Ontario's Provincial Parks System. Toronto, Ontario: Dundurn Press, in association with the Ontario Ministry of Natural Resources. 428 p.

Kirkwood, A. 1886. Algonkin, forest and park, Ontario: letter to Hon. T.B. Pardee, MPP Commissioner of Crown Lands. Toronto, Ontario: Warwick and Sons. 8 p.

Kirkwood, A. and J.T. Murphy. 1878. The Undeveloped Lands in Northern and Western Ontario: Information Regarding Resources. Toronto, Ontario: Hunter, Rose. 278 p.

Lambert, R.S. with P. Pross. 1967. Renewing Nature's Wealth: A Centennial History of the Public Management of Lands, Forests and Wildlife in Ontario, 1763-1967. Toronto, Ontario: Department of Lands and Forests. 647 p.

Lower, A.R.M. 1938. The North American Assault on the Canadian Forest: A History of the Lumber Trade between Canada and the United States. Toronto, Ontario: Ryerson Press. 404 p.

Lower, A.R.M. 1973. Great Britain's Woodyard: British America and the Timber Trade, 1763-1867. Montreal: McGill-Queen's University Press. 284 p.

MacKay, D. 1985. Heritage Lost: The Crisis in Canada's Forests. Toronto, Ontario: Macmillan. 270 p.

Marsh, G.P. 1965. Man and Nature. 1864. New York, New York: Charles Scribner and Cambridge, Massachusetts: Belknap Press of Harvard University Press. 501 p.

Martin, C. 1978. Keepers of the Game: Indian-Animal Relationships and the Fur Trade. Berkeley, California: University of California Press. 237 p.

Martin, C. 1980. Subarctic Indians and wildlife. In: C.M. Judd and A.J. Ray (editors). Old Trails and New Directions: Papers of the 3rd North American Fur Trade Conference, May, 1978, Winnipeg, Manitoba. Toronto, Ontario: University of Toronto Press. 73-81.

McCalla, D. 1993. Planting the Province: The Economic History of Upper Canada 1784-1870. Toronto, Ontario: University of Toronto Press. 462 p.

McCloy, B. 1986. Canadian-U.S. lumber trade. In: Tariffs and the Canadian Forest Industry: Proceedings of the 19th Annual Lakehead University Forestry Association Forestry Symposium, Thunder Bay, Ontario. Thunder Bay, Ontario: Lakehead University. 25-40.

McNab, D.T. 1991. "Principally rocks and burnt lands": Crown Reserves and the tragedy of the Sturgeon Lake First Nation in Northwestern Ontario. In: K. Abel and J. Friesen (editors). Aboriginal Resource Use in Canada: Historical and Legal Aspects. Manitoba Studies in Native History. VI. Winnipeg, Manitoba: University of Manitoba Press. 157-171.

Morris, A. 1880. The Treaties of Canada with the Indians of Manitoba and the North-West Territories. Toronto, Ontario: Belfords, Clarke and Company. 375 p.

Morse, L.A. 1984. White Pine: Ontario Celebrates its History. Toronto, Ontario: Ministry of Natural Resources. 48 p.

Mowat, F. 1990. Rescue the Earth: Conversations with the Green Crusaders. Toronto, Ontario: McClelland and Stewart. 282 p.

Nelles, H.V. 1974. The politics of development: forests, mines and hydro-electric power in Ontario, 1849-1941. Toronto, Ontario: Macmillan. 511 p.

Nute, G.L. 1941. The Voyageur's Highway: Minnesota's Border Lake Land. St. Paul, Minnesota: The Minnesota Historical Society. 106 p.

Oliver, P.N. 1977. G. Howard Ferguson: Ontario Tory. Toronto, Ontario: University of Toronto Press. 501 p.

Ontario Department of Lands and Forests. 1929. Report of the Minister of Lands and Forests of the Province of Ontario for the Year Ending October 31, 1928. Sessional Paper No. 3. Third Session of the Seventeenth Legislature of the Province of Ontario 1929. Toronto, Ontario: King's Printer.

Ontario Department of Lands and Forests. 1954. Suggestions for Program of Renewable Resources Development. Toronto, Ontario: Department of Lands and Forests. 43 p.

Ontario Department of Lands and Forests. 1967. Classification of Provincial Parks in Ontario 1963. Toronto, Ontario: Department of Lands and Forests, Parks Branch.

Ontario Economic Council. 1970. A Forest Policy for Ontario. Toronto, Ontario: Ontario Economic Council. 40 p.

Ontario Forest Industries Association. 1977a. Response to: Forest management in Ontario 1976 [by] Professor K.A. Armson. N.p.: Ontario Forest Industries Association. 28 p.

Ontario Forest Industries Association. 1977b. Forest management in Ontario. A brief to the Minister of Natural Resources. N.p.: Ontario Forest Industries Association. 20 p.

Ontario Independent Forest Audit Committee. 1992. K.W. Hearnden (chair). A Report on the Status of Forest Regeneration. Sault Ste. Marie, Ontario: Ontario Ministry of Natural Resources. 117 p.

Ontario Ministry of Natural Resources. 1972. Forest Production Policy Options for Ontario. Toronto, Ontario: Ministry of Natural Resources. 81 p.

Ontario Ministry of Natural Resources. 1978. Ontario Provincial Parks Planning and Management Policies. Manual. Toronto, Ontario: Ontario Ministry of Natural Resources. 195 p.

Ontario Ministry of Natural Resources. 1986. Action Plan on Forest Management. Toronto, Ontario: Ontario Ministry of Natural Resources.

Ontario Royal Commission on Forestry. 1947. Report. Toronto, Ontario: King's Printer. 196 p.
Ontario Royal Commission on Forestry Protection in Ontario. 1898. Report. Sessional Paper (1898-99), No. 35. Toronto, Ontario: Queen's Printer. 29 p.
Ontario Royal Commission on the Northern Environment. 1985. Letters from J.E.J. Fahlgren, Commissioner, to Premier William Davis (17 December 1982) and Minister of the Environment Keith Norton (17 December 1982 and 24 March 1983). In: Final Report and Recommendations. Appendices 4, 6 and 7. Toronto, Ontario: Ontario Ministry of the Attorney-General.
Outdoor Recreation Resources Review Commission. 1962. Outdoor Recreation for America. A Report to the President and to the Congress by the Outdoor Recreation Resources Review Commission. Washington, DC. 259 p.
Parr, J. 1990. The Gender of Breadwinners: Women, Men, and Change in Two Industrial Towns 1880-1950. Toronto, Ontario: University of Toronto Press. 321 p.
Pielou, E.C. 1991. After the Ice Age: The Return of Life to Glaciated North America. Chicago, Illinois: University of Chicago Press. 364 p.
Phipps, R.W. 1883. Report on the Necessity of Preserving and Replanting Forests Compiled at the Instance of the Government of Ontario. Toronto, Ontario: C. Blackett Robinson. 144 p.
Pomfret, R. 1976. The mechanization of reaping in nineteenth-century Ontario: a case study of the pace and causes of the diffusion of embodied technical change. Journal of Economic History 36(2): 399-415.
Pross, A.P. 1967. The development of professions in the public service: the foresters in Ontario, 1950-1970. Canadian Public Administration 10(3): 376-404.
Radforth, I. 1982. The mechanization of the pulpwood logging industry in Northern Ontario, 1950-1970. Unpublished paper presented to the Canadian Historical Association.
Radforth, I. 1987. Bushworkers and Bosses: Logging in Northern Ontario, 1900-1980. Toronto, Ontario: University of Toronto Press. 355 p.
Radforth, I. 1995. The shantymen. In: P. Craven (editor). Labouring Lives: Work and Workers in Nineteenth-Century Ontario. Toronto, Ontario: University of Toronto Press. 204-274.
Ray, A.J. 1974. Indians in the Fur Trade: their Role as Hunters, Trappers and Middlemen in the Lands Southwest of Thunder Bay. Toronto, Ontario: University of Toronto Press. 254 p.
Reed, F.L.C., and Associates. 1980. Wood Fibre Supply and Demand in Northern Ontario. Vancouver, British Columbia: F.L.C. Reed and Associates Ltd.
Reid, R.M., (editor). 1989. The Upper Ottawa Valley to 1855. N.p.: The Champlain Society with the Carleton University Press. 476 p.
Rich, E.E. 1966. Montreal and the Fur Trade. Montreal, Quebec: McGill University Press. 99 p.
Richardson, A.H. 1927. Forestry in Ontario. Toronto, Ontario: Ontario Department of Forestry. 73 p.
Rogers, E.S. 1986. Ecology, culture, and the subarctic Algonquians. *Anthropologica* 28(1): 1-2, 203-216.
Rosehart, B., O.B. Allen, J.R. Carrow, M. Hummel, G.R. Seed, and G. Sheehy. 1987. An Assessment of Ontario's Forest Resources Inventory System and Recommendations for its Improvement. Ontario Forest Resources Inventory Committee. Unpublished report. 48 p.
Rowe, J.S. 1972. Forest Regions of Canada. Publication No. 1300. Ottawa, Ontario: Canadian Forest Service. 172 p.
Russell, P.A. 1982. Upper Canada: a poor man's country? Some statistical evidence. Canadian Papers in Rural History 3: 129-47.
Russell, P.A. 1983. Forest into farmland: Upper Canadian clearing rates, 1822-1839. Agricultural History 57: 326-339.
Sanders, L. 1992. What forest audit didn't show. Thunder Bay, Ontario: *Chronicle-Journal*, December 24,1992. 1 and 5.
Saywell, J.T. 1991. "Just Call Me Mitch": The life of Mitchell F. Hepburn. Toronto, Ontario: University of Toronto Press. 640 p.
Schmalz, P.S. 1991. The Ojibwa of Southern Ontario. Toronto, Ontario: University of Toronto Press. 336 p.
Sharpe, J.F. and J.A. Brodie. 1931. The Forest Resources of Ontario 1930. Toronto, Ontario: Department of Lands and Forests. 60 p.
Sinclair, Peter W. 1987. The north and the north-west: forestry and agriculture. In: I. Drummond (editor). Progress Without Planning: The Economic History of Ontario from Confederation to the Second World War. Toronto, Ontario: University of Toronto Press. 77-90.
Sisam, J.W.B. 1961. Forestry Education at Toronto. Toronto, Ontario: University of Toronto Press. 116 p.
Small, H.B. 1884. Canadian Forests: Forest Trees, Timber and Forest Products. Montreal, Quebec: Dawson. 66 p.
Smith, A. 1990. Farms, forests and cities: the image of the land and the rise of the metropolis in Ontario, 1860-1914. In: D.R. Keane and C.F. Read. (editors). Old Ontario: Essays in Honour of J.M.S. Careless. Toronto, Ontario: Dundurn Press. 71-94.
Spears, J. 1979. Ontario stalling as vital forests devastated. *Toronto Star*, May 5,1979. C1.
Statutes of Canada. 1949. *Canada Forestry Act*. 13 George VI, c. 8.
Statutes of Ontario. 1878. *Act to Preserve the Forests from Destruction by Fire*. 41 Vict., c. 23.
Statutes of Ontario. 1893. *Algonquin Park Act*. 56 Vict., c. 8.
Statutes of Ontario. 1898. *Forest Reserves Act*. 61 Vict., c. 10.
Statutes of Ontario. 1906. *University of Toronto Act* . 6 Edw. VII, c. 55.
Statutes of Ontario. 1911. *Counties Reforestation Act*. 1 Geo. V, C. 74.
Statutes of Ontario. 1917. *Forest Fires Prevention Act*. 7 Geo. V, C. 54.
Statutes of Ontario. 1927. *Forestry Act*. 17 Geo. V, c. 12.
Statutes of Ontario. 1927. *Lakes and Rivers Improvement Act*. 17 Geo. V, c. 40; R.S.O. 1990, c. 3.
Statutes of Ontario. 1929. *Pulpwood Conservation Act*. 19 Geo. V, c. 13.
Statutes of Ontario. 1934. *Woodsmen's Employment Act*. 24 Geo. V, c. 66.
Statutes of Ontario. 1936. *Forest Resources Regulation Act*. 1 Edw. VIII, c. 22.
Statutes of Ontario. 1937. *Settlers' Pulpwood Protection Act*. 1 Geo. VI, c. 70.
Statutes of Ontario. 1947. *Forest Management Act* . 11 Geo. VI, c. 38.
Statutes of Ontario. 1954. *Provincial Parks Act*. 3 Elizabeth II, c. 75. c. 75.
Statutes of Ontario. 1959. *Wilderness Areas Act* . 7-8 Elizabeth II, c. 107.
Statutes of Ontario. 1975. *Environmental Assessment Act*. R.S.O. 1990, c. E18.
Statutes of Ontario. 1994. *Crown Forest Sustainability Act*. S.O., c. 25.
Statutes of Quebec. 1870. *Forest Fire Prevention Act*. 34 Vict., c. 19.
Statutes of Upper Canada. 1849. *Crown Timber Act*. 12 Vict., c. 30.
Swift, J. 1983. Cut and Run: The Assault on Canada's Forests. Toronto, Ontario: Between the Lines. 283 p.
Trigger, B.G. 1976. The Children of Aataentsic: A History of the Huron People to 1660. Kingston, Ontario and Montreal, Quebec: McGill-Queen's University Press. 921 p.

Trigger, B.G. 1985. Natives and Newcomers: Canada's "Heroic Age" Reconsidered. Kingston, Ontario and Montreal, Quebec: McGill-Queen's University Press. 410 p.

United Kingdom. 1791. *Constitutional Act*. 31 George III, c. 31.

United States of America. 1964. *Wilderness Act.* Congressional Record. Proceedings and Debates of the 88th Congress, Second Session. 17458-61.

Warecki, G. 1992. The people behind the parks: Ontario's wilderness conservationists. In: L. Labatt and B. Litteljohn (editors). Islands of Hope: Ontario's Parks and Wilderness. Willowdale, Ontario: Firefly Books. 75-82.

White, T.B. 1886. Arboriculture and Agriculture; Forestry and Farming in Ontario. Toronto, Ontario: Hunter, Rose. 16 p.

Wightman, W.R. and N.M. Wightman. 1997. The Land Between: Northwestern Ontario Resource Development, 1800 to the 1990s. Toronto, Ontario: University of Toronto Press. 566 p.

Wilson, A. 1968. The Clergy Reserves of Canada: A Mortmain. Toronto, Ontario: University of Toronto Press. 266 p.

Wright, J.V. 1995. A History of the Native People of Canada. Vol. I (10,000-1000 BC). Hull, Quebec: Canadian Museum of Civilization. 588 p.

Zaslow, M. 1971. The Opening of the Canadian North. Toronto, Ontario: McClelland and Stewart. 333 p.

Zavitz, E.J. 1908. Report on the Reforestation of Waste Lands in Southern Ontario. Sessional Papers, First Session of Twelfth Legislature of the Province of Ontario. Section 1919. No. 23. 28 p.

13 A New Foundation for Ontario Forest Policy for the 21st Century

DAVID L. EULER* and A. ERNEST EPP**

POLICY CULTURE AT THE END OF THE 20TH CENTURY

Four phases have been identified in the historical use of natural resources: the exploitation phase, the administrative phase, the ecological phase, and the social phase (Kimmins 1995). The exploitation phase began when humans first emerged as a species and used wild animals and plants for their needs without concern about the long-term consequences. As the number of people was small and their weapons poor, the impact on natural resources was negligible. The next period, the administrative phase, is defined by a growing realization that resources are not unlimited and that regulations must be imposed to sustain them. In North America, this phase began in the 19th century, with the first regulations concerning forests and wildlife.

The ecological phase had begun by the middle of the 20th century. By that time, knowledge of the ecology of the forest had developed, silviculture had become a mature science, and there was a vast body of knowledge about the relationship between tree growth and soil quality. The relationship of many wildlife species to forest habitats was also better understood; in fact, the impact of forest harvest on popular game animals was part of the forester's professional knowledge. In Ontario, advances in knowledge during the middle decades of the century increased the wealth of the province through the use of timber resources in the great building boom after the Second World War. Much of the economic well-being that the people of Ontario enjoy today is built on a foundation that came from the forests of the province.

By the present decade, the last of the 20th century, the social phase of forestry is in progress. Many people in Ontario and Canada have come to demand more than wood products from the forest, so choices must now be made among the various goals that can be achieved by forest management. In addition to recreation, tourism, and fish and wildlife, intangible values such as solitude and landscapes for contemplation are gaining as much legitimacy as the forest's economic benefits. The health of the forest is the overall concern, including such ecological processes as oxygen production and carbon sequestration. Rights and responsibilities in forest ecosystems are the themes that now dominate forest management plans.

THE ENVIRONMENTAL ASSESSMENT OF ONTARIO'S FOREST MANAGEMENT

Perhaps the best illustration of how the social phase of forestry has matured over the past several decades in Ontario is the *Environmental Assessment Act* and its impact on forestry in Ontario. In passing the Act in 1975, the government of Ontario set an international standard (Dunster and Gibson 1989). Events subsequently proved that it had also posed an enormous challenge for the provincial Ministry of Natural Resources. Although patterned on the US *National Environmental Policy Act* of 1970, Ontario's much more detailed legislation was intended to ensure that proposed plans and projects in Ontario were developed and evaluated publicly, in the light of clearly stated purposes, adequate consideration of alternatives, and careful evaluation of potential impacts on the social,

* Faculty of Forestry and the Forest Environment, Lakehead University, 955 Oliver Road, Thunder Bay, Ontario P7B 5E1
** Department of History, Lakehead University, 955 Oliver Road, Thunder Bay, Ontario P7B 5E1

economic, cultural, and biophysical environment (Gibson and Savan 1986).

The Act also applied to all provincial government activities, unless an exemption was obtained. From 1976 to 1985, the Minister of the Environment granted ten successive exemptions to give the Ministry of Natural Resources time to perform an internal environmental assessment of its various activities, termed "undertakings" in the Act, and to prepare a "class assessment" document. Unlike an individual assessment, which is concerned with a single activity, a class assessment is intended to examine the principles and ideas that govern an activity and allow the activity to occur if the principles are followed. For the Ministry of Natural Resources, the undertakings consisted of the forest management plans and operating plans, and included the building of forest access roads and the spraying of herbicides and insecticides for forest management purposes.

In older forest management plans, governed by the *Crown Timber Act*, the main purpose was to maintain an even flow of wood from the forest; that is, sustained yield. The planning process was intended to be "evergreen," meaning that the forest would always be composed of abundant healthy trees, and a supply of wood would always come from the forest. Plans were made for a 20-year period and reviewed every 5 years (see Epp 2000, this volume).

The challenge facing the Ministry can be appreciated from one part of the 1981 exemption, which required it to plan primary public forest access roads on Crown management units and emphasize the identification of alternative road locations; to consider the environmental effects of the alternatives; and to evaluate the rationale behind the selection made (Ontario Ministry of the Environment 1981). The Ministry was also required, from the spring of 1981, to give at least 30 days notice of the aerial spraying of herbicides or insecticides to assist in the regeneration or protection of the forest. At the same time, these requirements helped to focus the Ministry's attention on providing for timber production rather than on planning forest management.

A Royal Commission on the Northern Environment

While the Ministry was beginning to prepare for the environmental assessment of its activities, a Royal Commission on the Northern Environment was appointed to consider the request from a pulp and paper company, Reed Paper, for additional forest limits north of the 50th parallel of latitude. The company's record in dealing with the Ministry was poor and its environmental record abysmal (Wightman and Wightman 1997). In 1973, the discovery was made that the company's mill at Dryden in northwestern Ontario had poisoned the network of waterways known as the English-Wabigoon River system, ruined tourism on these rivers, and devastated the First Nations who depended on fish from the rivers for their sustenance (Shkilnyk 1985). The damage done illustrated in the starkest terms the conflict, which has repeatedly arisen in northern Ontario, between industrial activity and other uses of the forest.

It was consequently appropriate for the Minister of the Environment to seek the appointment of a commissioner to conduct an inquiry, under the guidance of the *Environmental Assessment Act*, into beneficial and adverse effects on the environment, such as those related to harvesting, supply and use of timber resources, mining, milling, smelting, oil and gas extraction, hydro-electric development, nuclear power development, water-use, tourism and recreation, transportation, communications, or pipelines north of the 50th parallel (Ontario Royal Commission on the Northern Environment 1985, Appendix I). The list of forest activities was comprehensive; the investigative challenge was enormous.

At that time, land use and resource management plans were being produced under a strategic land use planning initiative of the Ministry of Natural Resources. To determine whether Ministry-approved district land use plans could be adopted pending the outcome of the inquiry, the Commissioner sought to determine whether they were classified as "implementable" under the *Environmental Assessment Act*, or merely "conceptual/inventorial" (Ontario Royal Commission on the Northern Environment 1985, Appendix 6, p. 2). In a letter to the provincial premier on December 17, 1982, the Commissioner recommended that " ... all land use planning processes affecting Ontario north of 50° latitude be deferred ... and that the product of the Ministry of Natural Resources planning activities – land use plans – not be finalized until my findings and recommendations are released in the form of a public report and have been considered by your Government" (Ontario Royal Commission on the Northern Environment 1985, Appendix 4, p. 4).

The winter of 1982-83 proved to be a crucial time for the Ontario government, which failed the test it had set itself in the *Environmental Assessment Act*. In a letter to the Commissioner on January 18, 1983, the Natural Resources Minister stated that the land use plans were only guidelines intended for internal Ministry use and did not constitute a legal commitment to manage natural resources in any specific way. Some critics have noted that this approach downgraded the whole land use planning initiative (Dunster and Gibson 1989). In a letter of March 24, 1983, to the Environment Minister, the Commissioner stated that " ... this rationale, and now the apparent acceptance of it, gives rise to fundamental questions about the true nature, intent, status, substance, and implications" of the Natural Resources Minister's planning process, "about the appropriate locus of decision-making related to resource allocation, management and protection, about the effective injection of the Act's principles into the Ministry decision-making and planning processes, and about the continuing credibility of the Act itself"(Ontario Royal Commission on the Northern Environment 1985, Appendix 7, p. 1).

The conflicts that impeded the Ministry's use of environmental assessment to improve forest management became very clear in the Temagami region. More than a century had passed since Ontario officials had discovered the exclusion of the region's Tema-augama Anishnabai First Nation from the Robinson Huron Treaty, under which reserves had been established for other First Nations in that part of the province. The Tema-augama Anishnabai First Nation still had not been given a reserve. While negotiations continued, the Ministry attempted to plan activities in the region, and it did so despite the filing of a legal caution by the First Nation's lawyers. Unfortunately, background information on the Temagami district released by the Ministry in June 1980 indicated that "large volumes of white and red pine no longer exist [in the one-time pine reserve] due to past cutting practices, overmaturity and decadence" (Hodgins and Benidickson 1989, p. 276). This frank confession of the failure of Ontario's 80-year effort to achieve sustained-yield pine forestry in the former Temagami Forest led to desperate attempts to meet the timber requirements of nearby sawmills. At the same time, enjoyment of the area by cottage owners and tourists was increasing, and the Ministry's desire to create a northeastern Ontario wilderness park found a natural focus in this region. A conservation group known as the Alliance for the Lady Evelyn Wilderness supported this development and was later joined by other groups, including the International Union for the Conservation of Nature and Natural Resources.

Class Environmental Assessment of Forest Management

Between 1977 and 1987, the Ministry of Natural Resources made five attempts at a class environmental assessment of its forest management activities, the last three in public documents. The 1977 draft, circulated only within the Ministry, stated the goal to be "provid[ing] for an optimum continuous contribution to the economy by the forest based industries consistent with sound environmental practices; and to provide for other uses of the forest" (Dunster and Gibson 1989, p. 45). The Ministry suggested that there were only two options in forest management: "uncontrolled clearcut logging and partial high-grade logging," but it did allude to the following as alternative methods: "the clearcut silvicultural system, the shelterwood silvicultural system, the selection silvicultural system, and afforestation" (p. 45). These alternative methods were simply approaches that limited the number of trees that could be taken from the forest at any one time. The draft also offered new environmental quality guidelines developed from a pilot project in forest management.

Concerns expressed within the Ministry were sufficient to force a reworking of the draft over the next three years. The 1980 draft, reviewed by some staff of the Ministry of the Environment as well as some industry personnel, alluded to other uses of the forest than cutting for industry, but it continued to focus on logging and the associated road building and silvicultural activity. Among the main concerns of the reviewers was the lack of a clear planning process, making it difficult to see how forest management could be carried out as the *Environmental Assessment Act* required. It also appeared that the forest management objectives were predetermined and not open to question (Dunster and Gibson 1989).

When the Ministry of Natural Resources finally published an environmental assessment draft in September 1983, it defined forest management as "sustained yield timber production" and declared the Ministry's preoccupation to be "provid[ing] continuous, predictable wood supply sources, quantities and qualities" and "improv[ing] forest productivity through sound silvicultural practices [that] optimize returns on forest management investments" (Dunster and Gibson 1989, pp. 47-48). Secondary objectives, such as outdoor recreation and wildlife management, would have to be achieved as by-products of timber production. Areas might be reserved from timber production (in accordance with a century-old forest policy), and forest production might be subjected to specified conditions and limitations in some areas, but forest management was clearly timber harvesting as far as the Ministry of Natural Resources could see.

Criticized on all sides, the Ministry made a fourth attempt at an acceptable draft in December 1985. This fourth draft, presented to the Ministry of the Environment for review and approval as the *Class Environmental Assessment for Timber Management on Crown Lands in Ontario,* declared the Ministry's undertaking to be "timber management" designed to "provide a continuous and predictable supply of wood for Ontario's forest products industry" (Dunster and Gibson 1989, p. 56). Having been criticized for neglecting other objectives in the 1983 draft, the Ministry had dropped them entirely! When official reviewers within the Ministry of the Environment, other ministries, and federal agencies expressed their strong disapproval of this myopic attitude, the Ministry withdrew this proposal in December 1986. It made a fifth attempt in June 1987 with a revised *Class Environmental Assessment for Timber Management on Crown Lands in Ontario*, which included some recognition of Native land claims and Aboriginal rights and provided a brief discussion of heritage resources and archaeological values. The discussion of its timber production policy revealed, however, that the Ministry of Natural Resources remained preoccupied with the supply of timber to the forest companies.

Despite many years of preparation, the Ministry was not prepared for the hearing by the Environmental Assessment Board which Environment Minister Jim Bradley ordered in 1987. When the Board attempted to begin the hearing in May 1988, the Ministry could submit witness statements for only 4 of its 17 panels. As the Board concluded, "Other parties, therefore, did not know precisely what was in MNR's case, and could not prepare adequately" (Ontario Environmental Assessment Board 1994, p. 411). Adjournment of the hearing from mid-November 1988 to the end of January 1989 "to allow MNR to complete its witness statements" was only one of the reasons why the hearing was not concluded until October 1992 and the Decision not released until April 20, 1994.

The most significant fact for future forest policy, and a reality that affected everyone involved with the hearing, was the fact that the Ministry of Natural Resources proceeded to alter its operations while they were under review. Throughout the hearing, but particularly after a change of government in 1990, the Ministry announced numerous changes to its policies and practices (Ontario Environmental Assessment Board 1994). In fact, it was largely the progress made in forest management during the eight years after the Ministry published the third draft that enabled the Board to declare the class environmental assessment acceptable and to approve the undertaking of timber management planning (Ontario Environmental Assessment Board 1994, p. 411).

Results of the Environmental Assessment

The environmental assessment process, and the 115 Terms and Conditions on which the approval of the undertaking was based, moved Ontario forest management into a decisively new era. The Ministry of Natural Resources was now required to ensure that a timber management plan had been made for each forest management unit by a registered professional forester, in consultation with an interdisciplinary planning team and a local citizens' committee. Public notice was to be given annually of the timber management plans in preparation or being initiated, in order to facilitate public consultation. The Ministry was also to work with "native [*sic*] communities in or adjacent to the forest management unit ... [to] produce a Native Background Information Report for use in timber management planning" (Ontario Environmental Assessment Board 1994, p. 436).

Timber management was to be guided by implementation manuals reviewed at five-year intervals and updated "by suitably qualified people" in order to "reflect current scientific knowledge as it applies to Ontario" (Ontario Environmental Assessment Board 1994, p. 461). Not only was the Forest Resources Inventory to be regularly updated, but fisheries and wildlife inventory information was also to be made available to assist the planning effort. Each timber management plan was to include a "values map," or maps, to represent "known natural resource features, land uses and values which must be considered in timber management planning" (p. 436). An operations map was to be available for each forest management unit to "portray information about the locations of past, proposed and/or approved harvest operations for the two past five-year Timber Management Plans, the Plan currently in force and the proposed Plan" as well as "alternative road corridors for each forest management unit" (p. 436).

In item number 39 of its Terms and Conditions, the Environmental Assessment Board spelled out what the contents of a timber management plan should be. It provided a further 29 requirements regarding other timber management planning and operational matters, including negotiations at the local level with First Nations whose communities were in the management unit under study. The Reasons for Decision of the Environmental Assessment Board also included orders on a number of important items of provincial forest policy. These included approval of the use of herbicides to enable small trees to survive the early states of regeneration and an order to use chemical means only as a last resort in controlling severe infestations of spruce budworm and other insect predators. The Board's order that clear-cuts be no larger than 260 ha was designed to make forest harvesting operations more sensitive to the habitat needs of moose and other animals, although a regional office of the Ministry could approve larger clear-cuts. The Board also criticized the reductions in spending on regeneration efforts which the Ontario government had implemented during the severe economic recession of the early 1990s.

Finding on the Abundance of Timber

An important observation of the Environmental Assessment Board was that Ontario did not face a shortage of timber or wood fibre in the foreseeable future, although the Board conceded that planning was necessary to avoid over-cutting of the forest. This view contrasted sharply with warnings over several decades that shortages of timber and wood fibre loomed in the future. It also conflicted with internal studies that the Ministry of Natural Resources completed in 1991 and with *An Assessment of Ontario's Forest Resources* released by the Ministry in 1996 (C.A. Benson, pers. comm., January 1998; OMNR 1996a). The fuller utilization of the forest which had begun early in the 1980s was one of the factors influencing the supply of fibre to the mills that encouraged this new complacency.

FAILURE OF THE POLICY OF SUSTAINED YIELD

The class environmental assessment for timber management and subsequent events illustrated the failure of the policy of sustained yield, which had been the basis for most natural resource management in North American since the early 1950s. In its simplest expression, sustained yield means that some portion of any renewable natural resource can be harvested for use without harm to the system that produces the resource. Each year, by this principle, the forest grows more trees than are necessary to replenish the trees that die and more wildlife than can survive; therefore, removing some of these trees or animals is an effective way of using resources without harming the forested ecosystem (Meffe and Carroll 1997, p. 389).

While the sustained yield concept is sound in principle, implementing it has proven very difficult in the management of a wide range of resources. The collapse of Canada's Atlantic fishery for cod (*Gadus morhua)* is an example of the difficulty involved in blending the necessary social, ecological, and political forces into a sustained-yield management system that functions effectively over a period of years (Hutchings et al. 1997). Sometimes financial stress, as Clark (1990) argues, leads inevitably to over-exploitation. The over-exploitation of whales seems to spring from the inability of most whaling countries to follow the basic idea of sustained yield (Kellert 1996). It is far easier to hunt whales until they are gone than to pursue a policy of taking only as many as the population can replace through natural reproduction.

In Ontario, the failure of the sustained yield policy was made public by K.A. Armson when he suggested that cut rates continued to out-run the regenerative process, primarily because forest management for sustained yield was not working (Armson 1976; see also Epp 2000, this volume). By the mid-1970s there was general recognition in Ontario that sustained yield was not an effective management approach in forestry. The same conclusion had been reached in the United States. McQuillan (1994) pointed out that, despite almost a century of espousing sustained yield of timber for public lands (Parry et al. 1983) and nearly half a century of advocating the same goal for private lands (Dana and Fairfax 1980), there is an abiding discontent with the limited success of a sustained-yield policy in the western United States.

The pursuit of sustained timber yield led to some changes in Ontario's forested ecosystems which have had undesirable results. These included a reduction in the range of woodland caribou (*Rangifer tarandus*) and a fragmentation of the forest to favour wildlife species requiring open, semi-open, and early successional forests (Cummings 1997). In that case, applying the concept of sustained yield to one natural resource has created undesirable changes in the populations of other resources. Virtually all the failures of sustained-yield management have resulted not from a flaw in the concept, but from the difficulty of blending the ecological, social, and political forces that impact the management of the resource.

A NEW CONCEPT: ECOLOGICAL SUSTAINABILITY

The difficulties encountered in implementing the policy of sustained yield clearly pointed to the need for a new paradigm in forestry which would meet the needs of both people and ecosystems. This need gave rise to the concept of ecological sustainability. Beginning with the concept of "sustainable development" in the 1987 report of the Brundtland Commission, *Our Common Future* (World Commission on Environment and Development 1987), a mass of literature has appeared on the subject of managing natural resources in units termed "sustainable ecosystems."

The Brundtland Commission popularized the concept of sustainability by combining the word "development" with it. The Commission gave currency to the concept of "sustainable development" in a forum attended by many of the world's leaders in conservation and natural resource use. The purpose of this Commission was to report to the United Nations on how the use of natural resources could be managed to ensure that the needs of people could be met, both in the present and into the future. The Commission used the phrase "sustainable development" to call attention to the problem and to emphasize the idea that sustainability is important in any human endeavour.

The simple idea raised in "sustainable development" became the focus for a worldwide debate about how best to express the concept. The Brundtland Commission explained the term as focusing on an effort to use natural resources in such a way that future generations had the same choices as the current generation. Critics of the concept argued that the term was an oxymoron, because "development" implies growth, and growth cannot be sustained indefinitely. In natural resource management in North America, however, the ideas inherent in sustainable development have evolved into common approaches usually called ecosystem management or an ecological approach to management (Kimmins 1995; Gilmore 1997). This term, and others such as "ecological sustainability" and "sustainable forestry," reflect the goal of the Brundtland Commission and contain the same core ideas as "sustainable development."

Definitions of Ecosystem Management

An extensive literature has developed which explores ecosystem management in a forestry context. The concept of "New Forestry" (Franklin 1989) and "New Perspectives" advanced by the United States Forest Service (Salwasser 1990; Kessler et al. 1992; Thomas 1996) were the first attempts in North America to define and implement ecosystem management. "New Perspectives" stressed the fundamental changes needed to move from sustained yield to the concept of ecological sustainability, which included the way in which objectives are established, the role of public participation in resource management, and the role of science in resource management decisions. "New Forestry" likewise emphasized the need for ecosystem-level objectives and widespread involvement by the public in forest management planning. It called for management actions to be stated as scientific hypotheses that could be tested in a rigorous way.

There is no universally accepted definition for ecosystem management, although the key idea expressed by everyone is similar. Grumbine (1994) summarized much of the ecosystem management literature and suggested that ecosystem management was a process to integrate scientific knowledge of ecological relationships within a complex sociopolitical and values framework to protect native ecosystem integrity. The US Forest Service definition put forward in 1992 was "the use of an ecological approach to achieve multiple-use management of the national forests and grasslands by blending the needs of people and environmental values in such a way that the natural forest and grasslands represent diverse, healthy, productive and sustainable ecosystem" (Salwasser 1992, p. 469). More recent publications from the US Forest Service have elaborated the concept to include meeting the needs of people for forest products and maintaining healthy ecosystems (Blackwell et al. 1994; Salwasser 1994).

In Canada, ecosystem management is understood to be the management of human activities so that ecosystems, their structure, composition, and function, and the processes that shaped them, can continue at appropriate temporal and spatial scales (Canadian Biodiversity Working Group 1994). Kimmins (1995) reviewed changing paradigms in forestry and suggested that ecosystem management was an important concept in social forestry. He defined ecosystem management as "simply forest management that respects the ecological characteristics of local forested ecosystems and forested landscapes" (Kimmins 1995, p. 39).

An additional element was added to the definition by the Scientific Panel for Sustainable Forest Practices in Clayoquot Sound, British Columbia, which studied logging and forest management in an area of old-growth forest in that province. In a publication entitled *Sustainable Ecosystem Management in Clayoquot Sound*, the panel recommended "an ecosystem-based approach to planning in which the primary planning objective is to sustain the productivity and natural diversity of the Clayoquot Sound Region" (Clayoquot Sound Scientific Panel 1995, p. xiii). The import of the Panel's report is consistent with other policy documents under preparation by governments in North America which focus on ecosystem management (Thomas 1994). Booth et al. (1993) could as easily have been defining "ecosystem management" when they outlined the goal of "natural landscape management" to be maintenance of the long-term ecological integrity and capability of the forest landscape to serve all the functions and values dependent upon it.

Principles of Ecosystem Management

The main goal of ecosystem management, however it is defined, is to develop forest management plans that are ecologically sustainable, economically viable, and socially acceptable (Gilmore 1997). The fundamental principles of ecosystem management are still evolving, but a number of important themes are evident from various approaches taken to date. Slocombe (1993) suggested that implementing ecosystem management involves defining the management unit, developing an understanding of ecological processes, and creating planning and management frameworks in an ecosystem context. According to Grumbine (1994), the dominant themes of ecosystem management include a hierarchical context, appropriate ecological boundaries, adaptive management, and managing for the integrity of ecosystems. The essential elements of ecosystem management, according to Irland (1994), include the maintenance of biodiversity, establishment of ecosystem-level objectives, and management at the landscape level. Kimmins (1997) spoke of the need to maintain ecosystem health and integrity, to retain old-growth stages, to use low-disturbance harvesting systems, and above all else, to protect biodiversity. Booth et al. (1993) emphasized the need to maintain a continuing supply of all natural forest ecosystem types, the importance of basing forest management on sound knowledge of forest science, and the need to address a diverse range of interests in planning.

IMPLEMENTING ECOSYSTEM MANAGEMENT AND SUSTAINABILITY IN ONTARIO'S CROWN FOREST

The process of environmental assessment and new concepts of sustainability forced the Ministry of Natural Resources to consider new approaches to forest management. Largely on the basis of presentations made by professional foresters, the Environmental Assessment Board became convinced that integrated forest management was "a promising concept for timber management planning" and recognized that the Ministry's policy development was moving in that direction (Ontario Environmental Assessment Board 1994, p. 380). In the Board's view, the concept of ecological sustainability was captured by the term, "integrated forest management." The Ministry indicated that, although considerable progress would be made toward more comprehensive forest management by 1995-96, the transition to the new basis for forest management would continue throughout the nine-year term of the Board's class environmental assessment approval. The process was hastened by the Board's requirement that the Ministry "develop a conservation strategy, management directions and definitions for old-growth white pine and red pine by May 1995" (p. 387). It was strengthened by the Board's addition of the marten in the boreal forest and the pileated woodpecker in the Great Lakes-St. Lawrence Forest to be "featured species" in the Ministry's wildlife habitat management policy.

The Class Environmental Assessment hearing also considered "wildlife habitat supply modeling methodologies and landscape management methodologies as potential means of addressing biological diversity concerns in timber management" (Ontario Environmental Assessment Board 1994, p. 387). The Ministry asserted that its concern for biodiversity was evident in its recent publications and initiatives; it reported that its approach to conserving biodiversity had developed out of 1990 workshops on evaluating the effects of timber management on wildlife. By March

1991, the Ministry could recognize that, "although management for landscape diversity is more complex than the other alternatives examined, it would provide a number of significant benefits" based on the fact that recognition of biodiversity "contributes to multiple-use management of forest resources and that biodiversity is a key to maintaining future options and forest management flexibility" (p. 393). These conceptual advances clearly demonstrated the value of the process the Ministry had undergone through the *Environmental Assessment Act* of 1975.

In January 1990, the Deputy Minister of Natural Resources observed that the Ministry's goal was to "manage the forests on a sustainable basis" (Carrow 1990, p. 4). This source added that "there is more to management ... than simply understanding forest biology, including questions about land use policy and silvicultural practices, wilderness preservation and land ownership" (p. 5). Understanding the ecology of forests remains basic to sound management, however, and the Ministry launched an initiative known as the Forest Fragmentation and Biodiversity Project in 1991, within its new Sustainable Forestry Program. The studies carried out by the Ministry's Ontario Forest Research Institute initially focused on old-growth forests dominated by red and white pine, but they were expanded to include other studies such as the following: *A Pilot Landscape Ecological Model for Forests in Central Ontario* (Band 1993), *Managing the Land: A Medium-term Strategy for Integrating Landscape Ecology into Environmental Research and Management* (Merriam 1994), and *Ontario Fire Regime Model: Its Background, Rationale, Development and Use* (Li et al. 1996). These studies involved the use of data generated through the use of geographic information system (GIS) technology, which the Ontario Forest Resources Inventory Committee had urged the Ministry in September 1987 to incorporate into its planning (Rosehart et al. 1987).

Another initiative, launched in the fall of 1990 but funded within the Sustainable Forestry Program from 1991 to 1993, was the Forest Vegetation Management Task Force. The Task Force brought the Ontario Forest Industries Association and the Ministry of Natural Resources together with researchers seeking to develop a plan for future directions in forest vegetation management in Ontario. The plan was to be based on sound forest management principles, to recognize society's need for the sustainable production of a wide variety of forest resources, and to employ methods that were environmentally sound, socially acceptable, and economically viable (Wagner 1992).

The Ministry of Natural Resources responded to the new forest management challenges in November 1991 by appointing an Ontario Forest Policy Panel to develop forest policy for sustainable forests. The Panel consisted of the Chair of Forest Management and Policy at Lakehead University and three representatives from the private sector, including a regional planning consultant, a wildlife ecologist, and the regional vice-president of the Canadian organization of the International Woodworkers of America. The Forest Policy Panel published a discussion paper in June 1992 titled, *Our Future, Our Forests*, and invited responses from the interested public (Ontario Forest Policy Panel 1992). More than 3000 people responded in various ways with advice on a comprehensive policy framework for Ontario.

The policy framework proposed by the Forest Policy Panel was designed to secure forest sustainability. One of the five principles set down for achieving this end was that "forest practices, including clear cutting and other harvest methods, will emulate, within the bounds of silvicultural requirements, natural disturbances and landscape patterns" (Ontario Forest Policy Panel 1993, p. 5). The strategic objectives for forest sustainability were identified as the maintenance of biodiversity, natural heritage forest lands, and water, air, and soil quality; objectives with respect to social needs included the sustainability of communities and of resource use. It was hardly surprising that one of the chapters of the report was entitled, "Making the Difficult Choices." The Ministry of Natural Resources was challenged to focus its efforts over the next two years on developing a system of adaptive ecosystem management, a recommendation which marked the first time the term "ecosystem management" had appeared in Ministry policy documents. The report also urged the Ministry to encourage public participation in forest decision-making and to work on new policies relating to timber production on Crown lands, forest-based tourism, and forest investment and revenue.

Responding to the Panel's challenge, the Ministry of Natural Resources issued a *Policy Framework for Sustainable Forests* in April 1994 (almost simultaneously with the Decision of the Environmental Assessment Board) (OMNR 1994). Adopting the goal developed by the Panel, the Ministry stated that its mission was "to ensure the long-term health of forest ecosystems for the benefit of the local and global environments, while enabling present and future generations to meet their material and social needs" (OMNR 1994, p. 1). The principles for sustaining forests enunciated in the Panel's policy framework were also adopted, beginning with the statement that "forest practices, including all methods of harvesting, must emulate, within the bounds of silvicultural requirements, natural disturbances and landscape patterns" (p. 2). The Ministry also agreed both that "forest ecosystem types should not be candidates for harvest where this practice threatens or jeopardizes their long-term health and vigour" and that "forest practices must minimize adverse effects on soil, water, remaining vegetation, fish and wildlife habitat, and other values" (p. 2).

The Ministry of Natural Resources agreed that its strategic objectives included sustaining both resource-based communities and future resource use. The purpose was to encourage optimum levels and diversity of employment derived from Ontario's forests and to supply industrial and consumer wood needs, while maintaining forest sustainability. Another objective was to invest sufficiently to maintain forest sustainability – and to ensure that, over time, revenues from forest uses would meet investment requirements to maintain forest sustainability (OMNR 1994, p. 4).

THE *CROWN FOREST SUSTAINABILITY ACT*

The *Crown Forest Sustainability Act* (*CFSA*), drafted in the early months of 1994 and passed within that year, was intended to enshrine these new approaches to forestry in legislation. As a revision of the *Crown Timber Act* dating back through various amendments to the nineteenth century, this new Act was to "provide for the sustainability of Crown forests and, in accordance with that objective, to manage Crown forests to meet social, economic and environmental needs of present and future generations" (Part I, section 1). As Ministry of Natural Resources staff started to write the *CFSA*, the first problem they encountered was how to achieve a reasonable definition of sustainability that would be suitable for a law of the provincial parliament. Ecological concepts such as "forest complexity," "sustainability," and "forest health" were difficult to state in formal legislation; however, in preparation for the court challenges that would inevitably follow, it was in everyone's best interests to minimize ambiguity.

Part of the difficulty lay in the fact that many of the ideas from ecology require some element of judgment and resist empirical measurement. For example, one of the most celebrated principles for natural resource managers, from Aldo Leopold's *The Sand County Almanac* (1949, p. 262), is this: "A thing is right when it tends to preserve the integrity, stability and beauty of the biotic community. It is wrong when it tends otherwise." Statements of this nature may be appealing because they strike a chord of wisdom within everyone, but they are difficult to translate into legislation. Qualities such as "integrity," "stability," and "beauty" can rarely be identified or measured objectively; similarly, sustainability, complexity, and forest health are all multi-layered concepts that are understandable but difficult to measure.

In the *CFSA*, sustainability is defined as long-term forest health, so both of these concepts are needed to complete the Act's purpose. They must be defined and used in a way that can be monitored by the public, who own the forest, and by forest managers charged with implementing the law. "Forest health" means sustaining the complexity of forest ecosystems, while providing for the needs of people. The principle of sustaining ecosystem complexity is very similar to principles which insist that all ecosystem functions must be preserved. The *Policy Framework for Sustainable Forests* had stated that "maintaining ecological processes is essential" (OMNR 1994, p. 2), and the Act attempted to implement this principle. In general, the Act started from a base, not in timber production, as was the case with older legislation of this nature, but in the more modern idea of ecosystem management.

From a scientific point of view, however, the links among the diversity, health, and productivity of ecosystems are not well established. Ecosystems with low productivity and low diversity can be healthy, while ecosystems in similar ecological regions with high diversity and high productivity can also be healthy (Kimmins 1997). The challenge for forest managers is thus to implement the Act while recognizing that the scientific links between the ideas expressed in it are not always rooted in definitive scientific experimentation.

New Forest Management Mechanisms in the *CFSA*

The Act provides in section 24(1) that the Forest Management Agreements developed with forest companies are to be replaced by Sustainable Forest Licences "whenever the Minister is of the opinion that forest resources in a management unit should be made available for harvest or be used for a designated purpose." The Minister is required to "give public notice ... of the intention to make the resources available" and can only "grant a licence ... in accordance with a competitive process" (section 24[1] and [2]), although the Lieutenant Governor in Council can authorize another process. The transformation of types of tenure is spelled out in section 26, which authorizes "the Minister ..., with the approval of the Lieutenant Governor in Council, [to] grant a renewable licence to harvest forest resources in a management unit that requires the licensee to carry out renewal and maintenance activities necessary to provide for the sustainability of the Crown forest in the area covered by the licence." Such a licence can be granted for up to 20 years. It has to be reviewed every five years but it can then be renewed for an additional five years. Despite the similarity of terms between the Forest Management Agreements and the Forest Resource Licences, however, many industrial foresters believe that the security of wood supply that had previously justified forest management expenditures has been lost (Squires 1997).

Much more novel than the Sustainable Forest Licences is the Act's creation of two new trusts based on setting aside a portion of the stumpage fees for various kinds of regeneration activities. The first is a Forest Renewal Trust to "provide for reimbursement of silvicultural expenses incurred after March 31, 1994 in respect of Crown forests in which forest resources have been harvested and for such other matters as may be specified by the Minister" (section 48[3]). The Act declares that "the holder of a forest licence shall pay forest renewal charges to the Minister of Finance as required by the Minister of Natural Resources" and that these "forest renewal charges received by the Minister of Finance shall be held in a separate account in the Consolidated Revenue Fund [as] money paid to Ontario for a special purpose. This money can be used by the manager of the forest, whether it is a private company or the Ministry on Crown Management Units, to carry out regeneration work" (section 49[1]).

The Forestry Futures Trust was similarly established to provide for, *inter alia,* "the funding of silvicultural expenses in Crown forests where forest resources have been killed or damaged by fire or natural causes" and "the funding of intensive stand management and pest control in respect of forest resources in Crown forests" (section 51[3]). This trust, first created by an amendment to the *Crown Timber Act* in June 1994, became the responsibility of a Forestry Futures Committee composed of senior professional foresters, a former Member of the Provincial Parliament, and the regional planning consultant who had served as co-chair of the Ontario Forest Policy Panel. In its first two years of activity, the Committee received 165 applications and authorized spending in regard to 110 of them that amounted to more than $34 million of forest work, $21.1 million of it for remediation efforts, $10.6 million for stand improvement, and $2.4 million for forest protection (Naysmith 1997).

Ontario's *Forest Management Planning Manual*

Perhaps the key provision of the new Act was the requirement that forest managers follow a prescribed manual to write forest management plans. The planning manual instructs planners that sustainability is achieved by adhering to the following principles set out in Part I, section 2, subsection 3 of the Act:

1. Large, healthy, diverse, and productive Crown forests and their associated ecological processes and biological diversity should be conserved.

2. The long-term health and vigour of Crown forests should be provided for by using forest practices which, within the limits of silvicultural requirements, emulate natural disturbances and landscape patterns while minimizing adverse effects on plant life, animal life, water, soil, air, and social and economic values, including recreational values and heritage values.

Part II, section 9 of the Act stipulates, furthermore, that "The Minister shall not approve a forest management plan unless the Minister is satisfied that the plan provides for the sustainability of the Crown forest." The actual determination of sustainability is specified in Part VIII, section 68, subsection 3, as follows:

3. The Forest Management Planning Manual shall contain provisions respecting

(b) determinations of the sustainability of Crown forests for the purposes of the Act and the regulations in accordance with section 2;

(c) the requirement that management objectives in each forest management plan be compatible with the sustainability of the Crown forest; and

(d) the requirement that indicators be identified in each forest management plan to assess the effectiveness of activities in achieving management objectives and to assess the sustainability of the Crown forest.

Indicators of Sustainability

As the idea of sustainability is not yet amenable to direct measurement, to provide for implementation of the Act, measurable forest parameters must be identified to indicate whether sustainability has been achieved. If the system works as intended, at pre-selected values of such an indicator, some evidence will be available on the status of sustainability. Developing indicators of sustainability is a complex and technical task requiring considerable knowledge and understanding of the forest and of the measurement of elements within forested ecosystems. The task is best illustrated with an example using biodiversity as an indicator, although the *Forest Management Planning Manual* specifies several other indicators as well.

The Act is clear that biological diversity is an important component of a healthy forest and should be conserved. Conservation is not meant to imply that a low-diversity forest should be changed into a high-diversity forest, or that managers should strive to achieve a pre-selected level of biodiversity. As Kimmins (1997) points out, unmanaged forests exist in a wide range of biodiversity conditions. Many natural forests have low levels of biodiversity but are productive, healthy, and deliver important products to human society. In Ontario, large areas of jack pine (*Pinus banksiana)* and black spruce (*Picea mariana)* are common in much of the northern part of the province. Typically, these areas have low species diversity and relatively low productivity, but they are healthy forests that have existed this way for many decades. They harbour some wildlife species, for example, woodland caribou, that are unique or rare, but the total number of plants and animals in these areas is relatively low.

Other forests have high diversity and provide habitat for many species of wildlife and many plants. For example, Ontario's boreal mixedwood forest continues to be rich in species diversity and abundance. The intent of the Act is to manage the forest so as to keep the biodiversity values as close to "natural" values as possible, so the diversity of these forests must be considered and managed in different ways than the low-diversity forests. This implies, of course, that indicators of biodiversity would vary, depending on the forests and landscape being measured. There are, moreover, many measures of biodiversity, as has been pointed out by various authors (Noss 1983; Kimmins 1997), ranging from stand-level to landscape-level measurements. Any determination of sustainability must include an indicator of changes in biodiversity that occur as a result of the forest management plan, but it may be set at either the local or the landscape level.

Landscape-Level Measurements of Biodiversity

For the first time in Ontario, and perhaps in Canada, the *CFSA* requires that forest management planners prepare an explicit analysis of how forest management will affect the landscape. This provision is built into the planning process by the requirement to use indicators of sustainability, including at least one measure of diversity, at the landscape level. Each plan prepared under the authority of the *CFSA* must include the use of a model describing the landscape pattern and predicting how timber harvest will change the landscape pattern.

In addition to the conservation of biodiversity, the Act also directs forest managers to emulate natural disturbances and landscape patterns in forest harvest. The *Forest Management Planning Manual* specifies that the effects of forest harvest must be assessed by comparing the current landscape conditions with post-harvest landscapes created using a computer model. The guideline for acceptable changes is that they be "within bounds of natural variation." The Manual's criterion for accepting changes in either pattern or diversity is that they lie "within bounds of natural variation" (OMNR 1996b, p. A-63). This very broad guideline leaves considerable room for interpretation by the planners themselves. Such broad discretion is appropriate when the bounds of natural variation are not known, as is often the case, or when there are differences of opinion as to what constitutes natural variation. Finally, such a guideline may be the only realistic way of establishing a broad direction in a large bureaucracy, where a wide range of people are involved in a complex planning process.

One practical approach to landscape management that effectively implements the intent of the *CFSA* is described by Booth et al. (1993). Identified as "natural forest landscape management," this method involves designating a large area, such as a watershed, as the landscape management unit. Within this area, the objective of management is to maintain a combination of ecosystems, age classes, and stand types capable of providing the following benefits:

(a) Habitat for all plants and animals living in the area;

(b) An aesthetically pleasing forest landscape for recreation; and

(c) A healthy, vigorous forest to provide wood to meet the needs of society.

The basic tenet of natural forest landscape management is that long-term maintenance should not be compromised for short-term gains. This principle is followed by making maintenance of a continuing supply of ecosystems the primary management goal for the forest land base and overlaying other activities on that priority. This approach considers that the conservation of forested ecosystems is the fundamental biological basis for sustainability, and that keeping all natural ecosystems provides the best chance of maintaining the appropriate biodiversity for the landscape.

PUTTING THE *CFSA* TO THE TEST: A COURT CHALLENGE TO FOREST MANAGEMENT PLANS

The *Crown Forest Sustainability Act* made several provisions for a transition period of management planning, since it was clear that the Act could not be implemented immediately nor could the new concepts and procedures be incorporated instantly into forest management plans under preparation or in place. Two public-interest groups, however, the Algonquin Wildlands League and the Friends of Temagami, moved for a judicial review of six new management plans. Their request was based on the grounds that these plans were in violation of both the *Crown Forest Sustainability Act* and the *Environmental Assessment Act*. The applicants asked the Ontario Court of Justice (General Division) to declare the plans null and void on the grounds that "the Minister of Natural Resources did not determine the sustainability of the plans and work schedules in accordance with the Forest Management Planning Manual," and that "the plans violate some of the conditions imposed by the Environmental Assessment Board" (Court File No. 539/96, Ontario Court of Justice [General Division]).

After hearing extensive arguments, the Court ruled on February 6, 1998, that three of the plans were void for failure to comply with the *CFSA* and the *Forest Management Planning Manual*, and that the other three plans had significant problems that required attention before they could be implemented. The judgment included a review of many aspects of the Act, a discussion of how well the Ministry and its industrial partners had complied with the provisions of the Act, and an outline of the Crown's responsibility in implementing the Act.

A central point in the defence raised by the Ministry of Natural Resources, together with a dozen forest companies, was that the *Forest Management Planning Manual* required by the *Crown Forest Sustainability Act* was not mandatory but only directive. The Court declared (Campbell 1998) that this was not the position of the Environmental Assessment Board which, on October 10, 1995, approved the final draft of the manual dated May 18, 1995, subject to some amendments. The *Manual* was not approved by regulation and made available to the public, as required by the Act, until a year later. The complainants argued that the *Manual* had only been published at that time because of the legal action they had launched against the Ministry of Natural Resources. More significant to the case was the fact that major sustainability indicators required by the *Manual* were missing from the plans in question. In fact, none of the following criteria specified in the *Manual* had been addressed: landscape pattern or forest diversity indices, habitat for selected wildlife species, landscape processes, other forest resources, social and economic description, strategic direction and determination of sustainability, revenues and expenditures, and comparison of harvest, renewal, and tending forecast to selected management alternatives.

The Ministry of Natural Resources appealed the judgment on the grounds of errors in law. Its stated reason was that "the effect of the ruling goes far beyond the three forest management plans and has the potential to affect most forest companies" (OMNR 1998a). The judge had declared that "the short answer to the [Ministry's] argument, which was that the plans and work schedules can be saved because they comply with the overall general principles of sustainability, is that sustainability can only be measured by the Manual, compliance with which is a jurisdictional condition precedent to the validity of the plans" (Campbell 1998, p. 47). As the judge observed, "the whole point of the Manual, and the whole point of the new statute, is that sustainability will no longer be determined exclusively by the judgment of Ministry officials on the basis of vague statutory principles [but] sustainability must now be determined by the application of a public and concrete measurement standards [*sic*] based on the Manual" (p. 47).

The Ministry had argued further that the "objectively measurable yardsticks," as the judge described them, were only "words and forms" rather than (again in his words) "the indicators and parameters that are the concrete yardsticks against which the sustainability of the forest and the performance of the Ministry may be publicly measured and against which the Ministry may be held publicly accountable for its stewardship" (Campbell 1998, p. 47). Making Mr. Justice Campbell's judgment all the more serious was that he also found a number of violations of the conditions set by the Environmental Assessment Board when it approved the undertaking of timber management by the Ministry of Natural Resources.

In November 1998, the Divisional Court rejected the Ministry's appeal of the original decision and allowed the judgment to stand. The judgment has clarified the challenges that the Ministry of Natural Resources faces. The conditions of sustainability must be met in an open and public process in every part of Ontario. In addition, international demands continue to grow. The 1992 Rio de Janeiro gathering produced a *United Nations Convention on Biological Diversity*, and the Canadian and provincial governments have jointly signed *Canada's Biodiversity: A Commitment to its Conservation and Sustainable Use*. The Ontario government endorsed the Canadian policy on Conservation of Diversity in January 1996.

AN EXAMPLE OF ONTARIO'S NEW FOREST MANAGEMENT PLANS

Although the first plans prepared under the new Act were shown to be inadequate, management planners learned how important it is to follow the provisions of the *Crown Forest Sustainability Act*. Management plans are now more likely to be the responsibility of the forest management company, after the Ministry's recent reductions of staff and in-house operations. One of the first plans, prepared in 1997-98 for an area in northwestern Ontario called Nakina North (Nakina North Forest Management Plan), illustrates much promise concerning implementation of the new Act by modern, competent foresters.

In the Nakina North plan, a landscape-level modelling tool (FRAGSTATS) (McGarigal and Marks 1994) was used to estimate how the planned harvests would affect the landscape. The entire planning area, more than 100,000 ha, was defined as a landscape and broken into patches. Each patch was considered a homogenous unit or forest stand containing trees of a similar age and species composition. The composition and configuration of patches over the entire landscape area provided the description of landscape pattern required by the *Forest Management Planning Manual*.

The patches were classified into specific classes by the following: forest age, composition, and wildlife habitat units. With respect to age, the forest was divided into two classes: less than or equal to 40 years, and older than 40 years. Forest composition was defined as conifer (stands with more than a 70 percent component of coniferous species), hardwood (stands with more than an 80 percent component of deciduous species), and mixedwood (all other stands), all three broken into three age classes, 1 to 20 years, 21 to 80 years, and more than 80 years. Nine wildlife habitat units were established based on the species composition and age of the major tree species on the patch.

The indicators of biodiversity used in the landscape analysis included the total area of each class; the richness of types, measured by Shannon's equation; evenness, or the degree to which the various types were equally represented in the management unit; and various measures of the size, shape, and clustering of forest patches. These measures included edge density, total core area, mean core area per separate core, mean patch size, patch density, mean nearest-neighbour distance, and interspersion. Table 13.1 lists the general indicators and specific measures used in the Nakina North plan.

The impact of the planned harvest on the general indicators, and thus on the landscape pattern, was estimated for each of the parameters listed in Table 13.1. Each of the indicators changed as the proposed harvest plan was modelled. As the forest grows older, for example, the forest composition changes, as a result of both harvest and the passage of time; more young stands are created and stands that are not harvested grow older. Three of the ten indicators are discussed here: edge density, core area, and patch density.

Table 13.1

General indicators, specific measures, and formulae of landscape pattern measurements used in the Nakina North landscape pattern analysis.

		Landscape Classification				
General Indicator	*Specific Measure*	*Old Growth*	*Forest Age*	*Composition by Age*	*Habitat Unit*	*Habitat Unit by Age*
Forest Composition	Total Class Area	×	×	×	×	×
Forest Diversity	Shannon's Richness					×
	Shannon's Evenness					×
Forest Edge	Edge Density		×	×		
Forest Interior	Total Core Area	×	×	×	×	
	Mean Core Area per Separate Core	×			×	
Forest Fragmentation	Mean Patch Size	×			×	
	Patch Density					
Forest Isolation	Nearest Neighbour	×				
Forest Spatial Pattern	Interspersion	×				

Notes: × indicates landscape-specific measurements used.
Source: Nakina North Forest Management Plan 1994, Appendix 20, p. 7; M.Gluck, personal communication.

Edge Density

One indicator from Table 13.1 used in this modelling effort was "edge density." Edge is an important ecological concept that helps managers to understand how forest harvest may affect the ecological relationship of species living on the managed area. Many game animals such as moose (*Alces alces*) and white-tailed deer (*Odocoileus virginianus*) occupy habitats with an abundance of edge. Other species, such as woodland caribou, seem to require larger tracts of unbroken forest with relatively little edge. In this plan, edge density was defined as "the length of edge between patches relative to the area of a class in a landscape" (Nakina North Forest Management Plan 1994, Appendix 20, p. 11).

In the model representing the landscape of Nakina North after a proposed five-year timber harvest, there were significant increases in edge density for age classes up to 40 years (Figure 13.1). Edge density in the forest prior to harvest was approximately 1.5 m of edge per hectare in the younger age classes. By 2002, following the harvest, the edge density had increased to more than 2 m of edge per hectare of forest. The increase in edge density occurs because younger age classes are created by the harvest allocation, adjacent to the younger forest.

In this planning area, the edge density in younger forests after the harvest was significantly less than in older forests. The immediate result of the harvest may well be to create more habitat for edge-loving species, at the expense of species living in more unbroken areas; however, as the forest matures, the eventual result may be habitat for species that live in more unbroken forests.

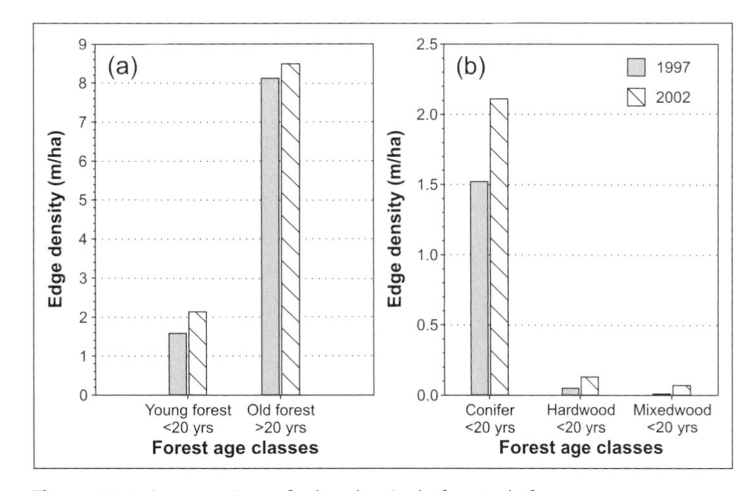

Figure 13.1 A comparison of edge density before and after timber harvest for forest classes in the Nakina North management area (Nakina North Forest Management Plan 1994, Appendix 20, p. 11).

Table 13.2

Summary of significant changes in total core area (ha) from the 1997 landscape to the 2002 modelled landscape of the Nakina North management area, by habitat unit, growth class, forest age, and forest composition.

		Total Core Area (ha) by Year					
Classification		*1997 (200m)*	*2002 (200m)*	*% Change*	*1997 (500m)*	*2002 (500m)*	*% Change*
Habitat Units	HU5				3164	2860	-10
	HU9				8	4	-50
Growth Classes	YGPj	2216	480	-78	32	0	-100
	OGPj	2000	4240	112	24	80	233
	YGSb				152	220	45
	OGSb				9336	11072	19
	YGPo	404	216	-47			
	OGLa	140	156	11			
	YGB	44	12	-73			
	OGB	572	648	13			
	OGBw	152	176	16			
Forest Age	1-40 yrs	1568	4496	187	4	356	8800
Composition by Age	Conifer >80 yrs				20500	17896	-13

Notes: The edge distance used to determine core area is shown in parentheses.
YGPj = young growth jack pine; OGPj = old growth jack pine; YGSb = young growth black spruce; OGSb = old growth black spruce; YGPo = young growth poplar; OGLa = old growth larch; YGB = young growth balsam fir; OGB = old growth balsam fir; OGBw = old growth white birch.
Source: Nakina North Forest Management Plan 1994, Appendix 20, p. 12; M. Gluck, personal communication.

Edge density in the older age classes, the age classes above 40 years, did not change significantly following the harvest, because relatively little harvest was planned for these units during the first five-year plan. The edge density in these areas was, therefore, based on the pattern of fire and other disturbances in the planning area over 40 years previous to this forest management plan.

Core Area

Core area is the area within a patch beyond a measured distance from the edge. Some species of wildlife, for example, American marten (*Martes americana)* and Connecticut warblers (*Oporornis agilis)*, seem to require larger areas of unbroken forest, even though they only use a portion in the "interior" for their habitat. These "forest interior species" are the opposite of edge-loving species, and thus require specific management action to ensure that they have habitat available. In the Nakina North plan, measurements of core areas were made to investigate how the plan would affect core species living in the area (Table 13.2).

Two categories of wildlife habitat, labelled HU5 and HU9, are described in Table 13.2. Both habitats are very similar, except for the underlying soil conditions, and consist predominately of jack pine and black spruce, with less than half of their area in hardwood species. These units are valuable to some passerine birds, provide winter cover for moose, and will eventually provide winter cover and feeding areas for woodland caribou. The harvest plan reduced the total coverage of these habitat types in the Nakina North area by amounts which the planners found acceptable: HU5 is reduced by 10 percent and, although HU9 is reduced by 50 percent, the small amount of land in that type of habitat makes it less significant.

Other changes in the landscape include increases in the amount of old-growth jack pine and old-growth black spruce in the 200-m core area and small increases in both species old growth in the 500-m core area.

Patch Density

Patch density is measured by the number of patches present, divided by the total area of a given class. The greater the number of patches, the more fragmented the landscape; therefore, an increase in patch density may call for consideration of whether the forest has been fragmented beyond acceptable levels.

A decrease in the patch density of young jack pine from about 0.03 per 100 ha to 0.02 per 100 ha occurred as a result of the planned harvest (Figure 13.2). A similar decrease occurred in poplar (*Aspen* spp.). Old-growth jack pine and old-growth poplar were affected in the opposite manner; in each case, patch density increased 0.01 per 100 ha (from 0.04 to 0.05 per 100 ha in jack pine and from 0.08 to 0.09 per 100 ha in poplar). The patch density of black spruce did not change, because the planned harvest would occur in blocks larger than the previous forest disturbances in this area.

After examining the natural disturbance pattern in the area and the landscape projections of the proposed timber harvest in the next five years, the planners concluded that, "Generally the results of this analysis show that the planned five year harvest allocations do not significantly change forest pattern any more than natural forest dynamics" (Nakina North Forest Management Plan 1994, Appendix 20, p. 20).

The Nakina North plan illustrates how far forest management has come in the last three decades. From an era of almost unregulated exploitation, forest harvests planned under the *Crown Forest Sustainability Act* must include a sophisticated analysis of the landscape and the impact of forest harvest on the ecosystem function of that landscape. Although this does not guarantee that forest harvest will have no undesirable effects, it does represent a significant advance in thinking about forest management and sustainability for the start of the twenty-first century. The Nakina North plan probably provides the best example at this time of how forest management planners should implement the *Crown Forest Sustainability Act*. Although future plans will improve, as people learn more about the planning process, the Nakina North plan represents a significant step towards full implementation of the Act.

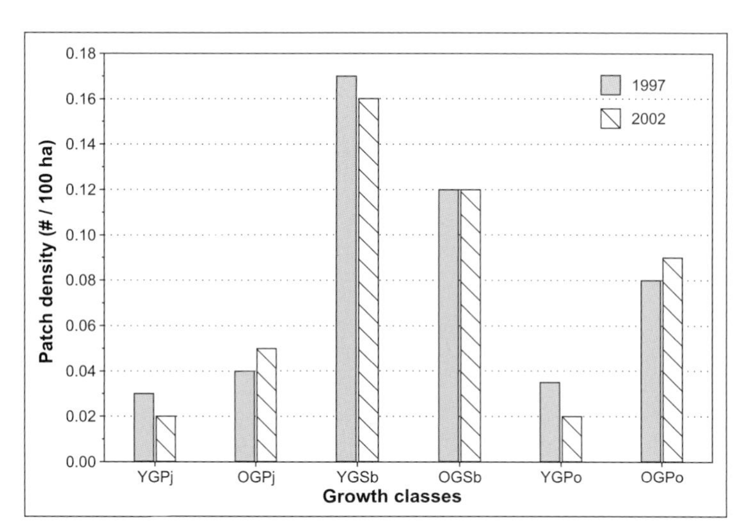

Figure 13.2 Patch density for young and old-growth jack pine, black spruce, and poplar in 1997 and 2002 in the Nakina North management area (Nakina North Forest Management Plan 1994, Appendix 20, p. 17).

Difficulties Inherent in Implementing the Act

The challenges the Ministry of Natural Resources now faces can also be appreciated through an examination of *Nature's Best: Ontario's Parks and Protected Areas: A Framework and Action Plan* (OMNR 1997b). This internal document on one of the "five core businesses – parks and protected areas" of the Ministry declares that "protected areas make a significant contribution to maintaining ecosystem health and safeguarding our natural heritage by conserving biological and geological diversity" (OMNR 1997b, p. 1). This document also reminds all concerned that "protected areas are a requirement of emerging international agreements on forest principles that will enhance access to foreign markets [and] are also a requirement of forest certification systems [such as the] Canadian Standards Association Sustainable Forest Management Standard" (p. 1).

The attitudinal change that has occurred in recent years is well expressed in the further assertion that a leading objective of the Ministry's business plan is "to ensure the long-term health of ecosystems by protecting and conserving our valuable soil, aquatic, forest and wildlife resources, as well as their biological foundations" (OMNR 1997b, p. 2). The objectives of the resulting Natural Heritage Areas Program include "protect[ing] a system of natural heritage areas ... foster[ing] land use planning and management in the intervening landscape that ensures ecological sustainability of a system of protected natural heritage areas ... [and] manag[ing] a system of protected natural heritage areas in order to retain and restore representative and special ecological and geological features, processes and systems" (p. 3). The reductions in staff that the Ministry of Natural Resources has experienced (together with other ministries) raise real concern, however, whether these fine words will be matched by action.

A FINAL PERSPECTIVE

The Canadian Shield and its forests have had a remarkable impact on the life of this province. On a continent marked by energetic agricultural settlement and aggressive industrial development, next to the world's pre-eminent capitalist country, Ontario has chosen to keep most of its lands and forests in public control. The Crown has the authority, as custodian for the people of this province, both to use the forest for present economic benefit and to protect it for future generations. This control may in the past have been honoured more in the breach than in the exercise, as vast areas of forest were conceded to the timber companies and the promoters of pulp and paper mills. Those responsible in law often hoped for the best while enjoying the revenues that forest exploitation brought them.

The establishment of Ministry control has been a struggle, and the application of scientific principles has only belatedly strengthened the hand of the professional foresters. The *Environmental Assessment Act*, however, has finally provided a context within which provincial, national, and international concerns about the forest can arouse a proper response in the Ministry of Natural Resources. In these circumstances, even the current provincial government, as preoccupied as it is with reducing expenditures and as receptive to the interests of the forest companies as it may be, finds itself speaking the language of sustainable development and respect for biodiversity.

The progress that has been made in understanding and appreciation of the forest does not mean that all is sweetness and light in the forest. The forest industries now find themselves under international pressure to cut carefully and regenerate thoroughly. The threat to their exports opens the door to a better understanding of the ecological needs of our time. It also lessens conflict for workers and their unions, who want to maintain their jobs and yet protect the forest that they enjoy between shifts. The forest is a vital endowment for the people of Ontario and much more than just a source of industrial raw materials. This realization, given new meaning by the scientific developments of our time, is reason for fresh optimism about the Ontario forest.

ACKNOWLEDGEMENTS

The authors gratefully acknowledge thoughtful reviews of sections of this chapter by J.F. Adderley, W.D. Addison, K.A. Armson, C.A. Benson, P. Duinker, J. Flowers, K.W. Hearnden, R. Klein, J. Naysmith, M.O. Nelson, M.F. Squires, and four anonymous reviewers.

REFERENCES

Armson, K.A. 1976. Forest Management in Ontario. Toronto, Ontario: Ontario Ministry of Natural Resources. 185 p.

Band, L.E. 1993. A Pilot Landscape Ecological Model for Forests in Central Ontario. Forest Fragmentation and Biodiversity Project Report No. 7. Sault Ste. Marie, Ontario: Ontario Ministry of Natural Resources, Ontario Forest Research Institute. 40 p.

Blackwell, C.J., U.C.S. Cotton, R.J. Domenick, Jr., B.W. Eddington, E.R. Eisenstadt, and C.M. Joy. 1994. Ecosystem Management: Additional Actions to Adequately Test a Promising Approach. Report to Congressional Requesters. General Accounting Office, Resources, Community and Economic Development Division. No. 94-111.

Booth, D.L., D.W.K. Boulter, D.J. Neave, A.A. Rotherham, and D.A.Welsh. 1993. Natural forest landscape management: a strategy for Canada. Forestry Chronicle 69: 141-149.

Campbell, J. 1998. Judgment between Algonquin Wildlands League and Friends of Temagami, Applicants; and the Minister of Natural Resources, E.B Eddy Forest Products Ltd., Agawa Forest Products Ltd., Grant Lumber Corporation, Elk Lake Planing Mill Limited, Algonquin Forestry Authority, Goulard Lumber (1971) Ltd., Midway Lumber Mills Limited, Birchland Veneer Limited, Grant Forest Products Corp., St. Mary's Paper Ltd., Tembec Inc., and Malette Inc., Respondents. Ontario Court of Justice (General Division) Divisional Court. Court File 539/96. 83 p. + appendices.

Canadian Biodiversity Working Group. Canadian Biodiversity Strategy. Canada's Response to the Convention on Biological Diversity. 1994. Hull, Quebec: Environment Canada. 52 p + appendices.

Canadian Forest Service. 1997. The State of Canada's Forests: Learning from History. Publication Fo-6/1997E. Ottawa, Ontario: Canadian Forest Service. 123 p.

Carrow, R. (editor). 1990. Old Growth Forests ... What are they? How do they work? Proceedings of a Conference on Old Growth Forests, January 20, 1990, Faculty of Forestry and the School of Continuing Studies, University of Toronto, and Ontario Ministry of Natural Resources, Toronto, Ontario: Canadian Scholars Press Inc. 1987 p.

Clark, C.W. 1990. Mathematical Bioeconomics. Second Edition. New York, New York: John Wiley. 396 p.

Clayoquot Sound Scientific Panel. 1995. Sustainable Ecosystem Management in Clayoquot Sound. Report No. 5. Victoria, British Columbia: Cortex Consultants Inc. 296 p.

Cummings, H. 1997. Don't cry wolf. Seasons 37 (2): 24-29.

Dana, S. and S. Fairfax. 1980. Forest and Range Policy. Second Edition. New York, New York: McGraw-Hill. 458 p.

Dunster, J.A. and R.B. Gibson. 1989. Forestry and Assessment: Development of the *Class Environmental Assessment for Timber Management* in Ontario. Toronto, Ontario: Canadian Institute for Environmental Law and Policy. 177 p.

Epp. A.E. 2000. Ontario forests and forest policy before the era of sustainable forestry. In: A.H. Perera, D.L. Euler, and I.D. Thompson (editors). Ecology of a Managed Terrestrial Landscape: Patterns and Processes in Forest Landscapes of Ontario. Vancouver, British Columbia: University of British Columbia Press. 237-275.

Franklin, J. F. 1989. Towards a new forestry. American Forester (November/December): 37-44.

Gibson, R.B. and B. Savan. 1986. Environmental Assessment in Ontario. Toronto, Ontario: Environmental Law Research Foundation. 422 p.

Gilmore, D.W. 1997. Ecosystem management – a needs driven, resource-use philosophy. Forestry Chronicle 73(5): 560-564.

Grumbine, R.E. 1994. What is ecosystem management? Conservation Biology 8: 27-38.

Hodgins, B.W. and J. Benidickson. 1989. The Temagami Experience: Recreation, Resources, and Aboriginal Rights in the Northern Ontario Wilderness. University of Toronto Press, Toronto, Ontario, Canada. 370 p.

Irland, L.C. 1994. Getting from here to there. Implementing ecosystem management on the ground. Journal of Forestry 92 (8):12-17.

Kellert, S.R. 1996. The Value of Life. Washington, DC: Island Press. 263 p.

Kessler, W.B., H. Salwasser, C.W. Cartwright, Jr., and J.A. Caplan. 1992. New perspectives for sustainable natural resources management. Ecological Applications 2(3): 221-225.

Killan, G. 1992. Ontario's provincial parks, 1893-1993: "We make progress in jumps." In: Islands of Hope: Ontario's Parks and Wilderness. Willowdale, Ontario: Firefly Books. 20-45.

Killan, G. 1993. Protected Places: A History of Ontario's Provincial Parks System. Toronto, Ontario: Dundurn Press and Ontario Ministry of Natural Resources. 431 p.

Kimmins, J.P. 1995. Sustainable development in Canadian forestry in the face of changing paradigms. Forestry Chronicle 71: 33-40.

Kimmins, J.P. 1997. Biodiversity and its relationship to ecosystem health and integrity. Forestry Chronicle 73: 229-232.

Leopold, A. 1949. A Sand County Almanac. New York, New York: Oxford University Press. 293 p.

Li, C., M. Ter-Mikaelian, and A. Perera. 1996. Ontario Fire Regime Model: Its Background, Rationale, Development and Use. Forest Fragmentation and Biodiversity Project Report No. 25. Sault Ste. Marie, Ontario: Ontario Ministry of Natural Resources, Ontario Forest Research Institute. 42 p.

McGarigal, K. and B.J. Marks. 1994. FRAGSTATS: Spatial Pattern Analysis Program for Quantifying Landscape Structure. General Technical Report PNW-GTR-351. Corvallis, Oregon: Oregon State University, Forest Science Department. 122 p.

McQuillan, A. 1994. National public tree farms. Journal of Forestry 92(1): 14-19.

Meffe, G.K. and C.R. Carroll. 1997. Principles of Conservation Biology. Second Edition. Sunderland, Maine: Sinauer Associates. 729 p.

Merriam, G. 1994. Managing the Land: A Medium-Term Strategy for Integrating Landscape Ecology into Environmental Research and Management. Forest Fragmentation and Biodiversity Project Report No. 13. Sault Ste. Marie, Ontario: Ontario Ministry of Natural Resources, Ontario Forest Research Institute. 59 p.

Monastyrski, J. 1998. Chiefs united in fight against the province. Sioux Lookout, Ontario: *Wawatay News*, July 30, 1998. 1.

Nakina North Forest Management Plan. 1994. Prepared under the authority of the *Crown Forest Sustainability Act* by the Nakina North Planning Committee. Available at the Ontario Ministry of Natural Resources, 435 James Street, Thunder Bay, Ontario P7E 6S3. >1000 p.

Naysmith, J., R. Carrow, M. Kershaw, J. Stokes, and M. Wanlin. 1997. Report of the Forest Futures Committee. Thunder Bay, Ontario: Forest Futures Committee. 11 p + appendices.

Noss, R.F. 1983. A regional landscape approach to maintain diversity. Bioscience 33: 700-796.

Ontario Environmental Assessment Board. 1994. Reasons for Decision and Decision. Class Environmental Asssessment by the Ministry of Natural Resources for Timber Management on Crown Lands in Ontario. Toronto, Ontario: Ontario Environmental Assessment Board. 561 p.

Ontario Forest Policy Panel. 1992. Our Future, Our Forests. Toronto, Ontario: Queen's Printer for Ontario. 13 p.

Ontario Forest Policy Panel. 1993. Diversity: Forests, People, Communities. Proposed Comprehensive Forest Policy Framework for Ontario. Toronto, Ontario: Queen's Printer for Ontario. 147 p.

Ontario Minister of the Environment. 1981. Order for Exemption – Ministry of Natural Resources – MNR-11/3. Toronto, Ontario: Ontario Ministry of Environment.

Ontario Ministry of Natural Resources. 1994. Policy Framework for Sustainable Forests. A statement of Ontario Government policy released April 6, 1994. Toronto, Ontario: Ministry of Natural Resources. 6 p.

Ontario Ministry of Natural Resources. 1996a. An Assessment of Ontario's Forest Resources. Sault Ste. Marie, Ontario: Ontario Ministry of Natural Resources. 133 p.

Ontario Ministry of Natural Resources. 1996b. Forest Management Planning Manual for Ontario's Crown Forests. Toronto, Ontario: Queen's Printer for Ontario. 452 p.

Ontario Ministry of Natural Resources. 1997a. Lands for Life: A Commitment to the Future. Toronto, Ontario: Ontario Ministry of Natural Resources. 6 p.

Ontario Ministry of Natural Resources. 1997b. Nature's Best: Ontario's Parks and Protected Areas: A Framework and Action Plan. Toronto, Ontario: Ontario Ministry of Natural Resources. 40 p.

Ontario Ministry of Natural Resources. 1998. MNR to appeal court ruling. Media release, February 19, 1998. Toronto, Ontario: Ontario Ministry of Natural Resources.

Ontario Royal Commission on the Northern Environment. 1985. Letters from J.E.J. Fahlgren, Commissioner, to Premier William Davis (17 December 1982) and Minister of the Environment Keith Norton (17 December 1982 and 24 March 1983). In: Final Report and Recommendations. Appendices 4, 6 and 7. Toronto, Ontario: Ontario Ministry of the Attorney-General.

Parry, T., H. Vaux, and N. Dennis. 1983. Changing conceptions of sustained-yield policy on the national forests. Journal of Forestry 81(3): 150-154.

Rosehart, B., O.B. Allen, J.R. Carrow, M. Hummel, G.R. Seed, and G. Sheehy. 1987. An Assessment of Ontario's Forest Resources Inventory System and Recommendations for its Improvement. Ontario Forest Resources Inventory Committee, Thunder Bay, Ontario. Unpublished report. 48 p.

Salwasser, H. 1990. Gaining perspective: forestry for the future. Journal of Forestry 88(11): 32-39.

Salwasser, H. 1992. From new perspectives to ecosystem management: response to Frissell et al. and Lawrence and Murphy. Conservation Biology 6(5): 469-472.

Salwasser, H. 1994. Ecosystem management: can it sustain diversity and productivity? Journal of Forestry 92(8): 12-17.

Shkilnyk, A.M. 1985. A Poison Stronger than Love: The Destruction of an Ojibwa Community. New Haven, Connecticut and London, UK: Yale University Press. 285 p.

Slocombe, D.S. 1993. Implementing ecosystem-based management. BioScience 43(9): 612-622.

Statutes of Ontario. 1994. *Crown Forest Sustainability Act*. S.O., c. 25.

Statutes of Ontario. 1990. *Crown Timber Act*. R.S.O., c. 51.

Statutes of Ontario. 1975. *Environmental Assessment Act*. R.S.O. 1990, c. E18.

Swift, J. 1983. Cut and Run: The Assault on Canada's Forests. Toronto, Ontario: Between the Lines. 283 p.

Thomas, J.A. 1994. Trends in forest management in the United States. Forestry Chronicle 70(5): 546-549.

Thomas, J.A. 1996. Forest service perspective on ecosystem management. Ecological Applications 6: 703-705.

United States. 1970. *National Environmental Policy Act*. Public Law 91-190. 42 United States Code 4321-4347.

Wagner, R.G. 1992. Vegetation Management Alternatives Program Program Prospectus. Sault Ste. Marie, Ontario: Ontario Ministry of Natural Resources, Ontario Forest Research Institute. 21 p.

Warecki, G. 1992. The people behind the parks: Ontario's wilderness conservationists. In: Islands of Hope: Ontario's Parks and Wilderness. Willowdale, Ontario: Firefly Books. 75-82.

White, A. 1957. A History of Crown Timber Regulations from the Date of the French Occupation to the Year 1899. Toronto, Ontario: Ontario Department of Lands and Forests. 284 p.

Wightman, W.R. and N.M. Wightman. 1997. The Land Between: Northwestern Ontario Resource Development, 1800 to the 1990s. Toronto, Ontario: University of Toronto Press. 575 p.

World Commission on Environment and Development. 1987. Our Common Future. New York, New York: Oxford University Press. 398 p.

14 Strategic Planning at the Landscape Level

GEORGE R. FRANCIS*

INTRODUCTION

This chapter explores some of the questions that arise from the need to undertake strategic planning for forest ecosystems at the landscape level, in part to incorporate current knowledge such as that summarized by earlier chapters in this volume. The new knowledge embraces such concepts as "ecosystem" and "landscape ecology," which are often interpreted as providing an applied-science basis for planning. Yet even these concepts are embedded in a cultural context. The need to be "strategic" implies that attention is to be directed to longer-term goals and objectives, that systems perspectives will be used to review forces and factors that bear on actions to be taken, and that spatial scales beyond local management units will be taken into account. But systems perspectives and related questions of spatial and temporal scale themselves raise a number of issues.

To explore those issues, this chapter begins with a note on the cultural contexts associated with ecosystems and landscape ecology. It then provides an overview of newer approaches to forest management planning that have emerged in Ontario over the past decade, and are still being developed. Planning issues raised by these approaches are then examined in the context of planning typologies and "rationalities," and from perspectives associated with the growing interest in complex systems. The chapter does not purport to provide answers, but may stimulate some insights among practitioners and students of resource management and others associated with the management of forest lands.

ECOSYSTEMS AND LANDSCAPE ECOLOGY AS CULTURAL CONCEPTS

Ecosystems

The term "ecosystem" was originally coined to place the study of natural history in a scientific context that emphasized the physics of material and energy flows. There are now several different schools of interpretation within ecosystem science, and the concept has also been grounded in certain cultural connotations (e.g., Francis 1995). This development has carried over into ecosystem management, which has been described as "inspired and informed by a mix of ideas drawn from 'ecophilosophy,' 'ecosystem science,' and 'political ecology.' The particular mix depends on the context within which ecosystem management is being discussed, especially with reference to the definition of purposes or problems for which ecosystem management is the solution" (Francis 1993, p. 317).

Concepts such as "ecosystem health" and "ecosystem integrity" are metaphors used to describe attributes and/or processes of ecosystem dynamics (e.g., Rapport 1992; Woodley et al. 1993). They have also been used as foundational concepts for environmental ethics (e.g., Westra 1994). Thus, the concept of an "ecosystem" encompasses connotations of science and human values.

The need to understand the dynamics of forests as ecosystems has been accepted for some time (e.g., Perera and Baldwin 2000, this volume; Galindo-Leal and Bunnell 1995; Kimmins 1995), especially as the diverse demands and expectations regarding forest

* Department of Environment and Resource Studies, University of Waterloo, Waterloo, Ontario N2L 3G1

management have grown. The emphasis given to fibre extraction as the central preoccupation of the forest industry is certainly a strong expression of values, but it is not the only one.

Broad principles sketched out for forest ecosystem management, such as the need to conserve biodiversity (e.g., Grumbine 1997), express other value preferences. One example is the application of "conservation biology," a normative applied science (Soule 1985), to issues of conserving biodiversity. Conservation in this sense goes beyond gap analyses undertaken in order to complete systems of parks and protected areas. It also calls for buffer areas around the protected core areas and linkages or networks of appropriate habitat types among core areas, in part to support viable metapopulations of certain species. In Ontario, these concepts of conservation biology have been drawn upon to develop conservation strategies for Carolinian Canada in southwestern Ontario (Allen et al. 1990; Reid and Symmes 1997) and to explore the notion of "greater ecosystems" in park-centred regions (e.g., Heritage Resources Centre 1997). Disputes associated with these approaches arise in part because of contested concepts such as the ecological reality of corridor functions and in part because of the values system underlying conservation biology, which accepts the intrinsic value of all living things regardless of their immediate usefulness to humans.

Landscape Ecology

Landscape ecology draws upon the concept of ecosystems to address issues concerning harmonious relationships among people in settled or other human-dominated landscapes (e.g., Moss 1988). An ecosystem, in this context, is described by components and functions that include humans and their activities, not just the other elements of nature; therefore, it is similar to the concept used in "an ecosystem approach" to water management (e.g., Allen et al. 1993) and the "ecosystem management" of forest or other terrestrial ecosystems. The emergence of these systems perspectives on the human use of resources and environment also necessarily relates questions of science to human values, a point noted by Baker (2000, this volume).

Cultural connotations and human values are especially evident in the term "landscape ecology" itself (e.g., Naveh 1991). The notion of landscape is fundamentally a cultural concept, expressed as much through the arts as it is through scientific analyses (e.g., Kemal and Gaskell 1993). The word "landscape" itself came into use as a technical term taken from Dutch and other painters by the late 16th century, and various genres of landscape art reflect different ways of viewing and relating to land and environments in changing cultural and historical contexts (Hirsch and O'Hanlon 1995). One example is the Group of Seven painters, who developed a distinctive Canadian school of landscape art from northern Ontario in the 1920s to 1940s.

The scientific understanding of landscape emphasizes the geomorphological and human causes of patterns on the land. It describes mosaics of different habitats with configurations of connectivity among various patches in some larger overall matrix of a dominant vegetation type, as discussed by Perera and Baldwin (2000, this volume). The human dimension is expressed in terms of values attached to different land use practices, various kinds of parks and gardens, and notions of wilderness. Dearden and Sadler (1989) reviewed the breadth of interest in landscape aesthetics from the perspectives of geography, culture, and human psychology. They note that questions of aesthetic quality in natural regions, the countryside, and urban areas, are important to many people, and that "bioaesthetics" underlies a number of environmental concerns. Beyond this, landscape can also be closely associated with a people's sense of place, their home, and, over time, with their very sense of identity. Many First Nations people have evolved this cultural component of home and identity further than more recent arrivals from Europe and elsewhere, but many of us, as individuals, can understand the relationship to favoured landscapes created by the experience of growing up in them.

The New Initiatives in Forest Ecosystem Planning

Following, and to a large extent during, an exhaustive, four-year class environmental assessment of timber management on Ontario Crown lands (Ontario Environmental Assessment Board 1994), the Ontario Ministry of Natural Resources (OMNR) considerably modified practices that were the focus of criticism during the hearings. In 1993, the Ontario government approved a Policy Framework for Sustainable Forests, and subsequently passed the *Crown Forest Sustainability Act* (1994) which came into effect in April 1995.

OMNR prepared operating manuals for a forest management planning process to incorporate the conditions specified by the 1994 Environmental Assessment Board report. The *Forest Management Planning Manual* issued in September 1996 is intended to be "the pivotal document which provides direction for all aspects of forest management planning for Crown lands in Ontario" (OMNR 1996, foreword). The historical context of these developments is described by Epp (2000, this volume).

The Regional Approach to Land Use Allocation

In February 1997, a new Ontario government launched a program known as "Lands for Life" (herein identified as "LfL"), which was billed as "a new comprehensive land use planning process that strikes a balance between the protection and use of Ontario's natural resources" (OMNR 1997a, p. iii). The spatial scale for which major decisions on land use allocations are to be made under LfL is impressive; it embraces about 451,000 km^2 across the province, some 86 percent of which is Crown land. The expressed intent of LfL is to complete a provincial system of parks and protected areas (OMNR 1997c), and provide "some certainty" (Ontario 1997c, p. iii) for resource-based industries and communities. This program would serve to update the regional strategic land use plans and district land use guidelines developed by OMNR during the early 1980s. If successful, LfL might also reduce the number and intensity of protracted land use disputes in the province, such as those associated with the Temagami region over the past 25 years or so (e.g., Hodgins and Benidickson 1989; Earthroots 1998). Under institutional arrangements with OMNR as the lead agency, strategic planning for forestry in Ontario is evolving to take two different approaches at different spatial scales, both beyond individual forest stands.

Basic land use allocations were to be made under LfL for three large planning regions defined by OMNR as Boreal West (198,349 km^2), Boreal East (148,416 km^2), and Great Lakes-St. Lawrence (104,545 km^2). For each of the three regions, OMNR appointed a 14-member round table with a chairperson. Appointees were residents in their respective regions and associated with natural resource industries, local community affairs, the tourism business, environmentalism, and First Nations. Within the framework of government policies for forest and resource management, resource-based tourism, and parks and protected areas, the round tables were to come up with recommendations for the Minister of Natural Resources on the spatial allocation of lands for these different purposes within each region.

OMNR laid out a framework for the LfL planning program to make sure it "will help meet the Ministry's goal and achieve the supporting corporate objectives. The goal of the Ministry of Natural Resources is to contribute to the environmental, social, and economic well-being of Ontario through the sustainable development of natural resources. Sustainable development depends on the continuing availability of natural resources. As steward of Ontario's natural resources, the Ministry has a mission to ensure ecological sustainability" (OMNR 1997b, p. 2).

Statements of objectives, desired outcomes, planning principles, and OMNR's role in planning were then spelled out for this planning program. OMNR stated that the program addressed "the short term need to establish strategic resource allocation and land use direction, and the longer term need to establish a process through which more comprehensive decisions are developed and outstanding natural resource issues are resolved" (OMNR 1997b, p. ii). LfL was a two-phase program. The first phase, scheduled for completion in mid-1998, had the objective of identifying "resource and land base requirements ... for forestry, natural heritage, resource-based tourism, fish and wildlife, and other important resources" (OMNR 1997 b, p. i). The second phase was intended to "address the full array of MNR's program responsibilities across 12 to 15 smaller planning units that are ecologically-based, and that cover all of Ontario."

In a guide for round table participants and others interested in the LfL process, OMNR (1997a) emphasized that the planning had to be "properly structured and supported to operate effectively" (p. 1). The guide outlined the general content of the strategies expected from the round tables; noted that "large parts of the planning area will be designated for multiple use" (p. 2); provided examples of the level of detail sought in statements of "directions and decisions" and in land use maps to be prepared; and laid out a suggested table of contents for the round tables' reports to the Ministry. Round tables had staff assigned by OMNR and other ministries to provide policy and planning expertise and administrative support to help them accomplish this duty.

By mid-1998, the round tables had received briefings from a number of people; they were debating issues among themselves; and they were outlining options and sketching scenarios with accompanying maps to show the possible spatial allocations of lands associated with different scenarios (e.g., Boreal East Planning Area 1998). These scenarios were put forward for more discussion in public meetings. At the same time, the round tables were still trying to resolve difficult questions for which, in a number of cases, little information was available. Individual interest groups were beginning to bypass the process to lobby directly at ministerial levels. OMNR called in the reports from the round tables and consolidated them into one public document having some 242 recommendations with varying degrees of specificity and mutual contradiction (OMNR 1998). The general public was then given 30 days during the month of November 1998 to comment on this document.

Outside of the round-table process, a group identified as the "Partnership for Public Lands" prepared a "protected areas vision" for all three regions based on a gap analysis for protected area planning and management, and called for "a new northern economy," one that should be based on diversifying local economic activities, increasing the value-added processing of forest products, and safeguarding large wilderness-based ecotourism opportunities (Partnership for Public Lands 1998). The question was also raised whether the government's decisions on land use allocations would be accompanied by major changes in land tenure for private-sector operations.

During the 30-day commentary period, the Partnership insisted that the round-table processes had failed to meet the government's commitment to parks and protected areas, and that the "business-as-usual" proposals that came from them for large-scale resource extraction were a recipe for economic disaster. Its "planning for prosperity" alternative called for some 20 percent of lands to be protected as parks or other kinds of protected areas, to support a more diversified tourism economy. This, along with enhanced forest management to ensure there would be no reductions in the total wood flow to mills, would serve to increase northern jobs and opportunities. The Partnership's campaign attracted considerable publicity, much of it favourable, along with petitions of support and hints of legal challenges, should the government accept the main thrust of the consolidated recommendations from the round tables.

The government deferred a final decision on LfL that had originally been promised by December 1998. In February 1999, the Minister of Natural Resources initiated discussions among the Partnership for Public Lands, the Ministry, and the forest industry to negotiate agreement on extending the number or extent of protected areas to reach a target of 12 percent of the total LfL planning region. In late March 1999, the Minister of Natural Resources announced that an agreement had been reached whereby 378 new parks and other protected areas totalling 2.4 million ha would be established in the LfL planning region to reach the 12 percent target (called "Ontario's Living Legacy"); there would be no logging, mining, or hydro-power development in these areas. An "Ontario Forest Accord" outlined an approach to addressing the needs of the forest industry. It included a $30 million fund and advisory board to help companies make re-adjustments in forest management operations, to provide enhanced access for hunting and fishing on Crown lands, and to make various improvements in fish and wildlife management (called a "Living Legacy Trust"). Procedures for implementing these decisions, including consultations with First Nations, are to be worked out.

Strategic planning to complete various systems of parks and protected areas is increasingly relying upon gap analyses to help identify important habitats or plant and animal associations that are not represented in existing parks or other protected areas within jurisdictions or particular ecoregions. Measures of protection could then be provided for these additional areas. This is one means through which the government could fulfill its commitments to complete the provincial system of protected areas (e.g., Statement of Commitment 1992) and help meet Canada's obligations under the 1992 Convention on Biological Diversity (Environment Canada 1995).

In Canada, gap analyses have been developed as pilot projects over the past several years (e.g., Canadian Council of Ecological Areas 1992, 1995; World Wildlife Fund 1996), and they have also been undertaken in Ontario by OMNR for a number of that Ministry's site districts. Related issues of conserving biodiversity in Ontario have also been addressed (e.g., Poser et al. 1993; Riley and Mohr 1994). The extent to which gap analyses contributed significantly to the

LfL process is not clear, although completion of some of these analyses has been paced by the schedule set for the LfL program. In identifying important gaps, much depends upon the spatial scales at which gap analyses are undertaken, the classification systems adopted to describe landscapes and associated biota, and the relative importance placed on the enduring features of landforms rather than types of vegetation or particular biota. The more detailed the classifications used, the more gaps that are likely to be discovered.

Forest Management Planning

The Crown lands allocated for forest management under various designation proposals suggested by the LfL will come under the provisions of The *Forest Management Planning Manual* (OMNR 1996). The *Manual* prescribes in great detail the procedures and processes that professional staff from OMNR are to follow, through four distinct phases, in the preparation of forest management plans for approval by the appropriate regional directors of the Ministry.

For example, the planning teams must use a "strategic forest management model" to analyze data about the land base of each forest management unit, the forest dynamics within it, and silvicultural options. There is to be a benchmark assessment of how the forest would naturally develop over time, if left undisturbed, to compare with projections of changes associated with at least three different management alternatives and objectives. Each management alternative "must include 160-year projections (in 20-year increments and five-year increments for the first 20-year period) for specific quantifiable [forest management] objectives" (OMNR 1996, p. A-61). A specified socio-economic impact model is to be used to identify the relative socioeconomic impacts of each alternative, for comparison with the socio-economic profile of the communities in the forest management area. The planning team and a local citizens' committee then select a "preliminary preferred management alternative." Extensive documentation must be prepared to explain the rationale and implications of choosing this alternative. Operational plans for the first five-year period of the preferred alternative must also be drawn up following detailed specifications. Additional public consultations are then held, after which the plan is submitted for official approval. The entire process for each forest management plan is expected to take just over two years.

Challenges of the New Planning Approaches

In an ideal world, the basic land use allocations at the scale of LfL would have incorporated the requisite landscape analyses, such as gap analyses, and identified the appropriate ecological boundaries of planning units for intensive forest management and multiple land uses. At the time of writing, the consequences of the LfL processes were still unfolding, the extent to which gap analyses will be used to decide on additional parks and protected areas remained uncertain, and the new forest management planning process was just getting underway on existing administrative units. The concepts of ecosystems and landscape ecology point to the need for understanding natural processes operating across a range of interrelated spatial and temporal scales. The planning processes themselves do not seem designed to reflect the cross-scale linkages.

Not surprisingly, the new forest management planning process has encountered some implementation problems, as noted by Euler and Epp (2000, this volume). There is also the question of performance monitoring. A legal challenge to OMNR from environmentalists led, in February 1998, to an Ontario Divisional Court order that OMNR bring three forest management plans into compliance within a year. The decision was appealed by OMNR, but the Divisional Court's ruling was upheld by the Ontario Court of Appeal in October 1998. This event raises the possibility of increased use of judicial review to help assure good forest management. Another approach would be third-party certification of forest management plans, along the lines proposed by the Canadian Standards Association (e.g., Elliott and Hackman 1996).

When placed in the longer-term perspective of the history of forest use and forestry in Ontario, however, these initiatives of the 1990s can be viewed as major efforts to adapt to new knowledge and new expectations about how forests in Ontario should be planned, managed, and responsibly used. The approaches taken towards planning for forested landscapes in Ontario do raise issues about planning itself. Different kinds of planning are more appropriate, or effective, under different circumstances. A brief overview of some planning issues will help to illustrate this principle.

Planning Typologies and Contexts

Planning itself has been defined in an introductory text as "the deliberate social or organizational activity of developing an optimal strategy of future action to achieve a desired set of goals, for solving novel problems in complex contexts, and attended by the power and intention to commit resources and to act as necessary to implement the chosen strategy" (Alexander 1986, p. 43).

Typologies of Planning

A variety of typologies for planning, and for the role of the planner, are represented among those who either do or teach planning. Different typologies are based on one or more considerations: what is being planned (e.g., natural resources, urban development, or social services); on whose behalf the planning is being done (e.g., government agencies, corporations, or other interest groups); the decision making techniques and processes adopted; and the institutional power and/or economic relationships that define the context in which plans are formulated and carried out.

A common theme among typologies is that planning is a question of how best to link knowledge to action, guided by some form of "rationality," and whether the political or social role performed through planning is in the service of social guidance or social transformation. "Rationality," as associated with planning, is commonly thought to be one of two kinds. Instrumental rationality refers to the application of particular methods, such as those from the applied sciences, to find the most effective means for achieving given ends. Communicative rationality is a deliberation within or among groups of people to decide upon ends as well as means, and resolve whatever conflicts of values arise in the process of reaching agreements. These two rationalities should be viewed more as a spectrum than as either/or alternatives. Main categories of planning theory that distinguish decision making approaches also represent different points in the spectrum of rationalities and different political or social roles for planning (e.g., Sager 1993). These categories are as follows: comprehensive, rational (termed "synoptic") planning; incremental planning; transactive planning; and advocacy planning.

The comprehensive, rational (synoptic) planning mode is the one most associated with instrumental rationality. This mode of planning can be applied most successfully by institutions that can exercise centralized decision making and top-down management control (e.g., Lawrence 1997) under quite stable social and environmental conditions. Problems to be addressed must be clearly defined, in terms such that expert scientific knowledge and analytical capability can be applied to resolve them with little uncertainty or ambiguity. The problem-solving situation has also to be one in which a comprehensive perspective can be used to identify and compare alternative possible solutions in order to select ones that would be the most efficient or cost-effective. Debates about synoptic planning criticize the underlying assumptions about comprehensiveness and rationality, and the political role of this type of planning in democratic societies. The political role is to allocate resources in a way that maintains, rather than changes, the basic social order. This kind of public sector planning within a dominant market economy has a relatively long tradition.

In situations where the nature of the problems are neither clear nor agreed upon, and the appropriate knowledge is largely unavailable or in dispute, synoptic planning is inappropriate. Instead, the situation requires some course of action to be negotiated among the stakeholders who are involved in it. Planning is then an "interactive communicative activity" and planners are "actors in the world rather than ... observers or neutral experts" (Innes 1995, p. 184). This entails some role of social guidance or social transformation, either in the conservative context of policy analysis or social learning, or in a more radical context of social reform or social mobilization (Friedmann 1987, 1995). The kinds and extent of social inequities or income disparities in a democratic society help to determine the contexts within which different types of planning emerge.

Funtowicz and Ravetz (1994) distinguish three kinds of problem-solving strategies, the success of which varies with key circumstances, such as the relative degree of uncertainty inherent in the situations being addressed and the consequences arising from the decisions to be made. The first of the three strategies is applied science, or mission-oriented research for problem-solving. This strategy succeeds best in situations which involve relatively low levels of uncertainty, most of which are in technical areas where they can be handled by standard procedures, and which have relatively low stakes, in terms of the decisions that use the research results. The second strategy lies in the realm of professional consulting, where innovative approaches may be required to deal with complex problems having greater uncertainties, in order to meet client needs.

The third kind of strategy addresses situations of what Funtowicz and Ravetz (1994) call "post-normal science." These are situations in which the inherent uncertainties are high; the consequences of the decisions required are also serious (either because of conflicting purposes among different groups of people or the many societal ramifications associated with the choices to be made); usable, scientific knowledge may be scarce; and the existing scientific models have limited application. Under these circumstances, Funtowicz and Ravetz suggest that the notion of peer groups must be extended beyond a narrow range of expertise, in order to understand and respond effectively to the new situations.

Cardinall and Day (1998) make similar observations about issues in resource and environmental management. For any given situation, two key attributes are "decision stakes" and "management uncertainty." The most challenging, but not uncommon, situations are ones in which issues are poorly understood (or denied), yet the decisions to be made have potentially irreversible consequences. The resulting uncertainties exceed the problem-solving capacities of decision making techniques, procedures, and institutions. Continued insistence upon the use of some technique (e.g., cost-benefit analysis) or procedure (e.g., negotiation among selected representative stakeholders) can only exacerbate the situation. Cardinall and Day (1998) note generally that current approaches to resource and environmental planning and decision making often reflect a paternalistic view, deemed to require a deficit model of planning under which professionals have to inform lay people whose local knowledge and experience are assumed to be limited, inferior, or not relevant to the situation at hand. Cardinall and Day suggest that a strong civic process of mutual dialogue about circumstances, visions of preferred futures, mobilization of resources, and actions to work towards better communities, may be required. Communicative rationality would be central to this dialogue.

Collaboration among individuals or among organizations to achieve strategic objectives may also rely primarily on the "planning mode," the "learning mode," and the "vision mode" (Westley 1995). These modes, too, have their strengths and weaknesses. The planning mode refers to some version of synoptic planning. When it is carried out by a particular organization, such as OMNR, the strength of the planning mode is that the organization can mobilize resources to carry out the planning. Given the institutional structures through which the planning has to be done, however, this mode can be weak when it comes to new problem definitions or to the acceptance of recommendations that vary from established procedures. The learning and vision modes assume communicative rationality. The learning mode describes processes whereby individuals or organizations come together voluntarily to collaborate on identifying the problems, setting goals, and taking action to achieve those goals. Difficulties include the processes of reaching consensus and the mobilization of resources to implement whatever is agreed upon. The vision mode recognizes that social or organizational change, in certain situations, can be brought about by creative individuals who have strong visions of what should be, combined with powers of persuasion to mobilize support from others. Charisma, however, is no substitute for the development of organizational capacity, especially to carry on when the original innovators no longer can.

Few real-world planning initiatives fit neatly or exclusively within any one box in these typologies. Some awareness of the typologies, however, helps reveal the relative strengths and weaknesses of particular approaches, their relevance to a situation at hand, and the basis for criticisms directed at them. It is not so much that some kinds of planning are right and others wrong. Rather, it is how well the assumptions underlying the chosen mode of planning mesh with the realities of the situation to which it is being applied. This consideration also raises some generic issues of science, policy, and reliable knowledge as discussed by Baker (2000, this volume).

SYNOPTIC PLANNING FOR FORESTS

The approaches taken to strategic planning for forests in Ontario – basic land use allocations, gap analyses, and forest management planning – exemplify a synoptic planning ideal. Even at the most operational level of the revised procedures for forest management plans spelled out in great detail by OMNR, the context in which these procedures are to be applied is problematic. Projections over two successive forest rotations (which, presumably, equates to the specified 160 years) may be plausible for slow-growing forests under stable environments, but strategic planning has also to consider socio-economic environments whose stability cannot be assumed over such time horizons. If, for example, one thinks back 160 years to Ontario around the year 1840, 27 years before Confederation, 65 years or so before the first forestry school was established, and long before industrial forestry at a global scale came into being, there would have been enormous scope for error in anticipating then the conditions for forest management now.

Gap analyses are expert-dependent and technique-oriented, and are conducted with an explicit set of values that may not be widely shared in the social situations in which they are applied. Their use in the LfL program was ostensibly to help implement a government policy which was treated as a given for the process. Despite this, the results of the process fell short of the prior commitment.

The LfL itself had to make much more use of communicative rationality, given the different values of its stakeholder participants, and the changing socio-economic contexts that underlie those values. The process had been highly structured, however, by the conditions imposed by the OMNR; that is, by the "givens" in terms of the Ministry's goal, objectives, policies, desired outcomes, and so on. At the early stages of the LfL program, the value conflicts inherent in the process were resolved satisfactorily through good will or skilled facilitation. But they had emerged in no uncertain terms by the middle of 1998, and remained unresolved by year's end in the face of a reported 14,000 public submissions received by the Minister during the 30-day commentary period on the LfL recommendations.

The challenges facing strategic planning for Crown lands in Ontario have some similarities to those faced by the shared decision making process undertaken in British Columbia to negotiate long-term sustainable uses of public forests and other resources. As the Commissioner of the former BC Commission on Resources and Environment (CORE) noted, representative government can be supplemented effectively with greater public participation by drawing on the best of both direct democracy and sectoral interest negotiation. The key is that such participation is open, so as to be responsible; balanced so as to be fair; and advisory, so as to leave decision making with accountable elected officials (Owen 1998).

Planners are not directly referred to here, but in the context of CORE they would have been playing a facilitating and supportive role, and would have the considerable "people skills" required to do so successfully. For as Sager (1995, p. 172) put it generally, "Problems have to be sorted out among people with very different professional backgrounds, contrasting ideologies, conflicting interests, and mutually off-putting personal peculiarities. The essential insight in this situation may be ... that there are several ways of being rational."

Synoptic planning is not meant to cope with such situations. From the perspective of Funtowicz and Ravetz (1994), forest management planning in Ontario seems to assume an enduring context somewhat like that noted above for successful applied science or consulting. So the question that arises is whether there are sufficient uncertainties and unknowns, especially in terms of the planning horizons (i.e., months to make basic resource use allocations, and multiple decades for forest management planning) to render synoptic planning less appropriate than other modes of planning that are based on different assumptions. That the answer appears to be "yes" becomes more evident when insights from complex systems perspectives are brought to bear on the question.

A "Complex Systems" Perspective

Emerging Perspectives on Emergent Complexity

Academic interest in complex systems has developed considerably in recent years. The phenomena of interest are self-organizing systems that regularly exhibit emergent properties, sudden discontinuous change, and various kinds of cycles or evolutionary spirals, all of which operate across a wide range of temporal and spatial scales. The theoretical interpretations of complex systems have been developed, in part, from new computer technologies that allow for a rich array of simulations. They have also developed from recent work to synthesize case study analyses of systems phenomena, in order to generate broad empirical generalizations and pose hypotheses for testing by both simulation and empirical research. General reviews of these developments, which can be viewed as elaborations of the general systems theory first developed by Ludwig von Bertalanffy and his colleagues in the 1930s, are given by Waldrop (1992), Lorenz (1993), Kauffman (1995), Bak (1996) and Ludwig et al. (1997). Applications of complex systems thinking to the study of ecosystem dynamics are discussed by Allen and Hoekstra (1992), Gunther and Folke (1993), Schneider and Kay (1994), and Regier and Kay (1996). Some implications for understanding social systems are discussed by Eve et al. (1997).

The application of complex systems thinking to human systems and ecosystems has emphasized self-organization phenomena operating over a wide range of different spatial and temporal scales, with cross-scale connections among them. The different scales convey images of hierarchies of systems-within-systems (or of systems with subsystems), but do not exhibit the strong top-down control usually implied by the concept of hierarchy. Boundaries of systems and subsystems are drawn quite arbitrarily, but not capriciously, by the observer, who for particular purposes needs to focus in more detail on some given system at a given scale. Observers also choose a system type that is of interest to them. System types can be categorized mainly as follows: abiotic systems, such as the geomorphology of watersheds, terrestrial ecoregions, and atmospheric/hydrological systems; primarily biotic systems, such as aquatic and terrestrial ecosystems; and primarily human/cultural systems, such as agricultural ecosystems, urban-industrial economies, and cultural-institutional systems.

There is no one correct way to classify types of systems and no correct scales for examining them. There are only helpful or less helpful ways, depending on the phenomena of interest. System types overlap; for example, managed forest ecosystems have abiotic, biotic, and human components, and the scale for observation depends on what the observer wants to find out.

Self-organizing systems develop organized structures and attributes over time in response to flows of high-quality energy (termed "exergy"), available resources, and information. For complex systems, "information" is encoded by genes in organisms, by biodiversity in ecosystems, and by rule systems in human social systems. A self-organizing system can maintain a particular steady-state configuration of structures and attributes for periods of time, as if it were held together by some attractor. For abiotic systems, the equilibrium points are the attractors. For ecosystems there appear to be two such attractors. One is associated with the survival and reproduction strategies of organisms, as they capture and dissipate the exergy flowing through food webs in the system. The other is associated with the amount of stored biomass available for renewal of the ecosystem. At some point, the slow release of biomass for re-use occurs suddenly on a much larger scale. In forest ecosystems, this sudden change can be caused by fires, storms, or insects, and generates the familiar patch dynamics phenomena. Holling (1986, 1994) describes a four-phase cycle of ecosystem dynamics that includes periodic change from one attractor to the other.

Self-organization is associated with positive feedback processes which drive a system in a given direction until constraints (negative feedbacks, or resource scarcities) come into play, to hold or reconfigure the system. Changes in access to energy, resources, and information can lead to rather sudden reconfigurations into some other steady-state associated with another attractor. These changes are inherently unpredictable. For human systems, this does not deny human agency. It only suggests that the collective striving of many individuals and organizations in society leads to collective results that are both unplanned and unexpected. Examples include the evolution of market systems, communication networks, and urbanization patterns.

Relevance of a Complex Systems Perspective for Strategic Planning

From the systems-within-systems perspective, forests can be viewed within some ecoregional classification system, such as the Hills (1959) classification or the ecoregions and ecodistricts defined under the Canadian ecological land classification system. The subsystems include the various forest habitats and stands, and their associated wetland habitats, lakes, and streams. This perspective on landscape is already in use.

The socio-economic system is also connected across scales. Direct employment or contract work in the forest and or mill operations contributes to the economic base of local communities, and the results produced are directly linked to corporate holders of the wood harvesting licences. In Canada, corporations in all sectors of the economy are frequently owned and controlled by other corporations, with various chains of ownerships combining to form extensive corporate empires ultimately owned by wealthy families or consortia of financial institutions. The picture of economic ownership and control is constantly shifting.

For example, at the time of writing, among the corporations with operations in Ontario, Noranda Forest

Inc. was tied quite directly into the Edward and Peter Bronfman group; and E.B. Eddy Paper Inc., long held by the George Weston group had recently been sold to Domtar. The families associated with these groups are among the most wealthy in Canada, owning and presiding over vast corporate holdings. Domtar was owned largely by a holding company controlled by a group of investment organizations and by the Quebec government. Avenor Inc., which had been owned largely by a group of financial intermediaries, such as Montreal Trust and the Investment Dealers Association of Canada, had recently sold major shares to Bowater Canada Inc., which is ultimately owned by a US parent company and the *Washington Post*. Donohue Inc. was owned by Les Placements Péladeau Inc. (Statistics Canada 1998). Others such as Tembec Inc. are widely held (*Financial Post* 1998). The merger of Abitibi-Price Inc. with Stone-Consolidated Corporation in 1997 to form Abitibi Consolidated Inc. reflects the increasing globalization of transnational corporations operating in world markets.

This changing corporate scene reflects "strategic planning" by the corporations. Corporations producing forest products seek profitable positions in their respective economic sectors by investing in labour-reducing technologies, expanding their market shares, developing specialty markets, improving productivity (sometimes aided by investments in environmental protection measures that reduce waste production), and engaging in cost-reduction measures at every opportunity (e.g., Canadian Pulp and Paper Association 1997). Investment funds serve to keep corporations focused on short-term financial returns for the debt and equity holdings of investors. Holding companies seek profitable returns by buying or selling assets from among the many corporations they own or control, or creating horizontal or vertical mergers and acquisitions of corporations in order to maintain an optimal mix of assets. Ownership arrangements can and do change suddenly, without notice. These changes may lead to the transfer of operations, layoffs, or expansions. These are all headquarters decisions that communities dependent on the forest industry may only find out about after the fact.

What self-organizational phenomena are in play that will affect the forests and the human communities in this system? For example, biophysical self-organizational phenomena include global atmospheric change, biogeochemical cycling of nutrients, forest and wetland successions and patch dynamics, and population cycles of particular biota. Human/societal self-organizational phenomena include an emerging global economy dominated by transnational corporations, population growth and urbanization, new technologies, especially for information and communication networks, and a complexity of institutional arrangements bordering on gridlock over many areas. Two examples from the global level illustrate the ramifications.

Global atmospheric change includes issues of climate change and variability, acidic precipitation, hazardous airborne pollutants, stratospheric ozone depletion leading to increased ultraviolet B radiation, and smog. There seems to be consensus that global warming is underway, and is due in part to human activities associated with the release of greenhouse gases by the combustion of fossil fuels, industrial activities, deforestation, and other land use practices. The impact on forest ecosystems has been explored through scenarios of possible shifts in the distribution of those ecosystems in Canada (e.g., Webb 1992; Lenihan and Neilson 1995). Munn (1996, p. 6) summarizes the findings as follows: "The general consensus is that the anticipated rate of forest ecosystem movement due to climate change will be about 200 km northwards in 50 years ... It has been estimated that the southern boundary of the Canadian Boreal Forest may shift 470-920 km as a result of a doubling of greenhouse gases, with a net losses of 100 million hectares due to atmospheric change ... Climate change is anticipated to be greater in the higher latitudes; therefore the northern frontiers of these ecosystems are likely to be affected first."

The ability of forests to adapt to the rates of change they may experience in the decades ahead is problematic. Each species has its own adaptive capabilities, so that new associations of trees should evolve into different forest ecosystems, much as they did in the postglacial period (Delcourt and Delcourt 1987). Stands at the southern edges of their current range may unravel first as growing conditions become too harsh and other plant species invade. The ability of particular species to establish outlier sites, as different forest species did during their postglacial range expansions (e.g., Davis and Zabinski 1992), is not known. Drier conditions and stressed forests are expected to increase the frequency of forest fires, which will constitute a positive feedback into atmospheric carbon dioxide.

The implications of climate change are that major environmental changes should be expected within the first few decades of the first rotation period under the new OMNR planning periods. In addition to the direct effects on stand structures, there are also other more indirect effects caused by changes in atmospheric composition that can affect the biophysical context within which the forests are growing, perhaps in ways that are foreign to their evolutionary and adaptive experience.

Other major changes in socio-economic conditions will become more evident over the next few decades. They will arise as the consequences of a single, global economy dominated by transnational corporations operating in multiple nation states (e.g., Korton 1995). While there are always elements of conjecture in anticipating the future, awareness of some of the major trends for change at the global level could help forest planners better anticipate the possibilities for the Ontario forest economy, as it may well evolve over the next few decades. For example, Marchak (1995) describes the kinds of structural changes that the forest industry in Canada has undergone, in terms of the replacement of labour by technological developments and the shift of new investments into higher-return plantation forestry in the southern hemisphere. Global markets also bring with them uncertainties about currency values, shifting trade patterns, and product demands that can change the business environment for forest and other resource industries quite suddenly.

Changes other than the two kinds noted above will also have their effects upon the forest ecosystems, the forest economy, and forestry-dependent communities. New technologies, especially in electronic communications, forest products using different sources of plant fibres (e.g., "tree-free" paper), and changing cultural values that cherish wilderness and forests, particularly old-growth forests, for the amenity value of pleasant landscapes, will all have their effects on human demands or expectations about forest management. Planning that is keyed to forest rotation periods over some eight decades to come is subject to considerable uncertainty and surprise. How might it respond?

Strategic Planning in a Complex Systems Context

Avoiding Traps

OMNR has embarked on strategic planning for forest use and management with a landscape focus. How much of this planning will be overtaken by events, and how rapidly that may happen, remains to be seen. There are other practical as well as philosophical implications arising from systems perspectives for understanding the natural and human worlds. Holling (1986, 1994) and Holling and Meffe (1996), for example, describe from a number of case studies a phenomenon referred to as the paradox or pathology of resource management, in which apparently successful practices triggered their own demise. The patterns of behaviours and responses underlying this paradox are generally as follows.

In the name of resource management, a designated agency seeks to control certain undesirable fluctuations or eliminate some undesirable components within ecosystems (e.g., fires and insect outbreaks, in the case of forests) in order to provide benefits from the set of conditions this brings about. They succeed, often over a time span of decades. They also become preoccupied with administering their chosen management methods and strive to increase efficiencies of these means, since the ends are so clearly a given. Local economic development takes place, based on the resources provided through these management actions. This fact leads to a strong economic dependency upon continuing the management practices. The ecosystems providing the resources evolve, but become less resilient to other stresses upon them, in part because they are already under stress from the continuing resource management practices.

Through this process, as Holling puts it (Holling 1986, 1994; Holling and Meffe 1996), the entire ecological-economic system eventually becomes an accident waiting to happen as ecosystems gradually become "brittle," management agencies become "rigid," and the local economy becomes dependent and vulnerable. Some unexpected event, usually of low probability but beyond the perceived normal range of variability, triggers a sudden change in the ecosystem, along with a collapse of the resource base and the economy dependent upon it. This is the most destructive phase in the four-phase cycle of ecosystem dynamics. The institutional implications of such cyclic phenomena are explored by Gunderson et al. (1995).

The challenge to forest planners and managers is not to get caught in this kind of situation as they plan for a steady course of action over many decades in the future. This caution may be especially important when synoptic planning is adopted without full awareness of its underlying assumptions, including those of the rationality and of the stability of the socioeconomic and environmental conditions.

This is where a much broader transdisciplinary systems perspective could be helpful, provided the institutional impediments to developing such a perspective could be overcome. For example, a general reiterative planning process classified as a "soft systems methodology" has been cited as the most appropriate for applications to ecosystems by Allen and Hoekstra (1992) and Allen et al. (1993). Participants in the planning process have to exchange their perceptions and understandings of the situations they confront, identify relevant systems and interrelationships, and decide what is desirable and feasible. Management actions are best structured and monitored for continual learning and adaptation, so that they will not be viewed as management failures whenever anticipated results do not occur (e.g., Lee 1993). Issues of active adaptive management, discussed by Baker (2000, this volume), become pertinent. The elimination or relaxation of the institutional, psychological, and other constraints that inhibit continual learning and adaptation is a major challenge itself (e.g., Michael 1993, 1995).

STRATEGIC PLANNING FOR SOCIAL FUTURES

As already noted, corporations have their own strategic planning processes for securing their futures in the operating environments of post-normal science. Many communities dependent on the forest industry, however, remain economically vulnerable, as they too face a future of uncertainty and change. While strategic planning for forest ecosystems at the landscape level is intended to help secure the economies of communities, this outcome cannot just be assumed. In many cases, it needs to be approached directly. Strategic planning for the well-being of communities is not so much about ecosystems and landscapes as it is about social futures. For the communities, the goal of strategic planning must be to reduce or minimize dependency on the forest industry for an economic base, rather than to plan how best to "lock into" it for some 80 years or more, knowing full well that everything, including the forests and especially the forest industries, could change dramatically over such time horizons. Humans will remain dependent on forest ecosystems, but primarily for their ecological functions.

Communities might do well to embark on their own strategic planning, without government domination of the process. Like most corporations, strategic planning by communities must accept the context of post-normal science, and this is what requires a different approach. Such planning must address questions about the kinds of futures that may be feasible in light of current circumstances, the prerequisites for sustainability, and the consultative processes through which the issues should be addressed and decisions taken. For communities, however they may be defined and bounded, desirable futures can be expressed through "visioning" exercises. Some ideals for the future will be widely shared; for example, ideals associated with decent levels of health and well-being, in pleasant surroundings, with strong community networks and a diversity of opportunities for work and fulfillment.

Part of the visioning process could be enhanced by depicting the different landscape patterns that could evolve from different technological or economic changes, using methods of "landscape impact analysis" developed in Scandinavia (Emmelin 1996). Achieving what is ideal in the vision is not a matter of forecasting futures so much as it is "backcasting"; that is, developing plausible scenarios, or "future histories," about how communities could get from the present situation to that ideal (e.g., Robinson 1996).

With a complex systems perspective comes the realization that sustainability is primarily about processes, not the perpetuation of attributes of systems. The essence of sustainability is to maintain the processes for natural renewal and the continued evolution of ecosystems, and the capacity for innovation and creativity in social systems. Forest ecosystems exhibit these processes through their patch dynamics and through longer-term adjustments to climatic or other abiotic changes. Social systems can also go through renewal after collapse (e.g., Tainter 1989), but only at considerable human cost. Strategic planning seeks alternatives for managing the transitions and adjustments required by change in a humanly decent way.

The consultation processes for this kind of planning require collective best judgments that draw upon a wide range of knowledge, experience, values, and preferences. Communicative rationality and the quality of the decision processes take precedence over instrumental rationality and reliance upon experts. Quality in decision processes refers to the inclusiveness of the consultations and the degree of consensus achieved. There are no blueprints about how to proceed. But there are many community-based initiatives underway in Canada and elsewhere that point to possibilities for creative, innovative change (e.g., Roseland 1992; Institute for Research on Environment and Economy 1996). They suggest organizational ground rules that differ considerably from those now specified for forest planning in Ontario.

For strategic planning in the world of post-normal science and continuous change, governing institutions must facilitate and support local community initiatives; issues of basic resource allocations for communities are open for consideration; stakeholder involvement must be as inclusive as possible; dialogue to achieve consensus is an important element of the process; the inevitability of disputes, uncertainty, and surprise must be accepted along the way; actions taken should be viewed more as learning opportunities than as management solutions; linkages of communication and cooperation across all manner of human-imposed boundaries must be developed and maintained; and effective leadership for a variety of agreed-upon new initiatives needs to be nurtured.

In thinking about the future of communities dependent on the forest industry, forestry professionals may well have a pivotal role to play in seeking alternatives, such as community forestry under local control, as one component of a more diversified economic base. It is the societal context and decision processes within which they contribute that will be refreshingly different.

REFERENCES

Alexander, E.A. 1986. Approaches to Planning: Introducing Current Planning Theories, Concepts, and Issues. London, UK: Gordon and Breach. 147 p.

Allen, G.M., P.F.J. Eagles, and S.D. Price (editors). 1990. Conserving Carolinian Canada. Waterloo, Ontario: University of Waterloo Press. 346 p.

Allen, T.F.H., B.L. Bandurski, and A.W. King. 1993. Ecosystem Approach: Theory and Ecosystem Integrity. Report to the Great Lakes Science Advisory Board, Great Lakes Office. Windsor, Ontario: International Joint Commission. 67 p.

Allen, T.F.H. and T.W. Hoekstra. 1992. Toward a Unified Ecology. New York, New York: Columbia University Press. 384.

Bak, P. 1996. How Nature Works: The Science of Self-Organized Criticality. New York, New York: Springer-Verlag. 212 p.

Baker, J.A. 2000. Landscape ecology and adaptive management. In: A.H. Perera, D.L. Euler, and I.D. Thompson (editors). Ecology of a Managed Terrestrial Landscape: Patterns and Processes in Forest Landscapes of Ontario. Vancouver, British Columbia: University of British Columbia Press. 310-322.

Boreal East Planning Area. 1998. Preliminary Land Use Options, Lands for Life Boreal East Planning Area. Boreal East Tabloid No. 3, March 1998. 24 p.

Canadian Council of Ecological Areas. 1992. Ecological Areas Framework for Developing a Nation-Wide System of Ecological Areas. Part 1: A Strategy. Ottawa, Ontario: Canadian Council of Ecological Areas. 39 p.

Canadian Council of Ecological Areas. 1995. Ecological Areas Framework for Developing a Nation-Wide System of Ecological Areas. Part 2: Ecoregion Gap Analysis. Ottawa, Ontario: Canadian Council of Ecological Areas.

Canadian Pulp and Paper Association. 1997. Paper's Changing World: Annual Review. Montreal, Quebec: Canadian Pulp and Paper Association.

Cardinall, D. and J.C. Day. 1998. Embracing value and uncertainty in environmental management and planning: a heuristic model. Environments 25(2 and 3): 110-125.

Davis, M.B. and C. Zabinski. 1992. Changes in geographic range resulting from greenhouse warming: effects on biodiversity in forests. In: R.L. Peters and T.E. Lovejoy (editors). Global Warming and Biological Diversity. New Haven, Connecticut: Yale University Press. 297-308.

Dearden, P. and B. Sadler (editors). 1989. Landscape Evaluation: Approaches and Applications. Victoria, British Columbia: University of Victoria Press. 305 p.

Delcourt, P.A. and H.R. Delcourt. 1987. Long-Term Forest Dynamics of the Temperate Zone: A Case Study of Late-Quaternary Forests in Eastern North America. New York, New York: Springer-Verlag.

Earthroots. 1998. Earthroots holds off logging in Temagami's ancient forests. Forest Bulletin, Spring/Summer.

Elliott, C. and A. Hackman. 1996. Current Issues in Forest Certification in Canada. Discussion Paper. Toronto, Ontario: World Wildlife Fund Canada. 63 p.

Emmelin, L. 1996. Landscape impact analysis: a systematic approach to landscape impacts of policy. Landscape Research 21(1): 13-35.

Environment Canada. 1995. Canadian Biodiversity Strategy: Canada's Response to the Convention on Biological Diversity. Biodiversity Convention Office, Ottawa, Ontario, Canada. 77 p.

Epp, A.E. 2000. Ontario forests and forest policy before the era of sustainable forestry. In: A.H. Perera, D.L. Euler, and I.D. Thompson (editors). Ecology of a Managed Terrestrial Landscape: Patterns and Processes in Forest Landscapes of Ontario. Vancouver, British Columbia: University of British Columbia Press. 237-275.

Euler, D.L. and A.E. Epp. 2000. A new foundation for Ontario forest policy for the 21st century. In: A.H. Perera, D.L. Euler, and I.D. Thompson (editors). Ecology of a Managed Terrestrial Landscape: Patterns and Processes in Forest Landscapes of Ontario. Vancouver, British Columbia: University of British Columbia Press. 276-294.

Eve, R.A., S. Horsfall, and M.E. Lee (editors). 1997. Chaos, Complexity, and Sociology: Myths, Models, and Theories. Thousand Oaks, California: SAGE Publications. 328 p.

Financial Post. 1998. Survey of Industrials. 72nd edition. Toronto, Ontario: Financial Post. 899 p.
Francis, G. 1993. Ecosystem management. Natural Resource Journal 33: 315-345.
Francis, G. 1995. Ecosystems. In L. Quesnel (editor). Social Sciences and the Environment. Ottawa, Ontario: University of Ottawa Press. 145-171.
Friedmann, J. 1987. Planning in the Public Domain: From Knowledge to Action. Princeton, New Jersey: Princeton University Press. 501 p.
Friedmann, J. 1995. Teaching Planning Theory. Journal of Planning Education and Research 14(3): 156-162.
Funtowicz, S.O. and J.R. Ravetz. 1994. Uncertainty, complexity and post-normal science. Environmental Toxicology and Chemistry 13(12): 1881-1885.
Galindo-Leal, C. and F.L. Bunnell. 1995. Ecosystem management: implications and opportunities of a new paradigm. Forestry Chronicle 71(5): 601-606.
Grumbine, R.E. 1997. Reflections on "What is ecosystem management?" Conservation Biology 11(1): 41-47.
Gunderson, L.H., C.S. Holling, and S.S. Light (editors). 1995. Barriers and Bridges to Renewal of Ecosystems and Institutions. New York, New York: Columbia University Press. 593 p.
Gunther, F. and C. Folke. 1993. Characteristics of nested living systems. Journal of Biological Systems 1(3): 257-274.
Heritage Resources Centre. 1997. Georgian Bay Islands National Park Ecosystem Conservation Plan. Waterloo, Ontario: Heritage Research Centre, University of Waterloo. 66 p.
Hills, G.A. 1959. A Ready Reference to the Description of the Land of Ontario and its Productivity. Preliminary Report. Maple, Ontario: Ontario Department of Lands and Forests. 142 p.
Hirsch, E. and M. O'Hanlon (editors). 1995. The Anthropology of Landscape: Perspectives on Place and Space. Oxford, UK: Clarendon Press. 268 p.
Hodgins, B.W. and J. Benidickson. 1989. The Temagami Experience: Recreation, Resources, and Aboriginal Rights in the Northern Ontario Wilderness. Toronto, Ontario: University of Toronto Press. 370 p.
Holling, C.S. 1986. The resilience of terrestrial ecosystems: local surprise and global change. In: W.C. Clark and R.E. Munn (editors). Sustainable Development of the Biosphere. Cambridge, UK: Cambridge University Press. 292-320.
Holling, C.S. 1994. Simplifying the complex: the paradigm of ecological function and structure. Futures 26(6): 598-609.
Holling, C.S. and G.K. Meffe. 1996. Command and control and the pathology of natural resource management. Conservation Biology 10(2): 328-337.
Innes, J. 1995. Planning theory's emerging paradigm: communicative action and interactive practice. Journal of Planning Education and Research 14(3): 183-189.
Institute for Research on Environment and Economy. 1996. Community Empowerment in Ecosystem Management. Ottawa, Ontario: Institute for Research on Environment and Economy.
Kauffman, S. 1995. At Home in the Universe: The Search for the Laws of Self-Organization and Complexity. Oxford, UK: Oxford University Press.
Kemal, S. and I. Gaskell (editors). 1993. Landscape, Natural Beauty and the Arts. Cambridge, UK: Cambridge University Press.
Kimmins, J.P. 1995. Sustainable development in Canadian forestry in the face of changing paradigms. Forestry Chronicle 71(1): 33-44.
Korton, D.C. 1995. When Corporations Rule the World. San Francisco, California: Berrett-Koehler. 374 p.
Lawrence, D.P. 1997. Renewing the EIA Planning Process [PhD thesis]. Waterloo, Ontario: University of Waterloo, School of Urban and Regional Planning.
Lee, K.N. 1993. Compass and Gyroscope: Integrating Science and Politics for the Environment. Washington, DC: Island Press. 243 p.
Lenihan, J.M. and R.P. Neilson. 1995. Canadian vegetation sensitivity to projected climatic change at three organizational levels. Climate Change 30: 27-56.
Lorenz, E.N. 1993. The Essence of Chaos. Seattle, Washington: University of Washington Press. 227 p.
Ludwig, D., B. Walker and C.S. Holling. 1997. Sustainability, stability, and resilience. Conservation Ecology 1(1): Article 7.
Marchak, L.P. 1995. Logging the Globe. Montreal, Quebec: McGill University Press; Kingston, Ontario: Queen's University Press. 404 p.
Michael, D.N. 1993. Governing by learning: boundaries, myths and metaphors. Futures 25(1): 81-89.
Michael, D.N. 1995. Barriers and bridges to learning in a turbulent human ecology. In: L.H. Gunderson, C.S. Holling, and S.S. Light (editors). Barriers and Bridges to Renewal of Ecosystems and Institutions. New York, New York: Columbia University Press. 461-485.
Moss, M.R. (editor). 1988. Landscape Ecology and Management. Montreal, Quebec: Polyscience. 240 p.
Munn, R.E. (editor). 1996. Atmospheric Change and Biodiversity: Formulating a Canadian Science Agenda. Toronto, Ontario: University of Toronto, Institute for Environmental Studies. 71 p.
Naveh, Z. 1991. Some remarks on recent developments in landscape ecology as a transdisciplinary ecological and geographical science. Landscape Ecology 5: 65-73.
Ontario. 1997. Lands For Life: A Commitment to the Future. Toronto, Ontario: Queen's Printer. 11 p.
Ontario Environmental Assessment Board. 1994. Reasons for Decision and Decision. Class Environmental Assessment by the Ministry of Natural Resources for Timber Management on Crown Lands in Ontario. Toronto, Ontario: Ontario Environmental Assessment Board. 561 p.
Ontario Ministry of Natural Resources. 1996. Forest Management Planning Manual for Ontario's Crown Forests. Toronto, Ontario: Queen's Printer for Ontario. 452 p.
Ontario Ministry of Natural Resources. 1997a. An Introduction to Regional Land Use Strategies. Lands for Life. Toronto, Ontario: Ontario Ministry of Natural Resources. 9 p.
Ontario Ministry of Natural Resources. 1997b. Lands for Life: A Land Use Planning System for Ontario's Natural Resources. Toronto, Ontario: Ontario Ministry of Natural Resources. 15 p.
Ontario Ministry of Natural Resources. 1997c. Nature's Best: Ontario's Parks and Protected Areas: The Framework and Action Plan. Toronto, Ontario: Ontario Ministry of Natural Resources. 37 p.
Ontario Ministry of Natural Resources. 1998. Lands for Life: Consolidated recommendations of the Boreal West, Boreal East, and Great Lakes-St. Lawrence Round Tables. Toronto, Ontario: Ontario Ministry of Natural Resources. 83 p.
Owen, S. 1998. Land use planning in the nineties: CORE lessons. Environments 25 (2 and 3): 14-26.
Partnership for Public Lands. 1998. It's Your Land: Completing Ontario's System of Parks and Protected Areas. Toronto, Ontario: Partnership for Public Lands [pamphlet]. 8 p.

Perera, A.H. and D.J.B. Baldwin. 2000. Spatial patterns in the managed forest landscape of Ontario. In: A.H. Perera, D.L. Euler, and I.D. Thompson (editors). Ecology of a Managed Terrestrial Landscape: Patterns and Processes in Forest Landscapes of Ontario. Vancouver, British Columbia: University of British Columbia Press. 74-99.

Poser, S.F., W.J. Crins, and T.J. Beechey (editors.) 1993. Size and Integrity Standards for Natural Heritage Areas in Ontario. Huntsville, Ontario: Ontario Ministry of Natural Resources. 138 p.

Rapport, D.J. 1992. Evaluating ecosystem health. Journal of Aquatic Ecosystem Health 1: 15-24.

Regier, H.A. and J.J. Kay. 1996. An heuristic model of transformations of the aquatic ecosystems of the Great Lakes - St. Lawrence river basin. Journal of Aquatic Ecosystem Health 5: 3-21.

Reid, R. and R. Symmes. 1997. Conservation Strategy for Carolinian Canada. London, Ontario: Carolinian Canada Steering Committee.

Riley, J. and P. Mohr. 1994. The Natural Heritage of Southern Ontario's Settled Landscapes: A Review of Conservation and Restoration Ecology for Land-Use and Landscape Planning. Aurora, Ontario: Ontario Ministry of Natural Resources. 78 p.

Robinson, J.B. (editor). 1996. Life in 2030: Exploring a Sustainable Future for Canada. Vancouver, British Columbia: University of British Columbia Press. 168 p.

Roseland, M. 1992. Towards Sustainable Communities: A Resource Book for Municipal and Local Governments. Ottawa, Ontario: National Round Table on the Environment and the Economy. 340 p.

Sager, T. 1993. Paradigms for planning: a rationality-based classification. Planning Theory 9: 79-118.

Sager, T. 1995. Teaching planning theory as order or fragments? Journal of Planning, Education and Research 14(3): 166-173.

Schneider, E.D. and J.J. Kay. 1994. Complexity and thermodynamics: toward a new ecology. Futures 24(6): 626-647.

Soule, M.E. 1985. What is conservation biology? BioScience 35(11): 727-734.

Statement of Commitment. 1992. A Statement of Commitment to Complete Canada's Network of Protected Areas. Ottawa, Ontario: Canadian Federal, Provincial and Territorial Ministers of Environment, Parks and Wildlife.

Statistics Canada. 1998. Inter-Corporate Ownership, 1998. Ottawa, Ontario: Statistics Canada. 1158 p.

Statutes of Ontario. 1994. *Crown Forest Sustainability Act*. S.O. 1994, c. 15.

Tainter, J.A. 1989. The Collapse of Complex Societies. Cambridge, UK: Cambridge University Press. 250 p.

Waldrop, M.M. 1992. Complexity: The Emerging Science at the Edge of Order and Chaos. New York, New York: Touchstone Books. 380 p.

Webb, T. 1992. Past changes in vegetation and climate: lessons for the future. In: R.L. Peters and T.E. Lovejoy (editors). Global Warming and Biological Diversity. New Haven, Connecticut: Yale University Press. 59-75.

Westley, F. 1995. Governing design: the management of social systems and ecosystem management. In: L.H. Gunderson, C.S. Holling, and S.S. Light (editors). Barriers and Bridges to Renewal of Ecosystems and Institutions. New York, New York: Columbia University Press. 391-427.

Westra, L. 1994. An Environmental Proposal for Ethics: The Principle of Integrity. Lanham, Maryland: Rowman and Littlefield. 235 p.

Woodley, S., J. Kay, and G. Francis (editors). 1993. Ecological Integrity and the Management of Ecosystems. Delray Beach, Florida: St. Lucie Press. 220 p.

World Wildlife Fund. 1996. A Protected Areas Gap Analysis Methodology: Planning for the Conservation of Biodiversity. Toronto, Ontario: World Wildlife Fund Canada. 68 p.

15 Landscape Ecology and Adaptive Management

JAMES A. BAKER*

INTRODUCTION

Science and tools developed in the field of landscape ecology are increasingly in use within land use and forest management planning in Ontario (OMNR 1996). The shift from single-species management to an ecosystem approach that has taken place in planning includes an emphasis on the analysis of landscape-level patterns and processes (Ostfeld et al. 1997; Peters et al. 1997; Wiens 1996, 1997). This broader approach to forest management has the goal of sustaining ecological processes, while also serving the multiple needs and values of society (Booth et al. 1993; Thompson and Welsh 1993). Ecosystem management at the landscape level has become a cornerstone of Ontario's forest management (Ontario Forest Policy Panel 1993).

As the concept of ecosystem management is relatively new, its ultimate success is uncertain; however, uncertainty about the outcomes of management is not a new problem for resource managers. Adaptive management was developed over 20 years ago as a rigorous process for reducing uncertainty in resource management (Holling 1978; Walters 1986). The process has been used with varying degrees of success to respond to problems in the management of fisheries, wildlife, forests, and whole ecosystems (Holling 1978; Baskerville 1985, 1995; Walters 1986, 1997; Nudds and Morrison 1991; Lee 1993; Gunderson et al. 1995; Lancia et al. 1996).

Briefly, adaptive management involves monitoring the implementation of policy through the use of scientific methods of investigation, so as to develop knowledge of policy effectiveness. The increasing expectations placed on resource managers to deliver a variety of goals and objectives, particularly in the area of forest management, have generated increasing interest in the use of adaptive management (Baskerville 1995; Taylor et al. 1997). Ontario has recognized the need to use "adaptive ecosystem management" to deal with the uncertainty of forest ecosystem management (Ontario Forest Policy Panel 1993; MacDonald et al. 1997).

Despite the history of adaptive management and increasing recognition of the need to apply it in Ontario, there is some confusion about both the basics of the process (Nyberg 1998) and the underlying scientific and policy assumptions. The purpose of this chapter is to describe the science and policy elements of adaptive management, the potential value of its use in conjunction with landscape ecology for forest ecosystem management, including sustainable forest management, and the power of this combination to convert management agencies into learning organizations.

THE CONTRIBUTION OF LANDSCAPE ECOLOGY TO ECOSYSTEM MANAGEMENT

Landscape ecology has served as catalyst in the recent shift to a new paradigm of ecosystems as "non-equilibrium systems." This new paradigm (Fielder et al. 1997) recognizes that ecosystems are dynamic and undergo various types of natural disturbance that create heterogeneity and complexity over a range of spatial and temporal scales (Holling 1992). One of the central challenges of an ecosystem approach to

* Ontario Ministry of Natural Resources, 300 Water Street, Peterborough, Ontario K9J 8M5

management is to maintain the processes that create and maintain heterogeneity and complexity. Although a number of landscape ecology principles relevant to this challenge have been derived from a variety of theoretical models (e.g., metapopulation models, population viability analysis, and island biogeography theory), few are widely tested for application to continuously forested landscapes. There is also no comprehensive theory of landscape ecology which can underpin long-term conservation policy (Ostfeld et al. 1997).

While concepts of ecosystem processes have changed, a variety of tools have developed, such as geographic information system (GIS) technology, which improve the capability to analyze landscape patterns (McGarigal and Marks 1995; Perera et al. 1998). These developments have provided a set of tools and information management systems with the potential to test assumptions about landscape management.

Assumptions about the negative effects of habitat fragmentation have been incorporated into conservation plans, for example, but many of the underlying assumptions about large-scale fragmentation effects (Schmiegelow 1992; Simberloff et al. 1992) have not been adequately tested. Doubt is emerging as to the validity of these assumptions (Haila et al. 1994; Fahrig 1997). Forest management plans in Ontario must now incorporate spatial analysis to reduce the potential impact of fragmentation (OMNR 1996). The analytical techniques available in landscape analysis software (McGarigal and Marks 1995; Perera et al. 1998) make these evaluations relatively easy. However, unless the models are used in their proper context, and unless the link between landscape patterns and landscape processes is critically evaluated, these techniques could eventually become just another descriptive tool having little or no added value for resource managers (Conroy 1993).

THE INTERACTION OF SCIENCE AND POLICY IN ECOSYSTEM MANAGEMENT

Science has long been a partner in the development of natural resource management policy for fisheries, wildlife, and forestry. Until recently, resource managers in these fields have used science primarily with the goal of increasing resource yields rather than understanding ecological processes (Grumbine 1994; Nudds 1999). One consequence of the broader mandate of ecosystem management outlined above has been to heighten the need for an improved understanding of ecological systems, because assumptions about the expected behaviour of ecosystems underlies all policy expectations (Booth et al. 1993; Thompson and Welsh 1993).

The result is a need for closer examination of the scientific rigour and methods employed to understand ecological processes. A lack of scientific rigour, or even general misunderstanding of the scientific method, can culminate in poor public policy and can mislead the public as to what their expectations should be (Walters 1997). Although the study of ecosystems presents many obstacles – including the often formidable difficulty posed by the complexity of the ecosystems and the practical difficulties of experimentation – these problems should not be used to excuse a lack of rigour in the application of science to an evaluation of resource management policies.

Ecosystems are composed of hierarchically nested subsystems at a range of scales, in which a few critical variables operate at each scale (Holling 1992). They are complex and exhibit non-linear behaviour which makes it extremely difficult to predict long-term changes. There is thus a high degree of uncertainty in the behaviour of ecological systems under management and a corresponding degree of uncertainty in the outcome of policies related to long-term sustainability. Uncertainty about the ultimate effect of policies is not usually part of public-policy statements or of the management programs that ultimately implement higher-level policy. However, an evaluation of the effectiveness of policy is a critical component of the cycle of policy development and evaluation (OMNR 1995).

Given the uncertainty underlying scientific understanding of ecological patterns and processes, and the complexity of the policy development process, it is obvious that the achievement of public-policy goals is uncertain. Consequently, the goals inherent in policy, legislation, and implementation procedures for natural resource management that are based on ecological theory should be considered much the same as hypotheses inherent in scientific theories. The uncertainty of the outcome of policy implementation is much the same as uncertainty whether a particular hypothesis is the appropriate explanation of an observed phenomenon (Nudds 1999). However, policy does not act like science in this regard: while alternative scientific hypotheses are considered and experimentally tested, natural resource management policy usually adopts one procedure and expects it to achieve the goals of the policy.

The task of evaluating the success or failure of a particular policy becomes difficult, because the criteria for whether and why it is effective are not usually embedded in the policy. The general goals of the policy usually reside elsewhere, either in the governing legislation or in lower-level policies. No explicit sets of observations are specified to indicate to the agency or to the public whether the policy is a success or a failure. A better understanding of expected outcomes on the part of both the agencies responsible for resource management and the general public which they serve would assist in the evaluation and improvement of policy. A lack of specifics in the higher-level legislation and strategic policies necessitates that regulatory agencies implement a dialogue among themselves and with the public on how to reduce the uncertainty of the expected outcomes of management (Ludwig et al. 1993). The policy cycle already includes an evaluation phase, and this chapter discusses the value of active adaptive management for achieving more rapid and reliable policy improvement.

ADAPTIVE MANAGEMENT FOR REDUCING UNCERTAINTY IN POLICY DEVELOPMENT

As a formal process for dealing with the uncertainties inherent in resource management policy-making, adaptive management conceives of management as an experiment which provides feedback, through monitoring, for improving the effectiveness of policies. Adaptive management has been classified into three different types: active, passive, and reactive (Hilborn 1992; MacDonald et al. 1997; Taylor et al. 1997). An active adaptive management process explicitly treats natural resource management policies as hypotheses to be tested in an experimental way, so as to provide reliable knowledge (Romesburg 1981) about the system under management and thus to reduce uncertainty (Lancia et al. 1993). Scientists are usually interested in rejecting hypotheses; whereas resource managers are more interested in identifying which management action is probably correct or which is more plausible (Bergerud and Reed 1998). Adaptive management can provide clues about which management or policy option is best for achieving a desired outcome.

To know which policy is likely correct, a manager needs to begin by knowing which policies to reject. Accordingly, a key requirement of active adaptive management is the development of alternative models of system response to potential policies, during the policy development phase. This step must be followed by the application of rigorous monitoring to learn what the system response actually is to alternative policies, thereby to reduce managerial uncertainty about that response. Attempting to evaluate policy using non-experimental, inductive approaches cannot provide reliable knowledge (Romesburg 1981; Nudds and Morrison 1991). Policies must be treated as hypotheses from the outset, however, not as an afterthought once the policy is implemented. Evaluating policies with adaptive management is equivalent to attempting to disprove a scientific hypothesis. Failure to disprove that a particular policy is successful provides evidence of its value, in terms of its consistency with underlying ecological principles and acceptability to society.

Active Adaptive Management

Active adaptive management can be conducted in three general ways. Ideally, a number of potential models are identified during the course of policy development about the predicted response of the system to alternative policies (Figure 15.1). Subsequently, each of the policies is implemented, using controls and replication simultaneously, to test which of the policy models achieves the desired outcomes, thus reducing uncertainty about how the system will respond to the policy options (Walters and Holling 1990; Ludwig et al. 1993). Examples include forest management

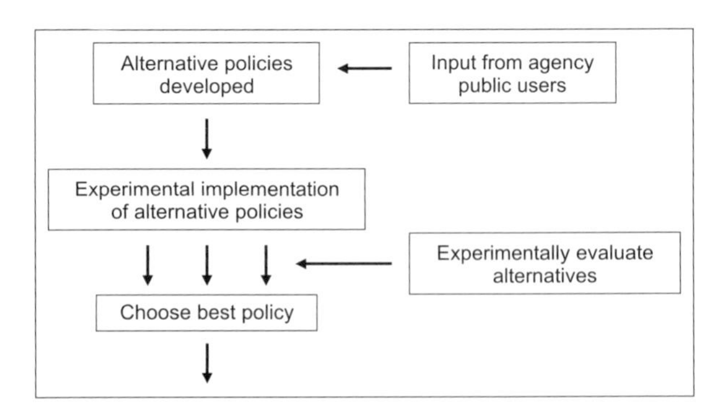

Figure 15.1 Active adaptive management using the ideal experimental form, in which alternative policy models are evaluated simultaneously.

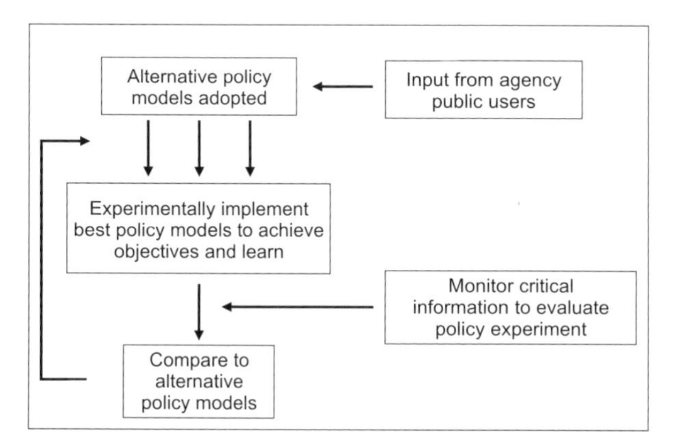

Figure 15.2 The second form of active adaptive management, in which it is impractical to implement alternative policy models simultaneously.

(Taylor et al. 1997), wildlife management (Gratson et al. 1993; Williams et al. 1996), and the implementation of alternative fisheries management programs (Walters and Holling 1990).

A second, less desirable method, but sometimes the only practical one, is to develop one or more alternative policy models, but implement only one of them. This process can reduce uncertainty if the results are monitored and subsequently used to determine whether one of the other potential policy models should be implemented in the future (Figure 15.2). This form of active adaptive management has been used in North America in waterfowl management (Williams and Johnson 1995; Williams et al. 1996; Johnson et al. 1997) and in fisheries management (Walters 1986; Walters and Holling 1990; and Lee 1993).

Comparative or retrospective studies (Smith 1998) can also be used to evaluate alternative policy models and should, therefore, be considered as a form of active adaptive management. Retrospective studies take advantage of unplanned perturbations, while active, experimental, adaptive management deliberately plans the perturbations. The evaluation of alternative policies would take the same form as illustrated in Figure 15.1 for active experimental adaptive management. However, because of the potential problems involved in accepting a postulate rather than a test, retrospective studies would need to be followed up by active experimental management. Through undertaking retrospective studies in this manner, reliable comparisons can be used in designing planned perturbations to test policy alternatives. This form of adaptive management is relevant to landscape ecology, in which it is often impractical to conduct large-scale experiments. It is also particularly relevant to the evaluation of forest management policies that attempt to emulate natural disturbance patterns.

Passive Adaptive Management

Passive adaptive management (Figure 15.3) is usually the form of adaptive management in use in agencies which proclaim that they already conduct or are going to conduct adaptive management. The cycle of passive adaptive management is usually defined as follows: plan, implement, monitor, and adjust; and then begin the cycle anew. This form of adaptive management is ineffective for reducing uncertainty about system response, because only one policy option is implemented and evaluated. Although the outcome of the policy is, indeed, monitored, there is no clear intent to collect data in such a way as to reduce uncertainty about the implemented policy and compare that policy to alternative policy models. Consequently, an agency may learn that the policy is ineffective, but will not understand why.

This passive form of adaptive management seems, on the surface, to be the same as the second form of active adaptive management identified above; however, the active form is distinguished by the deliberate design of its data collection program to reduce uncertainty about policy outcomes and to compare the policy with alternative models (Walters and Holling 1990; Hilborn 1992; Lee 1993; Williams and Johnson 1995; Johnson et al. 1997).

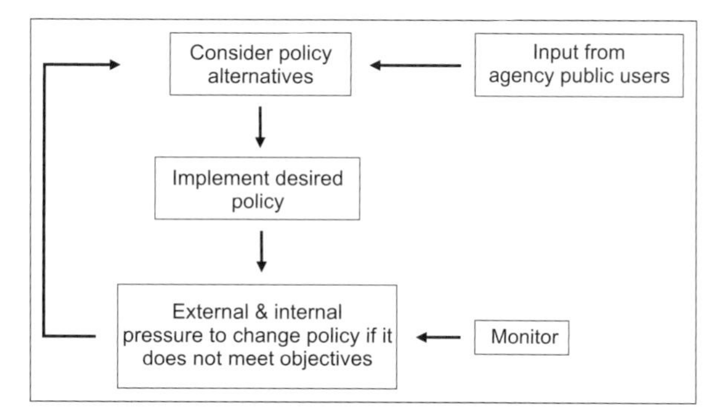

Figure 15.3 Passive adaptive management, in which only one policy is implemented and monitored. Although the success or failure of the policy is monitored, the monitoring protocol is not conducted experimentally.

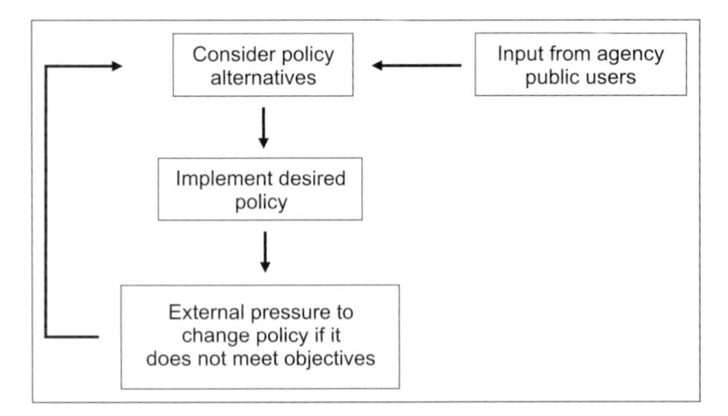

Figure 15.4 Reactive management, whereby a policy is implemented without any monitoring.

Reactive Management

Although reactive management (Figure 15.4) has been referred to as a type of adaptive management (Hilborn 1992; MacDonald et al. 1997), it should not be viewed as such, for the following reasons. Reactive management consists of implementing a policy without explicitly monitoring its effectiveness. The policy is changed only when pressure from outside the management agency forces a re-examination of the policy, sometimes through legal challenges (MacDonald et al. 1997), or when new knowledge becomes available. In most cases, the agency is simply responding to outside criticism and has no process in place to monitor the outcome of the policy, even if there may be good intentions to do so.

Adaptive Management and Reliable Knowledge

Active adaptive management, as outlined above and described by Walters (1986), Walters and Holling (1990), Lee (1993), Ludwig et al (1993), and Taylor et al. (1997), is a means of gaining reliable knowledge about resource management policy. The types of adaptive management represent a gradient of scientific rigour and reliability of knowledge (Figure 15.5).

Despite decades of experience in resource exploitation, in the study of natural history, and in fish and wildlife management, we still have only a limited ability to make reliable predictions about how and why

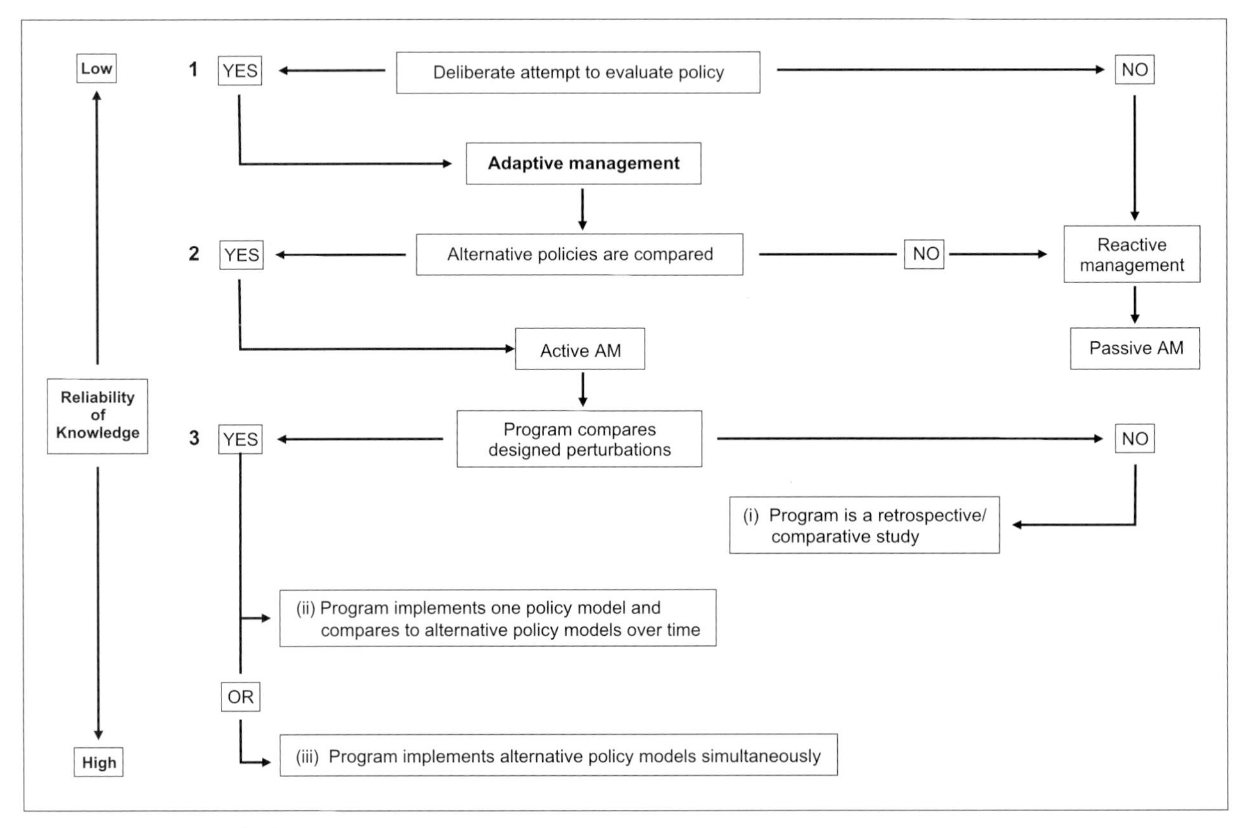

Figure 15.5 Key types of adaptive management.

natural systems respond to human intervention. In Holling's (1994) view, adaptive management is a means of "learning to live with the unexpected" and a process for dealing with "surprises." Since human intervention in natural systems, as approved within the context of natural resource management policies, changes those systems, we might as well learn from the intervention. The conventional method of policy development and implementation, which does not employ the concepts of control and replication, cannot provide reliable probes of ecosystems (Holling 1994).

CURRENT APPLICATIONS OF ADAPTIVE MANAGEMENT

Examples of the practical use of adaptive management are provided by Walters (1986), in the context of determining appropriate harvest policies for salmon. Adaptive management has been used in the Columbia River basin programs since 1984 (Lee 1993) to deal with the problems of hydroelectric power and salmon production in the basin. Adaptive harvest management is currently used to set limits on waterfowl harvests in North America (Williams et al. 1996; Johnson et al. 1997). Adaptive management is an integral part of forest management planning in the northwestern US states, where adaptive management areas have been established (Stankey and Shindler 1997). This management approach is used in Idaho to experimentally evaluate harvest rates for elk (Gratson et al. 1993) and in Missouri to evaluate long-term effects of forest management (Kurzejeski et al. 1993; Larsen et al. 1997).

Ontario has recently conducted a review of adaptive management examples in North America (MacDonald et al. 1997), and British Columbia has reviewed the efficacy of adaptive management for evaluating the effectiveness of various forest management practices (Taylor et al. 1997). Active adaptive management is being used to test the effectiveness of habitat guidelines for the provision of moose habitat in forest management planning and operations in Ontario (Rempel et al. 1997). An active adaptive management process is also being implemented to evaluate habitat guidelines for other wildlife species in the province (Other Wildlife Working Group 1995).

A key feature of active adaptive management is to simulate potential alternative models of the managed system and potential alternative responses to policies. Simulation models need not be overly complex (Walters 1986; Conroy 1993; McCarthy and Burgman 1995), but must be real enough to presuppose how a number of alternative policies will impact the system. Very few ecological models, regardless of their complexity, the genius of their design, or the empirical data used to verify them, can be confidently used to predict the future. Resource managers sometimes become disillusioned with models that are too complex or that are used to make positive predictions of the future (Conroy 1993; Johnson et al. 1997; Walters 1997). Good models, however, provide the means of selecting among alternative predictions.

Adaptive Management for Social Learning

The adaptive management process can be complex and challenging in the arena of public consultation. However, wider discussion of resource management issues is required, and conflicts among resource uses inevitably arise. Adaptive management should provide a comprehensive approach to the consultative process. Through an adaptive management approach, long-term risk of policy failure can be reduced (Holling 1995), including problems from inappropriate regulation. If reliable knowledge is lacking, the regulation of resource uses may be either too liberal or unnecessarily restrictive. In any case, resource users might accept a greater degree of regulation if the long-term consequences were evaluated as part of the policy.

Adaptive management has been described by Lee (1993) as a process for "social learning" because it includes society at large in the process of evaluating policy outcomes. Lee provides examples of agreements for the cooperative management of salmon, hydroelectric power, and irrigation water that were made to satisfy the needs of a variety of stakeholders. Central to effective public input into the process of policy development is the use of adaptive management to evaluate alternative management policies. The evaluation of policy alternatives through experimentation cannot be done as a technical experiment to discover ecological or other technical knowledge alone. If it is to be successful as a policy tool, it must also include an understanding of societal needs and expectations. The adaptive management process itself demands that interested parties disclose their desires and expectations frankly, and that they are kept informed of the outcomes at critical junctures of the experiment. In essence, adoption of the adaptive management process requires an understanding of societal expectations at the outset. Failure to include this consideration will likely result in accusations of an elitist, experts-only process.

Challenges in Implementing Adaptive Management

A review of adaptive management examples across North America (MacDonald et al. 1997) reveals that, where this management approach is being practised at all, it is largely passive. There are a number of reasons for the absence of more aggressive and reliable forms of adaptive management. There are at least five types of challenges to implementing an active adaptive management program: technical, economic, ecological, institutional, and political (Lee 1993; MacDonald et al. 1997).

Particularly in large-scale experiments, there are technical problems in deciding upon the design, including the appropriate controls, treatments, and replication (Walters 1997). Economics will dictate the level of staffing, data collection, data management, and data analyses that can be allocated over the long periods required to obtain reliable results. Walters (1997) takes the pessimistic view that most adaptive management applications fail when technical design and economic limitations lead to a disproportionate reliance on modelling, at the expense of experimentation. Institutionally, there are the additional problems of long-term commitments on the part of policy-makers, scientists, and operational staff, and the arrangements needed to manage these commitments. Moreover, policy-makers have a deep-seated reluctance to accept the potential perception of failure, if a particular policy is demonstrated to have failed (Campbell 1969; Lee 1993).

Science is a process of embracing critical thought in order to accelerate learning. As described by Lee (1993), conflict has a central role in the art of policy-making and policy evaluation. Societal learning can proceed with appropriate levels of managed conflict, but intense conflict can harm the process. Consequently, it is necessary to recognize the valuable role in accelerating learning that is played by conflict and its manifestation in criticism. In dealing with natural resource policy issues, adaptive management can readily embrace conflict and criticism in the effort to establish appropriate policies over the long term. For the process to be successful, however, it is necessary for the practitioners of adaptive management to eschew the role of advocate.

ACTIVE ADAPTIVE MANAGEMENT IN SUSTAINABLE FORESTRY

One fundamental problem in implementing sustainable forest management is to define and identify conditions that are sustainable and their "pass or fail point" (Kimmins 1997). A number of criteria and indicators (Canadian Council of Forest Ministers 1997) have been suggested. These indicators are simply data or pieces of information, but no thresholds have been established as benchmarks to identify a forest as sustainable. It is hoped that tracking these indicators will yield a better sense of what might or might not be sustainable, but to make that determination requires a baseline or reference point (Karr 1991; Nudds 1999) by which to compare current conditions to sustainable ones.

A fundamental difficulty in using adaptive management to develop and evaluate sustainable forest management policies in an experimental manner is the problem of experimentally manipulating the real world. This problem is even more critical when one is dealing with the question of ecological sustainability over periods of at least 100 years, which is the time scale of most forest management scenarios. Given the rate of human population growth and our collective rate of natural resource use, we cannot wait 100 years to determine if present-day policies are appropriate. The goal of managing for ecological sustainability is to create conditions which will keep the options open for at least two to three generations hence. We must try to evaluate current policies to obtain some reliable clues about the efficacy of current management policies that attempt to achieve this goal. Since landscape ecology promotes the integrated examination of many resource values, it can play an important role in active adaptive management approaches to addressing long-term questions about forest sustainability.

THE NEW MODEL OF SUSTAINABLE FOREST MANAGEMENT IN ONTARIO

In Ontario, the *Crown Forest Sustainability Act* (1994) established natural disturbance patterns as the baseline or control for sustainable forest management. Similar policies are being examined in jurisdictions such as Alberta, British Columbia, and Quebec (Hannon et al. 1996; Taylor et al. 1997; Bergeron and Harvey 1997). Forests are subjected to various natural disturbances caused by agents such as fire, insects, diseases, windstorms, gap phase replacement, and herbivory. Together, these natural disturbances create diversity in the forest systems over time and space (Holling and Meffe 1996). Fire is still a major source of disturbance in the boreal forest of Ontario. It is assumed that, since forests have existed continuously over long periods of time in the presence of disturbances, managing forests in a way that emulates natural disturbance should be the most reasonable course for sustaining them.

Since 1950, however, logging has accelerated the creation of landscape patterns consisting of small patches and an increased amount of edge (Perera and Baldwin 2000, this volume), likely as a result of a decrease in the size of clear-cuts. The increased amount of edge gives rise to an obvious concern that wildlife species, as well as other landscape processes, may be negatively affected. Habitat fragmentation is possibly the central issue in conservation biology (Wiens 1996, 1997) because of the potential for impact on wildlife (McGarigal and McComb 1995; Schmiegelow et al. 1997). Despite recent doubts about the impacts of fragmentation in continuous forests (Haila et al. 1994; Fahrig 1997), it might be prudent to require future forest management operations to increase patch sizes and reduce edge. While this change in policy is being implemented in Ontario, it might also be prudent to evaluate the impacts of this policy through an adaptive management approach. There are at least two other alternative hypotheses that might be considered and evaluated. These alternatives are to continue creating increasing amounts of edge by maintaining the status quo in forest policy, or to implement the patterns of intermediate patch size and shape that are produced by progressive, historical clear-cuts.

The key measure for comparing alternative policies is to be found in landscape processes. As fire is a chemical process and logging a mechanical one, the response of landscape processes to each could be different. It is not enough simply to assume that replicating patterns will replicate processes. However, given the difficulty involved in subjecting large landscapes (measuring at least 1000 km^2) to both natural disturbances and different forest management policies, it is not practical to compare these alternative policy models in real time using an active, experimental, adaptive management design with appropriate controls and replications.

The Role of Retrospective Studies in Evaluating Policy

Our current understanding of hierarchy theory, as applied to the types and rates of disturbance, can be depicted in a space-time continuum (Voigt et al. 2000, this volume, Figure 11.1). The spatial scale of landscape disturbance created by forest management activities, either selection cutting or clear-cutting, encompasses at least part of the range of patterns created by natural disturbances from fire and insects. A comparison of patterns and processes among different spatial scales of naturally disturbed and managed landscapes should provide insight into the similarities and differences between them. Such comparisons should provide clues for predicting the efficacy of the policy of emulating landscape patterns for future forest sustainability. In Ontario, as in other jurisdictions, notably the western provinces of Canada, there are large tracts of land that are not under management, as well as extensive areas in provincial and national parks that are still undergoing natural disturbance, either because they have not been logged or because no logging has been done for several decades. These areas can serve as baseline ecological controls.

The past range of forest management systems provides a range of managed landscape patterns for comparison to naturally disturbed landscape patterns. Such retrospective or comparative studies (Figure 15.5) can provide more reliable knowledge to guide the construction of retroductive policy hypotheses and to test landscape models, for example, of the effects of forest fragmentation. Retrospective studies can be easily dismissed on the grounds of a lack of detail in historical data (Lamont 1995); however, other more rigorous designs are impractical for whole systems (Allen and Hoekstra 1992; Holling 1995; Smith 1998). Large-scale comparisons of past and present can be performed with respect to landscape structure and composition, wildlife response at landscape scales, and other landscape processes. Landscape-scale studies of this type are presently underway in Ontario (Gurd 1996, 1997; Rempel et al. 1997; D. Voigt, pers comm.). These studies include comparisons of historical fire patterns and harvest patterns (Forsyth and McNicol, in press) and studies to model spatial fire dynamics over multiple, large spatial and temporal scales (Li et al. 1996, 1997; Boychuck and Perera 1997). All of these efforts are necessary to begin synthesizing some understanding of potential alternative policy models.

The scale of the perturbations studied will depend on the scale required to obtain results providing insight into the underlying processes of disturbance influence. The results must also be both practical and acceptable by society. Small-scale landscape experiments have been used to test hypotheses of wildlife response to different patch sizes and patterns (Schmiegelow and Hannon 1993; McGarigal and McComb 1995; Schmiegelow et al. 1997). Large-scale experiments are necessary, however: it is difficult to extrapolate from small to large spatial landscapes because of the non-linear differences in response among scales (Allen and Hoekstra 1992). In selecting which adaptive management method to use, the factors to be weighed include the desirability of perturbing the system deliberately, so as to obtain data quickly for revising policy models, and the reluctance of society to accept large-scale perturbations (Walters and Holling 1990; Johnson et al. 1997; Walters 1997).

Adaptive Management as a Cornerstone of Sustainable Forest Management

Holling (1995) suggests that the cornerstone of sustainable management of resources is the resiliency of managed ecosystems to recover from disturbance. Since disturbance from various natural processes is an internal property of forest ecosystems (Kimmins 1997), the real question is not whether humans should be disturbing these systems, but whether humans are adapting their management activities to the natural resiliency of these systems. Furthermore, Holling (1995, p. 32) proposed that both the resiliency of ecosystems and human opportunities to use those ecosystems can be maintained only through the development of "flexible, diverse, and redundant regulations; monitoring that leads to corrective responses; and experimental probing" of the dynamic ecosystems. Flexibility, multiplicity of options, monitoring, and experimentation are all hallmarks of adaptive resource management. In reviewing a number of situations around the world in which resource management was unsustainable, Holling (1986) found that none of these characteristics were evident in the associated resource management agencies.

MANAGEMENT AGENCIES AS LEARNING ORGANIZATIONS

If agencies charged with the responsibility of implementing ecosystem management are to be transformed into adaptive agencies, then internal organizational processes will need to change. Such change is needed to deal more effectively with the resource users and society as a whole (Grumbine 1994; Holling 1995). The focus of the changes should be the development of knowledge and the movement of that knowledge within and outside the agency. The same principles for using knowledge as a strategic and tactical resource that are now being used by many companies could be adopted by resource management agencies. Over the past few years, many companies have changed their organizational structures and internal processes, to become "learning organizations" (Senge 1990; Nonaka and Takeuchi 1995; Westley 1995; De Geus 1997). This transformation requires an organizational structure and a management process that encourages innovation and the flow of new knowledge throughout the organization. Organizations that achieve the required changes are responsive to new knowledge by being open to new stimuli. They maintain a clear link to action by taking advantage of and responding to both tacit and explicit knowledge (Nonaka and Takeuchi 1995).

Innovative Companies as Models for Ecosystem Management Agencies

Companies that use knowledge to develop innovative products could provide an appropriate model for agencies responsible for ecosystem-based resource management. An essential part of the model would be the transformation of the knowledge of resource users, resource managers, ecologists, biologists, foresters, naturalists, and members of society at large into designs for the experimental evaluation of policies. This process of design development would capture the reality of the policies required to achieve societal expectations. The success or failure of any particular policy turns on its acceptance by resource users and society in general. Just as companies develop products, management agencies would develop reliable knowledge about policies in which resource users and society at large would have confidence. The product of management would be policy built on the foundation of reliable knowledge.

In sustainable forest management, the move by individual companies toward the certification of sustainable forest products (Elliott and Hackman 1996; Kimmins 1997) beckons these direct users of the forest ecosystems to join regulators in developing reliable knowledge about forest sustainability. Forest product companies can no longer be just in the business of cutting trees and turning them into consumer products. They must also demonstrate stewardship of a much broader set of forest values, with the long-term purpose of sustainable development. Given the widespread public concern about forest management, it will be necessary for these companies to become learning organizations, as Alberta-Pacific Forest Industries in Alberta has done (Hebert 1995).

Accelerating Learning about Ecosystem Management

The future pace of accelerated learning will depend largely on the degree to which ecosystem management agencies need to become learning organizations. Although governments do not have the financial resources they once had, that fact should not be an excuse for slow learning. The development of learning organizations depends more on attitude than on funds, although funding will obviously be necessary. As De Geus (1997), Senge (1990), and others have pointed out, the change to learning organizations on the part of businesses is a deliberate attempt to become more competitive. Ecosystem management agencies do not have a competitive incentive, but they do have the incentive to manage potential political conflict. Such conflict can and likely will arise from attempts to make resource management decisions that require a balancing of various values, particularly in the face of accelerated resource development (Lee 1993). A commitment to learning is a promising means of avoiding conflict (Westley 1995). A move by agencies away from what has been called the "pathology of command and control" (Holling and Meffe 1996; 329) is needed for managers, users, and society to reach innovative solutions (Knight and Meffe 1997).

The incentive that exists for resource management agencies to become learning organizations is manifest in the speed at which management agencies are developing information management systems. Over the past few years, many agencies have dramatically increased their analytical capability by implementing GIS technology and developing decision-support models. All major forest companies have computer tools such as GIS software packages for use in a variety of forest management planning activities. Although these tools provide the capability to improve the information base of decision-making, this information is not necessarily translated into decisions based on more reliable knowledge. There is a real danger of mistaking masses of data and technological skills for knowledge. Consequently, the role of information systems, of the large databases incorporated into them, and of landscape analysis tools must be thoughtfully merged with the adaptive goal of a learning organization. A commitment to becoming a learning organization using an adaptive management approach has the potential to link science and policy for the long-term benefit of society and to improve the respective role of either element in public service.

SUMMARY

The principles and techniques of landscape ecology have provided for integration across major resource management disciplines and have thus permitted an ecosystem approach to management. This approach has the goal of sustaining ecological processes while also providing for the multiple needs of society. Pursuit of these goals requires both analyzing landscape spatial patterns and modifying them, as required, to maintain the integrity of landscape processes. Both land use and forest management planning have adopted spatial analytical tools for the purpose. The expected biological, social, and economic outcomes for ecosystem management policies remain to be evaluated in the harsh reality of real-life management, conflicts included.

Active adaptive management techniques, in which management is treated as an experiment, should be used for evaluating ecosystem management policies, together with the assumptions embedded in landscape planning techniques for implementing these policies. Although the baseline for sustainable forest management in Ontario is natural disturbance regimes, to experiment with natural and managed disturbances of 100 km^2 or more in the boreal forest will be challenging: it may take many years to convince society at large to allow experimental management at these scales. However, at least parts of the experiment may already be done or underway through forest management planning and operations, as Ontario and other jurisdictions in Canada have a history of natural and managed disturbances at such large scales. Retrospective or comparative studies, which are a subset of active management techniques, could be started immediately to evaluate current policies by analyzing differences between past and present landscapes.

Adopting an adaptive management process for policy evaluation will not likely be sufficient, however, unless management agencies can transform themselves into learning organizations. Management agencies may wish to emulate innovative companies that use knowledge as a strategic resource to provide a competitive edge. Management agencies do not have a competitive incentive, but they do have a need to reduce resource allocation conflicts. These conflicts might arise from the stated goal of ecosystem management to balance biological, social, and economic interests. It will be difficult to deliver all of these values unless society is included in the pursuit of knowledge about the effectiveness of strategic policy. Incorporating adaptive management into policy formation, and becoming a knowledge conduit to society at large concerning the effectiveness of ecosystem management policies, should provide management agencies with the means to provide open and meaningful public service.

ACKNOWLEDGEMENTS

I wish to thank John McNicol, Tom Nudds, and three anonymous reviewers for commenting on early drafts of this chapter.

REFERENCES

Allen, T.F.H. and T.W. Hoekstra. 1992. Toward a Unified Ecology. New York, New York: Columbia University Press. 384 p.

Baskerville, G.L. 1985. Adaptive management, wood availability and habitat availability. Forestry Chronicle 61: 171-175.

Baskerville, G.L. 1995. The forestry problem: adaptive lurches of renewal. In: L.H. Gunderson, C.S. Holling, and S.S. Light (editors). Barriers and Bridges to the Renewal of Ecosystems and Institutions. New York, New York: Columbia University Press. 37-102.

Bergeron, Y. and B. Harvey. 1997. Basing silviculture on natural ecosystem dynamics: an approach applied to the southern boreal mixedwood forest in Quebec. Forest Ecology and Management 92: 235-242.

Bergerud, W.L. and W.J. Reed. 1998. Bayesian statistical techniques. In: V. Sit and B. Taylor (editors). Statistical Methods for Adaptive Management Studies. Land Management Handbook No. 42. Victoria, British Columbia: British Columbia Ministry of Forests, Research Branch. 89-104.

Booth, D.L., D.W. K. Boulter, D.J. Neave, A.A. Rotherham, and D.A. Welsh. 1993. Natural forest landscape management. Forestry Chronicle 69: 141-145.

Boychuck, D. and A.H. Perera. 1997. Modeling temporal variability of boreal landscape age-classes under different fire disturbance regimes and spatial scales. Canadian Journal of Forest Research 27: 1083-1094.

Campbell, D.T. 1969. Reforms as experiments. American Psychologist 24: 409-429.

Canadian Council of Forest Ministers. 1997. Criteria And Indicators of Sustainable Forest Management in Canada. Technical Report. Ottawa, Ontario: Canadian Forest Service. 137 p.

Conroy, M.J. 1993. The use of models in natural resource management: prediction, not prescription. Transactions of the North America Wildlife and Natural Resource Conference 58: 509-519.

De Geus, A. 1997. The Living Company. Boston, Massachusetts: Harvard Business School Press. 214 p.

Elliott, C. and A. Hackman. 1996. Current Issues in Forest Certification in Canada. Discussion Paper. Toronto, Ontario: World Wildlife Fund. 62 p.

Fahrig, L. 1997. Relative effects of habitat loss and fragmentation on population extinction. Journal of Wildlife Management 61: 603-610.

Fielder, P.L., P.S. White, and R.A. Leidy. 1997. The paradigm shift in ecology and its implications for conservation. In: S.T.A. Pickett, R.S. Ostfeld, M. Shachak, and G.E. Likens (editors). The Ecological Basis of Conservation. New York, New York: Chapman and Hall. 83-92.

Forsyth, L. and J. McNicol. In press. Forest Management Guidelines for The Emulation of Fire Disturbance Patterns – Analysis Results. Thunder Bay, Ontario: Ontario Ministry of Natural Resources. 65 p.

Gratson, M.W., J.W. Unsworth, P. Zager, and L. Kock. 1993. Initial experiences with adaptive resource management for determining appropriate antlerless elk harvest rates in Idaho. Transactions of the 58th North American Wildlife and Natural Resources Conference 58: 610-619.

Grumbine, R.E. 1994. Introduction. In: R. E. Grumbine (editor). Environmental Policy and Biodiversity. Washington, DC: Island Press. 3-19.

Gunderson, L.H., C.S. Holling, and S.S. Light. 1995. Barriers broken and bridges built: a synthesis. In: L.H. Gunderson, C.S. Holling, and S.S. Light (editors). Barriers and Bridges to the Renewal of Ecosystems and Institutions. New York, New York: Columbia University Press. 489-593.

Gurd, D.B. 1996. Avian occurrence in Ontario boreal forests subjected to disturbance from timber harvest and fire [M.Sc. thesis]. Guelph, Ontario: University of Guelph. 105 p.

Gurd, D.B. 1997. Effects of Timber Management on Wildlife: Conceptual and Analytical Issues. Forest Fragmentation and Biodiversity Project Report No. 24. Sault Ste. Marie, Ontario: Ontario Ministry of Natural Resources, Ontario Forest Research Institute. 31 p.

Hannon, S., W. Ballard, S. Boutin, T. Diamond, R. Moses, J. Roland, P. Taylor, M. Villard, and D. Vitt. 1996. Influence of landscape structure on biodiversity and ecological processes in the boreal forest: a proposal. In: Sustainable Forest Management Network of Centres of Excellence. Conference Summaries: Forging a Network of Excellence, Sustainable Forestry Partnerships, March 8-10, 1996. Edmonton, Alberta: Sustainable Forest Management Network. 90-92.

Haila, Y., I.K. Hanski, J. Niemel, P. Punttila, and H. Tukia. 1994. Forestry and the boreal fauna: matching management with natural forest dynamics. *Annales Zoologici Fennici* 31: 187-202.

Hebert, D. 1995. Changing paradigms in forest ecosystem management. In: Proceedings of the Boreal Forest Ecosystem Management Conference and Workshop, January 29-31, 1995, Edmonton, Alberta. Edmonton: Alberta Pacific Forest Industries Inc. 67-72.

Hilborn, R. 1992. Institutional learning and spawning channels for sockeye salmon (*Onchorynchus nerka*). Canadian Journal of Fisheries and Aquatic Science 4: 1126-1135.

Holling, C.S. (editor). 1978. Adaptive Environmental Assessment and Management. London, UK: John Wiley. 377 p.

Holling, C.S. 1986. The resilience of terrestrial ecosystems: local surprise and global change. In: W.C. Clark and R. Munn (editors). Sustainable Development of the Biosphere. Cambridge, UK: Cambridge University Press. 292-317.

Holling, C.S. 1992. Cross-scale morphology, geometry and dynamics of ecosystems. Ecological Monographs 62: 447-502.

Holling, C.S. 1994. New science, and new investments for a sustainable biosphere. In: A. M. Jansson, M. Hanmer, C. Folke, and R. Costanza (editors). Investing in Natural Capital. Washington, DC: Island Press. 57-73.

Holling, C.S. 1995. What barriers, what bridges? In: L.H. Gunderson, C.S. Holling, and S.S. Light (editors). Barriers and Bridges to the Renewal of Ecosystems and Institutions. New York, New York: Columbia University Press. 4-34.

Holling, C.S. and G.K. Meffe. 1996. Command and control and the pathology of natural resource management. Conservation Biology 10(2): 328-337.

Johnson, F.A., C.T. Moore, W.L. Kendall, J.A. Dubovsky, D.F. Caithamer, J.R. Kelley, Jr., and B.K. Williams. 1997. Uncertainty and the management of mallard harvests. Journal of Wildlife Management 61: 202-216.

Karr, J.R. 1991. Biological integrity: a long neglected aspect of water resource management. Ecological Applications 1: 66-84.

Kimmins, H. 1997. Balancing Act: Environmental Issues in Forestry. Vancouver, British Columbia: University of British Columbia Press. 305 p.

Knight, R.L. and G.R. Meffe. 1997. Ecosystem management: agency liberation from command and control. Wildlife Society Bulletin 25: 676-678.

Kurzejeski, E.W., R.L. Clawson, R.B. Renken, S.L. Sheriff, L.D. Vangilder, C. Hauser, and J. Faaborg. 1993. Experimental evaluation of forest management: the Missouri Ozark forest ecosystem project. Transactions of the North American Wildlife and Natural Resources Conference 58: 599-609.

Lamont, B.R. 1995. Testing the effect of ecosystem composition/ structure on its functioning. *Oikos* 74: 283-295.

Lancia, R.A., C.E. Braun, M.W. Collopy, R.D. Dueser, J.G. Kie, C.J. Martinka, J.D. Nichols, T.D. Nudds, W.R. Porath, and N.G. Tilghman. 1996. ARM! for the future: adaptive resource management in the wildlife profession. Wildlife Society Bulletin 24: 436-442.

Lancia, R.A., T.D. Nudds, and M.L. Morrison. 1993. Adaptive resource management: policy as hypothesis, management by experiment? Opening comments: Slaying slippery shibboleths. Transactions of the North American Wildlife and Natural Resources Conference 58: 505-508.

Larsen, D.R., S.R. Shiley, F.R. Thompson III, B.L. Brookshire, D.C. Dey, E.W. Kurzejeski, and K. England. 1997. Ten guidelines for ecosystem researchers: lessons from Missouri. Journal of Forestry 95: 5-9.

Lee, K.N. 1993. Compass and Gryroscope: Integrating Science and Politics for the Environment. Washington, DC: Island Press. 243 p.

Li, C., M. Ter-Mikaelian, and A. Perera. 1996 Ontario Fire Regime Model: Its Background, Rationale, Development, and Use. Forest Fragmentation and Biodiversity Project Report No. 25. Sault Ste. Marie, Ontario: Ontario Ministry of Natural Resources, Ontario Forest Research Institute. 42 p.

Li, C., M. Ter-Mikaelian, and A. Perera. 1997. Temporal fire disturbance patterns on a forest landscape. Ecological Modelling 99: 137-150.

Ludwig, D., R. Hilborn, and C. Walters. 1993. Uncertainty, resource exploitation, and conservation: lessons from history. Science 260: 17 and 36.

MacDonald, G.B., R. Arnup, and R.K. Jones. 1997. Adaptive Forest Management in Ontario: A Literature Review and Strategic Analysis. Forest Research Information Paper No. 139. Sault Ste. Marie, Ontario: Ontario Ministry of Natural Resources, Ontario Forest Research Institute. 38 p.

McCarthy, M.A. and M.A. Burgman. 1995. Coping with uncertainty in forest wildlife planning. Forest Ecology and Management 74: 23-36.

McGarigal, K. and B. Marks. 1995. Spatial pattern analysis program for quantifying landscape structure. General Technical Report PNW-GTR-311. Portland, Oregon: USDA Forest Service. 122 p.

McGarigal, K. and W.C. McComb. 1995. Relationships between landscape structure and breeding birds in the Oregon coast range. Ecological Monographs 65: 235-260.

Nonaka, I. and H. Takeuchi. 1995. The Knowledge Creating Company. New York, New York: Oxford University Press. 284 p.

Nudds, T.D. 1999. Adaptive management and the conservation of biodiversity. In: R.K. Baydack, H.Campa III, and J. B. Haufler (editors). Practical Approaches to the Conservation of Biodiversity. Washington, DC: Island Press. 179-193.

Nudds, T.D. and M. L. Morrison. 1991. Ten years after "reliable knowledge": Are we gaining? Journal of Wildlife Management 55: 757-760.

Nyberg, B.J. 1998. Statistics and the practice of adaptive management. In: V. Sit and B. Taylor (editors), Statistical Methods for Adaptive Management Studies. Land Management Handbook No. 42. Victoria, British Columbia: British Columbia Ministry of Forests, Research Branch. 1-7.

Ontario Forest Policy Panel. 1993. Diversity: Forests, People, Communities. Proposed Comprehensive Forest Policy Framework for Ontario. Toronto, Ontario: Queen's Printer for Ontario. 147 p.

Ontario Ministry of Natural Resources. 1995. The Guide to Policy Development. Toronto, Ontario: Ontario Ministry of Natural Resources. 55 pages + appendices.

Ontario Ministry of Natural Resources. 1996. Forest Resources of Ontario. Toronto, Ontario: Queen's Printer for Ontario. 86 p.

Ostfeld, R.S., S.T.A. Pickett, M. Shachak, and G. E. Likens. 1997. Defining the scientific issues. In: S.T. A. Pickett, R. S. Ostfeld, M. Shachak, and G. E. Likens (editors). The Ecological Basis of Conservation. New York, New York: Chapman and Hall. 3-15.

Other Wildlife Working Group. 1995. Proposal for Research on the Impacts of Timber Management on Other Wildlife. Ontario Ministry of Natural Resources, Peterborough, Ontario. Unpublished manuscript. 25 p.

Perera, A. H., F. Schnekenburger, and D. Baldwin. 1998. LEAP II: A Landscape Ecological Analysis Package for Land Use Planners and Managers. Forest Research Report No. 146. Sault Ste. Marie, Ontario: Ontario Ministry of Natural Resources, Ontario Forest Research Institute. 88 p.

Perera, A.H. and D.J.B. Baldwin. 2000. Spatial patterns in the managed forest landscapes of Ontario. In: A.H. Perera, D.L. Euler, and I.D. Thompson (editors). Ecology of a Managed Terrestrial Landscape: Patterns and Processes in Forest Landscapes of Ontario.

Vancouver, British Columbia: University of British Columbia Press. 74-99.

Peters, R.S., D.M. Waller, B. Noon, S.T.A. Pickett, D. Murphy, J. Cracraft, R. Kiester, W. Kuhlmann, O. Houck, and W. J. Snape III. 1997. Standard scientific procedures for implementing ecosystem management on public lands. In: S.T.A. Pickett, R. S. Ostfeld, M. Shachak, and G.E. Likens (editors). The Ecological Basis of Conservation. New York, New York: Chapman and Hall. 320-336.

Rempel, R.S., P.C. Elkie, A.R. Rodgers, and M.J. Gluck. 1997. Timber-management and natural-disturbance effects on moose habitat: landscape evaluation. Journal of Wildlife Management 61: 517-524.

Romesburg, H.C. 1981. Wildlife science: gaining reliable knowledge. Journal of Wildlife Management 45: 293-313.

Schmiegelow, F.K.A. 1992. The use of atlas data to test appropriate hypotheses about faunal collapse. In: G.B. Ingram and M.R. Moss (editors). Landscape Approaches to Wildlife and Ecosystem Management in Canada. Montreal, Quebec: Polyscience. 67-74.

Schmiegelow, F.K.A. and S. Hannon. 1993. Adaptive management, adaptive science and the effects of forest fragmentation on boreal birds in northern Alberta. Transactions of the North American Wildlife and Natural Resources Conference 58: 584-598.

Schmiegelow, F.K.A., C.S. Machtans, and S.J. Hannon. 1997. Are boreal birds resilient to forest fragmentation? An experimental study of short-term community responses. Ecology 78: 1914-1932.

Senge, P.M. 1990. The Fifth Discipline. New York, New York: Doubleday. 423 p.

Simberloff, D., J.A. Farr, J. Cox, and D.W. Mahlman. 1992. Movement corridors: conservation bargains or poor investments? Conservation Biology 6: 493-504.

Smith, G.J. 1998. Retrospective studies. In: V. Sit and B. Taylor (editors). Statistical Methods for Adaptive Management Studies. Land Management Handbook No. 42. Victoria, British Columbia: British Columbia Ministry of Forests, Research Branch. 41-53.

Stankey, G.H. and B. Shindler. 1997. Adaptive Management Areas: Achieving the Promise, Avoiding the Peril. General Technical Report PNW-GTR-394. Portland, Oregon: USDA Forest Service. 21 p.

Statutes of Ontario. 1994. *Crown Forest Sustainability Act*. S.O. 1994, c. 25.

Taylor, B., L. Kremsater, and R. Ellis. 1997. Adaptive Management of Forests in British Columbia. Victoria, British Columbia: British Columbia Ministry of Forests, Forest Practices Branch. 93 p.

Thompson, I.D. and D.A. Welsh. 1993. Integrated resource management in boreal forest ecosystems: impediments and solutions. Forestry Chronicle 69: 32-39.

Voigt, D.R., J.A. Baker, R.S. Rempel, and I.D. Thompson. 2000. Forest vertebrate responses to landscape-level changes in Ontario. In: A.H. Perera, D.L. Euler, and I.D. Thompson (editors). Ecology of a Managed Terrestrial Landscape: Patterns and Processes in Forest Landscapes of Ontario. Vancouver, British Columbia: University of British Columbia Press. 198-233.

Walters, C. J. 1986. Adaptive Management of Renewable Resources. New York, New York: McGraw Hill. 374 p.

Walters, C.J. 1997. Challenges in adaptive management of riparian and coastal ecosystems. Conservation Ecology [online] 1(2): 1. www.consecol.org/vol 1/iss2/art 1

Walters, C.J. and C.S. Holling. 1990. Large-scale management experiments and learning by doing. Ecology 7: 2060-2068.

Westley, F. 1995. Governing design: the management of social systems and ecosystem management. In: L.H. Gunderson, C.S. Holling, and S.S. Light (editors). Barriers and Bridges to the Renewal of Ecosystems and Institutions. New York, New York: Columbia University Press. 391-427.

Wiens, J.A. 1996. Wildlife in patchy environments: metapopulations, mosaics, and management. In: D. R. McCullough (editor). Metapopulations and Wildlife Conservation. Washington, DC: Island Press. 53-84.

Wiens, J.A. 1997. The emerging role of patchiness in conservation biology. In: S.T.A. Pickett, R. S. Ostfeld, M. Shachak, and G.E. Likens (editors). The Ecological Basis of Conservation. New York, New York: Chapman and Hall. 93-107.

Williams, B.K. and F.A. Johnson. 1995. Adaptive management and the regulation of waterfowl harvests. Wildlife Society Bulletin 23: 430-436.

Williams, B.K., F.A. Johnson, and K. Wilkins. 1996. Uncertainty and the adaptive management of waterfowl harvests. Journal of Wildlife Management 60: 223-232.

16 Reflections on the Managed Forest Landscape

DAVID L. EULER,[*] IAN D. THOMPSON,[**] and AJITH H. PERERA[***]

THE CURRENT CONTEXT OF ONTARIO FOREST MANAGEMENT

Humans have always used forests to meet their needs for food, shelter, and recreation, but at the end of the 20th century there is an unprecedented demand for goods and services from the forest, a demand which is accompanied, at the global scale, by habitat loss and species extinction. In Ontario, the most difficult and compelling problem facing managers of the forests and their resources is how to meet this increasing demand for goods and services from the forest without degrading the ecological processes of forest landscapes. This focus on ecological processes is based on a growing understanding that forests are not an unchanging assemblage of trees and animals, and that the structure of the forest landscape does not remain static, either for relatively short periods of 50 to 100 years, or during millennia.

Ontario forest managers must address two central issues. The first is the real goal of forest sustainability? Most people intuitively understand what sustainability means, even though there is an abundance of definitions, some of which seem designed to confuse rather than help. How is a balance to be reached between sustaining the internal character of the forest (that is, the composition, structure, and function) and sustaining the external output of the forest (that is, products for consumption and non-consumptive benefits)? How can we ensure that the children of future generations have the same choices about the forest that we have? A currently emerging paradigm seems to be that if we attempt to emulate the forces of nature in our management activities, we at least have a reasonable chance of achieving sustainability. The second issue is whether the forest change caused by human activity lies within the range of variation that might be expected from natural ecological forces. In other words, are forest management practices simply substitutes for the disturbance forces of nature, or do they constitute an entirely different set of forces with different results?

The two major forest management activities that have had impacts on the forests of Ontario are logging and fire suppression. Views of the disturbance effects have changed over time. In the early years of European settlement, settlers considered the cutting of trees a necessary and socially responsible activity that would bring benefits to the entire community. Those local harvests did not always have profound or far-reaching effects, except for certain species such as white pine and red pine, which were considerably reduced in Ontario forests over two centuries. Timber harvest throughout most of this century was carried out by labour-intensive methods, often using horses to haul timber from the forest. Only during the last three or four decades have large areas been harvested with efficient machines that can clear trees rapidly over large areas. The increased mechanization of logging operations and development of the forest industry have meant that entire landscapes are affected by

* Faculty of Forestry and the Forest Environment, Lakehead University, 955 Oliver Road, Thunder Bay, Ontario P7B 5E1

** Canadian Forest Service, Great Lakes Forest Research Centre, 1219 Queen Street East, Sault Ste. Marie, Ontario P6A 5M7

*** Forest Landscape Ecology Program, Ontario Forest Research Institute, Ontario Ministry of Natural Resources, 1235 Queen Street East, Sault Ste. Marie, Ontario P6A 2E5 forest harvest during short time periods. Changes in wildlife habitat management policies and increased social pressure over clear-cutting have caused a dramatic reduction in average clear-cut sizes since the 1980s. This practice has resulted in smaller patches within the forest landscape and an increased extent of edge and ecotones between patches than before.

The area of forest harvested by clear-cut logging during the last five decades is more than three times the extent of forest fires during the same period (Carleton 2000, this volume; Li 2000, this volume). At a glance, it may seem that management is grossly exceeding the sustainable limits of disturbances. As forest fires have been very efficiently suppressed, however, the area burned during this period is not a good point of reference for comparison; in fact, the total area disturbed is probably about the same. In keeping with the view of wildfire as the cause of natural disturbance effects to be emulated, recent policy changes in Ontario have allowed fires to burn in some provincial parks and other areas where human life and property are not threatened.

Clearly, timber harvest practices and fire suppression have altered vegetation species composition over wide areas of the managed forest (Carleton 2000, this volume). Both types of disturbance have caused a decline in fire-tolerant, shade-intolerant conifer species, but clear-cutting has also promoted the development of less diverse, less stable plant communities dominated by broad-leaved trees and fire-intolerant understory plants. Even selective cutting has resulted in a decline in shade-tolerant tree species and a decline in certain conifer species. The nature of the disturbance is evidently very important to whether the forest ecosystems change or remain the same. The old, deterministic notions of forest succession evidently do not apply to forests where logging disturbance is a dominant factor and wildfire is suppressed. No definitive conclusion can be drawn from the evidence available at this time on the net effect wrought on the age-structure of Ontario's northern forests by both the combination of increased harvesting (which would tend to reduce the average age of the forests) and by fire suppression (which would tend to enable old forests to persist longer than they would have done under a natural fire regime).

While some comfort can be taken in the fact that forest management in Ontario has not resulted in large-scale destruction of populations of wild animals, it would be a mistake to believe that future management, continuing at current rates, will have the same limited impact. The net effect of changes in the patterns of forest patches in the managed forest landscape is to favour wildlife species that thrive in edge conditions, notably moose and deer, but also a wide variety of other species, from warblers to insects to wolves. Some specific changes in wildlife populations may be very significant, however, and perhaps irreversible. Although the information is not as clear as would be desirable, the preponderance of evidence suggests that populations of the woodland caribou have declined because of the fragmentation by logging of mature stands of the pioneer conifer species, jack pine and black spruce. Likewise, the widespread loss of pine as a result of many years of harvesting in the eastern North American continent has reduced the habitat of the red crossbill, a small songbird that feeds on the seeds in mature conifer cones. This species has probably suffered population declines as a result of habitat loss (Voigt et al. 2000, this volume).

On the other hand, many animal species seem adapted to the changes imposed by logging. For some species, particularly moose, logging and timber harvest have, in general, provided enhanced habitat. The evidence shows that moose populations have increased in areas where logging has occurred, if the hunting of these animals was not excessive (Voigt et al. 2000, this volume). Many other boreal and Great Lakes forest wildlife species have highly generalized habitat needs and are capable of surviving not only in many forest types, but also in a range of forest age-classes.

MAJOR CHANGES UNDERWAY IN ONTARIO FOREST MANAGEMENT

The roots of the current ecosystem approach to forest management planning, which was made law by the *Crown Forest Sustainability Act* (*CFSA*, 1994), lie in the environmental assessment of Ontario timber management conducted during the 1980s and early 1990s (Epp 2000, this volume; Euler and Epp 2000, this volume). Through the decision of the Environmental Assessment Board, issued in 1994, Ontario forest managers were for the first time required to meet an explicit, external standard for incorporating environmental scanning into plans for forest harvest and management. In the process, an awareness of the need to satisfy multiple uses of the forest, including a range of conservation values, has grown among policy-makers, resource managers, the forest industry, and the general public.

The guideline set by the *CFSA* and the *Forest Management Planning Manual for Ontario's Crown Forests* for any human intervention in the forest, especially for forest cutting operations, is that the effects produced should parallel those produced by natural disturbances (primarily wildfire). As noted above, the establishment of natural disturbance regimes as a standard is causing change in the old view that all fires should be suppressed, and has also given rise to an urgent new direction in forest research: the discovery of what natural disturbance patterns really are, or identification of a reasonable alternative to purely natural disturbance patterns among the current landscapes altered by human influences.

Awareness of a broader range of forest uses has been accompanied by a broader spatial perspective on forest land. While the unit of concern to traditional timber management was the individual forest stand or site, today's forest managers have begun to conceive of forest landscape units with emergent properties of their own. While research at the level of individual forest stands and sites will always be required, researchers now need to think in terms of much larger scales with respect to the emergent and cumulative effects of change on forest landscapes.

Ontario forest managers are adopting concepts from landscape ecology to help them find a balance between uses of the forest and maintenance of the integrity of forest ecosystems (Perera and Baldwin 2000, this volume). These concepts are based on a hierarchical framework of spatial scales, each one nested within the next. The fundamental principle of landscape ecology is that large-area questions must be studied at large scales, rather than through an aggregation of results from smaller scales. Although the results and effects of Ontario's Lands for Life initiative are still unfolding, this "big-picture" approach taken to strategic land-use planning in the province exemplifies application of the new perspective (Francis 2000, this volume).

Not only is a broader perspective on the Ontario forest becoming widely accepted, but a wide range of resources and tools for studying the forest at a landscape perspective has become available. Satellite remote sensing data and geographic information system technology provide effective digital means of storing and analyzing large spatial data sets. Models and programs are being developed for simulating various scenarios of forest disturbance (Li 2000, this volume) and net primary productivity (Band 2000, this volume), and for analyzing the spatial patterns that characterize landscapes and define landscape change.

In this growing wealth of data and tools, however, lies a potential trap. With respect to understanding the processes at work in forest landscapes and how these processes might change, we are already data-rich and information-poor. Landscape patterns detected by various measures mean little unless they are interpreted in the context of effects of management on flora and fauna, and thus on the sustainability of forest ecosystems in space and time. Although empirical knowledge and its extrapolation play a vital role, they are wholly inadequate to the endeavour of turning data into knowledge. It will be vital in coming years that null models be developed, against which current data on forest landscapes and landscape change can be interpreted, and that careful assessment of results from current data be carried out; otherwise, speculation may give rise to false and harmful conclusions.

As the Lands for Life program attests, another trend developing in forest land planning, in addition to the focus on ecosystem function and a large-scale approach, is the inclusion of a wide range of stakeholders, not least of all the general public, in the planning process. While promoting the representation of a wide range of interests and values, broad consultative processes may also serve to reduce excessive conflict over forest resource use.

FUTURE INFLUENCES ON ONTARIO'S FOREST LANDSCAPES

Global climate change will exert a long-term influence on Ontario's forest landscapes by redistributing species and changing the dynamics of fire, insect, and disease disturbance regimes (Flannigan and Weber 2000, this volume; Fleming et al. 2000, this volume; Li 2000, this volume). The direction of those changes is widely accepted, although various scenarios are offered for both the rate of climate change and the rapidity of the forest response.

Other future causes of change in Ontario forest management are found much closer to home. On the basis of trends already in evidence, including the growing scarcity of funds within government operations and reliance on the forest industry to conduct forest management planning, it is likely that a trend will develop toward the intensive management of selected areas rather than the extensive management of large areas. The increasingly effective integration of landscape-level analysis and monitoring into forest management planning will permit resource managers to select areas for intensive management. Additional remote sensing data sources, together with more versatile and accessible geographic information system packages, simulation models, and analysis programs, will enhance forest management planning and forest monitoring. One specific future event that definitely has the potential to change the future direction of Ontario forest management, and in ways unknown at present, is the next environmental assessment of forest management practices that is scheduled for 2003.

RESEARCH REQUIREMENTS FOR THE FUTURE

In order to take a true adaptive management approach, it is necessary to conduct forest management with a view to the design of experimentation (Baker 2000, this volume). Experimental studies need to focus on several spatial and temporal scales, so as to develop predictive models that incorporate scale-dependent processes of fauna and flora. The only scientifically valid hypothetical or deductive way to understand these processes is through the derivation of spatially explicit null models that are based on knowledge of the biology and dynamics of species.

While we have begun to understand the structure of forest landscapes and ecosystems, our knowledge of the functions and processes at work within these spatial units is much less advanced. Understanding how landscapes and ecosystems function will permit us to comprehend connections between structure and function, so that comparing scenarios of forest structure may also provide information on the influence of management options on functions that must be sustained. The pursuit of null models will provide a mechanistic understanding of landscape ecological patterns, processes, and change. This research goal will serve the evaluation of policies for forest sustainability. With the basic keys provided by null models and an understanding of the mechanisms by which landscape processes operate and change, the critical initiative can be advanced: conversion of data on landscape characteristics into knowledge of how forest landscapes are responding to management measures, other human activities, and other agents of change.

FUNDAMENTAL ISSUES UNDERLYING FOREST MANAGEMENT

In addition to the scientific issues, increased pressures on the forest by human populations raise many issues of values and ethics. Can scientific understanding provide adequate guidance to forest managers to avoid damaging ecosystems as timber is harvested? Can political processes find a consensus among the many values that people hold for forests and allow a fair distribution of the benefits that forests provide? One of the major dilemmas of forest landscape management concerns the amount of change deemed to be acceptable. Does it matter, for example, that some habitat for woodland caribou or red crossbills is lost?

As our understanding of the forested landscape improves, forest managers may learn to manage the forest within the bounds of natural variation. However, can the process of science and adaptive management keep pace with the need for forest products and the demands for economic benefits from the forest? This is a crucial question for the next century. The pressure for economic benefits and the need to sustain modern economies mean that forest managers will be asked to increase the flow of wood and other items a forest supplies.

As this century draws to a close, new information, much of it derived from landscape ecology, stimulates people to ask new and different questions. Most of these questions revolve around the concept of sustainability. The best hope for the future lies in the concept of adaptive management and in a rigorous process of planning and legislation that guides everyone in deriving benefits from the forest while maintaining the essential elements of the forested ecosystems. Adaptive management requires that the monitoring of policy results be established from the start and the possible responses to undesirable results formulated in advance. Despite the expressed intent, management agencies have great difficulty in actually implementing the concept of adaptive management, because doing so requires a political agency to confront its perceived failures, something that is very hard to do in the political arena. The tendency is for the political leaders of governments to emphasize success and not to discuss management initiatives that did not achieve their goals, as though only the former were useful for learning or worthy of discussion.

The risks associated with planning and forest management must also be considered. Silviculture and timber harvesting have made changes in the forest that could not have been anticipated when the plans were made for the forest management unit under consideration. No one envisioned 40 years ago, for example, that timber harvest would initiate a broad-scale process of conversion of forest types. Furthermore, changes in climate and ecological pressures may impose other, perhaps more far-reaching changes. The most important lesson from studies of the composition of the forest is that whatever actions are taken in forest management, the risk associated with the unexpected is always high.

In Ontario, forest management is at a critical point as the 21st century begins. A more sophisticated approach to forest management has been developed that includes a broader perspective on the forest as an assemblage of ecosystems instead of only a reservoir of wood waiting to be harvested. Effective legislation is in place to make forest management ecologically sustainable and economically beneficial to the people living in the province. The planning and policy approaches already established have the potential to make forest management truly sustainable. The ability of landscape ecology to provide a basis for management has improved dramatically over the last decade. As never before, forest managers understand how management has affected the forest landscape. The tools of adaptive management and synoptic assessment are relatively well developed and can be used to improve forest management in Ontario. Today's forest managers have the challenge and the opportunity of observing the results of past management policies and practices and interpreting changes in the forest landscape from that perspective.

What remains is to judge how well all of the knowledge of forested landscapes and management processes can be implemented to reach a sustainable future. The tools are available, the people who use them are ready, and only their future actions will determine how successful they will be.

Resource managers and planners on the "front lines" continue to face much uncertainty. How are the new forest ecological values to be defined, assessed, and balanced, in the specific case? How are the results of the new predictive analyses to be interpreted as a guide to action? The task of planning will not wait until all the research is done, so the temptation will be to mould prescriptions regarding the relationship between forest landscapes and management action from whatever descriptive evidence becomes available. The immediate challenge facing forest managers is to resist that temptation, to become adept at the new planning approach, and to collaborate with researchers in answering questions critical to fulfillment of the new mandate. We believe that this book will help to formulate some of these important questions.

REFERENCES

Baker, J.A. 2000. Landscape ecology and adaptive management. In: A.H. Perera, D.L. Euler, and I.D. Thompson (editors). Ecology of a Managed Terrestrial Landscape: Patterns and Processes of Forest Landscapes in Ontario. Vancouver, British Columbia: University of British Columbia Press. 310-322.

Band, L.E. 2000. Forest ecosystem productivity in Ontario. In: A.H. Perera, D.L. Euler, and I.D. Thompson (editors). Ecology of a Managed Terrestrial Landscape: Patterns and Processes of Forest Landscapes in Ontario. Vancouver, British Columbia: University of British Columbia Press. 163-177.

Carleton, T.J. 2000. Vegetation responses to the managed forest landscape of central and northern Ontario. In: A.H. Perera, D.L. Euler, and I.D. Thompson (editors). Ecology of a Managed Terrestrial Landscape: Patterns and Processes of Forest Landscapes in Ontario. Vancouver, British Columbia: University of British Columbia Press. 179-197.

Epp. A.E. 2000. Ontario forests and forest policy before the era of sustainable forestry. In: A.H. Perera, D.L. Euler, and I.D. Thompson (editors). Ecology of a Managed Terrestrial Landscape: Patterns and Processes of Forest Landscapes in Ontario. Vancouver, British Columbia: University of British Columbia Press. 237-275.

Euler, D.L. and A.E. Epp. 2000. A new foundation for Ontario forest policy for the 21st century. In: A.H. Perera, D.L. Euler, and I.D. Thompson (editors). Ecology of a Managed Terrestrial Landscape: Patterns and Processes of Forest Landscapes in Ontario. Vancouver, British Columbia: University of British Columbia Press. 276-294.

Flannigan, M.D. and M.G. Weber. 2000. Influences of climate on Ontario forests. In: A.H. Perera, D.L. Euler, and I.D. Thompson (editors). Ecology of a Managed Terrestrial Landscape: Patterns and Processes of Forest Landscapes in Ontario. Vancouver, British Columbia: University of British Columbia Press. 103-114.

Fleming, R.A., A.A. Hopkin, and J.-N. Candau. 2000. Insect and disease disturbance regimes in Ontario's forests. In: A.H. Perera, D.L. Euler, and I.D. Thompson (editors). Ecology of a Managed Terrestrial Landscape: Patterns and Processes of Forest Landscapes in Ontario. Vancouver, British Columbia: University of British Columbia Press. 141-162.

Francis, G.R. 2000. Strategic planning at the landscape level. In: A.H. Perera, D.L. Euler, and I.D. Thompson (editors). Ecology of a Managed Terrestrial Landscape: Patterns and Processes of Forest Landscapes in Ontario. Vancouver, British Columbia: University of British Columbia Press. 295-309.

Li, C. Fire regimes and their simulation with reference to Ontario. 2000. In: A.H. Perera, D.L. Euler, and I.D. Thompson (editors). Ecology of a Managed Terrestrial Landscape: Patterns and Processes of Forest Landscapes in Ontario. Vancouver, British Columbia: University of British Columbia Press. 115-140.

Ontario Environmental Assessment Board. 1994. Reasons for Decision and Decision. Class Environmental Assessment by the Ministry of Natural Resources for Timber Management on Crown Lands in Ontario. Toronto, Ontario: Ontario Environmental Assessment Board. 561 p.

Ontario Ministry of Natural Resources. 1996. Forest Management Planning Manual for Ontario's Crown Forests. Toronto, Ontario, Canada. 452 p.

Ontario Ministry of Natural Resources. 1997. Lands for Life: A Land Use Planning System for Ontario's Natural Resources. Toronto, Ontario: Ontario Ministry of Natural Resources. 15 p.

Perera, A.H. and D.J.B. Baldwin. 2000. Spatial patterns in the managed forest landscape of Ontario. In: A.H. Perera, D.L. Euler, and I.D. Thompson (editors). Ecology of a Managed Terrestrial Landscape: Patterns and Processes of Forest Landscapes in Ontario. Vancouver, British Columbia: University of British Columbia Press. 74-99.

Statutes of Ontario 1994. *Crown Forest Sustainability Act*. S.O. 1994, c. 25.

Voigt, D.R., J.A. Baker, R.S. Rempel, and I.D. Thompson. 2000. Forest vertebrate responses to landscape-level changes in Ontario. In: A.H. Perera, D.L. Euler, and I.D. Thompson (editors). Ecology of a Managed Terrestrial Landscape: Patterns and Processes of Forest Landscapes in Ontario. Vancouver, British Columbia: University of British Columbia Press. 198-233.

Glossary

ANPP (annual net primary productivity) - Annual amount of carbon produced by an ecosystem, minus the amount used in respiration.

Autogenic succession - The replacement of one community by another as a result of intrinsic changes in the environment brought about by the community itself. *Allogenic succession* is caused by extrinsic changes in the environment.

AVHRR (Advanced Very High Resolution Radiometer) - A remote sensor carried on NOAA meteorological satellites that provides reflectance information on the Earth's surface at resolutions from 1 km^2 to 65 km^2 across a 2000-km-wide swath.

Biome - A major ecosystem with a distinct assemblage of vegetation, animals, microbes, and physical environment, often reflecting a certain climate and soil.

Colonization:extinction - The ratio between the rates of a species occupying a habitat patch and its disappearance. This ratio is an important component of *island biogeography theory* and *metapopulation* concepts.

Contagion - A *landscape* characteristic that refers to the degree of dispersion of cover types. A clumped landscape has high contagion values compared to a dispersed landscape.

Core area - see *interior*

Discriminant analysis - An exploratory statistical technique that groups communities on the basis of their multivariate characteristics and assigns communities to previously defined groups.

Disturbance - A discrete event, either natural or human-induced, that changes the existing condition of an ecological system. An *autogenic disturbance* results from interactions among plants and animals within the ecological system; an *allogenic disturbance* is the result of external factors such as fire, drought, and wind. Landscape diversity is greater with large, infrequent disturbance; whereas stand diversity and ecosystem diversity are greater with small, frequent disturbance.

Diversity - A measure of an ecological system that indicates the relative *richness* and *evenness* of its components. A landscape with many cover types and/or a high degree of evenness is more diverse than a landscape with fewer cover types and/or a low degree of evenness. Also referred to as heterogeneity.

Eco-regionalization - The zonation of the earth's surface based on integrated environmental factors, usually at scales ranging from global, to continental, regional, and local, producing hierarchical classes such as eco-zones, eco-regions, eco-districts, and eco-sites.

Ecotone - An area of transition between two or more types of ecosystems.

Edge - The perimeter length of a patch that borders other cover types. The degree of contrast offered by the edge implies the degree of ecological dissimilarity between the patch and the surrounding cover types.

Evenness - The relative degree of apportionment of components in an ecological system. Within a landscape, evenness refers to proportionate distribution of area among the cover types; within an ecosystem, the proportionate distribution of individuals or biomass among species.

Featured species - A species of fish or wildlife for which specific management guidelines have been written; a species chosen to represent the needs of a group of like species.

Fire cycle - The number of years required to burn an area equal to the entire area of interest.

Fire frequency - The average number of fires that occur per unit time at a given point.

Fire regime - The fire activity or patterns that characterize a given area; important elements include *fire cycle* or *interval*, fire season, and the number, type, and intensity of fires.

Fragmentation - The relative degree to which a landscape or a given cover type is "broken up" by *patches* of smaller size and/or by the isolation of some patches from others. Fragmentation is a disruption of continuity and should be distinguished from habitat loss.

Gap analysis - The systematic technique used to determine geographic priorities in conserving biological diversity.

Guild - A group of species that share the same habitat, use the same resources, or use resources in the same manner.

Interior - The area of a *patch* that is left after the transition or 'buffer' zone of a specified width (e.g., 50 m) along the patch perimeter or *edge*. *Interior* is a species-specific concept.

Interspersion - Indicates the relative degree of spatial intermixing of *patches* of different cover types in a *landscape*.

Island biogeography - The field of study of species migration and extinction in relation to the spatial characteristics of habitat.

Isolation - An interpretation of the distance between two like patches (e.g., greater distance = higher degree of isolation).

Landsat TM - The thematic mapper optical remote sensor carried on Landsat satellite that provides reflectance information on the Earth's surface at a 900 m^2 resolution across a 180-km-wide swath.

Landscape - A heterogeneous ecological system composed of a cluster of interacting ecosystems. A forest landscape is an ecological system that constitutes a mosaic of forest cover types.

Landscape ecology - The study of the structure, function, and change in a defined *landscape*.

Metapopulation - A system of populations of a given species in a landscape linked by balanced rates of extinction and colonization. A key concept of metapopulation is the constant turnover of local populations through extinctions and emigration from existing sources.

NDVI (Normalized Difference Vegetation Index) - Computed as an index based on the surface reflectance in the red and near-infrared wavelengths, NDVI values indicate the relative "greenness" of the Earth's surface.

NEP (net ecosystem productivity) - Gross ecosystem productivity minus biomass used in respiration.

Ordination - A non-spatial statistical technique that orders communities along multi-dimensional environmental gradients.

PAR (photosynthetically active radiation) - The portion of the light spectrum that can be used in photosynthesis. FPAR is the fraction of PAR that a specified canopy group can use in photosynthesis. Combining FPAR and PAR produces APAR, which is the photosynthetically active radiation absorbed by a specified canopy group.

Patch - The basic component of a landscape, denoted by a contiguous area of a cover type that is bounded by area(s) of other cover type(s). A patch is equivalent to an *ecosystem*.

Richness - Describes the variety of component types in an ecological system. In a *landscape*, richness refers to the number of cover types, and in an *ecosystem*, the number of species.

Scale - Refers to physical dimensions of observed entities and phenomena. The modifiers *micro* and *macro* refer to relatively small and large physical dimensions, respectively, and *meso* denotes intermediate dimensions. Similarly, the modifiers *large* and *coarse* refer to relatively large physical dimensions; whereas *small* and *fine* refer to relatively small physical dimensions.

Shape index - The perimeter:area ratio of a *patch*, which is an indicator of its relative amount of *interior*.

Spatial complexity - A generalized notion of the spatial pattern in a *landscape*. A landscape is deemed relatively more complex when the patches are complex (e.g., convoluted) and have a higher degree of *interspersion*.

Spatial correlation - The degree to which a measured value at a given spatial point in a *landscape* is related to measured values at surrounding spatial points.

Spatial scale - Refers to the spatial dimension of *scale* in two components: *spatial resolution* refers to the minimum size of identifiable information; *spatial extent* to the magnitude of the space represented.

Succession - A series of directional changes in ecosystem composition, structure, and function over time, which results in the replacement of one group of organisms by another. For example, the changes in a plant community following *disturbance* are *seral stages*, in which each stage represents a different community.

Temporal scale - Refers to the temporal dimension of *scale* in two components: *frequency* refers to the minimum interval for which information is represented, and *term* to the length of the period represented.

Toposequence - A sequence of ecosystem conditions that differ from one another primarily as a result of topography and its influence on soil.

List of Contributors

James A. Baker, Science, Development and Transfer Branch, Ontario Ministry of Natural Resources, 300 Water Street, Peterborough, Ontario K9J 8M5

David J.B. Baldwin, Forest Landscape Ecology Program, Ontario Forest Research Institute, Ontario Ministry of Natural Resources, 1235 Queen Street East, Sault Ste. Marie, Ontario P6A 2E5

Lawrence E. Band, Department of Geography, University of North Carolina, CB# 3220 Chapel Hill, North Carolina, USA 27599

Jean-Noël Candau, Forest Landscape Ecology, Ontario Forest Research Institute, Ontario Ministry of Natural Resources, 1235 Queen Street East, Sault Ste. Marie, Ontario P6A 2E5

Terence J. Carleton, Faculty of Forestry and Department of Botany, University of Toronto, 33 Willcocks Street, Toronto, Ontario M5S 3B3

Joseph R. Desloges, Department of Geography, University of Toronto, 100 St. George Street, Toronto, Ontario M5S 3G3

A. Ernest Epp, Department of History, Lakehead University, 955 Oliver Road, Thunder Bay, Ontario P7B 5E1

David L. Euler, Faculty of Forestry and the Forest Environment, Lakehead University, 955 Oliver Road, Thunder Bay, Ontario P7B 5E1

Michael D. Flannigan, Canadian Forest Service, Northern Forest Research Centre, 5320 - 122nd Street, Edmonton, Alberta T6H 3S5

Richard A. Fleming, Canadian Forest Service, Great Lakes Forest Research Centre, 1219 Queen Street East, Sault Ste. Marie, Ontario P6A 5M7

George R. Francis, Department of Environment and Resource Studies, University of Waterloo, Waterloo, Ontario N2L 3G1

Anthony A. Hopkin, Canadian Forest Service, Great Lakes Forest Research Centre, 1219 Queen Street East, Sault Ste. Marie, Ontario P6A 5M7

Chao Li, Canadian Forest Service, Northern Forest Research Centre, 5320 - 122nd Street, Edmonton, Alberta T6H 3S5

Ajith H. Perera, Forest Landscape Ecology Program, Ontario Forest Research Institute, Ontario Ministry of Natural Resources, 1235 Queen Street East, Sault Ste. Marie, Ontario P6A 2E5

Robert S. Rempel, Centre for Northern Forest Ecosystem Research, Ontario Ministry of Natural Resources, Lakehead University, 955 Oliver Road, Thunder Bay, Ontario P7B 5E1

Ian D. Thompson, Canadian Forest Service, Great Lakes Forest Research Centre, 1219 Queen Street East, Sault Ste. Marie, Ontario P6A 5M7

Dennis R. Voigt, 1457 Heights Road, Lindsay, Ontario K9V 4R3. Formerly: Ontario Ministry of Natural Resources, 300 Water Street, Peterborough, Ontario K9J 8M5

Michael G. Weber, Canadian Forest Service, Great Lakes Forest Research Centre, 1219 Queen Street East, Sault Ste. Marie, Ontario P6A 5M7

Index

Set in Charter and Myriad by Artegraphica Design Co.

Printed and bound in Canada by Friesens

Copy editor: Susan Smith

Proofreader: Fran Aitkens

Indexer: Elizabeth Bell

Designer: Artegraphica Design Co.